Immunobiology and Immunopharmacology of Bacterial Endotoxins

University of South Florida International Biomedical Symposia Series

IMMUNOBIOLOGY AND IMMUNOPHARMACOLOGY OF BACTERIAL ENDOTOXINS
Edited by Andor Szentivanyi, Herman Friedman, and Alois Nowotny

VIRUSES, IMMUNITY, AND IMMUNODEFICIENCY
Edited by Andor Szentivanyi and Herman Friedman

A Continuation Order Plan is available for this series. A continuation order will bring delivery of each new volume immediately upon publication. Volumes are billed only upon actual shipment. For further information please contact the publisher.

Immunobiology and Immunopharmacology of Bacterial Endotoxins

Edited by
Andor Szentivanyi
and
Herman Friedman
University of South Florida College of Medicine
Tampa, Florida

and
Alois Nowotny
University of Pennsylvania
Philadelphia, Pennsylvania

PLENUM PRESS • NEW YORK AND LONDON

Library of Congress Cataloging in Publication Data

University of South Florida International Symposium in the Biomedical Sciences on
Immunobiology and Immunopharmacology of Bacterial Endotoxins: Basic and
Clinical Aspects (1985: Tampa, Fla.)
Immunobiology and immunopharmacology of bacterial endotoxins.

   (University of South Florida international biomedical symposia series)
   "Proceedings of the University of South Florida International Symposium in the
Biomedical Sciences on Immunobiology and Immunopharmacology of Bacterial En-
dotoxins: Basic and Clinical Aspects, held January 14–16, 1985, in Tampa, Florida"
— T.p. verso.
   Includes bibliographies.
   1. Endotoxins — Physiological effect — Congresses. 2. Immune response — Regula-
tion — Congresses. 3. Immunopharmacology — Congresses. I. Szentivanyi, Andor. II.
Friedman, Herman, 1931–   . III. Nowotny, A. (Alois), 1922–   . IV. Title. V.
Series. [DNLM: 1. Bacteria — immunology — congresses. 2. Endotoxins — immunol-
ogy — congresses. 3. Endotoxins — pharmacodynamics — congresses. QW 630 U58i
1985]
QP632.E4U55   1985   616'.014   86-25372
ISBN-13:978-1-4612-9319-4   e-ISBN-13:978-1-4613-2253-5
DOI:10.1007/978-1-4613-2253-5

Proceedings of the University of South Florida International
Symposium in the Biomedical Sciences on Immunobiology and
Immunopharmacology of Bacterial Endotoxins: Basic and Clinical
Aspects, held January 14–16, 1985, in Tampa, Florida

© 1986 Plenum Press, New York
Softcover reprint of the hardcover 1st edition 1986

A Division of Plenum Publishing Corporation
233 Spring Street, New York, N.Y. 10013

PREFACE

Endotoxins are constituents of all gram negative bacteria, as well as many other microorganisms. Since their original discovery and study at the beginning and middle parts of this century, many investigations have been performed concerning their immunochemistry and physicochemistry, as well as their pharmacologic activities and physiologic effects on the host. It became widely recognized during the beginning of this century that the pyrogenicity of many microbial infections may be associated with endotoxins. Furthermore, some 80 years ago, attempts were begun to "treat" a variety of illnesses including neoplasia, with such "pyrogens", i.e., bacterial endo- toxins. Inconclusive results were observed including some detrimental ones as well as, in some cases, beneficial ones. It became widely accepted that during infections with many gram negative organisms the fever occurring in patients, as well as many of the untoward pathophysiological effects of the infections, seemed to be due to the endotoxin the bacteria contained or released. In this regard, septic shock has been studied in detail by many clinicians, physiologists and pharmacologists and attempts have been made to relate the devastating effects of infection on metabolic and physiologic alterations caused by endotoxins. Recently, however, many beneficial effects of endotoxin have also been studied. It is now widely recognized that most, if not all mediators of immune responses, including interferon induction as well as interleukins, colony stimulating factor, tumor necrosis factor, and other soluble mediators of immunity, may have beneficial effects on a wide variety of immunologic and physiologic events, including resistance to neoplastic conditions. Thus, it seemed of value to the organizers of the International Symposia Series at the University of South Florida College of Medicine to bring together experts on the subject of bacterial endotoxin to focus attention on these important components.

This volume is based on the proceedings of the International Symposium on the Immunobiology and Immunopharmacology of Bacterial Endotoxins, Basic and Clinical Aspects. The speakers and participants provided newer informa- tion concerning fundamental and clinical aspects of endotoxin research conducted over the last half decade or so. Advances have been made in understanding the structure and nature of the endotoxin molecules(s) and their effects on a wide variety of cellular and subcellular parameters of immunity, metabolism and physiology, both in vivo and in vitro. This publication provides a permanent collection of up-to-date research and review articles concerning endotoxin biology and chemistry presented by outstanding chemists, microbiologists, immunologists, pharmacologists and physiologists in this important and ever-expanding area of biology. In addition, the editors of this volume believe that the subject of the immuno- biology and immunopharmacology of endotoxins provides an important focal point for continued investigation on the dynamics of the host/parasite relationship. It is anticipated that publication of this volume will provide a stimulus for further increased study concerning both detrimental and beneficial effects of endotoxins on the host and the interested reader with new perspectives in this rapidly evolving field.

We take this opportunity to express our most profound gratitude to Mrs. Christine Abarca for her outstanding editorial assistance in the preparation of this book.

Andor Szentivanyi
Dean of the College of Medicine
Deputy Vice President for
  Medical Affairs
Professor of Pharmacology and
  Internal Medicine
Chairman of the Department of
  Pharmacology and Therapeutics
University of South Florida
Tampa, Florida

Herman Friedman
Professor and Chairman
Department of Medical Microbiology
  and Immunology
University of South Florida
College of Medicine
Tampa, Florida

CONTENTS

IV. EFFECTS OF ENDOTOXINS ON CELLS OF THE IMMUNE SYSTEM

V. MODULATION OF THE IMMUNE RESPONSE BY ENDOTOXINS

INTRODUCTORY SECTION

The Biology and Physiopharmacology of Bacterial Lipopolysaccharide
Endotoxins and Their Role in Immunoregulation -
An Introductory Preview
Andor Szentivanyi and Herman Friedman

Retrospective and Prospective View of Endotoxin Research
L. Joe Berry

THE BIOLOGY AND PHYSIOPHARMACOLOGY OF BACTERIAL LIPOPOLYSACCHARIDE

ENDOTOXINS AND THEIR ROLE IN IMMUNOREGULATION - AN INTRODUCTORY PREVIEW

Andor Szentivanyi and Herman Friedman

Departments of Pharmacology and Therapeutics and
Medical Microbiology and Immunology
University of South Florida College of Medicine
Tampa, Florida

The topic of the immunobiology and immunopharmacology of bacterial endotoxins has attracted the attention of many biomedical investigators, including immunologists, microbiologists and biochemists, as well as physiologists,
pharmacologists and clinical practitioners of medicine. This has occurred
mainly because of the realization that these important bacterial components
affect many normal physiologic events and metabolic reaction of a host, including the immune system. Endotoxins have been studied for many decades in
terms of their chemistry and chemical reactivity. Although constituting a
diverse group of distinct chemical moieties on the bacterial cell surface,
as indicated in the preface, bacterial endotoxins may have different effects
on different target organs and cells. This volume provides a focus for presentation of pertinent newer findings concerning the immunologic and pharmacologic effects of endotoxins and to interrelate these effects with physiologic and immunologic alterations which may occur, both in disease states as
well as in specific situations in which endotoxins may be used for therapy,
especially in viral and cancer pathology.

In the introductory section, the first chapter is based on the keynote
address by L. Joe Berry of the University of Texas in Austin, Texas, who is
one of the pioneers of endotoxin research. Dr. Berry reviews previous
research concerning endotoxins over the past few decades and proposes some
newer areas which should be investigated. The remainder of the book is
divided into five sections dealing with various aspects of the subject.

The first section is concerned with the chemical characteristics of
endotoxin. Alois Nowotny of the University of Pennsylvania Health Science
Center in Philadelphia, presents an elaborate discussion on the theoretically
possible molecular explanations of the beneficial effects of endotoxins
including a review of old and new data relevant to some of the assumed possibilities. This is followed with an exposition of the methodological and
other problems encountered in attempting to identify beneficially active
structural subunits. Michael Apicella of the State University of New York
at Buffalo describes the antigenic structure of one of the well-studied
endotoxin-like molecules from an important pathogen which has heretofore not
been fully examined in terms of endotoxemia, i.e., Neisseria gonorrhoeae.
It is shown that the antigenic expression of gonococcal lipooligosaccharide
is complex and regulated by a number of factors including pyocin susceptibility, growth conditions, divalent cation interactions and association with

3

Lipid A. Since the lipooligosaccharide on the gonococcal cell surface is a key target for human bactericidal antibody, the factors which orchestrate this antigenic expression will have to be further elucidated for the understanding of the immunobiology of the gonococcus. Otto Luderitz represented the group from Germany that has made and continues to make major contributions concerning Lipid A and the structure and biologic activity of endotoxins. In their current chapter they give an excellent account of the investigations of the relationships between the chemistry and biology of Lipid A for the past three decades up to the point of the recent success in the chemical synthesis of complete E. coli Lipid A and the demonstration of its chemical and biological identity with natural Lipid A. In the next chapter of this section, in order to establish the validity of earlier immunochemical observations, and to provide potential therapeutic agents in gram-negative sepsis, James W. Larrick and his associates of Cetus Laboratories in California demonstrate the feasibility of producing human monoclonal antibodies that recognize cross-reactive determinants of the LPS core region, including those related to the biologically active Lipid A moiety. Arnold A. Peterson and Estelle J. McGroarty describe the physical properties of short-chain and long-chain fractions of LPS as they relate to the packing arrangement or aggregate structure of endotoxins in the outer membrane of bacteria and consequently to their barrier function in the bacterial cell. The section closes with the chapter of Chao-Ming Tsai and Carl E. Frasch of the Food and Drug Administration in Bethesda discussing their experiences on the analysis and characterization of bacterial endotoxins by SDS-polyacrylamide gel electrophoresis followed by silver stain. They describe considerable heterogeneity among LPS preparations from different species, many of the nonenteric bacterial species producing only low molecular weight LPS. This technique offers a simple yet powerful method for the chemical characterization of bacterial endotoxins, and the high sensitivity of the silver stain also permits the procedure to be used to detect endotoxin impurities in purified gram-negative bacterial cell components.

The second section deals with the physiological and pharmacological effects of endotoxins on the host. John Spitzer and his associates of Louisiana State University Medical School of New Orleans, describe in detail some of the metabolic alterations affecting both the disappearance and appearance rates of glucose. Although both of these parameters are affected, it appears that the increased rate of disappearance of glucose is the primary response to endotoxin as far as carbohydrate homeostasis is concerned. While the increased rate of disappearance is due to elevated glucose utilization of extrahepatic tissues, the increased hepatic glucogenolysis and gluconeogenesis are responsible for the augmented rate of appearance of glucose. The absolute rate of gluconeogenesis is elevated despite of the impaired gluconeogenic ability of the liver because of the altered hormonal environment (predominance of catabolic over anabolic hormones) and the increased rate of delivery of the gluconeogenic precursors to the liver. This glucose dyshomeostasis during endotoxicosis is further discussed from a unique perspective in the chapter of Michael Yelich and associates from the Loyola University of Chicago. This chapter shows that in the process of development of the profound and progressive hypoglycemia which is the ultimate pathophysiologic conclusion of endotoxicosis, the reticuloendothelial system plays a central role. Of particular importance are the hepatic sinusoidal macrophages which phagocytize and detoxify endotoxin. The latter may be processed by endocytosis or may escape RES processing. Endotoxin which is processed by the RES, triggers the production and release of macrophage-derived monokines that affect metabolic processes which control glucose homeostasis. On the other hand, the endotoxin that escapes processing by the RES affects directly the metabolic processes which are of primary importance for glucose homeostasis. The two macrophage-derived monokines (MILA = Macrophage Insulin-Like Activity; and MIRA = Macrophage Insulin Release-potentiating Activity) act in distinctly different ways: MILA acts directly

on the utilization of glucose by extrahepatic tissues, while MIRA acts on
the β-cells of the endocrine pancreas resulting in systemic hyperinsulinemia.

While observations are consistent with the concept that endotoxin is
important in the pathology of the shock state, they fail to establish an
endotoxin role in the hypermetabolic or "high flow" state of sepsis. Using
an animal model in which a non-lethal dose of endotoxin is infused continuously, that is a design that mimics clinical realities with greater fidelity
than does a bolus injection, Judy A. Spitzer of the Louisiana State University of New Orleans, examines in her chapter a wide range of homeostatic
mechanisms which are overwhelmed in lethal models of endotoxic septic shock.
It is pointed out that this low dose endotoxin infusion model does in fact
simulate some cardinal features of the metabolic profile of the septic
patient, and provides evidence that the perturbations in gluconeogenesis
implicate metabolic lesions in mitochondrial and cytoplasmic compartments of
isolated hepatocytes. At the same time, the endoplasmic reticulum, rather
than mitochondria appears to be the site of endotoxin-induced interference
with intracellular Ca homeostasis. In the next chapter on endotoxin toxicity
to hepatocytes and its modulation by macrophages, Patricia S. Latham and
Susan B. Sepelak of the University of Maryland in Baltimore, indicate that
macrophages are not necessary for the response of hepatocytes to endotoxin,
but they do appear to modify that response. Another highlight of this
chapter is the demonstration that endotoxin injury of hepatocytes includes a
condensation of the mitochondrial matrix as one of the early signs of toxic
injury, and a striking increase in the formation of plasma membrane blebs.
Another analysis of the endotoxin-macrophage interaction in the context of
the liver is presented in the chapter of Joseph F. Williams and Andor
Szentivanyi. In this discussion, the effects of endotoxin on the hepatic
microsomal mixed-function oxidase system are explored. This enzyme system
is responsible for the biotransformation of many drugs and certain endogenous
substances including steroids, fatty acids, and prostaglandins. Of the
various immunomodulators that have been found in the past decade to depress
this enzyme system, endotoxin is one of the most potent. The authors have
recently shown that injection of endotoxin-tolerant animals with supernatant
fluids from cultures of peritoneal macrophages incubated with endotoxin will
depress the hepatic microsomal mixed-function oxidase system suggesting that
a macrophage-derived product and not endotoxin per se may be the effector
agent.

Myocardial performance and adrenergic modulation of cyclic AMP following
endotoxin administration is discussed by Raymond E. Shepherd and associates
of the Louisiana State University in New Orleans. The authors show that
myocytes from endotoxin-treated rats are less responsive to beta adrenergic
activation and stimulation by forskolin, indicating that the locus of myocardial dysfunction in endotoxicosis may be within the guanine nucleotide
regulatory protein or within the adenylate cyclase itself. Although increased catecholamine release in endotoxicosis may produce adrenoceptor desensitization, no difference is demonstrable in receptor density or affinity, and that the basic abnormality is a post-receptor event is further suggested by sodium fluoride's inability to activate adenylate cyclase in cell
membranes of endotoxin-treated rats. John W. Hadden and his associates from
the University of South Florida in Tampa, point out in their chapter that
endotoxins are likely candidates for positive regulators of the development
and differentiation of both classes of lymphocytes, and of macrophages and
various granulocyte populations. The evidence suggesting non-toxic physiological roles for endotoxin is compelling, and indicates a role for cyclic
AMP to induce precursor cell differentiation, and for cyclic GMP to promote
proliferative or secretory functions of mature lymphocytes and macrophages.
In addition, endotoxin-induced mediators such as CSF, IFN, IL-1, and IL-2
appear to involve cyclic nucleotide mediation as part of their action.

In the field of resistance to bacterial infections, specific antibody formation has its limitations as a defense mechanism. A primary immune response is the end-product of a sequence of events that occurs too late to rescue the host in many septicemias, a circumstance, most evident in neonatal resistance. Indeed, antibody formation is probably one of the highest evolutionary forms of protein synthesis, whereas other major defense mechanisms such as phagocytosis or elevation of temperature appear to be more powerful and more anciently established in phylogenetic studies. For these reasons, the topic of the third section of this volume is the immunogenecity and related non-specific effects of endotoxin starting out with the chapter of Louis Chedid and his associates of the Pasteur Institute of Paris. Following a general review of the non-specific effects of endotoxin on bacterial infections, the authors point out that fever enhances non-specific resistance essentially by two mechanisms: 1) inhibition of proliferation of thermosensitive organisms, and 2) production of monokines and other immunostimulatory agents by exogenous pyrogens. Of these, IL-1 possess both pyrogenic and lymphocyte-activating properties that can be selectively produced. Pyrogenic IL-1 lowers plasma iron concentration which in turn inhibits bacterial proliferation during fever. Since non-specific resistance can be enhanced by agents and conditions that do not produce fever or hypoferremia, there likely exist other factors capable of influencing non-specific resistance. One of these is the tumor necrosis factor which can be shown to enhance resistance to bacterial infections. In contrast, the subsequent chapter by A. C. Rodloff and associates of the Free University of Berlin describes a reduction in non-specific resistance in "synergistic" infections between enterobacteria and different Bacteroides species, that is situations in which the Bacteroides increased the endotoxin susceptibility of experimental animals.

Whereas mechanisms involved in the induction of endotoxin tolerance and non-specific host defense to bacterial infection by endotoxin may be similar, the radioprotective effect of endotoxin has been related to the early hematopoietic recovery, thus enabling the host to resist the consequences of gut-derived bacteremia and endotoxemia following x-ray exposure. This problem is discussed by Renate Urbaschek and associates of the University of Heidelberg in Mannheim in their chapter on the role of post-endotoxin serum components from BCG infected mice in the protection of compromised hosts. It is demonstrated that BCG/ET serum is capable of transferring stimulation of granulopoiesis as well as an increase in splenic granulocyte macrophage precursor cells, and an increase in serum levels of colony-stimulating activity (CSA). A great variety of mediators are present in BCG/ET serum also referred to as TNS (tumor necrosis serum) because of its high level of tumor necrosis factor that has been isolated and characterized from such sera including interferon, IL$_1$, and many other mediators. The critical factor or factors in such serum responsible for the enhancement of non-specific resistance have not been identified as yet.

Experiences with a number of gram-negative infections suggest the involvement of antigenetically similar cell wall components in resistance to infection. These observations are complemented by findings showing that antibodies against "core" glycolipids of enterobacteria protect against gram-negative sepsis. This issue is discussed by B. W. Fenwick and associates of the University of California at Davis in their chapter arranged around their studies on the protection against lethal Haemophilus pleuropneumoniae infection in swine by antibodies to LPS core antigens. The mechanism by which this protection is achieved is as yet undetermined but the study indicates that antibodies to core antigens are most effective when bacterial growth is unrestricted, as during the early stages of an infection or in a compromised host. There are two companion chapters in this section by the same group of workers: 1) discusses information on active immunization with E. coli J5 and its protective effects from endotoxic shock in calves, and 2) presents an analysis of the protective function of neutrophils during experimental endotoxic shock. The first chapter describes a study designed to determine the antibody classes which provide the protective effects of E. coli LPS core antigens during experimental endotoxic shock in calves, and concludes that antibodies of the IgG class are associated with this protective effect. The second chapter explores the relationship between the disappearance of neutrophils during endotoxic and septic shock and the role of this event in the pathogenesis of these conditions. Significance of the nature of this relationship lies in the fact that two of the most consistent hemostatic aberrations in the shock state are a rapid decrease in circulating neutrophils and consumptive coagulopathies. Neutrophils as inflammatory cells have the potential of damaging host tissues and thus may contribute to the limitation of these events. However, the results of their elegantly designed studies indicate that neutrophils provide a degree of protection from the effects of endotoxin rather than contribute to the severity of the host response. Although the mechanism of this protective activity is unknown the speculation is advanced that the neutrophils may be acting by way of their superior affinity to endotoxin thereby sequestering it and preventing interactions between endotoxin and more biologically active cellular targets.

In another chapter Toby K. Eisenstein of Temple University of Philadelphia and his associates present studies in which Salmonella endotoxins were tested along with other Salmonella antigens and vaccines, for ability to protect mice against virulent Salmonellae, and for ability to induce antibody. Strains of mice differing in their genetically determined responsiveness to LPS and their innate susceptibilities to Salmonella infection were used, and the findings show that the capacity of nonviable preparations to protect mice against Salmonella depends on the strain, and correlates with the innate susceptibility of the mouse to Salmonella infection, but not with the histocompatibility type, nor with the capacity to respond to endotoxin physiologically or immunologically. Jeanne Becker and associates of the University of South Florida in Tampa describe the central role of macrophages in general host resistance to microbial infections using the model of Legionella pneumophila, a facultative intracellular pathogen that grows within monocytes and macrophages. Proliferation of this organism within macrophages occurs even in the presence of serum antibody and complement, and it appears that L. pneumophila evades destruction within phagocytes by inhibiting phagosome-lysosome fusion. Thus, the prevention of phago-lysosome fusion impairs this normal macrophage function which can lead to decreased host resistance. The chapter describes the ability of these bacteria to influence two additional macrophage functions. One is the inhibition of macrophage spreading, the other is a transient inhibition of phagocytosis, and when supplied with corticosteroids, a potentiation of the steroid-induced suppression of phagocytosis, also in a transient manner. Samuel B. Salvin and Pamela B. Renda of the University of Pittsburgh present a general overview of the effects of endotoxin on migration inhibitory factor and interferon including their own experiences in this field. They conclude that endotoxin can stimulate release of migration inhibitory factor and of interferon to varying degrees with optimum release depending on a variety of factors such as the strain of the animal, the degree of exposure to endotoxin, and the nature and source of the lipopolysaccharide. Finally, Joseph G. Sinkovics of the University of South Florida and Baylor College of Medicine presents a comprehensive, state of the art, review on the clinical recognition, pathophysiology, and treatment of endotoxinemia.

The fourth section of this volume concentrates on the effects of endotoxins on cells that participate in the organization of the immune response starting out with the chapter of Fritz Melchers from the Basel Institute for Immunology on the regulation of the cell cycle of murine B lymphocytes by endotoxins. Normally, the cell cycle of murine B lymphocytes is controlled by antigen [acting via surface-bound immunoglobulin (IgG)], by macrophages (producing α-factor-type lymphokines), and by helper T lymphocytes (producing β-factor-type lymphokines). For both types of lymphokines, B cells display so far uncharacterized specific receptors. Endotoxins act in three ways at three restriction points within the B cell cycle. Early after mitosis, endotoxins act to excite B cells to susceptibility for α- and β-factors, replacing the requirements for the occupancy of surface Ig by either antigen and MHC-restricted helper T cells, or by Ig-specific antibodies. They then indirectly activate A cells (macrophages) to α-factor production that control the B cell cycle three to five hours after mitosis within the G1 phase. Finally, they replace the action of helper T cell-derived β-factors late in the cell cycle, two to four hours before mitosis in the G2 phase of the cell cycle. Masayasu Nakano and Toshimasa Nitta of Jichi Medical School in Tochigiken, discuss the synergistic effect of endotoxin and Concanavalin A on DNA synthesis in lymphocytes in cultures in relation to productions of IL-1 and 2, and demonstrate that the synergy is mainly due to an elevation of IL-2 production from cells co-stimulated with endotoxin and Concanavalin A. In a subsequent chapter, David C. Morrison and associates of the University of Kansas Medical Center in Kansas City examine the antigenic and biochemical basis for the genetic defect in the C3H/HeJ mouse which results in a phenotypic functional unresponsiveness of B lymphocytes to mitogenic stimulation by LPS. On the basis of intensive immunizations of C3H/HeJ mice with congenic LPS responder lymphocytes, it appears that if this mutation results in an altered gene product expressed on the B cell surface, it is significantly less immunogenic than a minor histocompatibility antigen. These considerations are in agreement with biochemical studies using a photo-activatable radiolabelled and cleavable LPS probe to detect differences in LPS binding sites on responder and non-responder lymphocytes. With poly-acrylamide gel electrophoresis of LPS binding targets, indistinguishable patterns are obtained for C3HeB/FeJ and C3H/HeJ lymphocytes. Data are also presented which confirm that C3H/HeJ lymphocytes are not refractory to a variety of LPS preparations with R-chemotype structures, and such activity is not due to contaminating lipid A-associated protein but represents a true LPS-induced proliferative signal. This combination of findings lends strong support for the concept that the genetic defect in the C3H/HeJ mouse which is manifest in an inability to respond mitogenically to some LPS preparations may not be due to the absence of a membrane receptor specific for LPS.

Suzanne M. Michalek and associates of the University of Alabama in Birmingham discuss the effects of gut endotoxin on the cells of gut-associated lymphoreticular tissue and on subsequent IgA responses. They show that endogenous LPS can influence both T-dependent antigen-specific and polyclonal IgA responses in a manner which may occur following gut LPS stimulation of cells of the gut-associated lymphoreticular tissue. In studies designed to assess human IgA subclass responses to gut LPS, it appears that IgA subclass responses to LPS differ in serum and in external secretions, and that IgA1 in serum and IgA2 in secretions are the predominant anti-LPS antibodies found implying that serum and secretory IgA responses are under separate regulatory control. The principal theme of the chapter of Edgar Pick of the Sackler School of Medicine of Tel-Aviv University, is that immunological activation of macrophages, the essential elements of which are evident in macrophages exposed to lymphokines in vitro, involves major changes in the production of oxygen radicals. The essential function of these is to mediate cytostasis or killing of phagocytosed pathogens while causing minimal damage to adjacent cells. It is emphasized that this metabolic change is not synonymous with macrophage activation and against both intra- and extracellular targets exist. The two most common explanations for this multiplicity of effector pathways are: macrophage heterogeneity and the involvement of more than one lymphokine. D. K. Blanchard and associates of the University of South Florida discuss the induction of gamma interferon by endotoxin in "aged" murine splenocyte cultures, and point out that this effect of endotoxin is abrogated by T cell depletion indicating that production of gamma interferon in response to LPS is dependent on an aged population of T cells. Since

IL-2 is found to increase in culture supernatant fluids upon aging of unstimulated splenocytes, it is possible that in vitro activation of T cells
by IL-2 is responsible for gamma interferon production by LPS-stimulated
lymphocytes.

Robert L. Duncan and associates of Emory University in Atlanta discuss
the fate of E. coli LPS after the uptake and catabolism of the whole bacterium by macrophages. It is shown that LPS are exocytosed slowly from the
macrophage, and those remaining within the macrophage at 72 hours, and the
LPS released from the macrophage over that period retain endotoxic activity.
The LPS remaining within the macrophage at 48-72 hours is associated with
phagocytic vacuoles, and both the retained and the exocytosed LPS have a
significantly enhanced capacity to stimulate splenocytes. As assessed by
SDS-PAGE, macrophage processed LPS is enriched for higher molecular weight
subunits resulting in enhanced immunostimulatory activity. Thus, macrophage
processing of LPS may contribute to the local amplification of immunologically relevant events.

For many years the laboratories of R. C. Butler (Arlington Hospital,
Virginia), H. Friedman (University of South Florida), and A. Nowotny
(University of Pennsylvania) have collaborated in studying a variety of the
immunostimulatory effects of LPS, and conversely the immunosuppressive
effects of the Friend leukemia virus (FLV). In their current chapter, the
authors discuss the mechanism by which FLV suppresses the antibody response,
and they conclude that it is the suppression of macrophage function by the
direct infection of macrophages of FLV. While this infection does not
affect macrophage viability or obvious cell morphology, it does inhibit both
the spontaneous and the LPS-induced production of antibody response helper
factor(s) by the macrophage. In accord with this conclusion is the demonstration that the development of antibody responses by normal splenocytes is
greatly enhanced by the production of antibody helper factor(s) by macrophage. In a subsequent chapter, Stefanie N. Vogel and Gary S. Madonna of
the Uniformed Services University of the Health Sciences in Bethesda discuss
the nature of the two phases of "endotoxin tolerance." The "early endotoxin
tolerance" that is the transient period of LPS hyporesponsiveness that occurs
within the first few days following initial exposure to LPS is independent
of anti-O-specific antibodies, transferable with spleen cells, and is associated with the appearance of increased numbers of macrophage progenitor
cells in the bone marrow. In contrast, "late phase tolerance" is delayed,
persistent, and antibody-mediated. Dov H. Pluznik and Stephan E. Mergenhagen
of the National Institute of Dental Research in Bethesda discuss the synergistic relationship between LPS and tumor promoting phorbol esters (TPA) in
the induction of colony-stimulating factor generation by murine bone marrow
cells. The cooperation between LPS and TPA is effective only when both
agents are added simultaneously to the bone marrow cells. Since the calcium
ionophore A23187 could replace LPS in cooperating with TPA in stimulating
bone marrow cells to generate colony-stimulating factor, the hypothesis is
put forward that LPS could be involved in calcium mobilization which is required for many cellular activities and in the production of various soluble
mediators.

In general, the final section deals with modulation of the immune response by endotoxins. The first chapter by Edgar Ribi and associates from
Ribi ImmunoChem Research, Inc., in Hamilton, Montana, establishes the structural relationship between the toxic diphosphoryl lipid (DPL) and the nontoxic monophosphoryl lipid A (MPL) which are chemically derived from the
native LPS molecule. MPL retains many of the beneficial properties of the
parent endotoxin such as being able to stimulate humoral and cell-mediated
immune responses, while being significantly less active in causing fever,
shock, and death in animals which are highly susceptible to the toxic
effects of endotoxin. Moreover, MPL used in combination with chemically-

defined adjuvants from mycobacteria acts synergistically in mediating tumor and antimicrobial immunity. Thus, MPL appears to have the potential for a wide range of clinical applications. Kathyrn Nixdorff and Sigrid Schell discuss the selective induction of particular subclasses of IgG antibody-producing cells specific for LPS and produced by complexing LPS with bacterial outer membrane components used as modulators of the immune response. T cells are involved in this selective modulation, but they are apparently not specific for the modulators. Furthermore, LPS molecules of different structure also effect a selective induction of particular subclasses of IgG antibody-producing cells. LPS molecules having short O-polysaccharide chains show a tendency to induce IgG1 and IgG2 PFC while LPS molecules possessing long O-polysaccharide chains induce mainly IgG2 and IgG3 PFC. It appears, that a change in the physical character or the conformation of the LPS molecule occurs upon complex formation with modulators, which in some manner not only allow enhanced IgG production but also determines the IgG subclass induced. Complementary to the preceding material is the chapter of Barnet M. Sultzer and associates of the Downstate Medical Center of the State University of New York that also discusses endotoxin associated membrane proteins and their polyclonal and adjuvant activities. For their studies, the authors adopted both in vivo and in vitro model systems to measure the immune response to cholera enterotoxin and sheep erythrocytes. Endotoxin associated membrane protein preparations from Salmonella typhi, Vibrio cholerae and Bordetella pertussis were used for comparison purposes. It is of great interest that in these model systems it was the endotoxin associated membrane protein preparation derived from B. pertussis that was found to be the most potent adjuvant. This probably contributes to the known adjuvant effect of the whole cell vaccine previously attributed solely to the endotoxin component of the bacterial membrane, and it is in accord with the series of experiments by Szentivanyi and Fishel in the 1960's and early 1970's indicating that these pertussis-derived endotoxin associated membrane protein complexes have a powerful inhibitory effect on the beta adrenergic receptor-cyclic nucleotide system. The authors conclude that the maturation and differentiation of B lymphocytes in the host in response to gram-negative bacterial antigens is dependent, at least in part, on the immunomodulating components of the bacterial outer membranes as they interact with these host cells directly or indirectly through the activation of accessory immunocompetent cells. Stephen A. Stimpson and associates of the University of North Carolina at Chapel Hill describe in their chapter their analysis of the LPS-induced recurrence of arthritis initiated by peptidoglycan-polysaccharide. They hypothesize that LPS, although usually associated with acute toxic reactions, might play a role in chronic arthritis through the exacerbation and perpetuation of inflammation initiated by certain peptidoglycan-containing polymers. The mechanism by which LPS induces a recurrence of arthritis in joints previously exposed to PG-PS is not known, but may involve the shared biological activities of lipid and peptidoglycan. Important in this regard may be the synergistic interaction of muramyl dipeptide and endotoxin observed previously by Ribi et al. in an in vivo model of antitumor activity and by Butler and Friedman in an in vitro adjuvanticity assay.

The two subsequent chapters of W. and K. Roszkowski, J. Jeljaszewicz, G. Pulverer, and associates discuss the effects of some selected antibiotics on immunity, the endogeneous intestinal microflora, and tumor growth. Among these antibiotics, the most dramatic and long-lasting immunosuppressive effect was caused by mezlocillin. This agent, furthermore, produces an indirect anti-tumor effect which may be related to the elimination of the endogeneous intestinal bacterial flora. In the apparent correlation between inhibition of tumor growth and the intestinal flora, E. coli endotoxin does not seem to be involved, and the speculation is advanced that the common denominator between these phenomena may be the elimination of the proliferative stimulus of the intestinal flora on normal (i.e., plasma cells) and malignant cells. Mauro Bendinelli and his associates of the University of

Plea present information on the mechanisms involved in the immunorestorative effect of endotoxin in the immunodeficiency state induced by the two viral components of the Friend leukemia complex. It is demonstrated that macrophages are deeply involved in the immunorestorative effect of endotoxin. This involvement is supported by the following considerations: 1) macrophage-rich cell populations derived from normal mice, themselves incapable of reversing the immunologic deficit of infected cultures, became capable of doing so once pretreated with LPS; 2) the latter restores the ability of infected macrophages to help nonadherent cells from normal mice in the generation of an antibody response; and 3) supernatants conditioned from LPS-stimulated cells which were over 95% pure macrophages proved very effective in enhancing the responsiveness of infected cells. In view of the failure of purified IL-1 to reproduce the effects of the LPS-conditioned supernatants, and since infected macrophages produce normal or enhanced levels of this mediator following in vitro LPS stimulation, it appears likely that other soluble mediator(s) of LPS-activated macrophages by themselves, or in concert with IL-1, mediate the immunorestorative action of LPS. In the ensuing chapter, Herman Friedman and Andor Szentivanyi discuss the nontoxic polysaccharide enriched derivative of endotoxin from gram-negative bacteria as an immunoadjuvant in acquired immunodeficiency induced by retroviruses. Both intact endotoxin and the polysaccharide-rich derivative are immunoenhancing in both normal as well as Friend leukemia virus-infected splenocytes. The Lipid A free, polysaccharide-rich extract of endotoxin also augments the activities of BCG and/or purified synthetic MDP in increasing antibody formation by normal and leukemic splenocytes and stimulates the generation of cell-free factors which are immunopotentiating. The final chapter by James P. Nolan of the State University of New York at Buffalo deals with the clinical relevance of endotoxemia associated with gram-negative bacterial infections and various endotoxemic states without bacteremia. It is pointed out that the concept of the existence of endotoxemia in the absence of circulating bacteria is one that is gaining in acceptance, but still is unfamiliar to many clinicians. Origin of the endotoxins in these clinical conditions appears to be the cell walls contained in large quantities in the intestine. By the mechanism of increased absorption through a break in the usually efficient mucosal barrier, or by a failure of hepatic removal and detoxification mechanisms, LPS gains entry into the circulation. While the most extensive and convincing demonstration of the clinical importance of endotoxemia lies in liver disease, clinical states implicating endotoxins as key pathophysiologic mediators are also examined before the chapter concentrates on their role in hepatic injury. These extrahepatic manifestations include acute renal failure, adult respiratory distress syndrome, neonatal endotoxemia, and gastrointestinal disease.

In conclusion, it appears from the broad range of topics discussed in this volume that there is a rapid increase of interest in the immunobiology and immunopharmacology of endotoxins. It is anticipated that publication of this volume will stimulate further investigative work in this exciting area of biomedicine.

RETROSPECTIVE AND PROSPECTIVE VIEW OF ENDOTOXIN RESEARCH

L. Joe Berry

Department of Microbiology
University of Texas
Austin, Texas

This chapter will, more or less in chronological order, describe work done over a number of years with endotoxin. So many contributions have been made to each topic that to include them all is impossible. The selections here represent works that are considered highly significant since they serve as milestones in the study of this fascinating and challenging substance. Not all important observations can be covered within the limits of this chapter thus, if recognition is not given where it is due, the reader will recognize the difficulties inherent in the author's task. First we shall look backward on endotoxin research and then try to project the direction it will take in the future.

ANTITUMOR EFFECT OF ENDOTOXIN

There are few, if any bacterial products that have attracted as much research interest as endotoxin. About 90 years ago, Pfeiffer (99) introduced the term endotoxin, perhaps mistakenly, to distinguish this class of substances from the exotoxins that are released by the bacteria into the environment. No significant studies were initiated just because endotoxin was given a name but at about the same time Coley (28), in New York, attracted considerable attention when he noted that certain bacterial infections or the administration of a mixture of Serratia marcescens and streptococci resulted in a decrease in the size of some tumors. After a brief flurry of interest, little was done for several decades.

Research started again after the end of World War II. Murray Shear at NIH had published a report in 1941 (110) showing that a preparation derived from Serratia marcescens had potent tumor necrotizing activity. A spate of publications ensued when hostilities ended. One of the active contributors, Hugh Creech of the Cancer Research Institute of Philadelphia was a part-time colleague of mine at Bryn Mawr College. One of Creech's major interests was the immunological response of laboratory animals to endotoxin and how this might contribute to its progressive loss of efficacy against tumors (31). Hugh was the first to arouse my interest in endotoxin even though my initial research with it was several years away. Murray Shear deserves credit for getting Maurice Landy involved in its study. Landy, as much as anyone in this country, helped to keep endotoxin at the forefront of research for a number of years and attracted important scientists as his collaborators, two of whom are contributors to this volume, Arthur Johnson and Edgar Ribi.

Table 1. Anti-Tumor Effects

Year	Authors	Observation
1893	Coley	Antitumor effect of some bacteria
1941	Shear & Associates	Tumor necrotizing activity of LPS
1949	Creech & Associates	Immunity to endotoxin
1974	Yang and Nowotny	Adoptive transfer of spleen cells protects against tumors
1975	Carswell et al.	Established that a postendotoxin serum factor (tumor necrosis factor, TNF) had activity
1976	Green et al.	Purified TNF
1979	Mannel et al.	TNF found in supernatant of endotoxin treated macrophages
1980	Hammerling et al.	Multiple activity of TNF

Recognition of the loss of the ability of endotoxin to provide effective therapy with repeated injections resulted in a lull in this type of research. During the present decade renewed interest has been generated as a result of several observations. Prior or concurrent administration of endotoxin was found by Yang and Nowotny (137) to protect mice against tumor challenge when an adoptive transfer of spleen cells from these endotoxin-treated donors to untreated recipients was made. Carswell et al. (25) reported soon thereafter that a factor which appeared in serum of endotoxin-treated mice also provided protection against tumors in recipient animals. Green et al. (50) purified the tumor necrosis factor (TNF) and found it to have multiple activities including properties similar to those of interleukin I and others (54). Mannel et al. (83) were able to obtain tumor necrotizing activity in the supernatants from endotoxin-treated activated macrophages.

TOLERANCE

As investigators became increasingly aware that the efficacy of endo-toxin in the treatment of malignancies declined with repeated injections, other fascinating observations were being made at the same time. Paul Beeson at Yale (9) recognized the loss in pyrogenic response to endotoxin that accompanied second or third injections into the same animal. Supposedly, the editor of the journal to which Beeson submitted his paper dealing with these observations did not approve the word Beeson had used, and changed it to tolerance. Whatever its origin, it is now part of the vocabulary of endo-toxinologists. Beeson, incidentally, suggested that tolerance was due to an activation of the reticuloendothelial system (RES) by endotoxin, which resulted in its more rapid clearance and elimination (10,11). This concept was dispelled when Benacerraf et al. (14) showed that mice were sensitized to endotoxin and not tolerant to it when their RES was activated by Zymosan. Suter et al. (125) had found the previous year that mice vaccinated with BCG were greatly sensitized to endotoxin even though their RES was activated. These animals were killed by not more than 0.001 LD_{50}. Chedid and Parant (27) associated tolerance with an accelerated clearance and presumably destruction of endotoxin from the circulation of tolerant animals. This left

Table 2. Tolerance

Year	Authors	Observation
1946	Beeson	Tolerance to fever detected
1947	Beeson	Tolerance due to activated RES
1955	Stetson	Host responses to endotoxin due to immune phenomena
1958	Suter et al.	BCG vaccination sensitizes mice to endotoxin. RES is activated.
1959	Benacerraf et al.	RES activated with Zymosan sensitizes mice to endotoxin
1959	Freedman	Tolerance passively transferred with serum
1961	Chedid & Parant	Tolerance associated with rapid clearance of endotoxin from blood
1963	Watson & Kim	Tolerance due to specific antibody
1963	Greisman et al.	Tolerance due to specific antibody
1970	Greisman & Woodward	Tolerance associated with depleted mediators
1978	Goodrum & Berry	Minimal levels of GAF in tolerant mice

the nature of tolerance in limbo, however, until somewhat contradictory observations were made. Henry Freedman (39) made rabbits passively tolerant by injecting them with serum from tolerant animals. This established that some humoral factor, but not likely specific antibody, was involved. Watson and Kim (131), on the other hand, believed that some specificity existed in tolerance to endotoxin fever. In fact, they maintained that hypersensitivity played a role in the primary response to endotoxin. This point of view was changed later, however. It did agree with Stetson (122) who emphasized that host responses to endotoxin were probably immunological. Greisman and his associates also believed tolerance was due to specific antibody (51). They later (52) established that mediators, such as endogenous pyrogen, fail to be released into the blood of tolerant animals, hence tolerance represents a phase when endotoxin elicits minimal responses. Goodrum and Berry (49) found supporting evidence for this concept when barely detectable levels of glucocorticoid antagonizing factor (GAF) were present in the serum of tolerant mice. Normal mice similarly manipulated had elevated levels of GAF.

THE SANARELLI-SHWARTZMAN PHENOMENA

Another observation that generated a number of studies was made in the 1930's by Shwartzman (112). He noticed that successive injections of typhoid vaccine could result in the development of a necrotic lesion at the site of the initial injection. He also found that in some cases, following two intravenous injections, serious illness or even death resulted from renal

Table 3. Sanarelli-Shwartzman Phenomena

Year	Authors	Observation
1894	Sanarelli	Skin lesion at site of first vaccine injection after a second (iv) injection
1930s	Shwartzman	Confirmed above and found renal cortical necrosis after 2 iv injections
1951	Stetson	Similarity betwen local and Arthus reactions
1952	Good & Thomas	Thorotrast or trypan blue substituted for first LPS injection in generalized reaction
1955	Thomas et al.	Fibrinoid deposits in vascular bed in generalized reaction
1956	Thomas	Epinephrine could replace first injection in localized reaction
1958	McKay & Shapiro	Followed changes in blood coagulability in generalized reaction
1960	Brunson & associates	Rabbits tolerant to epinephrine still show local reaction
1962	Brunson	In generalized reaction, first LPS injection "blocks" RES with fibrin and second blocks glomerular capillaries

damage. These effects became known respectively as the local and generalized Shwartzman reactions. Sanarelli, in 1894, (106) had made similar observations and his name is sometimes and should always be used along with Shwartzman's in referring to these phenomena.

These findings were pursued by a number of pioneering investigators. Thomas (126) was able to substitute an injection of epinephrine as the priming dose for the localized lesion and Thomas et al. (127) described the fibrinoid deposits found in the generalized Shwartzman. Good and Thomas (48), McKay and Shapiro (81), Stetson (121), Brunson and associates (34), and Leung Lee (72) were prominent in analyzing the mechanisms of these reactions. The fibrin-like deposits resulting from the first intravenous injection of endotoxin are believed to be cleared by cells of the RES but when the second injection (the eliciting dose) is given, additional deposits cannot be successfully eliminated and their accumulation in the glomeruli results in renal cortical necrosis, the major characteristic of the generalized reaction.

The basis for the local Shwartzman reaction is not well understood. Since epinephrine can replace endotoxin as the priming dose, Evers and Brunson (34) made rabbits tolerant to the hormone without diminishing the response to two injections of endotoxin. Moreover, the accumulation of leukocytes at the site of the lesion suggested that these cells might be responsible. Animals depleted of leukocytes by treatment with nitrogen mustard responded normally. Since fibrin was deposited in the vessels in and around the lesion, localized ischemia might play a role in necrosis.

Table 4. Enhanced Nonspecific Resistance

Year	Authors	Observation
1956	Rowley	LPS elevates nonspecific resistance to infection
1956	Dubos & Schaedler	Confirmed Rowley's observation
1956	Landy & Pillemer	Attributed elevated resistance to properdin increase
1958	Shilo	LPS blocks infection-enhancing effects of levans and dextrans
1959	Gledhill	LPS increases resistance to ectromelia virus infection in mice
1962	Fukui et al.	Requires prior exposure to specific bacterial antigen
1963	Kampschmidt et al.	LPS lowers serum iron
1964	Mulholland & Cluff	Emphasized role of leukocytes in resistance
1975	Kampschmidt et al.	Passively transferred resistance with LEM
1979	Galelli et al.	Transfer of resistance to infection with bone marrow cells

NONSPECIFIC RESISTANCE TO INFECTION

In 1956, Derrick Rowley (105) was the first to show that an injection of endotoxin made mice more susceptible to infection during the first three to four hours thereafter, but following an interval of 15-18 hours, a highly significant increase in resistance ensued. Dubos and Schaedler (32) and Landy and Pillemer (70) made similar observations the same year. The latter workers attributed the resistance to elevated serum levels of properdin. Other scientists who pursued these findings include Shilo (111), Alan Gledhill (45), Fukui et al. (41), Mulholland and Cluff (93), and others. It was generally believed that resistance was due to some influence on body phagocytes, probably to a combination of activation and mobilization, and/or to elevated levels of antibody. There were suggestions that a humoral factor was also involved. There seems to be general agreement today, however, that enhanced resistance to a broad spectrum of pathogens is probably related to the endotoxin-induced decrease in concentration of available iron in the serum of the mammalian host. Kampschmidt and Arrendondo (63) were the first to show that endotoxin lowers plasma iron and this was confirmed by Baker and Wilson (8) and Eaves and Berry (33). The factor responsible for lowered iron has been identified by Kampschmidt as leukocytic endogenous mediator (LEM) since it also produces fever, lowers serum zinc and elevates serum copper, and acts as a leukocytosis inducing factor. It also increases by passive transfer nonspecific resistance to infection with S. typhimurium (64). These effects occur in response to a partially purified factor. The term LEM was introduced because this substance is considered to have a broader spectrum of activity than does endogenous pyrogen. It has not been purified to homogeneity, however. Since all cells require iron and must

Table 5. Protection Against X-Irradiation

Year	Authors	Observation
1953	Mefferd et al.	Endotoxin protects against x-rays
Late 1950s	Smith & associates	LPS stimulates leukocyte recovery in irradiated mice
1959	Kind & Johnson	LPS enhances antibody synthesis after irradiation
1977	Urbaschek et al.	LPS fails to protect C3H/HeJ mice
1980	Behling et al.	Passive protection with postendotoxin serum against irradiation
1981	Addison & Berry	Protected C3H/HeJ mice with post-endotoxin serum

compete for it in whatever environment they encounter, the slower growth of pathogens in an iron-depleted animal gives the body defenses a chance to successfully eliminate the microorganism. These well established observations are finding new admirers in those now studying the genetics of iron acquisition in a variety of bacterial pathogens. Low iron would not readily account for the passive transfer of resistance to infection by bone marrow cells (43) unless they are responsible for the release of an iron-binding substance.

PROTECTION AGAINST X-IRRADIATION

In 1953, Mefferd et al. (84) reported the ability of endotoxin to protect against lethal x-irradiation. This was studied in depth during the late 1950's by Willie Smith and her colleagues (115-118). The severe leukopenia that follows lethal irradiation was not eliminated but leukopoiesis and a return to normal levels of circulating white blood cells was found to be accelerated. Kind and Johnson (66) demonstrated that endotoxin enhanced the restoration of antibody synthesis in x-irradiated rabbits, an effect somewhat related to the above work, at least in light of present-day knowledge about the cells involved in immunity. Little work was directed toward further analysis of this problem for nearly two decades until Urbaschek et al. (129) failed to protect nonresponder C3H/HeJ mice against x-irradiation with endotoxin. There was not colony stimulating activity found in the serum of those given endotoxin alone. Behling et al. (13) in Nowotny's laboratory found that they could passively protect mice against lethal irradiation using postendotoxin serum. Addison and Berry (1) confirmed these findings and were able also to protect the C3H/HeJ mice as well as tolerant mice with postendotoxin serum. Whether there are substances in serum other than colony stimulating factor that are required to protect animals against lethal irradiation must await studies with purified compounds.

ISOLATION AND CHEMISTRY OF ENDOTOXIN

In 1933, Boivin and Mesrobeanu (21) described the use of trichloroacetic acid (TCA) for the extraction of a toxic material from gram-negative bac-

Table 6. Isolation and Chemistry of Endotoxin

Year	Authors	Observation
1933	Boivin & Mesrobeanu	Use TCA to extract endotoxin
1937	Morgan	Used organic solvents and water to prepare LPS
1945	Goebel	Used organic solvents and water to prepare LPS
1952	Westphal & Luderitz	Developed phenol-water method of extracting LPS
1963	Osborn	Identified 2-keto-3-deoxycotonate as constituent of LPS
1964	Westphal el al.	Maintained liped moiety was active component in endotoxin
1964	Ribi et al.	Maintained polysaccharide moiety was active component in endotoxin
1966	Luderitz et al.	Reviewed structure of endotoxin
1971	Nowotny	Heterogeneity and endotoxin and its responsible moieties
1973	Luderitz et al.	Reviewed isolation and structure of lipid A
1975	Nowotny et al.	Polysacchairde moiety stimulates immune response in irradiated animals
1982	Ribi et al.	Nontoxic modified lipid A protects against tumors

teria. There were several publications from this group over the next few years. Lydia Mesrobeanu maintained her interest in endotoxin until her death under rather sad circumstances a few years ago in her native Romania. Nearly 20 years later, a major milestone was reached when Westphal and Luderitz (134) published their first paper on the phenol-water method of preparing endotoxin. This was their initial contribution to the field of endotoxinology and they and their associates at the Max Planck Institute in Freiburg are responsible as much as any other group for establishing endotoxin as a viable and worthwhile research field. Morgan in London in 1937 (91) and Goebel in New York in 1945 (47) had used mixtures of water and organic solvents to extract materials that contained polysaccharides, lipids, and proteins, similar to Boivin's preparation but the phenol-water preparation was proteinfree.

From the late 1950's until about the middle 1960's there was a difference of opinion between the Freiburg group of Westphal and Luderitz and Edgar Ribi and associates at the Rocky Mountain Laboratories and Maurice Landy and his colleagues at NIH as to the precise role lipids play in causing the biological effects of endotoxin. The two positions are well summarized in the articles that appeard in the book that grew out of the Rutgers meeting in 1963 (103,135). Nowotny (96) has consistently maintained that both lipids and polysaccharides are involved, which for selected activities, is generally accepted today. The detailed structure of bacterial lipopolysaccharide was summarized in an outstanding review article by Luderitz, Staub and Westphal (74). Another review of value by this group appeared a few years later (75). Mary Jane Osborn (98) had made an earlier contribution in which she identified 2-keto-3-deoxyoctonate as part of the LPS molecule. This unusual compound was later found to link the core polysaccharide to the lipid component. The isolation of lipid A and the establishment of its chemical structure and biological activity is summarized in a paper by Luderitz et al. (76). Most biological effects of endotoxin are due to the lipid A moiety but nontoxic polysaccharide constituents of LPS have the ability to stimulate recovery of the immune response in irradiated animals (97), protect against tumors (97), while nontoxic monophosphoryl lipid A will also necrotize tumors (104).

INFLUENCE OF ENDOTOXIN ON IMMUNE PHENOMENA

A voluminous literature on the influence of endotoxin on various immunological phenomena exists. These include the relationship betwen hypersensitivity and biological responses to endotoxin, endotoxin's antigenic properties, its influence on complement, its adjuvanticity, its ability to modulate the immune response, and its use or one of its constitutent parts as a protective antigen against heterologous gram negative pathogens. There are, no doubt, other facets of this area not included here but this gives a general idea of how broad the topic is. Only a few comments will be made about each of these dimensions below.

Stetson (123), as mentioned previously, and many others, such as Landy and Weidanz (71) believed that many of the biological effects of endotoxin were due to some manifestation of hypersensitivity. Kim and Watson (65), however, demonstrated that gnotobiotic colostrum-deprived miniature piglets devoid of detectable immunoglobulins in their blood were susceptible to the lethal effects of endotoxin. We found (15) neonatally thymectomized mice also normally susceptible to the lethal action of endotoxin. Repeated injections of cortisone, which result in strongly suppressed immunity, sensitized mice to the lethal action of endotoxin, according to the observations of Chedid and Boyer (26). Thus, it seems apparent that many, if not all, host reponses to endotoxin are not dependent on hypersensitivity.

Neter et al. (95) developed a technique for coating erythrocytes with alkali-treated endotoxin so that hemagglutinin titers or plaque assays could be used for measuring the immune response to it. Of course, O antigens and others could be evaluated in this way. Braude et al. (23) were able to prevent the generalized Shwartzman reaction in rabbits by pretreating them with antiserum derived from animals vaccinated with side-chain deficient mutants of Salmonella typhimurium. The protection was possible against heterologous LPS. McCabe (78) protected mice against several pathogens using the Re mutant of S. minnesota as vaccine but Bruins et al. (24) found poor protection in rabbits using lipid A as the vaccine when results were compared to those in animals that received the Re mutant vaccine. Lipid A can be solubilized as a BSA complex (42) but its usefulness in protecting animals is not dramatic, possibly because it is masked within the LPS structure.

Mergenhagen and associates (87) have made major contributions to our understanding of the interaction of endotoxin with the complement system. A number of biological effects of endotoxin can be related to specific components of complement or to products of their cleavage. Endotoxin is also known to activate the alternate pathway of the complement cascade (44).

MacLean and Holt (77) in a 1940 publication described the enhanced immune response to diphtheria or tetanus toxoid when it was incorporated with typhoid vaccine. It was shown by Johnson et al. (61) and Condie et al.

Table 7. Influence of Endotoxin on Immune Phenomena

Year	Authors	Observation
1940	McLean & Holt	Adjuvant effect of typhoid vaccine
1952	Chedid & Boyer	Repeated injections of cortisone into mice sensitized them to endotoxin
1956	Neter et al.	Developed the hemagglutination test for endotoxin
1956	Johnson et al.	Adjuvant effect of endotoxin
1964	Stetson	Host response due to hypersensitivity
1964	Landy & Weidanz	Biological effects due to antigen-antibody interaction
1964	Berry	Thymectomized neonatal mice sensitive to endotoxin
1966	Kim & Watson	Immunoglobulin-free piglets sensitive to endotoxin
1968	Gewurz et al.	LPS activates alternate C pathway
1971	Galanos et al.	Solubilized lipid A serves as better antigen
1972	McCabe	Protected mice against several gram-negative pathogens with an Re mutant
1973	Braude et al.	Prevented generalized Shwartzman with antiserum against rough mutant of S. minnesota
1973	Mergenhagen et al.	Reviewed interaction of endotoxin with complement
1973	Ziegler et al.	Protected agranulocytic rabbits with serum from a rough vaccine
1977	Bruins et al.	Lipid A was a poor vaccine to protect rabbits
1979	Morrison & Ryan	Effect of LPS on cell types in the immune response - a review
1981	McGhee et al.	Review of work on adjuvant effect of LPS on the IgA response
1982	Ziegler et al.	Treated human beings with bacteremia with anti-serum against rough E. coli
1983	Friedman et al.	Immunomodulatory effect of LPS

(29) about 15 years later that endotoxin was the component in typhoid vaccine that had the adjuvant effect. This early work is summarized in papers by Johnson (59,60). McGhee and his collaborators (80) have focused on the influence of endotoxin on the secretory antibody response in experimental animals and man, especially as it relates to dental caries and periodontal disease. The influence of endotoxin on various cell types involved in the immune response has been reviewed by Morrison and Ryan (92) but this issue needs much more work before it is better understood. Friedman et al. (40)

21

Table 8. Metabolic Effects

Year	Authors	Observation
1923	Menten & Manning	Depletion of carbohydrates
1923	Zeckwer & Goodell	Depletion of carbohydrates
1948	Kun	Blocks entry of glucose in cells and inhibits glycogenesis
1959	Berry & Smythe	Proved blockage of glyconeogenesis
1960s – present	John & Judy Spitzer and associates	Analyzed nature of hypoglycemia and lipid metabolism in endotoxemia
1963 –	Berry & associates	Discovered GAF and its mode of action
1970s – present	Filkins and associates	Studied parallelism in carbohydrate metabolism in endotoxemia and with insulin
1970s – present	Mela and associates	Effect of LPS on mitochondria
1972	Agarwal	LPS blocks glucocorticoid receptors
1981	Lowitt et al.	Enzyme induction and mixed function oxidases in endotoxemia
1982	McCallum & Stith	LPS blocks glucocorticoid receptors

have recently described the immunomodulatory effect of endotoxin and a non-toxic polysaccharide moiety derived from it in both in vivo and in vitro experiments.

As mentioned previously, McCabe (78) successfully immunized mice with an Re mutant of S. minnesota and protected them against a variety of gram-negative pathogens. Ziegler et al. (139) protected agranulocytic rabbits against pseudomonas bacteremia with serum from animals immunized with a rough strain of E. coli. Direct immunization with E. coli 0111 was not protective. Clinical trials using immune serum against the rough E. coli strain gave promising results in patients with gram-negative bacteremia (140).

METABOLIC EFFECTS OF ENDOTOXIN

Menten and Manning (86) and Zeckwer and Goodell (138) sixty years ago vaccinated rabbits with a typhoid vaccine and observed an initial hyperglycemia followed by a progressively severe hypoglycemia. Throughout this time, liver glycogen reserves were steadily declining. Kun observed in rats that endotoxin prevented the conversion of glucose to liver glycogen (67) and he also noted that the entry of glucose into the isolated rat diaphram was suppressed in poisoned animals (68). Berry and his associates, over a period of years, confirmed these original findings (16) and extended them to show that the synthesis of certain key hepatic enzymes under the regulatory control of glucocorticoid hormones is blocked by endotoxin (19). This, in turn, is believed to interfere with gluconeogenesis in the liver. This work has been reviewed in several articles (16-18) and will not be elaborated at this time. The current evidence suggests very strongly that the induction of phosphoenolpyruvate carboxykinase (PEPCK), the rate limiting enzyme in gluconeogenesis is blocked, not by endotoxin but by a mediator, GAF, which interferes with either transcription, translation, or posttranslational events (19). The Spitzers, in a recent review (119), agree with the sequence of changes in carbohydrate metabolism elicited by endotoxin. They attribute the initial hyperglycemia to an elevated sympathetic discharge and an increased hepatic output of glucose fueled by lactate and alanine as gluconeogenic precursors. The severe hypoglycemia is attributed to the inability of the liver to maintain gluconeogenesis at the same rate as sugar is demanded. Impaired blood flow to adipose tissue is believed to account for the net decrease in release of free fatty acids. Moreover, a reduction in lipoprotein lipase is thought to explain the hypertriglyceridemia.

Filkins and his students (37) have emphasized the contributions of insulin or insulin-like activity to the altered carbohydrate metabolism in endotoxicosis. The parallelisms they discern are unarguable. One of the challenges they face is to identifiy the factor responsible for the commonly observed changes.

It is difficult to be cursory in discussing a topic that has occupied so much of my research attention. I shall, therefore, mention briefly the work of some of my former students who were at least trained to be independent thinkers since some of their findings differ from mine. Agarwal (2) and McCallum and Stith (79) attribute the inhibition by endotoxin of glucocorticoid induction of PEPCK to an interference with binding of the hormone to its receptors in the cell. Shackleford in my lab (19) could not confirm their observations so the reasons for the difference in results is not yet explained. Mela and her associates have published a number of papers dealing with the effect of endotoxin on mitochondrial function. Various levels of oxidative metabolism are interrupted as she reveals in a recent review (85). Lowitt et al. (73) have studied the response of several regulatory enzymes to endotoxin. Williams and Szentivanyi will discuss some of this work in another chapter of this book.

Table 9. Interaction of Endotoxin with Membranes

Year	Authors	Observation
1971	Shands	Reviewed physical properties of endotoxins
1972	Anderson et al.	Mitogenic effect of endotoxin
1973	Shands	Reviewed interaction of endotoxin with membranes
1975	Springer & Adye	Isolated from rbc lipoglycoprotein believed to be the LPS binding substance
1975	Skidmore et al.	Lipid A with associated protein in mitogenic for C3H/HeJ spleen cells
1977	Kabir & Rosensteich	C3H/HeJ spleen cells bind LPS normally

Soon after Neter et al. (95) established that endotoxin attaches to erythrocyte membranes Shands and his associates initiated a series of studies on the physical properties of the bacterial substance and related this to its affinity for membranes. This work is well summarized in two review articles (108,109). With electron photomicrographs, Shands could detect particles of various shape, including discs, ribbons, and doughnuts. All were bilayers. The particles attached by their edges to membranes. This attachment was believed to allow the lipid moiety to interact with and possibly become incorporated into membrane lipids. Springer (120) has also made significant observations regarding endotoxin receptor sites on cell membranes. Partial chemical characterization of the erythrocyte receptor has been described.

With the discovery that endotoxin acts as a potent mitogen for certain cells of the RES (3), a number of publications soon appeared. It was generally assumed that attachment to, for example, B lymphocytes resulted in the initiation of their division. Kabir and Rosenstreich (62), however, reported that spleen cells from the nonresponder C3H/HeJ mice bound as much ^{14}C-labeled LPS as cells from the responder C3H/HeN strain. The nonresponder mice were found to react more or less normally to preparations of endotoxin to which a lipid A-associated protein was present (114). This complicated relationship will be more fully elucidated in this volume by Drs. Morrison and Sultzer.

PHARMACOLOGICAL ASPECTS OF ENDOTOXEMIA

One of the editors of this book, Professor Szentivanyi, is one of the earliest investigators of the pharmacology of host responses to endotoxin. He and the late Charles Fishel (38) focused their efforts on the sensitization of mice to histamine and serotonin following the administration of

Table 10. Pharmacological Aspects of Endotoxemia

Year	Authors	Observation
1960	McLean & Berry	Protected against lethal effect of LPS with a drug that blocks inhibitor effects of epinephrine
1963	Szentivanyi et al.	Interrelationships between pertussis-induced pharmacologic hypersensitivity and glucose metabolism
1963	Fishel & Szentivanyi	Involvement of beta adrenergic mechanisms in the sensitization of mice to histamine and serotonin by pertussis vaccine
1964	Hinshaw	Role of biogenic amines in endotoxemia
1975	B. & R. Urbaschek	Protected microcirculation with a glucofuranoside
1980	Faden & Holaday	Maintained blood pressure in endotoxemic animals with naloxone

Bordetella pertussis vaccine. They suggested that this might result from
the formation of a substance that blocks the beta-adrenergic receptors on
cells thereby leaving the alpha-adrenergic receptors unbalanced. This type
of functional imbalance was not believed to be limited to the effects of B.
pertussis but to represent a far broader manifestation of host-parasite
interaction. Many of the earliest detectable responses to endotoxin are
believed to be initiated by the release of biogenic amines and subsequent
changes in the physiology of a poisoned animal may reflect some of these
early influences (57). To test this possibility, McLean and Berry (82)
tested drugs that exerted various forms of adrenergic inhibition and found
one, 3,4-dichloroisoproterenol, capable of providing statistically significant protection against lethality in mice. The drug is known to block inhibitor and vasodilator effects of epinephrine and related catecholamines.

The Urbascheks have carried out detailed studies on various pharmacological factors involved in the altered microcirculation of several animals.
Pretreatment with detoxified endotoxin prevented these changes and protected
against the microcirculatory alterations but not death. These experiments
are summarized in a brief review (128).

Faden and Holaday (35) have successfully prevented the hypotension that
develops in rats and dogs by treating them with the endorphin agonist, naloxone. Even though blood pressure was more or less normal, survival was not
consistently increased. Larger doses of naloxone can protect rats against
endotoxin lethality but without any evidence of other physiological improvement. The relationship between endotoxin shock and endorphins needs additional work for clarification. Not only does this particular aspect of a
possible pharmacological intervention in the the manifestations of endotoxemia need further study but other uses of drugs in altering the severity of
host response to endotoxin should be investigated as well.

GENETICS OF THE RESPONSE TO ENDOTOXIN

In 1940, Hill et al. (56) developed a strain of mice that was unusually
resistant to endotoxin lethality. It was selected for this purpose alone.
This strain was lost as a result of World War II but prior to its destruction
one final experiment proved that these animals were normally susceptible to
mouse typhoid. Heppner and Weiss (55) found that inbred A mice were

Table 11. Genetics of the Response to Endotoxin

Year	Authors	Observation
1940	Hill et al.	Developed inbred mice resistant to LPS
1965	Heppner & Weiss	Inbred A mice unusually susceptible to LPS
1969	Sultzer	C3H/HeJ mice unusually resistant to LPS
1973	Vas et al.	Found other resistant inbred strains
1974	Watson & Riblet	Single gene found to control mitogenic and immune response to endotoxin
1976	Glode et al.	Spleen cell transfer betwen irradiated C3H/HeJ or C3H/HeN mice resulted in characteristic response of donor animal

unusually susceptible to endotoxin while Sultzer was the first to report the unusually high resistance to endotoxin of C3H/HeJ mice (124). Vas et al. (130) made a similar observation soon thereafter. This last named strain has been a boon to endotoxin research and has made possible numerous studies otherwise impossible. Watson and Riblet (132), for example, were able to show that a single gene controls both the mitogenic and immune response to LPS in the C3H mouse while Glode et al. (46) transferred spleen cells from the susceptible HeN strain into irradiated HeJ mice and rendered them more susceptible. The reverse transfer of spleen cells from HeJ mice into irradiated HeN mice gave a more resistant chimera. These observations leave no doubt that endotoxin lethality is cell-mediated. No one has succeeded in identifying a lethal mediator, however, except by implication. A short review by Watson et al. (133) of the genetic control of the response to endotoxin appeared in 1980.

SELECTED MEDIATORS OF ENDOTOXIN EFFECTS

There is ample reason to believe that many of the biological and metabolic effects of endotoxin are mediated (16). Endotoxin was first reported by Beeson (10) to elicit a biphasic febrile response in rabbits. The first rise in body temperature occurred only after a delay of 15 to 30 minutes. Beeson (12) then noted that a pyrogen released from leukocytes produced a similar response and the delay was shortened or eliminated. Atkins and Wood (5) were also actively investigating fever and how endotoxin initiated the response. While endogenous pyrogen was originally thought to be produced by granulocytic leukocytes [for reviews see Atkins and Atkins and Bodel (4,7)], it subsequently was shown by two groups, Atkins et al. (6) and Hahn et al. (53) to come from macrophages. It has been purified to homogeneity (94). Whether or not endogenous pyrogen acts directly on the thermo-regulatory center in the hypothalamus or acts through the release of another messenger is not resolved. There is considerable evidence suggesting that a product of arachidonic acid, possibly PGE_2, is the final mediator (30,113). The effectiveness of aspirin as an antipyretic supports this role of a prostaglandin.

It has been known for some years that endotoxin induces the release or synthesis of interferon. This topic has been effectively reviewed by Ho (58) and his paper should be consulted for detailed references. Gamma interferon is probably the type formed in response to endotoxin since its properties differ from those of alpha and beta interferons. Now that recombinant interferons are available for investigation, this interesting and potentially valuable therapeutic substance is destined for current detailed study.

In 1966, Bradley and Metcalf (22) introduced a technique for the in vitro culturing of myeloid precursor cells in mouse bone marrow. Their procedure was based on one used by Pluznik and Sachs (101) for growing mast cells in vitro. Colony formation by these cells is dependent upon the constant presence of a stimulatory glycoprotein called colony stimulating-factor (CSF) (102). The release of CSF from a variety of cell types may be spontaneous and appear in serum and urine or induced in lymphoid cells by antigens, mitogens and endotoxin (for a recent review of the topic, see Pluznik) (100). CSF has been purified to the point that a useful radio-immunoassay has been developed by Shadduck and Waheed (107). In addition, recent studies by Moore et al. (89) have revealed a complex feedback regulation of CSF-induced macrophage proliferation by endogenous E prostaglandins and interferon alpha/beta.

Other mediators of great current interest are interleukins 1 and 2 (IL1 and IL2). Briefly, interleukin 1 was formally called lymphocyte activating

Table 12. Selected Mediators of Endotoxin Effects - Fever

Year	Authors	Observation
1946	Beeson	Biphasic fever response
1948	Beeson	Identified leukocytic pyrogen
1955	Atkins & Wood	Pathogenesis of fever
1960	Atkins	Review of early work on fever
1967	Atkins et al.	Macrophages as origin of endogenous pyrogen (EP)
1967	Hahn et al.	Macrophagic origin of EP
1974	Murphy et al.	Purified EP to homogeneity
1980	Cranston	PGE_2 as fever mediator
1980	Skarnes & McCracken	PGE_2 as fever mediator

factor (LAF) and hence stimulates antibody production by activating T cells. The mechanism of activation remains unclear. Endotoxin stimulates the release of IL1 through its action on macrophages. This probably accounts, at least in part, for the adjuvanticity of endotoxin. These relationships are reviewed in two recent publications (69,88).

Table 13. Selected Mediators of Endotoxin Effects - CSF, Interleukin 1 and 2, GAF

Year	Authors	Observation
1965	Pluznik & Sachs	Grew mast cells in vitro
1966	Bradley & Metcalf	Cultured myeloid cells in bone marrow in vitro
1966	Pluznik & Sachs	Identified a glycoprotein necessary for colony growth (CSF)
1977	Berry	Review of glucocorticoid antagonizing factor (GAF)
1979	Shadduck & Waheed	Purified CSF
1980	Mizel	Review of interleukin 1
1980	Farrar et al.	Review of interleukin 2
1983	Pluznik	Recent review of CSF
1983	Lachman	Review of interleukin 1
1983	Berry & Gaska	Review of GAF
1984	Moore et al.	Identified a complex feedback regulation of macrophage proliferation
1985	Moore et al.	Review of GAF

Interleukin 2 is also involved in the adjuvanticity of endotoxin since it acts as a T-cell growth factor. It stimulates thymocyte mitogenesis.

This topic is briefly reviewed by Farrar et al. (36) is further discussed by Nakano and Nitta in Section IV, Chapter II of this book.

It has been known for decades that adrenal cortical hormones diminish or suppress a number of the biological effects of endotoxin. For reviews see Berry (16), Berry and Gaska (18) and Moore et al. (90). This interaction is due to a macrophage factor produced in response to endotoxin that antagonizes the hormone effects. The factor has been called glucocorticoid antagonizing factor (GAF). GAF has only been partially purified and is believed to have a molecular weight similar to that of TNF, in excess of 150,000. When serum rich GAF is administered to nonresponder C3H/HeJ mice or to tolerant mice they behave as if they were normally susceptible. One of the effects GAF has (and also endotoxin when given to typical responder animals) is to block the hormonal induction of selected hepatic regulatory enzymes. Phosphoenolpyruvate carboxykinase (PEPCK) is an example and its behavior has been extensively studied. It appears as if GAF and indirectly endotoxin does not interfere with the entry of hydrocortisone into liver cells and their binding to the receptor protein and its entry into the nuclei but rather to an interference with PEPCK-mRNA synthesis. The details are yet to be elucidated.

PROSPECTIVE VIEW OF ENDOTOXIN RESEARCH

The only prospective event that can be predicted is another endotoxin symposium or conference to be held somewhere in the world before the end of

Table 14. Schedule of Endotoxin Conferences

Year	Place	Organizers
1959	Freiburg, W. Germany	Westphal & Luderitz
1961	Bryn Mawr, PA	Berry, Landy, and Brown
1963	Wakulla Springs, FL	Suter and Shands
1963	Rutgers Univ., NJ	Landy and Braun
1965	Chicago, IL	Ritts
1965	New York, NY	Nowotny
1967	Paris, France	Chedid
1968	Heidelberg, W. Germany	B. and R. Urbaschek
1972	Airlie House, VA	Wolff and Kass
1973	Vienna, Austria	B. and R. Urbaschek and Neter
1976	Austin, TX	Berry, Mergenhagen, and Neter
1981	Heidelberg, W. Germany	B. and R. Urbaschek
1982	Kansas City, KS	Anderson and Unger
1983	Walter Reed Hospital, Washington, D.C.	Alving, Mattsby-Balzar, and Morrison
1985	Tampa, FL	Szentivanyi, Nowotny and Friedman

1987. So far there have been sixteen since the first meeting in 1959, twenty-six years ago, and there have been six satellite meetings since 1973, the last one, this past summer in Quebec (see Tables 14 and 15). For awhile it looked as if the observations that gave impetus to much of the interest in endotoxin, namely its tumor necrotizing activity, was no longer a fertile area for study but now that TNF has been purified, the details of its mode of action can now be explored. The human TNF gene had been cloned and large quantities will soon be available for clinical tests. There is a good possibility that tolerance will not develop to TNF and a valuable antitumor drug will hopefully be added to the war on cancer. Moreover, Ribi and his associates and Nowotny and his have pursued the isolation of an LPS component that is nontoxic and still acts as an antitumor agent.

Very little is being done today on the febrile response to endotoxin or to its mediator, endogenous pyrogen, but since the role of prostaglandins and endorphins in this process is poorly understood, some imaginative neurophysiologist within the next few years is likely to explain what now appears to be a complex relationship.

Studies of the Shwartzman phenomena and of tolerance have virtually disappeared from the literature. Unanswered questions remain but they will not be solved easily. Because of the difficulty of the problem and probably a low priority for funding, there is not likely to be much activity here.

Publications concerned with nonspecific resistance to infection have appeared steadily throughout the years. The basis of this effect, in addition to the lowering of serum iron, will continue to attract attention. The host factors involved in resistance and the components of LPS responsible for eliciting this change will continue to be studied. Bertok (20) recently reviewed work in which detoxified endotoxin has been used successfully in stimulating nonspecific resistance.

There does not seem to be much prospect for research concerned with protection against ionizing radiation. Should the world experience a nuclear war, the prospects for survival for any of a wide variety of reasons are not sanguine, regardless of precautionary measures taken. Questions do, however, remain. The protective mediator has not been purified unless someone has shown it to be CSF. If it has been done by direct experimentation, and not by inference it has escaped my attention. It would be important to know whether repeated injections of the protective mediator would continue to be effective. Would tolerance be by-passed by this means?

We are now entering an era in which structure-function relationships of LPS moieties is under active investigation, and work of this type will no

Table 15. Schedule of Satellite Meetings

Year	Place and Occasion	Organizer
1973	Amsterdam - European Immunology Society	Bertok
1978	Babolna, Hungary - European Immunology Society	Bertok
1980	Paris - International Congress of Immunology	Agarwal
1983	Tokyo - International Congress of Immunology	Agarwal
1983	Budapest - European RES Society and Tissue Culture Society	Bertok
1984	Quebec - International Congress of Endocrinology	Agarwal

doubt continue for at least another decade. This approach has tremendous potential for helping us understand many of the puzzles about endotoxin that remain unanswered.

In addition to predicting another meeting on endotoxin by the end of 1987, I feel most secure in projecting a steady outpouring of papers dealing with the various ways in which LPS and/or its fragments modify the immune response. In fact, endotoxin has provided and will continue to serve as a probe that helps us understand the intricacies of cell interactions in immunity. This work will continue for several years and will play an important role in their elucidation. Moreover, the unanswered question of whether endotoxin contributes to the pathogenesis of gram-negative infections also will continue to be explored. To be sure, the normal or even greater suscep- tibility to infection with gram-negative organisms in inbred strains of mice known to be highly resistant to endotoxin may not be as conclusive an argu- ment against a pathogenic role of endotoxin as some investigators would make of it. To extrapolate in vivo events during an infection to those that follow the administration of a bolus of endotoxin or even its infusion over time is hardly a rigorous analysis.

When work started in my laboratory on the effects of endotoxin on host metabolism this was a lonely area. Mark Woods and associates (136) were among the few who were also engaged in this type of study and through the years the numbers have slowly but steadily increased. There are far more questions remaining than answers provided so one can predict with some confidence that this will continue to grow in importance in the years ahead. Inbred strains of animals (mostly mice) should prove of value in these inves- tigations and the approach should and will become increasingly molecular. Even at a ripe old age the investigator has been forced to learn and apply the techniques of molecular biology in order to understand the specific site where GAF blocks enzyme synthesis. It is an exciting prospect and I wish Judy and John Spitzer, James Filkins, Joseph Williams, and Andor Szentivanyi all the best in their pursuit of this problem.

The underlying mechanism of a mitogenic stimulus has never been ex- plained at the molecular level. Obviously, endotoxin and any other mitogen must first interact with the membrane of the target cell and this interaction must, in some at presently unknown manner, activate the cell cycle. Just as Shands (108,109) and Springer (120) have tackled this problem in the past, we have in this book chapters indicating continued progress in this area (Jacobs and Morrison).

The pharmacological approach to an analysis of biological and metabolic influences of endotoxin is presented in chapters by Drs. Szentivanyi and Williams and Shepherd and associates. As mentioned previously, this is a fertile field for future investigation and will surely prosper.

The identification of a number of mediators of endotoxemic effects has opened new vistas of research. Some, but not all, have been purified and that effort must continue. My lab is delinquent in not having purified GAF. Once pure mediators are available, antibody can be raised against them so as to pinpoint more precisely the role that mediator plays following an injec- tion of endotoxin. Not very much has been done in this way and it surely awaits the near future. Purified mediators should make it feasible to ana- lyze their effect(s) at the molecular level which now seems so promising in our effort to understand the role of GAF in blocking PEPCK induction by hydrocortisone. GAF might offer a means of combating hyperactivity of the adrenal cortex and the interleukins should prove valuable in correcting some types of immune deficiency. These examples are not all-inclusive but serve to illustrate the point.

ACKNOWLEDGMENTS

The work reported from the author's laboratory was supported by a research grant (AI-100) from the National Institue of Allergy and Infectious Diseases, Department of Health and Human Services.

REFERENCES

1. P. D. Addison and L. J. Berry, J. Reticuloendothelial Soc. 30:301 (1981).
2. M. K. Agarwal, Int. J. Biochem. 3:408 (1972).
3. J. Anderson, G. Mollen, and O. Sjoberg, Cell. Immunol. 4:381 (1972).
4. E. Atkins, Physiol. Rev. 40:580 (1960).
5. E. Atkins and W. B. Wood, Jr., J. Exp. Med. 101:519 (1955).
6. E. Atkins, P. Bodel, and L. Francis, J. Exp. Med. 126:357 (1967).
7. E. Atkins and P. Bodel, N. Eng. J. Med. 286:27 (1972).
8. P. J. Baker and J. B. Wilson, J. Bacteriol. 90:903 (1965).
9. P. B. Beeson, Proc. Soc. Exp. Biol. Med. 61:248 (1946).
10. P. B. Beeson, J. Exp. Med. 86:29 (1947).
11. P. B. Beeson, J. Exp. Med. 86:39 (1947).
12. P. B. Beeson, J. Clin. Invest. 27:524 (1948).
13. U. H. Behling, P. H. Pham, F. Madani, and A. Nowotny, in: "Microbiology 1980," D. Schlessinger, ed., American Society of Microbiology, Washington, D. C. (1980).
14. B. Benacerraf, G. J. Thorbecke, and D. Jacoby, Proc. Soc. Exp. Biol. and Med. 100:796 (1959).
15. L. J. Berry, in: "Bacterial Endotoxin," M. Landy and W. Braun, eds., Rutgers University Press, New Brunswick, NJ (1964).
16. L. J. Berry, in: "Microbial Toxins," S. Kadis, G. Weinbaum, and S. J. Ajl, eds., Academic Press, New York (1971).
17. L. J. Berry, CRC Critical Rev. Toxicol 4:239 (1977).
18. L. J. Berry and J. E. Gaska, in: "Beneficial Effects of Endotoxin," A. Nowotny, ed., Plenum Press, New York (1983).
19. L. J. Berry and G. M. Shackleford, in: "Adrenal Steroid Antagonisms," M. K. Agarwal, ed., de Gruyter and Company, Berlin (1984).
20. L. Bertok, in: "Beneficial Effects of Endotoxins," A. Nowotny, ed., Plenum Press, New York (1983).
21. A. Bouvin and L. Mesrobeanu, C. R. Soc. Biol. 112:611 (1933).
22. T. R. Bradley and D. Metcalf, Aust. J. Exp. Biol. and Med. 44:287 (1966).
23. A. I. Braude, H. Douglas, and C. E. Davis, J. Infect. Dis. 128S:149 (1973).
24. S. C. Bruins, R. Stumacher, M. A. Johns, and W. R. McCabe, Infect. Immun. 17:16 (1977).
25. E. A. Carswell, L. J. Old, R. L. Kassel, S. Green, N. Fiore, and B. Williamson, Proc. Natl. Acad. Sci. USA 72:3666 (1975).
26. L. Chedid and F. Boyer, Compt. Rend. Soc. Biol. 146:239 (1952).
27. L. Chedid and M. Parant, Ann. Inst. Pasteur 101:170 (1961).
28. W. B. Coley, Amer. J. Med. Sci. 105:487 (1893).
29. R. M. Condie, S. J. Zak, and R. A. Good, Proc. Soc. Exp. Biol. Med. 90:355 (1955).
30. W. I. Cranston, in: "Microbiology 1980," D. Schlessinger, ed., American Society of Microbiology, Washington, D.C. (1980).
31. H. J. Creech, R. F. Hankwitz, Jr., and D. R. A. Wharton, Cancer Res. 9:159 (1949).
32. R. J. Dubos and R. W. Schaedler, J. Exp. Med. 104:53 (1956).
33. G. N. Eaves and L. J. Berry, Am. J. Physiol. 211:800 (1966).
34. C. G. Evers and J. G. Brunson, Am. J. Pathol. 37:55 (1960).
35. A. I. Faden and J. W. Holaday, J. Infect. Dis. 142:229 (1980).

36. J. J. Farrar, S. B. Mizell, J. Fuller-Bonar, M. L. Hilfiker, and W. L. Farrar, in: "Microbiology 1980," D. Schlessinger, ed., American Society of Microbiology. Washington, D.C. (1980).
37. J. P. Filkins, Circulatory Shock 9:269 (1982).
38. C. W. Fishel, A. Szentivanyi, and D. W. Talmage, in: "Bacterial Endotoxins," M. Landy and W. Braun, eds., Rutgers University Press, New Brunswick, NJ (1964).
39. H. H. Freedman, Proc. Soc. Exp. Biol. Med. 102:504 (1959).
40. H. Friedman, S. Specter, and R. C. Butler, in: "Beneficial Effects of Endotoxin," A. Nowotny, ed., Plenum Press, New York (1983).
41. G. M. Fukui, G. Chandlee, P. Klapsogeorge, and F. M. Berger, Fed. Proc. 21:277 (1962).
42. C. Galanos, O. Luderitz, and O. Westphal, Eur. J. Biochem. 24:116 (1971).
43. A. Galelli, Y. LeGarrec, and L. Chedid, Infect. Immun. 23:232 (1979).
44. H. Gewurz, H. S. Shin, and S. E. Mergenhagen, J. Exp. Med. 128:1049 (1968).
45. A. W. Gledhill, Brit. J. Exp. Pathol. 40:195 (1959).
46. L. M. Glode, S. E. Mergenhagen, and D. L. Rosenstreich, Infect. Immun. 14:626 (1976).
47. W. F. Goebel, F. Binkley, and E. Perlman, J. Exp. Med. 81:315 (1945).
48. R. A. Good and L. Thomas, J. Exp. Med. 96:625 (1952).
49. K. J. Goodrum and L. J. Berry, Proc. Soc. Exp. Biol. Med. 159:359 (1978).
50. S. Green, A. Dohrjansky, E. A. Carswell, R. L. Kassel, L. J. Old, N. Fiore, and M. K. Schwartz, Proc. Natl. Acad. Sci. USA 73:381 (1976).
51. S. E. Greisman, F. A. Garozza, Jr., and J. D. Hills, J. Exp. Med. 117:663 (1963).
52. S. E. Greisman and B. DuBuy, Proc. Soc. Exp. Biol. Med. 148:675 (1975).
53. H. Hahn, D. C. Char, W. B. Postel, and W. B. Wood, Jr., J. Exp. Med. 126:385 (1967).
54. U. Hammerling, L. J. Old, E. a. Carswell, J. Abbott, H. F. Oettgen, and M. K. Hoffmann, in: "Microbiology 1980," D. Schlessinger, ed., American Society of Microbiology, Washington, D.C. (1980).
55. G. Heppner and D. W. Weiss, J. Bacteriol. 90:696 (1965).
56. A. B. Hill, J. M. Hatswell, and W. W. C. Topley, J. Hyg. 40:538 (1940).
57. L. B. Hinshaw, in: "Bacterial Endotoxins," M. Landy and JW. Braun, eds., Rutgers University Press, New Brunswick, NJ (1964).
58. M. Ho, in: "Beneficial Effects of Endotoxins," A. Nowotny, ed., Plenum Press, New York (1983).
59. A. G. Johnson, in: "Bacterial Endotoxins," M. Landy and W. Braun, eds., Rutgers University Press, New Brunswick, NJ (1964).
60. A. G. Johnson, in: "Beneficial Effects of Endotoxins," A. Nowotny, ed., Plenum Press, New York (1983).
61. A. G. Johnson, S. Gaines, and M. Landy, J. Exp. Med. 103:225 (1956).
62. S. Kabir and D. L. Rosenstreich, Infect. Immun. 15:156 (1977).
63. R. F. Kampschmidt and M. I. Arrendondo, Proc. Soc. Exp. Biol. Med. 113:142 (1963).
64. R. F. Kampschmidt and L. A. Pulliam, J. Reticuloendothelial Soc. 17:162 (1975).
65. Y. B. Kim and D. W. Watson, Ann. NY Acad. Sci. 133:727 (1966).
66. P. Kind and A. G. Johnson, J. Immunol. 82:415 (1959).
67. E. Kun, J. Biol. Chem. 174:761 (1948).
68. E. Kun, Proc. Soc. Exp. Biol. Med. 68:496 (1948).
69. L. R. Lachman, in: "Beneficial Effects of Endotoxins," A. Nowotny, ed., Plenum Press, New York (1983).
70. M. Landy and L. Pillemer, J. Exp. Med. 104:838 (1956).
71. M. Landy and W. P. Weidanz, in: "Bacterial Endotoxins," M. Landy and

W. Braun, eds., Rutgers University Press, New Brunswick, NJ (1964).

72. L. Lee, J. Exp. Med. 115:1065 (1962).
73. S. Lowitt, A. Szentivanyi, and J. F. Williams, Biochem. Pharmacol. 30:1999 (1981).
74. O. Luderitz, A. M. Staub, and O. Westphal, Bacteriol. Rev. 30:192 (1966).
75. O. Luderitz, O. Westphal, A. M. Staub, and H. Nikaido, in: "Microbial Toxins IV Bacterial Endotoxins," G. Weinbaum, S. Kadis, and S. J. Ajl, eds., Academic Press, New York (1971).
76. O. Luderitz, C. Galanos, V. Lehmann, M. Nurminen, E. T. Rietschel, G. Rosenfelder, M. Simon, and O. Westphal, J. Infect. Dis. 128:17 (1973).
77. I. J. MacLean and L. B. Holt, Lancet 2:581 (1940).
78. W. McCabe, J. Immunol. j108:601 (1972).
79. R. E. McCallum and R. D. Stith, Circ. Shock 8:220 (1982).
80. J. R. McGhee, S. M. Michalek, H. Kiyono, J. L. Babb, M. P. Clark, and L. M. Mosteller, in: "Mechanisms in Mucosal Immunity," L. A. Hanson and J. W. Strober, eds., Raven Press, New York (1981).
81. D. G. McKay and S. S. Shapiro, J. Exp. Med. 107:353 (1958).
82. R. A. McLean and L. J. Berry, Proc. Soc. Exp. Biol. Med. 105:91 (1960).
83. D. N. Mannel, D. L. Rosenstreich, and S. E. Mergenhagen, Infect. Immun. 24:573 (1979).
84. R. B. Mefferd, D. T. Hendel, and J. B. Loeffer, Proc. Soc. Exp. Biol. Med. 85:53 (1953).
85. L. Mela, in: "Cellular Biology of Endotoxin," L. J. Berry, ed., Elsevier, Amsterdam, in press (1985).
86. M. L. Menten and H. M. Manning, J. Med. Res. 44:675 (1924).
87. S. E. Mergenhagen, R. Snyderman, and J. K. Phillips, J. Infect. Dis. 128S:78 (1973).
88. S. B. Mizel, in: "Microbiology 1980," D. Schlessinger, ed., American Society of Microbiology, Washington, D.C. (1980).
89. R. N. Moore, F. J. Pitruzello, H. S. Larsen, and B. T. Rouse, J. Immunol. 133:541 (1984).
90. R. N. Moore, G. M. Shackleford, and L. J. Berry, in: "Cellular Biology of Endotoxin," L. J. Berry, ed., Elsevier, Amsterdam, in press (1985).
91. W. T. J. Morgan, Biochem. J. 31:2003 (1937).
92. D. C. Morrison and J. L. Ryan, Adv. Immunol. 28:293 (1979).
93. J. H. Mulholland and L. E. Cluff, in: "Bacterial Endotoxins," M. Landy and W. Braun, eds., Rutgers University Press, New Brunswick, NJ (1964).
94. P. A. Murphy, P. J. Chesney, and W. B. Wood, Jr., J. Lab. Clin. Med. 83:310 (1974).
95. E. Neter, O. Westphal, O. Luderitz, E. A. Gorznyski, and E. A. Eichenberger, J. Immunol. 76:377 (1956).
96. A. Nowotny, in: "Microbial Toxins, IV," G. Weinbaum, S. Kadis, and JS. J. Ajl, eds., Academic Press, New York, (1971).
97. A. Nowotny, U. H. Behling, and H. L. Chang, J. Immunol. 115:199 (1975).
98. M. J. Osborn, Proc. Nat. Acad. Sci. USA 50:499 (1963).
99. R. Z. Pfeiffer, Hyg. Infektkv. 11:393 (1893).
100. D. H. Pluznik, in: "Beneficial Effects of Endotoxins," A. Nowotny, ed., Plenum Press, New York (1983).
101. D. H. Pluznik and L. Sachs, J. Cell. Physiol. 66:319 (1965).
102. D. H. Pluznik and L. Sachs, Exp. Cell Res. 43:553 (1966).
103. E. Ribi, R. L. Anacker, K. Fukushi, W. T. Haskins, M. Landy, and K. C. Milner, in: "Bacterial Endotoxins," M. Landy and W. Braun, eds., Rutgers University Press, New Brunswick, NJ (1964).
104. E. Ribi, K. Amano, J. L. Cantrell, S. M. Schwartzman, R. Parker, and K. Takayama, Cancer Immunol. Immunother. 12:91 (1982).

105. D. Rowley, Brit. J. Exp. Pathol. 37:233 (1956).
106. J. Sanarelli, Ann. Inst. Pasteur 8:193 (1894).
107. R. K. Shadduck and A. Waheed, Blood Cells 5:421 (1979).
108. J. W. Shands, Jr., in: "Bacterial Toxins, Vol. IV," G. Weinbaum, S. Kadis, S. J. Ajl, eds., Academic Press, New York (1971).
109. J. W. Shands, Jr., J. Infect. Dis. 128S:189 (1973).
110. M. J. Shear, Cancer Res. 1:731 (1941).
111. M. Shilo, Brit. J. Exp. Pathol. 39:652 (1958).
112. G. Shwartzman, "Phenomenon of Local Tissue Reactivity," Hoeber, New York (1937).
113. R. C. Skarnes and J. A. McCracken, in: "Microbiology 1980," D. Schlessinger, ed., American Society of Microbiology, Washington, D. C. (1980).
114. B. J. Skidmore, D. C. Morrison, J. M. Chiller, and W. O. Weigle, J. Exp. Med. 142:1488 (1975).
115. W. W. Smith, I. M. Alderman, and R. E. Gillespie, Am. J. Physiol. 191:124 (1957).
116. W. W. Smith, I. M. Alderman, and R. E. Gillespie, Am. J. Physiol. 191:124 (1957).
117. W. W. Smith, R. Q. Marston, and J. Cornfield, Blood 14:737 (1959).
118. W. W. Smith and I. M. Alderman, Radiation Res. 17:594 (1962).
119. J. A. Spitzer and J. J. Spitzer, in: "Beneficial Effects of Endotoxin," A. Nowotny, ed., Plenum Press, New York (1983). 120.
G. F. Springer and J. C. Adye, Infect. Immun. 12:978 (1975).
121. C. A. Stetson, J. Exp. Med. 94:347 (1951).
122. C. A. Stetson, J. Exp. Med. 101:421 (1955).
123. C. A. Stetson, in: "Bacterial Endotoxins," M. Landy and W. Braun, eds., Rutgers University Press, New Brunswick, New York (1964).
124. B. M. Sultzer, J. Immunol. 103:32 (1969).
125. E. Suter, G. E. Ullman, and R. G. Hoffman, Proc. Soc. Exp. Biol. Med. 99:167 (1958).
126. L. Thomas, J. Exp. Med. 104:865 (1956).
127. L. Thomas, R. T. Smith, and R. von Korff, J. Exp. Med. 102:263 (1955).
128. B. Urbaschek and R. Urbaschek, in: "Gram-Negative Bacterial Infections," B. Urbaschek, R. Urbaschek, and E. Neter, eds., Springer-Verlag, New York and Vienna (1975).
129. R. Urbaschek, S. E. Mergenhagen, and B. Urbaschek, Infec. Immun. 18:860 (1977).
130. S. I. Vas, R. S. Roy, and M. G. Robson, Can. J. Microbiol. 19:767 (1973).
131. D. W. Watson and Y. B. Kim, J. Exp. Med. 118:425 (1963).
132. J. Watson and R. Riblet, J. Exp. Med. 140:1147 (1974).
133. J. Watson, K. Kelly, and C. Whitlock, in: "Microbiology 1980," D. Schlessinger, ed., American Society of Microbiology, Washington, D.C. (1980).
134. O. Westphal, O. Luderitz, and F. Bister, Z. Naturforsch 7B:148 (1952).
135. O. Westphal, I. Beckmann, U. Hammerling, B. Jann, K. Jann, and O. Luderitz, in: "Bacterial Endotoxins," M. Landy and W. Braun, eds., Rutgers University Press, New Brunswick, NJ (1964).
136. M. Woods, M. Landy, D. Burk, and T. Howard, in: "Bacterial Endotoxins," M. Landy and W. Braun, eds., Rutgers University Press, New Brunswick, NJ (1964).
137. C. Yang and A. Nowotny, Infect. Immun. 9:95 (1974).
138. I. S. Zeckwer and H. Goodell, J. Exp. Med. 42:43 (1925).
139. E. J. Ziegler, H. Douglas, J. E. Sherman, C. E. Davis, and A. I. Braude, J. Immunol. 111:433 (1973).
140. E. J. Ziegler, J. A. McCutchan, J. Fierer, M. Glauser, J. C. Sadoff, H. Douglas, and A. I. Braude, N. Eng. J. Med. 307:1225 (1982).

SECTION I. CHEMICAL AND MOLECULAR CHARACTERISTICS OF ENDOTOXINS

Beneficially Active Structural Entities in Endotoxin Preparations
Alois Nowotny

The Antigenic Structure of the Lipooligosaccharides of Neisseria Gonorrhoeae
Michael A. Apicella

Lipid A: Relationships of Chemical Structure and Biological Activity
Otto Luderitz, Chris Galanos, Ernst Th. Rietschel,
and Otto Westphal

Production of Human Monoclonal Antibodies Recognizing Cross-Reactive
Determinants on Lipopolysaccharides
James W. Larrick, Mark Jahnsen, George Senyk, Stefan Weiss,
and Karen Watson

Physical Properties of Short-Chain and Long-Chain Fractions
of Lipopolysaccharide
Arnold A. Peterson and Estelle J. McGroarty

Analysis and Characterization of Bacterial Endotoxins (Lipopolysaccharides)
by SDS-Polyacrylamide Gel Electrophoresis Followed by Silver Stain
Chao-Ming Tsai and Carl E. Frasch

BENEFICIALLY ACTIVE STRUCTURAL ENTITIES IN ENDOTOXIN PREPARATIONS

Alois Nowotny

University of Pennsylvania
Center for Oral Health Research
Philadelphia, Pennsylvania

INTRODUCTION

The effects of endotoxin preparations can be divided into two large groups. One consists of reactions which are harmful and may lead to irreversible vascular collapse, shock and death. The other covers those which are beneficial, such as enhanced immune responsiveness, elevated non-specific resistance to microbial infections including fungi and viruses, increased leukopoiesis, activation of lymphocytes and of the functions of macrophages as well as of the entire reticuloendothelial system. Finally, one should mention the anti-tumor effects of endotoxins on some established animal malignancies and the resistance enhancement to transplantable experimental tumors induced by endotoxin. Table 1 shows some of those endotoxic effects which may be considered as beneficial.

While all investigators in this field seem to accept that the relative integrity of the lipid moiety of endotoxin is a prerequisite for its harmful or toxic effects, some reports indicate that the structural requirements for the elicitation of beneficial endotoxin effects are different.

In this review we intend to elaborate first on the theoretically possible molecular explanations of the beneficial effects. This will be followed by a review of old and new data relevant to some of the assumed possibilities. The next section deals with the problems encountered in attempting to identify beneficially active structural subunits, and the last part of this chapter will discuss the present status of our understanding of this problem, including its ambiguities as well as the pitfalls of attempts to find simple answers in such an immensely complex field.

THEORIES

1. The very first possibility we have to entertain may be summarized by stating that there are no separate sites in the endotoxin structure for toxic and beneficial effects, rather both are elicited by the same. If this is so, the lipid moiety of endotoxin should be considered as the sole active center. Whether it will elicit beneficial or toxic effects depends entirely on the dosage applied, as is the case for practically all drugs and other therapeutic modalities.

Table 1. Beneficial Effects of Endotoxin

Immunogenicity
Fibrinolysis
Radioprotection
Non-specific Protection Against Several Infectious Agents
Immune Adjuvant Effect
Release of IL-1 and other Mediators
Release of Interferon
Stimulation of Leukopoiesis
Suppression of GVH Reaction
Activation of Macrophages
Hemorrhagic Necrosis of Some Tumors
Enhancement of Resistance to Transplantable Tumors

---

2. The second possible molecular explanation of beneficial and toxic effects is that although both are within the lipid moiety, the most essential structural elements for the elicitation of toxic reactions are different from those required to elicit beneficial effects. One can assume that toxic reactions require a more or less intact, native lipid moiety, with all the long chain carboxylic acids and the phosphorylations in the right position on the proper carbohydrate backbone, but skeletal remains of this moiety are also sufficient to elicit beneficial reactions. In that case the intact and toxic molecule can have both toxic and beneficial effects and this depends only on the dose applied, while some fragments can still elicit beneficial reactions without being toxic.

3. The third theory assumes that while the lipid moiety is to be blamed for the toxic effects, the beneficially active site resides in other parts of the very complex structure. There is a large polysaccharide moiety in smooth endotoxins, the carbohydrate content of rough mutants is less but still significant. Although often ignored, there are amino acids present in the hydrolysates of endotoxin preparations. These may come from non-endotoxic contaminants but there are convincing data which make us believe that they might be covalently linked to the endotoxin structure (34,40). The possibility that these two or other constituents are the loci of beneficially active structural subunits cannot be dismissed.

4. The fourth assumption could have been considered first, since it deals with the heterogeneity of endotoxin preparations. As we have repeatedly elaborated on this point, we have to recognize that there is an extrinsic and an intrinsic heterogeneity in endotoxin preparations. The extrinsic heterogeneity is caused by the incomplete removal of co-extracted, non-endotoxic components of the bacterial cell. The intrinsic heterogeneity is an ever-present property of biological macromolecules, where the synthesis, the assembly and the degradation of the large structure occurs in multiple steps. This unavoidably results in the simultaneous presence of precursors, unassembled or incomplete subunits with the fully constructed macromolecules, together with their decaying split products (39,41,45,47). It is possible that some of the contaminants will elicit beneficial effects without being endotoxic. It is equally possible that among the precursors or partially degraded products of endotoxins, one can find some with reduced activity in one biological effect while their activity in other reactions is unimpaired.

5. We should also include a fifth theory which refers to the often observed additive or synergistic effect of multiple components in various biological systems. The simultaneous presence of certain components in

the preparation can cause end results quite different from the effects of
the isolated components alone. Similarly, the absence of one or more participants can drastically alter the final outcome of physiological or pathological reactions. If this is applicable to endotoxins, it is possible
that certain combinations of the multiple components present in the conventional endotoxin preparations are the cause of toxic effects while another
combination or proportion of components will induce beneficial reactions.

DATA

1. The largest volume of published findings is supportive of the first
point in discussion which emphasizes the pivotal role of the lipid moiety
(most often called Lipid A) in both toxic and beneficial reactions. This
theory considers the lipid moiety as the only active site in the endotoxin
structure and assumes that a small dose of it will induce beneficial effects
since it will only stimulate the host's natural defenses, while a large
dose will overpower the defenses and lead to manifestations of toxemia.

It is hard to debate this argument, particularly because so many laboratories have found that the Lipid A preparations they prepared or obtained
from other laboratories are qualitatively active in both toxic and beneficial
reactions. Probably the effect of mild acidic hydrolysis and other chemically induced structural changes on the biological properties comprise the
data which argue most effectively against the first hypothesis. If a given

Table 2. Biological Activities of Endotoxins and Endotoxoids

Sample	Mouse LD_{50}	CE LD_{50}[1]	NSR[2]
Endotoxin	250 – 500 µg	$10^{-2} - 10^{-1}$	2.04×10^{-3}
Endotoxoid[3]	>4000	>10	3.63×10^{-3}
Endotoxin	250 – 500	$10^{-3} - 10^{-2}$	1.66×10^{-3}
Endotoxoid	>4000	>100	1.44×10^{-2}
Endotoxin	ND[4]	$10^{-4} - 10^{-3}$	4.36×10^{-4}
Endotoxoid	ND	>10	8.12×10^{-4}

[1]CE LD_{50} = chick embryo LD_{50} in micrograms.

[2]Nonspecific resistance inducing capacity as measured by the dose
(in micrograms) which would give 50% protection against a
Salmonella typhi 0 901 challenge.

[3]Endotoxoid = endotoxin detoxified by 0.1 M CH_3OK under anhydrous
conditions at 60°C for 60 minutes.

[4]ND = not determined.

structural entity were the elicitor of both toxic and beneficial effects, then changes in this structure should influence both effects equally. One would expect that loss of toxicity would lead to loss of other biological activities if the same structural requirements elicit all these reactions. This does not seem to be the case. We as well as numerous other laboratories showed that chemical alterations of the structure can lead to loss of toxicity while some of the beneficial effects are retained almost quantitatively. We called such preparations "endotoxoids" (26,35).

The most critical point in such studies is the correct comparison of the biological activities on a strictly quantitative basis. By quantitation we mean not only the measurement of the extent of the biological response at a given dose of the preparations, but the comparison of the dose-response curves at a rather wide dose range. The aim of such determinations is to find the dose of each preparation which elicits an approximately 50% response, and compare these findings after testing the preparations.

We carried out such determinations measuring the LD_{50} of toxic and chemically detoxified (anhydrous CH_3OK treated) preparations. We also determined the non-specific resistance enhancing effect of these samples by measuring the dose which reduces the mortality of mice by 50%, if they are challenged 24 hours later by virulent pathogens (44). Table 2 summarizes some of the findings. Accordingly, while the toxicity of the samples was reduced by a factor of 100 or more, their capacity to elevate non-specific resistance was practically unchanged (59).

Another way to investigate whether the toxic or beneficial effects both depend on the same structural entity is to follow the fate of these properties during slow and controlled hydrolysis of the structure. We carried out such studies, and the parameters were chick embryo lethality for toxicity, CSF generation and immune adjuvant effects for beneficial

Table 3. Effect of Acidic Hydrolysis on the Toxicity, CSF
Generating and Immune Adjuvant Activities of
Serr. Marcescens TCA Endotoxin

Time of Hydrolysis[1]	Toxicity[2]	CSF[3]	ADJ[4]
0'	125	53	15.3
30'	8	43	4.2
60'	0	22	6.3
90'	0	7	8.2
120'	0	3	8.8
180'	0	3	1.4

[1] Hydrolysis in 0.2 N acetic acid in 100°C waterbath.
[2] Toxicity = number of LD_{50} units per microgram material injected i.v. to 11 day old chick embryos.
[3] CSF = colony stimulating factor determined as described earlier (48). Numbers indicate the colonies developed from 10^5 bone marrow cells.
[4] ADJ = immune adjuvant effect, expressed as stimulation index. Immunogen was 10^7 SRBC given i.p. four days before the determination of the number of plaque forming cells by the Jerne method.

properties. In all these experiments we observed that while the toxicity diminishes rather rapidly, the beneficial effects remain high for a longer time. For example, in one of these experiments, we subjected TCA extracted Serratia marcescens endotoxin to mild acidic hydrolysis for various lengths of time. This resulted in a precipitate (Lipid A) and a supernatant which we call PS, because it was rich in polysaccharides. By determining the CSF-inducing capacity of the supernatant, we found that the toxic properties of the supernatant diminish rapidly as hydrolysis progresses. The CSF inducing structural subunits present in the PS preparation had 80% of the CSF activity of the starting endotoxin, while its toxicity (as measured by chick embryo lethality and local Shwartzman test) was reduced to less than ten percent of the original. Table 3 shows such results.

Summarizing the findings of the experiments relevant to the first hypothesis, we believe that toxicity and beneficial effects are not elicited by the same structures. Partial degradation leads to rapid loss of toxicity, as measured by the very sensitive chick embryo lethality test. At the same time these non-toxic degradation products showed beneficial effects in the same dose range as the starting endotoxin preparation. These facts are in full accord with our experimental findings made 23 years ago when we showed that some chemical attacks on selected functional groups of endotoxic macromolecules can detoxify the preparations while maintaining their beneficial effects (26,35). Our conclusion was that the structural requirements of toxic reactions are different from those needed to elicit beneficial effects (49).

2. The second working hypothesis assumes that while both toxic and beneficial reactions reside in the lipid moiety, the elicitation of toxic effects requires a more or less intact lipid moiety. The beneficial effects, on the other hand, can be still elicited by partially degraded structural remains of the lipid moiety. We arrived at this conclusion (37,38,40,68) after studying the mechanism of the chemical detoxification which we have already discussed above. At first sight one is inclined to believe that the activities of chemically detoxified preparations provide supporting evidence for the second hypothesis, since the cleavage of some ester linkages eliminated toxicity while it maintained some beneficial properties. This could be used as supporting evidence for this hypothesis only if the lipid moiety were the starting material. Since we and several other laboratories (64) detoxified whole endotoxin, it is possible that the beneficial effect of the detoxified samples may reside in parts of the structure different from the lipid moiety (see third hypothesis), which structural entities were not affected by the chemical alteration.

More convincing support for the second hypothesis comes from studies of the Lipid precipitates. The assumption that the lipid moiety of the endotoxin is the carrier of toxicity and eventually of all other biological effects comes from early studies (11,71,72). These investigators used the lipid-rich precipitate obtained after mild acidic hydrolysis of endotoxin, called Lipid A. Since such a precipitate was found to be composed of several hydrophobic split products (27,36,42) we initiated systematic studies to isolate the biologically active components from the Lipid A. Using preparative TLC systems we isolated several chromatographically pure components and the majority of these were inert. Some were non-toxic but could induce hemorrhagic necrosis of Sarcoma 37 (42). Others were highly toxic but could also induce tumor necrosis. These latter preparations were called RESI (being the residue of the lipid A preparation after acetone and chloroform extraction). Fraction No. 2 of RESI, the most active component of this preparation, was subjected to chemical analysis that showed that fraction 2 is similar in its chemical composition to the endotoxic glycolipid we isolated from the hepatoseless Re mutant of S. minnesota (15). The conclusion of these experiments was that among the partially degraded split products of the lipid moiety one can find compounds which are fully toxic and also active in some beneficial reactions, and one can isolate others which are no longer toxic, but still active in antitumor reactions (15,42). Ribi and co-workers used RESI preparations successfully in combination with trehalose dimycolate to induce complete regression of guinea pig line 107 hepatocarcinomas. For these studies we used Serratia marcescens TCA extracted endotoxins (55).

Ribi and associates took S. minnesota R595, a rough mutant bacterium, and extracted from it a mutant endotoxin. This is believed to consist almost entirely of the lipid moiety without the O-specific and core polysaccharides. The amino acid content of these preparations is very low and is often undetectable. This as well as similar preparations were subjected to mild acidic hydrolysis and the products were tested for the anti-tumor effects and analyzed chemically. Ribi and co-workers found that hydrolysis removes one phosphoric acid radical from the diphosphorylated D-glucosamine disaccharide backbone of the lipid moiety but no other changes in the chemical composition were observed. Most importantly, this change produced preparations with no toxicity but full activity in the tumor necrosis assay (56,57,66).

Ribi and associates assume that the presence of a phosphoric acid radical in a given position renders the lipid structure toxic (57). Luderitz and co-workers as well as a number of other investigators, including the author, believe that the presence of certain long chain carboxylic acids in very critical positions on the D-glucosamine backbone of the lipid structure is responsible for the toxic manifestations (35,37,40,67,68,72). In the search for the carboxylic acid which is essential for the toxic reactions, we found that some carefully controlled chemical reactions which detoxify endotoxins cleave palmitic acid amongst exclusively from the lipid moiety (68). At this meeting Luderitz presented convincing evidence that the presence of palmitic acid ester bound to the 3-OH group of amide bound 3-OH myristic acid is the critical component on synthetic lipid A models.

While the above results strongly indicate that the second working hypothesis may be correct, they do not rule out the possibility that there are additional structural elements or features which also contribute to the beneficial effects of endotoxins.

3. The third hypothesis says that not only the lipid but other parts of the structure can also have beneficial effects. Relevant results were first published by us (14,48). We found that the supernatant of the hydrolyzed endotoxin still has activity in CSF generation, in protection against lethal irradiation and in vitro and in vivo tests as an immune adjuvant (4,5,6,19,48). This supernatant contains the soluble split products of the endotoxin, it is free of long-chain fatty acids, it contains peptides and it is rich in oligo- and polysaccharides. We named this preparation PS.

The PS preparations obtained from TCA extracted endotoxin samples of smooth gram negative bacteria were particularly active. Rough mutant endotoxins deficient in their polysaccharide moiety did not yield active PS, except those which still possessed a significant portion of their core structure. For example, S. minnesota R595 or R5 or R7 endotoxins did not release active PS fragments while R345 or the smooth 1114 endotoxins did (unpublished findings). As already discussed, we studied the optimal conditions for CSF inducing PS isolation. It was found that in measuring the CSF activity of the PS samples, short duration of mild hydrolysis was needed to eliminate toxicity and maintain the CSF generative activity (Table 3). Longer hydrolysis, even with 0.2 N acetic acid, rapidly destroyed the CSF activity of the PS. It is interesting to compare the immune adjuvant effect of the PS samples with their CSF activity. The supernatant of the acid hydrolyzed endotoxin rapidly lost its CSF activity, while the adjuvant effect was unchanged at that time. These observations lead us to believe that different structural requirements exist for CSF generation and for immune potentiation.

All these results indicate that PS preparations contain beneficially active fragments. Not only our laboratories, but several others confirmed this (22,23,24,52,70). The major question is whether they came from the polysaccharide moiety of the endotoxin or from other structural elements, such as the lipid moiety or from the peptides, which latter compose five to ten percent of TCA extracted endotoxin, but less than one percent of rough mutant endotoxins. It is feasible that the hydrolysis liberates some polar constituents from the lipid moiety so that they can be found among the hydrophilic components of the PS. The fact that very rough mutants, such as S. minnesota R595, R5 or R7 did not yield active PS preparations indicate that it is unlikely that the active fragments would come from the lipid moiety, because these mutants consist mainly of this part of the endotoxin structure. On the other hand they do not contain the complete polysaccharide part and their amino acid content is very low. The two likely possibilities are that the active fragments in the PS are either constituents of the polysaccharide or of the peptide constituents. It is a less likely possibility that the lipid moiety of smooth endotoxins has side chains or other hitherto unindentified substituents, which are absent in very rough mutant endotoxins. If so, these may be the elicitors of the reactions observed in various beneficial assays.

At any rate, the final identification of the origin of these active fragments in the PS will be possible only after their isolation and chemical structural analysis. The major stumbling block towards this aim has been the lack of the proper separation technique. We think that we solved this problem recently, developing a TLC and high voltage electrophoresis combina-

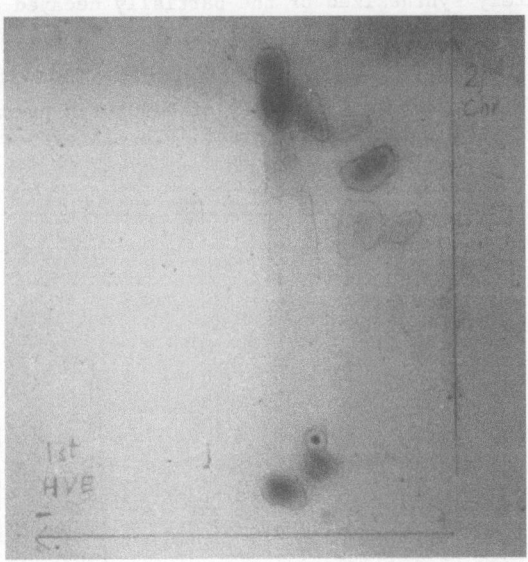

Figure 1. Separation of PS components by high-voltage electrophoresis in the first dimention at pH 5.4 in pyridine acetic acid buffer followed by chromatography in the second dimension. This solvent was isopropanol:water: cc. acetic acid - 5:5:1. The separation was carried out on Baker TLC plates coated with silicic acid. For the detection of the components of TLC plate was sprayed with orcinol-$FeCl_3$-H_2SO_4 and heated at 110°C for 10 min.

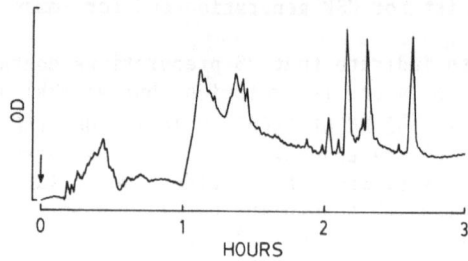

Figure 2. Chromatographic separation of PS on ion exchange resin. Effluent was monitored for carbohydrate using a Technicon analyzer system. The procedure was described earlier (41).

tion carried out on silica gel covered glass plates (Kovats and Nowotny, manuscript in preparation). For the staining of the separated components we used orcinol-$FeCl_3$-H_2SO_4 spray which detects carbohydrate containing components (62). Figure 1 shows such a separation. The preparative fractionation of the PS preparations is in progress. Promising results were obtained using weak anion exchange column chromatography, monitoring the column effluent for carbohydrates with a continuous Technicon autoanalyzer (41). Figure 2 illustrates the elution profile, which as expected, indicates the presence of many split products in the PS sample.

4. The fourth working hypothesis assumes that since the endotoxin preparations are heterogenous, it is possible that some of the non-endotoxin components have beneficial effects, or they contribute to the beneficial reaction elicited by endotoxins. Furthermore, it is logical to assume that either the incompletely synthesized or the partially decayed endotoxin derivatives may have a reduced toxicity, but, if hypothesis No. 2 is correct, they may be beneficially active.

We do not think that the heterogeneity of endotoxin preparations which we reported first in 1966 needs further proof. The latest review on this

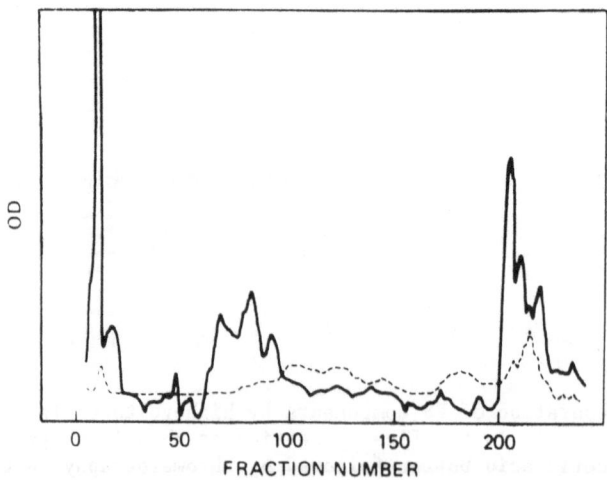

Figure 3. Column chromatography of TCA extracted endotoxin of Serr. marcescens on ion exchange resin. The effluent was monitored for carbohydrate (——) and UV absorption at 260 (----) as described earlier (41).

44

Table 4. Biological Activities of the Fractions Obtained by Ion
Exchange Chromatography of S. marcescens TCA Endotoxin

Table 4. Biological Activities of the Fractions Obtained by Ion
Exchange Chromatography of S. marcescens TCA Endotoxin

Fraction Number	Toxicity[1]	Shwartzman[2]	CSF[3]	NSF[4]
1 - 25[5]	3	0	42	3.6×10^{-1}
55 - 80	134	3.5	34	8.4×10^{-3}
81 - 100	74	3.8	40	5.5×10^{-3}
200 - 210	2	2.1	3	4.8×10^{-2}
211 - 250	2	3.1	4	4.0×10^{-2}

[1] Number of chick embryo LD_{50} units per µg material.
[2] Area of hemorrhagic skin lesion on rabbits measured on cm^2 (44).
[3] Number of colony forming cells from 10^5 bone marrow cells (48).
[4] Dose in µg which give 50% protection from viable S. typhi 0901
exchange (44).
[5] Fraction numbers (see Figure 3) pooled and tested.

topic appeared quite recently (45). For the sake of completeness, however,
we repeat the most relevant point of this old observation, namely the
column chromatographic isolation of non-endotoxic contaminants from TCA
extracted endotoxins which were active either as enhancers of non-specific
resistance to bacterial infection, or as generators of CSF (41,45,47).
Figure 3 shows the chromatographic profile and the fractionation while
Table 4 gives the biological activities of the fractions.

Since these observations, several laboratories showed extensive
heterogeneity of endotoxin preparations and the isolation of non-endotoxic
but beneficially active components from endotoxin preparations, as well as
from whole bacterial cells (9,10,16,17,29,31,49,52,53,65,73).

Anacker and co-workers (2) and Rudbach and associates (61) isolated
incompletely synthesized polysaccharides from the protoplasm of E. coli and
S. enteritidis bacteria which appear to be precursors of the endotoxic
lipopolysaccharide. Beside the reactivity of these so called "native" or
"protoplasmic haptens" with specific O-antisera, no activities were found
in the conventional assays of endotoxicity. The results with the biological
activities of the PS preparations made us interested in the protoplasmic
haptens, which were generously provided for our studies by Dr. E. Ribi. We
found that these are active as CSF inducers and they could potentiate the
antitumor effect of BCG infection (13).

Another precursor of the endotoxin molecule was isolated from a pbsB
gene deficient mutant E. coli strain. It was observed by Nishijima and
Raetz (33) that this strain accumulates phosphomuco-lipids. Fractionation
of these lipids lead to the isolation of a 2,d-diacyl-D-glucosamine-1-phos-
phate, called Lipid X. Testing the biological properties of Lipid X it was
found to be mitogenic for murine spleen cells and it also possessed some
other biological activities of endotoxins (12,54,75). Raetz and associates,
who established the complete structure of Lipid X, found that the essential
functional group is a 3-OH C14 acid ester-bound to C3 of the D-glucosamine-1-
phosphate (51). Furthermore, the group reported that Lipid X can be also
found in wild type E. coli cells which may contain 4000 molecules of Lipid
X in each cell. The pbsB deficient mutant contains 100 to 300 times more
than the wild strain (12).

45

It should be evident that precursors like the protoplasmic hapten of Ribi and associates and the Lipid X of Raetz and co-workers might contaminate conventional endotoxin preparations and their amount in these will depend upon the bacterial strain extracted and by the method of extraction, as we discussed above. It is also obvious that these precursors will not have the entire spectrum of diverse biological activities of endotoxins, but might be able to elicit some, thus contributing the overall effect of the heterogenous endotoxin sample.

Another set of evidence can be brought up at this point to prove the existence of nonendotoxic bacterial products with beneficial effects. White (74) subjected whole bacterial cells to mild acetic acid hydrolysis and used the supernatant obtained as an antigen preparation for serological identification of gram negative bacteria. Freeman and Philpot (20) continued this work in more extensive studies and such extracts became known as Freeman polysaccharides. No toxicity or other biological activity of these prepara-tions were reported until Urbaschek noted that these polysaccharides, like our PS preparations, are activators of granulopoiesis and can protect against lethal irradiation (70). We pursued these results and observed that such polysaccharide-rich samples, which we call White-type polysaccharides or WPS, elicit a large number of biological activities originally thought to be inducible by endotoxins (60). Chemical analysis of the WPS samples re-vealed that they contain the O-specific and the core structural fragments of the endotoxin, but do not have detectable amounts of the characteristic components of the lipid moiety, namely fatty acids, glucosamine or phos-phorous. Furthermore, active WPS could be obtained from bacterial residue, which did not contain endotoxins due to their previous removal from chloro-form-methanol (S. minnesota R595) or with cold five percent TCA (Serratia marcescens) (Nowotny, Pham and Kovats, unpublished).

These results showed that components of gram negative bacteria, which are not identical with endotoxins, are fully capable of eliciting beneficial reactions. It seems to be assumable that several of these non-endotoxic but beneficially active preparations might be co-extracted with endotoxins, particularly if harsh methods are used to liberate the endotoxin layer from the bacterial cell surface.

That conventional endotoxin preparations contain a series of endotoxin molecules was shown by using SDS-PAGE stained by the Tsai-Frash procedure (69). Goldman and Leive (21) found more than 40 components and this hetero-geneity was attributed to the various numbers of repeating units in the polysaccharide moiety. They assumed that the fast migrating endotoxin mole-cules have only a few of these, while the slowly migrating ones have many repeating units. Hitchcock and Brown (25) as well as others reached similar conclusions.

We isolated three major zones from the SDS-PAGE separated components of S. minnesota phenol-water extracted endotoxins. The top zone contained the slowest and the bottom zone the fastest migrating components. Deter-mining the biological activities of the fractions it was found that the bottom fraction had the highest activity on a weight basis in polyclonal B cell activation and it also induced significant enlargement of the spleen of mice. In other assays, the three zones were indistinguishable (50). We also determined the serological reactivity (antigenicity) of the SDS-PAGE separated components of TCA extracted Serratia marcescens endotoxins by the immunoblotting method. As Figure 4 shows, the bands in the upper zone were quite reactive with O-antiserum while the fast migrating, supposedly low molecular weight endotoxins were not (Kovats and Nowotny, unpublished). This intrinsic heterogeneity shows clearly that in the bacterial cells not only one molecular type of endotoxin, but a variety of endotoxin molecules exists. So far we assumed that they differ only in their molecular weights, but based on the above results, we have to think that they could differ in their chemical composition and in their potency to elicit biological reactions as well.

5. There is little experimental evidence for the fifth working hypothesis applied to endotoxins. Our laboratory has frequently observed that purification of endotoxin preparations by column chromatography, which separated the various fractions, did not result in the expected increase of activity as determined on a weight basis (41). Quite often a crude TCA extracted endotoxin is more active, particularly in beneficial effects, than the carefully prepared and purified fractions thereof. If the extraction procedure was even milder, such as water at 80° or the use of detergents (46) or hydrolytic enzymes (28), crude preparations were obtained which were more active in CSF generation than the most highly purified, TCA or phenol-water extracted endotoxins from the same bacterial strains. Table 5 illustrates this point.

The explanation of this phenomenon usually blames the excessive manipulation of the fractions during the isolation. They may become denatured or they lose solubility, due to the removal of inactive but solubility-enhancing components from the mixture. Recombination of these with the active material sometimes restores the activity to a certain degree, as we also observed working with column chromatographically separated endotoxin fractions (41).

The explanation may be much more complex, particularly if one considers the possible pathways through which one can reach a beneficial end result,

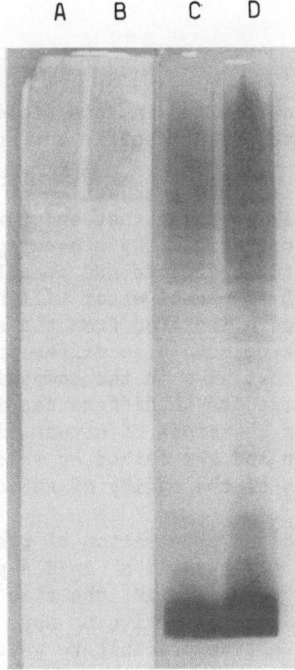

Figure 4. Immunoblot of Serratia marcescens endotoxin separated on SDS polyacrylamide gel. A = 10 μg and B = 30 μg endotoxin, transferred to nitrocellulose and reacted with hyperimmune rabbit O-anti serum to S. marcescens. C = 10 μg and D = 30 μg of above endotoxin, separated in the same run, but stained by the Tsai-Frasch method.

Table 5. Comparisons of the CSF Activities of Various
Crude Endotoxin Preparations from S. Marcescens

Preparation Extracted By	CSF Activity[1]
80° water[2]	106
CTB detergent[2]	81
Hydrolytic enzymes[3]	76
Phenol-water[4]	12
TCA[5]	38

[1] Number of colonies formed by 10^5 bone marrow cells
[2] For procedures see (46)
[3] see (28)
[4] see (71)
[5] Modified Boivin procedure (44)

for example, enhanced immune responsiveness or enhanced resistance to infections or to transplantable tumors. We assumed that several chain reactions are involved in the so-called endotoxic effects, they may run simultaneously and they are probably triggered by different fractions present in most endotoxin preparations. They may set in motion independent physiological events which add to or synergistically amplify the overall symptoms (41).

PROBLEMS

While great progress has occurred in this field, many problems are still overlooked or are treated with insufficient scrutiny.

One of these problems lies in the selection of the endotoxin preparations we are going to use. It is known that endotoxin preparations differ from each other by multiple criteria. Their heterogeneity depends on a number of extrinsic and intrinsic factors and it follows that endotoxins isolated by detergents or TCA or phenol-water will have quite different "contaminants" even if they were isolated from the same strain of bacteria (41,45,46). The endotoxin structures also differ not only in the composition of their polysaccharide but also in the composition of their lipid moiety (30,32,41,58). This results in differences in their biological activities (32,43,73). It is therefore of utmost importance to identify not only the bacterial strain and the method by which the endotoxin was obtained, but also the status of the purity of the chosen endotoxins.

The nature and method of the preparation of the "lipid A" samples requires even closer scrutiny. If this is a Lipid A precipitate obtained after mild acidic hydrolysis of endotoxin, the time of treatment will determine whether the preparation will be active or not. Short and very mild hydrolysis will yield a very active precipitate which contains incompletely degraded endotoxins (15). The choice of the starting material used to prepare Lipid A precipitate is even more important. Wild-type endotoxins with a complete polysaccharide moiety require a relatively long and strong hydrolysis to split the macromolecule and to precipitate its lipid-rich zone. Such lipid A preparations are extremely heterogenous due to extensive degradation of the lipid moiety, and it follows that their activities will also be much less than the activities of the parent structure. On the other hand, if one takes a rough mutant endotoxin, with deficiencies in the hydrophilic polysaccharide moiety, very weak acidic hydrolysis will precipitate the endotoxin which had very limited solubility to begin with. In this case, the precipitate will contain large quantities of almost unchanged endotoxins. None of the above preparations can be considered as true representatives of the native lipid moiety of the endotoxin.

Another source of problems is the improper use of biological assays for the comparison of the potency of various preparations. The difficulties here are numerous, and the first to mention is the right choice of the biological reaction. Toxicity is by no means a reaction which is specific for endotoxins only. Pyrogenicity is a more characteristic property of endotoxins, but its quantitation is cumbersome and again, not only endotoxins but some products of gram positive bacteria, or even simpler organic compounds such as LSD, can also elicit febrile responses. From our experience, probably the local Shwartzman reaction (63) and the numerous variations thereof are of the most applicable for the determination of endotoxicity. The reaction is rather characteristic for endotoxins and the quantitation turned out to be easier and more accurate than expected (44).

While the local Shwartzman reaction is what we use most often as a measure of endotoxicity, to select a beneficial reaction which is characteristic for endotoxins and can be quantitated is considerably more difficult. Undeniably none of the reactions shown in Table 1 can be considered typical for endotoxins alone, although evidently the various endotoxin preparations are the most potent among the natural products which can elicit them. Most often the specific interest of the laboratories determine which one of these will be studied. As beneficial reaction, our laboratories most often measure the immune potentiation, macrophage activation by phagocytosis and cytotoxicity, and generation of colony stimulation factor (CSF).

There is one more word of caution to be said at this point and that is that the various endotoxic reactions are not interchangeable. If a preparation is very active in one parameter, it may be less active in another. We showed that first using toxicity measurements (18,40) later by comparing toxic and some beneficial reactions (1). As we have already discussed, there is only one logical explanation for this, namely that the elicitation of the various endotoxic reactions does not require the same structural components and/or features.

Probably the most important is to carry out extensive dose-response studies for the quantitation of the response. Comparison at one dose range may not give a reliable and true reflection of the relative potencies. Quite often, one has to establish first the dose range where increasing amounts elicit increasing reactions, since an overdose might diminish the responses or completely inhibit its expression, being already toxic in vitro or in vivo. Further difficulty is experienced when the dose range where the changes in the response are linear with the changes of dose applied is very narrow.

Even if the proper dose range has been established and pure preparations have been obtained and applied, additional complications may be generated by the timing of the application. While this seems to be less critical for the measurement of toxic effects, it can be manifested in several beneficial reactions elicited by endotoxins. We observed that while endotoxin given to mice one day before lethal irradiation protects them, the same dose given five days before irradiation does not (5). Similar time-dependency of beneficial effects could be observed in immune adjuvant tests and in measuring the resistance of mice to transplantable tumors. More extensive studies of this phenomenon established that a single injection of endotoxin elicits periods where the host manifests beneficial effects, but these periods are interrupted with negative phases. For example, these alternating and almost sinusoidally changing conditions of the host's organism produce time periods where the host is resistant to lethal irradiation or to transplantable tumors but this is soon followed with another phase where the host is more susceptible to these than are normal, untreated controls (3,5). The possible mechanisms of these oscillations were discussed elsewhere (7,8).

The significance of this phenomenon in relation to the topic we are discussing in this chapter is evident. The use of an improper time interval between the endotoxin injection and the test of its effect on the host can determine whether the outcome is beneficial or harmful. While this may not be the case for all beneficial reactions, one has to keep this possibility in mind and experiments must be designed to optimize not only the dose range but the timing of endotoxin applications as well.

CONCLUSION

Several theoretically possible explanations were entertained in this chapter with the aim of shedding some light on the molecular basis of beneficial endotoxic reactions. Those listed here do not exhaust all the working hypotheses one might conceive and set up, only the most evident five were discussed and their likelihood illuminated by experimental data.

In our minds, it is beyond doubt that among the multiple components present in conventional endotoxin preparations one can find non-toxic substances which are either directly beneficial or which facilitate the development of a beneficial overall effect in the recipient.

Similarly, there is little doubt that the "pure" endotoxin molecules have a beneficial effect, but the extreme sensitivity of human subjects to the toxic effects of endotoxin leaves only a very narrow dose range where the beneficial effects could be utilized. The point elaborated in the second working hypothesis might help to extend this dose range. If non-toxic structural residues of the endotoxin's lipid moiety are capable of eliciting some of the beneficial effects, we might consider developing a therapeutically useful preparation. According to previous findings, the chances of doing so are excellent. Recent results coming from several independent laboratories justify this optimistic outlook.

The third possibility, namely, that not only the lipid moiety but other subunits of the endotoxin structure can also elicit beneficial effects, finds support in the results we and a few other laboratories obtained using various PS preparations. The fact that those rough mutant endotoxins which almost completely lack the polysaccharide moiety cannot yield an active PS preparation is the major argument for the validity of the third hypothesis. The same rough mutants are believed to possess a complete lipid moiety, and judging from their high toxicity, this lipid moiety is fully active in the toxicity tests. It is known that the PS preparations consist of many split products of the hydrolyzed endotoxin. While the carbohydrates are the major constituents, there are components present in the PS samples which might be cleaved off from other subunits of the macromolecular structure. These components of the PS preparation are unlikely to be split products of the lipid moiety, because if they were, the rough mutants would also release them since they have a complete lipid moiety. This is not the case, therefore we have to assume at this point that sites other than the lipid moiety can yield acid soluble breakdown products which have reduced but definitely measurable activities in some, but not all, beneficial reactions. At the time this review is being written, the third theory remains a working hypothesis, in obvious need of isolation followed by chemical analysis of the active component(s) in the PS preparation.

In the final statement, we urge all editors of scientific periodicals
to request sufficient information from the authors regarding the source,
the extraction method and the status of the purity of the endotoxin or lipid
A or other samples they used in their experiments. If the samples were
purchased or gifts from other laboratories, the authors still should request
and provide a full description of the preparations. Discrepancies in the
findings of various laboratories very often can be caused by the use of
non-identical preparations which wear identical designations, as it was
found in the case of different samples, all called "Lipid A" (15). No
results which come from the use of heterogenous mixture are acceptable for
structure and function relationship studies, since neither the structure
nor the function can be properly identified in mixtures of various entities.
This applies equally to conventional endotoxin or Lipid A preparations
as well as to other mixtures of components such as PS or WPS.

We think that this conclusion spells out rather clearly the several
times emphasized necessity for the isolation of the beneficially active
entities in chemically pure form. This also means that the major part of
our research efforts are still ahead of us and the endotoxin field, instead
of winding down, offers more exciting possibilities than ever before.

REFERENCES

1. A. M. Abdelnoor, H. L. Chang, P. H. Pham, and A. Nowotny, Lack of
    relationship between toxicity and bone marrow cell colony stimulating
    activity of endotoxin preparations, Proc. Soc. Exp. Biol. Med. 163:15
    (1980).
2. R. L. Anacker, R. A. Finkelstein, W. T. Haskins, M. Landy, K. C.
    Milner, E. Ribi, and R. W. Stashak, Origin and properties of naturally
    occurring hapten from Escherichia coli, J. Bacteriol. 88:1705 (1964).
3. U. H. Behling, Alterations of survival and immunity in normal and
    irradiated mice treated with bacterial endotoxin and its derivatives,
    Ph.D. Thesis, Temple University, Philadelphia, Pennsylvanyia (1975).
4. U. H. Behling and A. Nowotny, Immune adjuvancy of lipopolysaccharide
    and a nontoxic hydrolytic product demonstrating oscillating effects
    with time, J. Immunology 118:1905 (1977).
5. U. H. Behling and A. Nowotny, Long-term adjuvant effect of bacterial
    endotoxin in prevention and restoration of radiation-caused
    immunosuppression, Proc. Soc. Exp. Biol. Med. 157:348 (1978).
6. U. H. Behling, P. H. Pham, F. Madani, and A. Nowotny, Components of
    lipopolysaccharide which induce colony stimulation, adjuvancy and
    radioprotection, in: "Microbiology 1980," D. Schlessinger, ed.,
    American Society of Microbiology, Washington, D.C. (1980).
7. U. H. Behling and A. Nowotny, Cyclic changes of positive and negative
    effects of single endotoxin injection, in: "Bacterial Endotoxins and
    Host Response," M. K. Agarwal, ed., Elsevier/North-Holland Biomedical
    Press (1980).
8. U. H. Behling and A. Nowotny, Bacterial endotoxins as modulators of
    specific and non-specific immunity, in: "Regulatory Implication of
    Oscillatory Dynamics in the Immune Response," J. Hiernaux and C.
    DeLisi, eds., CRC Press (1982).
9. F. M. Berger, The effect of endotoxin on resistance to infections and
    disease, Adv. Pharmacol. 5:19 (1967).
10. F. M. Berger, G. M. Fukui, B. J. Ludwig, and J. P. Rosselet, Increase
    of non-specific resistance to infection by Protodyne, a protein
    component derived from bacterial protoplasm, Proc. Soc. Exp. Biol. Med.
    127:556 (1968).
11. A. Boivin, J. Mesrobeanu, and L. Mesrobeanu, Extraction d'un complexe
    toxique et antigenique a partir du bacille d' Aertrycke, Compt. Rend.
    Soc. Biol. 114:307 (1933).

12. C. E. Bulawa and C.R.H. Raetz, The biosynthesis of gram negative endotoxin: identification and function of UDP-2,3-Diacylglucosamine in Escherichia coli, J. Biol. Chem. 259:4846 (1984).

13. R. C. Butler and A. Nowotny, Combined immunostimulation in the prevention of tumor take in mice using endotoxins, their derivatives and other immune adjuvants, Cancer Immunol. Immunother. 6:255 (1979).

14. H. Chang, J. J. Thompson, and A. Nowotny, Release of colony stimulating factor (CSF) by non-endotoxic breakdown products of bacterial lipopolysaccharides, Immunol. Commun. 3:401 (1974).

15. C. M. Change and A. Nowotny, Relation of structure to function in bacterial O-antigens. VII. The biological activity of "lipid A" preparations, Immunochemistry 12:19 (1975).

16. J. Choay and M. Sakouhi, Fraction capable of inducing in vivo a resistance to bacterial infections, process for obtaining said fraction from bacteria and drugs containing said fraction, U.S. Patent 4,148,877.

17. M. J. Crutchley, D. G. Marsh, and J. Cameron, Biological studies on free endotoxin and a non-toxic material from culture supernatant fluids of Escherichia coli 078K80 (1968).

18. K. R. Cundy and A. Nowotny, Quantitative comparison of toxicity parameters of bacterial endotoxins, Proc. Soc. Exptl. Biol. Med. 127:999 (1968).

19. S. Frank, S. Specter, A. Nowotny, and H. Friedman, Immunocyte stimulation in vitro by nontoxic bacterial lipopolysaccharide derivatives, J. Immunol. 119:855 (1977).

20. G. G. Freeman and J. St. L. Philpot, The preparation and properties of a specific polysaccharide from Bact. typhosum Ty$_2$, Biochem J. 36:340 (1942).

21. R. C. Goldman and L. Leive, Heterogeneity of antigenic side chain length in LPS from Escherichia coli 0111 and Salmonella typhimurium LT2, Eur. J. Bioch. 107:145 (1980).

22. N. Haeffner-Cavaillon, J. M. Cavaillon, and L. Szabo, Macrophage dependent polyclonal activation of splenocytes by Bordetella pertussis endotoxin and its isolated polysaccharide and lipid A region, Cell Immunol. 74:1 (1982).

23. N. Haeffner-Cavaillon, J. M. Cavaillon, M. Moreau, and L. Szabo, Interleukin secretion by human monocytes stimulated by the isolated polysaccharide region of Bordetella pertussis endotoxin, Mol. Immunol. 21:389 (1984).

24. N. Haeffner-Cavaillon, J. M. Cavaillon, and L. Szabo, Cellular receptors for endotoxin, in: "Handbook of Endotoxin," Volume 3, L. J. Berry, ed., Elsevier-North Holland Publishers, Amsterdam and New York (1985).

25. P. J. Hitchcock and T. M. Brown, Morphological heterogeneity among Salmonella lipopolysaccharide chemotypes in silver-stain polyacrylamide gels, J. Bacteriol. 154:269 (1983).

26. A. G. Johnson and A. Nowotny, Relationship of structure to function in bacterial O-antigens. III. Biological properties of endotoxoids, J. Bact. 87:809 (1964).

27. N. Kasai and A. Yamano, Studies of the lipids of endotoxins. Thin layer chromatography of lipid fractions, Jap. J. Exp. Med. 34:329 (1964).

28. S. Lehrer and A. Nowotny, Isolation and purification of endotoxin by hydrolytic enzymes, Inf. and Imm. 6:928 (1972).

29. D. G. Marsh and M. J. Crutchley, Purification and physico-chemical analysis of fractions from the culture supernatant of Escherichia coli 078K80: free endotoxin and a non-toxic fraction, J. Gen. Microbiol. 47:405 (1967).

30. H. Mayer and J. Weckesser, "Unusual" lipid A's structures, taxonomical relevance and potential value for endotoxin research, in: "Handbook of Endotoxin," Vol. 1, E. Rietschel, ed., Elsevier Publishing, Amsterdam and New York (1984).

31. D. C. Morrison, S. J. Betz, and D. M. Jacobs, Isolation of a lipid A bound polypeptide responsible for "LPS-initiated" mitogenesis of C3H/HeJ spleen cells, J. Exp. Med. 144:840 (1976).

32. A. K. Ng, C. M. Chang, C. H. Chen, and A. Nowotny, Comparison of the chemical structure and biological activities of the glycolipids of Salmonella minnesota R595 and Salmonella typhimurium SL1102, Inf. and Immun. 10:938 (1974).

33. M. Nishijima and C.R.H. Raetz, Membrane lipid biogenesis in Escherichia coli: identification of genetic loci for phosphotidylglycerol, J. Biol. Chem. 254:7837 (1979).

34. A. Nowotny, Chemical structure of a phosphomucolipid and its occurrence in some strains of Salmonella, J. Am. Chem. Soc. 83:501 (1961).

35. A. Nowotny, Endotoxoid preparations, Nature 197:721 (1963a).

36. A. Nowotny, Relation of structure to function in bacterial O-antigens. II. Fractionation of lipids present in Boivin-type endotoxin of Serratia marcescens, J. Bact. 85:427 (1963b).

37. A. Nowotny, Chemical detoxification of bacterial endotoxins, in: "Bacterial Endotoxins," M. Landy and W. Braun, eds., Rutgers University Press, New Brunswick, NJ (1964).

38. A. Nowotny, Relation of chemical structure to pathologic activity of endotoxins, in: "Shock and Hypotension," L. J. Mills and J. H. Moyer, eds., Grune and Stratton, Inc., New York (1965).

39. A. Nowotny, Heterogeneity of endotoxic bacterial lipopolysaccharides revealed by ion exchange column chromatography, Nature 210:278 (1966).

40. A. Nowotny, Molecular aspects of endotoxic reactions, Bact. Rev. 33:72 (1969).

41. A. Nowotny, Chemical and biological heterogeneity of endotoxins, in: "Microbial Toxins," Vol. IV, Weinbaum et al., eds., Academic Press, New York (1971a).

42. A. Nowotny, Relationship of structure and biological activity of bacterial endotoxins, Naturwissenschaften 58:397 (1971b).

43. A. Nowotny, Relation of structure to function in bacterial endotoxins, in: "Microbiology 1977," D. Schlessinger, ed., American Society of Microbiology, Washington, D.C. (1977).

44. A. Nowotny, "Basic Exercises in Immunochemistry," 2nd Edition, Springer-Verlag, New York, Heidelberg, Berlin (1979).

45. A. Nowotny, Heterogeneity of endotoxins, in: "Handbook of Endotoxins," Vol. 1, E. T. Rietschel, ed., Elsevier, Amsterdam and New York (1984).

46. A. M. Nowotny, S. Thomas, O. S. Duron, and A. Nowotny, Relations of structure to function in bacterial O-antigens. I. Isolation methods, J. Bact. 85:418 (1963).

47. A. Nowotny, K. Cundy, N. Neale, A. M. Nowotny, R. Radvany, S. Thomas, and D. Tripodi, Relation of structure to function in bacterterial O-antigens. IV. Fractionation of the components, Ann. N.Y. Acad. Sci. 133:586 (1966).

48. A. Nowotny, U. H. Behling, and H. L. Chang, Relation of structure to function in bacterial endotoxins. VIII. Biological activities in a polysaccharide-rich fraction, J. Immunol. 115:199 (1975).

49. M. Ohta, M. Mori, T. Hasegawa, F. Hagase, I. Nakashima, S. Naito, and N. Kato, Further studies of the polysaccharide of Klebsiella pneumoniae possessing strong adjuvanticity. I. Production of the adjuvant polysaccharide by non-capsulated mutant, Microbiol. Immunol. 25:939 (1981).

50. M. Ohta, J. Rothmann, E. Kovats, P. H. Pham, and A. Nowotny, Biological activities of lipopolysaccharides fractionated by preparative acrylamide gel electrophoresis, Microbiol. Immunol. 29:1 (1985).

51. C.R.H. Raetz, S. Purcell, and K. Takayama, Molecular requirements for B-lymphocyte activation by Escherichia coli lipopolysaccharide, Proc. Natl. Acad. Sci. USA 80:4624 (1983).

52. D. Raichvarg, C. Brossard, and J. Agneray, Preparation of a nontoxic and immunogenic polysaccharide fraction from a Haemophilus influenzae

phenol-water extract, Infec. and Immun. 1:171 (1980a).

53. D. Raichvarg, M. Guenounou, C. Brossard, J. E. Alouf, and J. Agneray, Endotoxin-like substances in bacterial cell-outer membrane: correlation between structure and biological activities of Haemophilus influenzae type A endotoxin, in: "Bacterial Endotoxins and Host Response," M. K. Agarwal, ed., Elsevier/North-Holland Biomedical Press (1980b).

54. B. L. Ray, G. Painter, and C.R.H. Raetz, The biosynthesis of gram negative endotoxin. Formation of lipid A disaccharides from monosaccharide precursors in extracts of Escherichia coli, J. Biol. Chem. 259:4852 (1984).

55. E. Ribi, R. Parker, S. M. Strain, Y. Mizuno, A. Nowotny, K. von Eschen, J. L. Cantrell, C. A. McLaughlin, K. M. Hwang, and M. B. Goren, Peptides as requirement for immunotherapy of the guinea-pig line-10 tumor with endotoxin, Cancer Immunol. Immunother. 7:43 (1979).

56. E. Ribi, K. Amano, J. Cantrell, S. Shwartzman, R. Parker, and K. Takayama, Preparation and antitumor activity of non-toxic lipid A, Cancer Immunol. Immunother. 12:91 (1982).

57. E. Ribi, J. L. Cantrell, K. Takayama, K. Amano, Enhancement of anti-tumor resistance by mycobacterial products and endotoxin, in: "Beneficial Effects of Endotoxins," A. Nowotny, ed., Plenum Press, New York and London (1983).

58. E. T. Rietschel, S. Hase, M.-T. King, J. Redmond, and V. Lehmann, Chemical structure of lipid A, in: "Microbiology 1977," D. Schlessinger, ed., American Society of Microbiology, Washington, D.C. (1977).

59. N. Rote, The role of long chain carboxylic acid in endotoxin, Ph.D. Thesis, Temple University, Philadelphia, Pennsylvania (1974).

60. J. Rothman, A. G. Johnson, H. Friedman, E. Kovats, P. H. Pham, F. Sanavi, A. M. Nowotny and A. Nowotny, Biological effects of White-type polysacharides of gram negative bacteria, J. Biol. Res. Mod. 4:169 (1985).

61. J. A. Rudbach, R. L. Anacker, W. T. Haskins, K. C. Milner, and E. Ribi, Physical structure of a native protoplasmic polysaccharide from Escherichia coli, J. Immunol. 98:1 (1967).

62. J. Samu, E. Kovats, T. Keler, and A. Nowotny, TLC and high-voltage TLC of endotoxins and their derivatives, Abstract, Fifth American-Eastern European Symposium on Liquid Chromatography, Hungarian Chemical Society, Budapest, Hungary (1985).

63. G. Shwartzman, Studies on Bacillus typhosus toxic substance. I. The phenomenon of local reactivity to B. Typhosus culture filtrate, J. Exp. Med. 48:247 (1928).

64. B. M. Sultzer, Chemical modification of endotoxin and inactivation of its biological properties, in: "Microbial Toxins," Vol. 5, Kadis et al., eds., Academic Press, New York (1971).

65. B. M. Sultzer and G. W. Goodman, Endotoxin protein: A B-cell mitogen and polyclonal activator of C3H/HeJ lymphocytes, J. Exp. Med. 144:821 (1976).

66. K. Takayama, E. Ribi, and J. L. Cantrell, Isolation of a nontoxic Lipid A fraction containing tumor regression activity, Cancer Res. 41:821 (1981).

67. D. Tripodi, Active sites of Serratia marcescens endotoxin preparations, Ph.D. Thesis, Temple University, Philadelphia, Pennsylvanyia (1965).

68. D. Tripodi and A. Nowotny, Relation of structure to function in bacterial O-antigens. V. Nature of active sites in endotoxic lipopolysaccharides of Serratia marcescens, Ann. N.Y. Acad. Sci. 133:604 (1966).

69. C. M. Tsai and C. E. Frash, A sensitive silver stain for detecting lipopolysaccharide in polyacrylamide gels, Anal. Biochem. 119:115 (1982).

70. R. Urbaschek, Effects of bacterial products in granulopoiesis, Adv. Exp. Med. Biol. 121B:51 (1980).

71. O. Westphal and O. Luderitz, Chemische Erforschung von Lipopolysac-
     chariden gram negativer Bakterien, Angew. Chem. 66:407 (1954).
72. O. Westphal, A. Nowotny, O. Luderitz, H. Hurni, E. Eichenberger, and G.
     Schonholzer, The significance of the lipid component (lipid A) in the
     biological activities of bacterial endotoxins (lipopolysaccharides),
     Pharmeceutica. Acta. Helvetica (Germany) 33:401 (1958).
73. O. Westphal, O. Luderitz, E. T. Rietschel, and C. Galanos, Bacterial
     lipopolysaccharide and its lipid A component: some historical and some
     current aspects, Biochem. Soc. Trans. 9:191 (1980).
74. P. B. White, Notes on intestinal bacilli with special reference to
     smooth and rough races, J. Path. and. Bact. 32:85 (1929).
75. P. D. Wightman and C.R.H. Wightman, The activation of protein kinase C
     by biologically active lipid moieties of lipopolysaccharide, J. Biol.
     Chem. 259:10048 (1984).

THE ANTIGENIC STRUCTURE OF THE LIPOOLIGOSACCHARIDES

OF NEISSERIA GONORRHOEAE

Michael A. Apicella

Division of Infectious Diseases, Department of Medicine
State University of New York at Buffalo, School of Medicine
Buffalo, New York

INTRODUCTION

The importance of lipooligosaccharide (LOS) (this term which will be
used in this chapter in place of "lipopolysacharide" was suggested by
Schneider et al. (29) for the lipid A containing glycolipids of pathogenic
Neisseria because of the oligosaccharide nature of the saccharide moiety in
these organisms) in the pathogenesis and immunobiology of Neisseria
gonorrhoeae is unquestioned. Studies by Ward et al. (32), Rice and Kasper
(27) and Schneider and co-workers (28) indicate the LOS is a key target on
the gonococcal cell surface for human bactericidal antibody. Thus, an understanding of the physicochemical and antigenic structure of gonococcal LOS
as well as its characteristics in vivo and in vitro will be important in
comprehending the immunobiology of the gonococcus.

Isolation of gonococcal LOS was first published by Tauber and Garson
in 1959 using the phenol-water method (31). Little was known about this
material until Maeland and associates (13-20) published a series of studies
concerning the chemical and antigenic nature of gonococcal LOS. He has
described two major antigenic determinants in various crude endotoxin preparations and has designated these the alpha and beta antigens. Maeland believed that these antigens were components of the lipid A carbohydrate complex of gonococcal endotoxin and proteins from the outer membrane. His
studies indicate that the beta antigen was a protein. The alpha determinant
is a carbohydrate and has been shown to lose antigenicity after oxidation
with sodium metaperiodate. This antigen is unaffected by pronase digestion.
Maeland had assumed that the alpha antigen may be analogous to the O somatic
antigen of the enterobacterial lipopolysaccharides. Using a hemagglutination
inhibition system with cross absorption, he has demonstrated that there may
be as many as six alpha antigen factors and has revealed a potential tool
for serotyping the gonococcus.

Perry et al. (23), Stead et al. (30), and Wiseman and Caird (33) studied
the chemical composition of gonoccal LOS isolated by phenol-water extraction.
These investigators found glucose, galactose, glucosamine hepatose and KDO
present as the major components. Stead et al. were the first to recognize
that gonococcal LOS lacked the repeating O-side chains of the enterobacteriaceae and clearly showed that colonial type had no bearing on LOS chemical
structure.

The studies of Maeland et al. (13-20) are the foundations upon which more recent efforts in studying gonococcal LOS have been based. Apicella and co-workers confirmed and extended his observations and have isolated a series of six immunologically distinct acidic polysaccharides from the phenol-water LOS (2-9). These antigens are the carbohydrate component of gonococcal LOS and are isolated from the phenol-water extract of gonococci after NaOH digestion of the phenol-water extract followed by pronase digestion and ion exchange chromatography (2-4). The resulting product is a series of oligosaccharides linked by the amide linked lipid which retains the antigenic properties of the native LOS. It contains less than one percent protein or fatty acid and less than 0.5% nucleic acids (6). Studies have indicated that the antigenic structure of gonococcal LOS is complex with each LOS containing a common determinant (6). An additional antigen, termed the variable antigen, is present on three (types 1,3, and 4) of the six LOS serotypes. These studies have been confirmed in part by the work of Rappuoli who demonstrated two distinctive "core" antigens on one gonococcal LOS (25-26). A second gonococcal LOS was shown to contain only a single "core" antigen. In addition, our own studies have shown that each LOS contains its respective serotype antigen (6,22).

The terminology "rough and smooth" has been applied to Neisserial LOS by many investigators (6,25,26). Such terminology probably has little meaning when considered in the same context as enterobacterial LPS. Unlike enterobacterial LPS, gonococcal LOS has an intricate antigen structure without a series of repeating O side chains (10-12). Sizing of gonococcal LOS oligosaccharide chains indicates chain length no greater than the enterobacterial Ra mutant by both SDA-PAGE (11) and molecular sieve chromatography (28). Thus, while gonococcal LOS has specific serotype determinants similar to enterobacterial O-antigens, it lacks the corresponding chain length of these enterobacterial LPS. A different mechanism for expression of these antigens must be operative and probably is a steric arrangement of a limited repertoire of sugars. Studies by Allen and co-workers (1) and those of others (24) have indicated that phenol-water extracted gonococcal LOS contains glucose, galactose, glucosamine, galactosamine, and ethanolamine. The hexosamine content of the six serotype LOS preparations is quantitatively similar (1) but qualitatively different, with molar ratios of glucosamine to galactosamine varying from 1:1 (LOS type 1) to 4:1 (LOS types 4 and 5). Studies in a number of laboratories indicated that the O-D-galactopyranosyl (1-4)-D-glucopyranose (lactose) moiety may be an important constituent of the gonococcal LOS antigen structure (8,15). Using a monoclonal antibody, 3F11, which recognizes a region in the oligosaccharide portion of all six gonococcal LOS serotypes, Apicella and co-workers have identified its specificity for a D-galactosamine-O-D galactopyranosyl-(4-4)-glucopyranose moiety. In five of the six gonococcal LOS serotypes studied, isolation of the pyocin resistant variant resulted in a strain which lacked serotype antigen expression and also lacked the ability to bind 3F11 monoclonal antibody (8,21). This would suggest that presence or expression of this trisaccharide moiety while common to all gonococcal LOS studied, is related in some way to the serotype antigen expression.

The antigenic structure of gonococcal LOS is important immunobiologically, is complex and regulated by a number of diverse factors. The following results outline recent studies with this intriguing macromolecule.

Utilizing the serum resistant strain JW31 and its serum sensitive pyocin resistant variant, JW31R, bactericidal studies were undertaken to determine the role of the different human immunoglobulin types in the bactericidal response. The LOS of these strains differ since JW31 LOS contains the Gc_4 serotype antigen, the variable antigen and the common antigen, while JW31R LOS contains only the LOS common antigen (21). Thus, these bactericidal studies allow analysis of the effect of changes in the

58

Table 1. Bactericidal Studies with Affinity Purified Human Anti-LPS Immunoglobulins Isolated from Pooled Normal Serum

Anti-LPS immunoglobulin	dilution	% Killing at 1 Hour Strain	
		JW31	JW31R
IgA	undiluted	13.2%	0
IgM	1:10	2.1%	100%
	1:50	--	100%
IgG	undiluted	93.7%	0
	1:2	85.4%	--

LOS on the bactericidal response since these variants differ only in the antigenic expression of their LOS.

Naturally occurring antibody from pooled normal human serum to lipopolysaccharide was isolated by an affinity column to which E. coli J-5 LPS had been conjugated. ELISA studies using J-5 LPS showed the eluate contained IgM, IgG and IgA LPS antibodies at titers of 1:6400, 1:1600 and 1:400, respectively.

This affinity purified anti-J-5 LPS antibody was then fractionated into the respective immunoglobulin fractions by passage over affinity columns to which anti-human heavy chain specific antisera had been covalently linked. The isolated immunoglobulin anti-J-F LPS antibody fractions were tested in ELISA and showed reactivity only with this respective immunoglobulin. A bactericidal system was established with these immunoglobulin fractions, hypogammaglobulinemic sera as a complement source and either gonococcal strain JW31 or JW31R. Table 1 gives the results of these studies. The IgM fraction was bactericidal at a titer of 1:50 for strain JW31R while the IgA and IgG fractions failed to kill this strain. The IgG fraction was bactericidal for strain JW31 at 1:2 dilution and the IgA and IgM fractions failed to kill this strain. Blocking experiments showed that IgA or IgG did not prevent IgM killing of JW31R while IgA blocked IgG killing of JW31.

These studies indicate that there are at least two gonococcidal sites on gonococcal LOS. One site is acted on by the IgM anti-LOS antibody and is based in a deep core site of JW31R LOS. The second site is acted on by IgG and potentially by IgA. This site appears to be more peripheral and involves either the gonococcal LOS serotype or variable antigen.

Monoclonal antibodies have begun to elucidate the antigenic structure of gonococcal LOS. Using monoclonal antibodies 3F11 and 1-1-M, studies have demonstrated that specific determinants in the LOS may be markers for serum sensitivity and resistance. As seen in Table 2, monoclonal antibody 3F11 recognizes a site on 60% of serum sensitive strains and only 9% of serum resistant strains. The epitopes recognized by these antibodies are present on distinct bands after SDS-PAGE and Western Blot analysis indicating that these chains represent distinct physicochemical and antigenic moieties rather than a series of polymeric structures. As monoclonal antibodies with additional specificities for gonococcal LOS are identified, it is feasible that an LOS serovar type system could be developed.

Using pyocin selection, a number of investigators have shown that isogenic variants of a variety of gonococci can be selected and that these

Table 2. Reactivity of Gonococcal LOS Specific Monoclonal Antibodies in Dot Assay for Serum[s] and Serum[r] Strains

| serum sensitivity | # strains | Monoclonal reactivity in dot assay | | | |
| | | 3F11 | | 1-1-M | |
		+	−	+	−
Ser[s]	20	19	1	12	8
Ser[r]	11	11	0	1	10

pyocin resistant strains produce lipooligosaccharides which are physicochemically (11, 22) or both antigenically and physicochemically different than their parents. Morse and Apicella (21) studied strain JW31 and its pyocin resistant mutant JW31R. Auxotype and outer membrane proteins of these strains were identical. Immunodiffusion and an ELISA assay showed that the pyocin resistant strain JW31R lost both the LOS serotype and variable antigens while retaining at least a portion of the common determinant. The use of monoclonal antibody 3F11 indicated that the LOS from strain JW31R and pyocin 611 131 resistant strains of other LOS serotypes lack a D-galactosaminyl-D-galactopyranosyl-D-glucose moiety. Studies with pyocin resistant variants from LOS serotypes 2, 3, 4, 5 and 6 indicate that these variants failed to inhibit the 3F11 ELISA at concentrations as high as 1000 μgm/ml (Table 3). It is interesting that the Gc$_1$ strain 1342 was resistant to pyocin 611 131. Whether this is only characteristic of this strain or is a function of the serotype is unknown. The LOS derived polysaccharide from strain JW31 polysaccharide exhibits a markedly reduced affinity. Additionally, in the presence of 25 percent normal human serum, 99 percent of strain JW31R was killed within 20 minutes and strain JW31 was resistant to this bactericidal activity. This suggests that the LOS structure plays an important role in determining serum susceptibility of these strains. Similar observations have been made by Guymon (11) (see below).

Studying the same strains, Connelly and Allen showed similar changes in LOS antigens after pyocin selection and demonstrated changes in migration in the two LOS preparations in SDA-PAGE (10). Guymon and co-workers identi-

Table 3. Inhibition of Monoclonal 3F11 ELISA* by LOS Isolated from Pyocin Sensitive and Pyocin Resistant Strains

| Prototype strain | Gc type | μgm/ml for 50% inhibition | |
		Pyocin[s] LOS	Pyocin[r] LOS
1291	2	8.6 ± 3.2	1000
4505	3	5.7 ± 2.7	1000
8551	4	3.6 ± 2.5	1000
PID2	5	9.2 ± 3.6	1000
3893	6	6.3 ± 3.7	1000

* 4505 LOS well coat at 5 μgm/ml and 3F11 at 1:4000 as antiserum

Table 4. Inhibition of Gc_4 Serotype Specific ELISA* by
Lipooligosaccharides (LOS), Oligosaccharides
(OS) and Lipid A (LA).

Inhibitor	Concentration for 50% inhibition
Gc_4 LOS	2.5 µgm ± 1.4
Gc_4 oligosaccharide	210 µgm ± 20.6
Gc_4 lipid A	265 µgm ± 41.7
Gc_1 LOS	>1000
Gc_1 oligosaccharide	>1000
Gc_1 lipid A	>1000

Well coat was 8551 LOS at 25 µgm/ml and Gc_4 antiserum was
diluted at 1:4000

fied three classes of gonococcal LOS mutants utilizing pyocin selection
(11). The LOS of one lacked galactose while a second lacked the typical
heptose found in the parental LOS and was reduced in glucose, galactose,
and N-acetylglucosamine of the parent strain. The strains lacking galactose
were as resistant as the parent to normal human serum bactericidal activity
while the other two variants were serum sensitive. Additionally, the pyocin
selected variants were three to four times more sensitive to polymyxin B
than the parent strain.

Divalent cations appear to be important in the antigenic expression of
gonococcal LOS. Studies using monoclonal antibodies 3F11 and 1-1-M indicate
that these monoclonal antibodies fail to interact with the LOS which is
bound to nitrocellulose unless Mg++ is present in the buffers in which the
immunologic interaction occurs. Studies with radiolabelled LOS indicate
that the action of the Mg++ is not to enhance LOS binding. From these experiments it would appear that divalent cation Mg++ is important in the antigenic expression of gonococcal LOS potentially through steric effects on
the oligosaccharide portion of the LOS.

The interaction of the lipid A and oligosaccharide portion of the gonococcal LOS appears to be important in the antigenic expression of the LOS.
Studies in ELISA inhibition utilizing polyvalent antisera to gonococcal
strain 8551 and strain 8551 LOS as the well coat clearly indicate that the
inhibitory activity of the lipid A and oligosaccharide are at least 80 to
100 times less than the intact LOS (Table 4).

Studies with monoclonal 3F11 which has specificity for the oligosaccharide region of gonococcal LOS demonstrates the importance of the relationship
between the oligosaccharide and the Lipid A. Table 5 shows the results of
a 3F11 ELISA inhibition using native LOS from the prototype strains of the
six LOS serotypes and a meningococcal group B strain 986.

As can be seen the oligosaccharides fail to inhibit at concentrations
as high as 1000 µgm/ml. The lipid A inhibited at 20 (Gc_1) to 100 (Gc_4)
fold less that of the native LOS. Aqueous extraction of the lipid A after
chloroform solubilization to remove residual oligossacharide resulted in
approximately 50 percent further decrease of the inhibitory capacity of the
lipid A. It is assumed that the remaining inhibition was due to unhydrolyzed
LOS contaminating the lipid A preparation. These studies would suggest
that the antigenic expression of gonococcal LOS as regards both polyclonal

Table 5. Inhibition of Monoclonal 3F11 ELISA* by Lipooligo-
Saccharides (LOS), Oligosaccharides (OS) and Lipid A (LA)

Strain	Gc LOS Serotype	μgm/ml for 50% inhibition		
		LOS	OS	LA
1342	1	17.1 ± 4.6	>1000	262 ± 42
1291	2	8.6 ± 3.2	>1000	300 ± 78
4505	3	5.7 ± 2.7	>1000	352 ± 75
8551	4	3.6 ± 2.5	>1000	420 ± 60
PID2	5	9.2 ± 3.6	>1000	250 ± 110
3893	6	6.3 ± 3.7	>1000	460 ± 153
MnB-986	-	255 ± 47	>1000	1000

*4505 LOS 5 μgm/ml as well coat and 3F11 at 1:4000 as antiserum.

antiserum and an oligosaccharide directed monoclonal antibody is dependent
upon the lipid A portion of the macromolecule for complete expression. It
is interesting to hypothesize that the lipid A exerts steric constraints on
the oligosaccharide by hydrophobic interactions which are important in antigenic expression.

CONCLUSIONS

The antigenic expression of gonococcal LOS is complex and regulated by
a number of factors including pyocin susceptibility, growth conditions,
divalent cation interaction and association with lipid A. Because the LOS
is an important target for human bactericidal antibody, the factors which
orchestrate this antigenic expression must be elucidated and understood
before LOS can be incorporated into a vaccine.

ACKNOWLEDGMENTS

This work was supported in part by Grant AI 18384 from the National
Institutes of Health.

REFERENCES

1. P. Z. Allen, M. C. Connelly, and M. A. Apicella, Interactions of
   lectins with Neisseria gonorrhoeae, Canadian J. Microbiol. 26:468
   (1980).
2. M. A. Apicella and J. C. Allen, Isolation and characterization of the
   beta antigen of Neisseria gonorrhoeae, Infect. and Immun. 7:315
   (1973).
3. M. A. Apicella, Antigenically distinct populations of Neisseria
   gonorrhoeae: isolation and characterization of the responsible
   determinants, J. Infect. Dis. 130:619 (1974).
4. M. A. Apicella, Serogrouping of Neisseria gonorrhoeae: identification
   of four immunologically distinct acidic polysaccharides, J. Infect.
   Dis. 134:377 (1976).
5. M. A. Apicella, J. F. Breen, and N. C. Gagliardi, Degradation of the
   polysaccharide component of gonococcal lipopolysaccharide of
   gonococcal and meningococcal sonic extracts, Infect. and Immun.
   20:228 (1978).

6. M. A. Apicella and N. Gagliardi, Antigenic heterogeneity of the non-serogroup antigen structure of the lipopolysaccharides of Neisseria gonorrhoeae, Infect. and Immun. 26:870 (1979).

7. M. A. Apicella, Lipopolysaccharide-derived serotype polysaccharides from Neisseria meningitidis group B, J. Infect. Dis. 140:62 (1979).

8. M. A. Apicella, K, M. Bennet, C. A. Hermerath, and D. E. Roberts, Monoclonal antibody analysis of lipopolysaccharide from Neisseria gonorrhoeae and Neisseria meningitidis, Infect. Immun. 34:751 (1981).

9. J. F. Breen and M. A. Apicella, Immunogenicity of gonococcal Gc$_2$ polysaccharide: comparative studies with pneumonococcal type III polysaccharide, and Salmonella typhosa Vi antigen, Infect. and Immun. 22:195 (1978).

10. M. C. Connelly and P. Z. Allen, Chemical and immunochemical studies on lipopolysaccharides from pyocin 103 sensitive and resistant Neisseria gonorrhoeae, Carbohydrate Research 120:171 (1983).

11. L. F. Guymon, M. Esser, and W. M. Shafer, Pyocin-resistant lipopolysaccharide mutants of Neisseria gonorrhoeae: alterations in sensitivity to normal human serum and polymyxin B, Infect. and Immun. 36:541 (1982).

12. B. Jann, K. Reski, and K. Jann, Heterogeneity of lipopolysaccharides. Analysis of polysaccharide chain lengths by sodium dodecylsulfate-polyacrylamide gel electrophoresis, Eur. J. Biochem. 60:239 (1975).

13. J. A. Maeland, Antigenic properties of various preparations of Neisseria gonorrhoeae endotoxin, Acta. Path. Microbiol. Scan. 73:413 (1968).

14. J. A. Maeland, Antigenic determinants of aqueous ether extracted endotoxin from Neisseria gonorrhoeae, Acta. Path. Microbiol. Scan. 76:475 (1969).

15. J. A. Maeland, Antibodies in human sera against antigens in gonococci, demonstrated by a passive hemolysis test, Acta. Path. et Microbiol. Scand. 67:102 (1966).

16. J. A. Maeland, Serological properties of antisera to Neisseria gonorrhoeae antigens, Acta. Path. Microbiol. Scand. 69:145 (1967).

17. J. A. Maeland, Immunochemical characterization of aqueous ether extracted endotoxin from Neisseria gonorrhoeae, Acta. Path. Microbiol. Scand. 76:484 (1969).

18. J. A. Maeland, Serological cross-reactions of aqueous ether extracted endotoxin from Neisseria gonorrhoeae strains, Acta. Path. Microbiol. Scand. 77:505 (1969).

19. J. A. Maeland and T. Kristoffersen, Immunochemical investigations on Neisseria gonorrhoeae endotoxin. I. Characterization of phenol-water extracted endotoxin and comparison with aqueous ether preparations, Acta. Path. Microbiol. Scand. Sec. B:226 (1971).

20. J. A. Maeland, T. Kristofferson, and T. Hofstad, Immunochemical investigations on Neisseria gonorrhoeae endotoxin. II. Serological multispecificity of other properties of phenol-water preparations, Acta. Path. Microbiol. Scand. Sec. B:233 (1971).

21. S. A. Morse and M. A. Apicella, Isolation of a lipopolysaccharide mutant of Neisseria gonorrhoeae: an analysis of the antigenic and biologic differences, J. Infect. Dis. 145:206 (1982).

22. S. A. Morse, C. S. Mintz, S. K. Sarafran, L. Bartenstein, M. Bertram, and M. A. Apicella, Effect of dilution rate on lipopolysaccharide and serum resistance of Neisseria gonorrhoeae grown in continuous culture, Infect. and Immun. 41:74 (1983).

23. M. D. Perry, V. Dauost, B. B. Diena, E. Ashton, and R. Wallace, The lipopolysaccharides of Neisseria gonorrhoeae colony types 1 and 4, Can. J. Biochem. 53:623 (1975).

24. M. D. Perry, V. Dauost, V. G. Johnson, B. B. Diena, and F. E. Ashton, Gonococcal R-type lipopolysaccharides, in: "Immunobiology of Neisseria Gonorrhoeae," ASM Publications, Washington, D.C. (1978).

25. R. Rappuoli, Purification of gonococcal R Type lipopolysaccharide not bearing the lactose-like antigen, *Giorn. Batt. Virol. Immun.* 74:191 (1981).

26. R. Rappuoli, Production and characterization of high titer rabbit antigonococcal R-type lipopolysaccharide serum, *Microbiologica* 5:225 (1982).

27. P. A. Rice and D. Kasper, Characterization of gonococcal antigens responsible for induction of bactericidal antibody in disseminated infection, *J. Clin. Invest.* 60:1149 (1977).

28. H. Schneider, J. M. Griffis, G. D. Williams, and G. B. Pier, Immunological basis of serum resistance to Neisseria gonorrhoeae, *J. Gen. Microbiol.* 128:13 (1982).

29. H. Schneider, T. L. Hale, W. D. Zollinger, R. C. Seid, C. A. Hammack, and J. M. Griffis, Heterogeneity of molecular size and antigenic expression within lipooligosaccharide of individual strains of Neisseria gonorrhoeae and Neisseria meningitidis, *Infect. and Immun.* 45:544 (1984).

30. A. Stead, S. S. Main, M. E. Ward, and P. J. Walt, Studies on lipopolysaccharides isolated from strains of Neisseria gonorrhoeae, *J. Gen. Microbiol.* 88:123 (1975).

31. H. Tauber and W. Garson, Isolation of lipopolysaccharide endotoxin, *J. Biol. Chem.* 235:961 (1959).

32. M. E. Ward, P. R. Lambden, J. E. Heckels, and P. J. Ward, The surface properties of Neisseria gonorrhoeae: determinants of susceptibility to antibody complement killing, *J. Gen. Micro.* 108:205 (1978).

33. G. M. Wiseman and J. D. Caird, Composition of the lipopolysaccharide of Neisseria gonorrhoeae, *Infect. and Immun.* 16:550 (1977).

LIPID A: RELATIONSHIPS OF CHEMICAL STRUCTURE AND BIOLOGICAL ACTIVITY

Otto Lüderitz, Chris Galanos, Ernst Th. Rietschel, and Otto Westphal

Max-Planck Institut für Immunobiologie, Freiburg
and Forschungsinstitut Borstel, Borstel
Federal Republic of Germany

Thirty years ago in 1954, we summarized our knowledge then of the chemistry and biology of endotoxins (73). At that time, the phenol/water extraction procedure had been established (74), and the extracts from a number of bacterial strains studied chemically and biologically. These investigations had led to the identification of the extracted products as the O antigens and endotoxins of these bacteria, resembling products isolated earlier by Boivin and Mesrobeanu (2), Morgan et al. (44), Goebel et al. (15), Shear et al. (62), and others. It had been recognized that endotoxins are composed of a polysaccharide component linked covalently to a lipid component, and thus were chemically lipopolysaccharides. The lipid had been termed lipid A (73) and found to contain D - glucosamine, phosphate, and long-chain fatty acids in ester and amide linkages. It had been shown that the linkage of the polysaccharide to lipid A could be cleaved by hydrolysis with dilute acetic or hydrochloric acid, leading to free water-soluble polysaccharide and free water-insoluble lipid A. Neither free polysaccharide nor free insoluble lipid A expressed endotoxic activities. Nevertheless, it has been anticipated that lipid A was the biologically active part and that activity would result from the binding of lipid A to the polysaccharide component, the latter functioning as a hydrophilic carrier for lipid A. Evidence for this was obtained by demonstrating that cleavage of lipopolysaccharides by mild acid in the presence of a non-endotoxic protein (casein, albumin) leads to the formation of a water-soluble lipid A-protein complex which expresses pyrogenicity and lethal toxicity. This procedure, however, was tricky and our results were confirmed by some but not by all laboratories.

This was what we knew in 1954 on lipid A biology and chemistry. For about one decade, no significant progress was made in lipid A chemistry until 1965 when analytical work was resumed (13). At this time the backbone structure of lipid A from Salmonella minnesota was recognized as a 1,4'-biphosphorylated β1,6-linked D-glucosamine disaccharide. Results of periodate oxidation seemed to indicate that the polysaccharide chain was linked to position 3' of the nonreducing glucosaminyl residue of the backbone (14), a conclusion which recently had to be revised (see below).

In 1972, Galanos, et al. introduced new reproducible methods for the preparation of defined endotoxic lipid A/albumin complexes (11), and in 1977, they succeeded in solubilizing lipid A directly (without carrier) by conversion into the triethylammonium salt form (6). These findings allowed the unequivocal demonstration that lipid A represents the endotoxic principle of lipopolysaccharides. They greatly stimulated new interest into the chemical structure of lipid A, especially with the aim to identify subregions in the molecule essential for distinct activities, and to study the possibility to attribute individual beneficial (e.g., immune-modulating, therapeutic) and toxic properties to defined structures. Lipid A analysis (performed mainly on Salmonella minnesota R-forms) was encountered with many kinds of difficulties. The number of fatty acid residues demonstrable in lipid A increased with the improvement of the techniques of analysis and finally came near to seven (S. minnesota R595 lipid A), i.e. to a value exceeding the number (five) of hydroxyl and amino groups available for acyl substitution on the disaccharide backbone (36,56). It was then found that one ester-linked (R)-3-hydroxymyristic acid was esterified at its hydroxyl group by another fatty acid to form a 3-acyloxyacyl residue while a second 3-hydroxymyristic acid carried a free 3-hydroxyl group (56). Later, it turned out that also amide-bound 3-hydroxy fatty acids can be 3-O-acylated to form 3-acyloxyacyl residues (75) and that in fact, only 3-hydroxymyristic acid units are directly linked to the lipid A backbone. It became possible to define the identity and exact positions of the amide-linked acyl and 3-acyloxyacyl residues (19,57,76), but it took longer to find and establish the techniques allowing the determination of the positions of each of the ester-bound 3-hydroxy- and 3-acyloxy-acyl residues. Methylation techniques in neutral medium allowed the detection of free hydroxyl groups at position 4 in the Re lipopolysaccharide and at positions 4 and 6' in free lipid A (64), the hydroxyl group at position 6' representing the attachment site of the polysaccharide component. Two-dimensional nuclear magnetic resonance, fast atom bombardment, and laser desorption mass spectrometry finally led to the definite structure of lipid A of Salmonella and Escherichia coli. These techniques in the last phases of investigations have been applied by several groups. They all came to identical results (52,58,59,61,66). Enterobacterial free lipid A contains a β1,6-linked D-glucosamine disaccharide which is substituted by two phosphate groups one being linked in an α-glycosidic linkage to C1 of the reducing glucosamine unit (GlcN I), the other being ester-linked to position 4' of the non-reducing glucosamine unit (GlcN II).

In lipid A, this so-called backbone structure can be substituted by polar head groups. Thus, in E. coli, the phosphate group at position 1 is (partly) substituted by a phosphate residue, in S. minnesota, it carries (also in non-molar amounts) a phosphorylethanolamine unit. In E. coli, the phosphate group at position 4' is unsubstituted. In S. minnesota it is partly substituted by 4-amino-L-arabinose in an α-glycosidic linkage (in Figure 1 the polar groups have been omitted).

The backbone of enterobacterial lipid A carries four 3-hydroxymyristic acid residues which are linked to positions 2, 3, 2', and 3' and which may be substituted, at their 3-hydroxyl group, by a non-hydroxylated fatty acid. Thus, in E. coli and Salmonella, the hydroxymyristoyl residues at positions 2' and 3' are esterified by a lauroyl and a myristoyl residue, respectively. S. minnesota lipid A contains additional palmitoyl residues which partly substitute the 3-hydroxymyristic acid at position 2. The backbone, therefore, contains two free hydroxy groups at positions 4 and 6', the latter representing the attachment site of the polysaccharide component in lipopolysaccharides (Figure 1).

Investigations of other genera of Enterobacteriaceae as well as of lipid A's from strains of other families (58) have revealed that most lipid A's resemble each other in their backbone structure, which therefore represents the most conservative region of lipopolysaccharides (for exceptions see 42,43,72). Lipid A's of different origin, however, may differ in the nature and linkages of the substituents of the backbone (42,43,52,72). The results of these investigations allowed the first comparative studies on

66

structure/activity relationships, which were performed on enterobacterial lipopolysaccharides of S and R forms, on free lipid A, on degradation and modification products thereof, as well as on lipid A's derived from bacterial families remote from <u>Enterobacteriaceae</u>. These studies led to the following conclusions (36):

The biological effects of lipid A are dependent on solubility and its state of aggregation (5,7).

Different lipid A effects are separable and it seems, therefore, that different activities are dependent directly or indirectly on the presence of distinct substructures of lipid A (36,68,69).

Polar substituents of the backbone (such as ethanolamine and 4-amino-L-arabinose) do not contribute directly to endotoxic activity. In some cases substitutions may rather inhibit biological activities (36).

Whether phosphate groups as such play a role in some activities is not clear. They do counterbalance the lipophilic properties of the molecule and confer solubility in water. Therefore, at least indirectly, they are important at least for acute endotoxic activity.

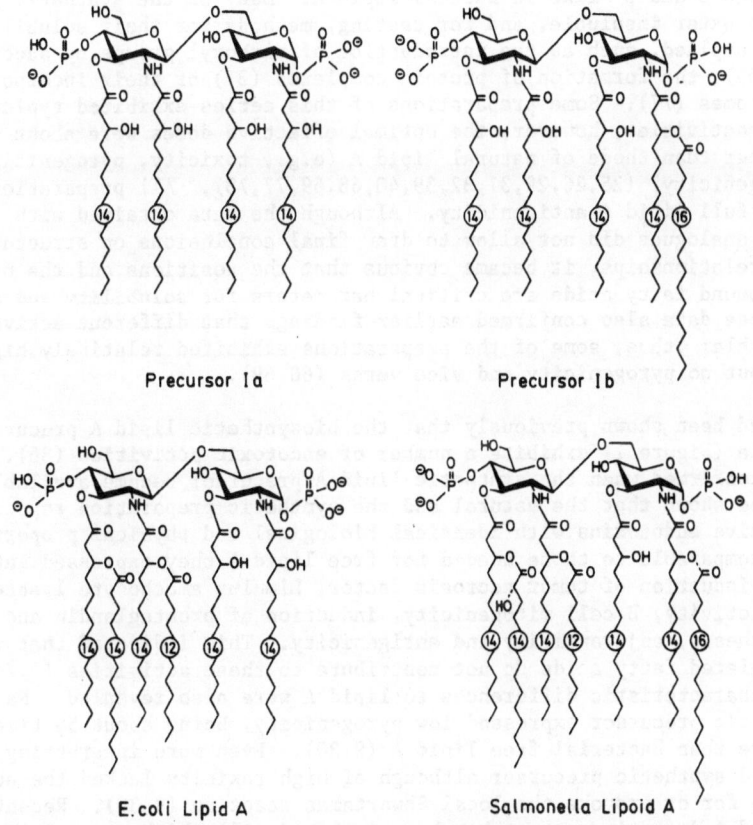

Figure 1. Anticipated structures for precursors Ia and Ib, and lipid A of <u>E. coli</u> and <u>Salmonella</u>. Numbers in circles indicate number of carbon atoms in the chain. Substituents of phosphate groups of lipid A have been omitted. Dotted linkage indicates incomplete substitution.

Lipid A's containing the Salmonella backbone exhibit strong serological cross reactions (8). Finally, these investigations have shown that certain gram-negative bacteria (e.g. members of the family of phototrophic Rhodospirillaceae) produce lipopolysaccharides containing lipid A's which are structurally distinct from enterobacterial lipid A, which are endotoxically inactive (42,43,72), and which are antigenically distinct from Salmonella lipid A (8).

Simultaneously with the structural work important steps of the biosynthesis of lipid A have been elucidated. Monomeric (50,54,67) and dimeric (18,34,35) precursor molecules have been isolated from mutant strains, their structures evaluated, and their role in lipid A biosynthesis studied (45,46,-50,51,55).

Finally, several groups of carbohydrate chemists have undertaken the chemical synthesis of lipid A analogues (1,3,22,23,27,47,63), lipid A precursor molecules (21,28,33,41), and very recently, complete lipid A (20).

These achievements represent important progress regarding possibilities to study structure/activity relationships of lipid A on a molecular basis. The first synthetic compounds chemically resembled enterobacterial lipid A in that they contained the lipid A backbone carrying amide-linked 3-hydroxyl or 3-acyloxyacyl residues. The ester-bound acyl and 3-acyloxyacyl residues in the synthetic preparations were linked to positions 3,4, and 6' and not to positions 3 and 3' like in natural lipid A. Many of the synthetic preparations were water insoluble, and for testing, methods for their solubilization had to be applied, such as the introduction of carboxyl groups by succinylation (68,69), the formation of protein complexes (31) or their incorporation into liposomes (77). Some preparations of this series exhibited typical endotoxic activities, however, the optimal effective doses were about 500 times higher than those of natural lipid A (e.g., toxicity, pyrogenicity, B cell mitogenicity) (25,26,29,31,32,39,40,68,69,77,78). All preparations expressed full lipid A antigenicity. Although the data obtained with this series of analogues did not allow to draw final conclusions on structure/-activity relationships, it became obvious that the positions and the number of ester-bound fatty acids are critical parameters for solubility and activity. These data also confirmed earlier findings that different activities are separable: thus, some of the preparations exhibited relatively high toxicity but no pyrogenicity and vice versa (68,69).

It had been shown previously that the biosynthetic lipid A precursor molecule Ia (Figure 1) exhibits a number of endotoxic activities (36). This was confirmed when the synthetic lipid A precursor became available. It could be shown that the natural and the synthetic preparation represented highly active endotoxins with identical biological and physical properties. In doses comparable to those needed for free lipid A they expressed lethal toxicity, induction of tumor necrosis factor, Limulus amoebocyte lysate gelation activity, B cell mitogenicity, induction of prostaglandin and interferon synthesis, adjuvanticity and antigenicity. This indicated that the nonhydroxylated fatty acids do not contribute to these activities (9,24,30). However, characteristic differences to lipid A were also revealed. Natural and synthetic precursor expressed low pyrogenicity, being about 50 times less active than bacterial free lipid A (9,30). Even more interesting, natural and synthetic precursor although of high toxicity lacked the ability to prepare for or provoke the local Shwartzman reaction (9,30). Recently, a new lipid A-defective mutant has been isolated (17) which accumulates in addition to precursor Ia, a second precursor Ib in about equal amounts (17,18). Structurally, precursor Ib occupies a position intermediate between precursor Ia and lipid A since it contains one amide-linked 3-acyloxyacyl residue (Figure 1). It was found that precursor Ib is capable of inducing and provoking the local Shwartzman reaction (12). This finding was surprising and may indicate that Shwartzman reactivity is not bound to a specific position of the additional fatty acyl residue (palmitic acid), but is rather dependent on the degree of hydrophobicity of the molecule. This is concluded from the fact that E. coli lipid A which contains two 3-acyloxyacyl residues, however neither of these at position 2, but contains instead a nonsubstituted 3-hydroxymyristic acid at this position, does exhibit strong Shwartzman reactivity. We would expect that one acyloxyacyl residue on either position 2' or 3' would fulfill the structural requirements for Shwartzman reactivity. The synthesis of positional isomers of precursor Ib is planned.

In one property natural and synthetic precursor behaved completely adverse. Synthetic precursor, as expected, stimulated B cells from lipopolysaccharide-responder mice but was inactive towards B lymphocytes from lipopolysaccharide-nonresponder mice (9). In contrast, the natural precursor preparation stimulated B cells from both mouse strains (70, and C. Galanos et al., unpublished data).

The reason for the discrepancy regarding the different mitogenic activities of the natural and synthetic precursor towards B lymphocytes of lipopolysaccharide-nonresponder mice (strain C3H/HeJ) was investigated. It was found (16 and C. Galanos, unpublished data) that the mixture of precursor Ia and Ib, as obtained after phenol/chloroform/petroleum ether-extraction, was contaminated with bacterial protein (up to 10%). After purification by silica gel thin layer chromatography and separation of precursor Ia and Ib, the protein content of the preparations was reduced to less than 0.2%. Comparison of the mitogenic activities of the compounds at different steps of purification showed that with reduction of the protein content the response of B cells from nonresponder mice decreased while the response of B cells from the responder strain remained unchanged. Obviously, a protein is coextracted with the precursor, which exhibits strong mitogenicity also with lipopolysaccharide-nonresponder mice cells, possibly in a synergistic way with the precursor. This question is under investigation. Very recently Shiba and Kusumoto and their colleagues (20) succeeded in synthesizing the structure anticipated for lipid A of E. coli. Like lipid A from other bacteria, E. coli lipid A exhibits structural microheterogeneity. This may be demonstrated by thin layer chromatography, whereby the lipid A is resolved into several bands (48,49), one of which predominates. The synthetic preparation corresponds to the major band of E. coli lipid A. Natural and synthetic lipid A were found to be identical in all physicochemical properties tested so far, such as solubility, NMR and IR spectrometry (Shiba et al., Seydel et al., unpublished data). Furthermore, we could show (10) that synthetic and natural lipid A were identical in all biological test systems applied, i.e., they exhibited strong lethal toxicity, pyrogenicity, Shwartzman reactivity, cross tolerance to lethality, tumor necrotizing activity, B cell and macrophage stimulating activity, antigenicity, and gelation of Limus amoebocyte lysate. These results show that synthetic lipid A prepared according to the formula evaluated for E. coli and other enterobacterial lipid A's exhibits all endotoxic activities in the expected doses.

The first synthesis of a partial structure of a lipopolysaccharide was performed in 1960. It contained the immunodominant sugar, colitose, of the O antigen of E. coli 0111 coupled to a carrier protein (38). This artificial antigen cross-reacted with anti- E. coli 0111 antiserum and induced the formation of anti-colitose antibodies. In recent years, a number of artificial antigens and immunogens containing immunodeterminant structures of enterobacterial O antigens have been synthesized (for examples, see 37). The corresponding highly specific antisera represent valuable tools in clinical diagnosis. Also, the synthesis of determinant structures of the core region has been started (71). These antigens are investigated as non-toxic vaccines for their possible capacity to protect against infection with a wide range of gram-negative pathogens.

The recent success in the chemical synthesis of complete E. coli lipid
A and the demonstration of its chemical and biological identity with natural
lipid A signals the beginning of a new era in endotoxin research. The synthesis of lipid A part structures and analogues (e.g., radio-labeled at
will) will provide the possibility to study mechanisms of endotoxin action,
to investigate structure/activity relationships, and to elucidate the minimal
structural requirements for individual lipid A effects. This may lead ultimately to the preparation of nontoxic products exhibiting defined beneficial
activities.

REFERENCES

1. L. Anderson and M. A. Nashed, The convergent approach to the synthesis
    of lipid A and its analogs, Am. Chem. Soc. Symp. Series 231:255
    (1983).
2. A. Boivin and L. Mesrobeanu, Recherches sur les antigenes somatiques et
    sur les endotoxines des bacteries. 1. Considerations generales et
    expose des techniques utilisees, Rev. Immunol. 1:553 (1935).
3. D. Charon, C. Diolez, M. Mondange, S. R. Sarfati, L. Szabo, P. Szabo,
    and F. Trigalo, Synthetic studies on structural elements of the
    hydrophobic region present in bacterial endotoxins, Am. Chem. Soc.
    Symp. Series 231:301 (1983).
4. R. Christian, G. Schulz, and F. M. Unger, Zur Struktur der
    3-desoxyoctuloson-(KDO)-region des lipopolysaccharids von Salmonella
    minnesota R595, Tetrahyd. Lett. 25:3433 (1984).
5. C. Galanos, Physical state and biological activity of
    lipopolysaccharides. Toxicity and immunogenicity of the lipid A
    component, Z. Immun. Forsch. 149:214 (1975).
6. C. Galanos, M. Freudenberg, S. Hase, F. Jay, and E. Ruschmann,
    Biological activities and immunological properties of lipid A, in:
    "Microbiology 1977," D. Schlessinger, ed., American Society of
    Microbiology, Washington, D.C. (1977).
7. C. Galanos, M. A. Freudenberg, O. Lüderitz, E.Th. Rietschel, and O.
    Westphal, Chemical, physicochemical and biological properties of
    bacterial lipopolysaccharides, in: "Biomedical Applications of the
    Horseshoe Crab (Limulidae)," Alan R. Liss, Inc., New York (1979).
8. C. Galanos, M. A. Freudenberg, F. Jay, and D. Nerkar, Immunogenic
    properties of lipid A, Rev. Infect. Dis. 6:546 (1984).
9. C. Galanos, V. Lehmann, O. Lüderitz, E.Th. Rietschel, O. Westphal, H.
    Brade, L. Brade, M. A. Freudenberg, T. Hansen-Hagge, Th. Lüderitz,
    G. McKenzie, U. Schade, W. Strittmatter, K. Tanamoto, U. Zähringer,
    M. Imoto, H. Yoshimura, M. Yamamoto, T. Shimamoto, S. Kusumoto, and
    T. Shiba, Endotoxin properties of chemically synthesized lipid A
    part structures. Comparison of synthetic lipid A precursor and
    synthetic analogues with biosynthetic lipid A precursor and free
    lipid A, Eur. J. Biochem. 140:221 (1984).
10. C. Galanos, O. Lüderitz, E.Th. Rietschel, O. Westphal, H. Brade, L.
    Brade, M. Freudenberg, U. Schade, M. Imoto, H. Yoshimura, S.
    Kusumoto, and T. Shiba, Synthetic and natural Escherichia coli free
    lipid A express identical endotoxic activities, Eur. J. Biochem. in
    press (1985).
11. C. Galanos, E.Th. Rietschel, O. Lüderitz, O. Westphal, B. Kim, and D.
    W. Watson, Biological activities of lipid A complexed with bovine
    serum albumin, Eur. J. Biochem. 31:230 (1972).
12. C. Galanos, T. Hansen-Hagge, V. Lehmann, and O. Lüderitz, Comparison of
    two lipid A precursor molecules in their capacity to express the
    local Shwartzman phenomenon, Infect. Immun. in press (1985).
13. J. Gmeiner, O. Lüderitz, and O. Westphal, Biochemical studies on
    lipopolysaccharides of Salmonella R mutants 6. Investigations on the
    structure of the lipid A component, Eur. J. Biochem. 7:370 (1969).

14. J. Gmeiner, M. Simon, and O. Lüderitz, The linkage of phosphate groups and of 2-keto-3-deoxyoctonate to the lipid A component in a Salmonella minnesota lipopolysaccharide, Eur. J. Biochem. 21:355 (1971).

15. W. F. Goebel, F. Binkley, and E. Perlman, Studies on the Flexner group of dysentery bacilli I. The specific antigens of Shigella paradysenteriae (Flexner), J. Exp. Med. 81:315 (1945).

16. Th. Hansen-Hagge, Entwicklung und Anwendung neuer Selektionsmethoden zur Anreicherung von Salmonella Lipoid A Mutanten. Isolierung und Charakterisierung einer neuen Vorstufe, Thesis University of Freiburg (1984).

17. Th. Hansen-Hagge, V. Lehmann, and O. Lüderitz, Free flow electrophoresis as a tool for enrichment of mutants with temperature dependent lethal mutations in lipid A synthesis, Eur. J. Biochem. in press (1985).

18. T. Hansen-Hagge, V. Lehmann, U. Seydel, B. Lindner, and U. Zähringer, Isolation and structural analysis of two lipid A precursors from a KDO deficient mutant of Salmonella typhimurium which differ in their hexadecanoic acid content, Arch. Microbiol. in press (1985).

19. M. Imoto, S. Kusumoto, T. Shiba, H. Naoki, T. Iwashita, E.Th. Rietschel, H. W. Wollenweber, C. Galanos, and O. Lüderitz, Chemical structure of E. coli lipid A: linkage site of acyl groups in the disaccharide backbone, Tetrahedron Lett. 24:4107 (1983).

20. M. Imoto, H. Yoshimura, S. Kusumoto, and T. Shiba, Total synthesis of lipid A, the active principle of bacterial endotoxin, Proc. Japan Acad. 60B:285 (1984).

21. M. Imoto, H. Yoshimura, M. Yamamoto, T. Shimamoto, S. Kusumoto, and T. Shiba, Chemical synthesis of phosphorylated tetraacyl disaccharide corresponding to a biosynthetic precursor of lipid A, Tetrahedron Lett. 25:2667 (1984).

22. M. Inage, H. Chaki, S. Kusumoto, and T. Shiba, Synthesis of liposaccharide corresponding to fundamental structure of Salmonella-type lipid A, Tetrahedron Lett. 21:3889 (1980).

23. M. Inage, H. Chaki, S. Kusumoto, and T. Shiba, Chemical synthesis of phosphorylated fundamental structure of lipid A, Tetrahedron Lett. 22:2281 (1981).

24. S. Kanegasaki, Y. Kojima, M. Matsuura, J. Y. Homma, A. Yamamoto, Y. Kumazawa, K. Tanamoto, T. Yasuda, T. Tsumita, M. Imoto, H. Yoshimura, M. Yamamoto, S. Shimamoto, and T. Shiba, Biological activities of analogues of lipid A based chemically on the revised structural model: comparison of mediator inducing, immunomodulating and endotoxic activities, Eur. J. Biochem. 143:237 (1984).

25. S. Kanegasaki, T. Yasuda, T. Smumita, T. Tadakuma, J. Y. Homma, S. Kusumoto, and T. Shiba, Biological activities of synthetic lipid A analogues in artificial membrane vesicles, in: "Bacterial Endotoxins: Chemical, Biological, and Clinical Aspects," J. Y. Homma, S. Kanegasaki, O. Lüderitz, T. Shiba, and O. Westphal, eds., Verlag Chemie, Weinheim (1984).

26. N. Kasai, K. Egawa, J. Mashimo, T. Shiba, and S. Kusumoto, Shwartzman activities of synthetic lipid A analogues and their effects on hepatic enzyme activities, in: "Bacterial Endotoxins: Chemical, Biological, and Clinical Aspects," J. Y. Homma, S. Kanegasaki, O. Lüderitz, T. Shiba, and O. Westphal, eds., Verlag Chemie, Weinheim (1984).

27. M. Kiso and A. Hasegawa, Synthetic studies on lipid A and related compounds, in: "Bacterial Endotoxins: Chemical, Biological, and Clinical Aspects," J. Y. Homma, S. Kanegasaki, O. Lüderitz, T. Shiba, and O. Westphal, eds., Verlag Chemie, Weinheim (1984).

28. M. Kiso, H. Ishida, and A. Hasegawa, Synthesis of biologically active, novel monosaccharide analogs of lipid A, Agric. Biol. Chem. 48:251 (1984).

29. S. Kotani, H. Takada, M. Tsujimoto, T. Ogawa, Y. Mori, N. Sakuta, A. Kawasaki, M. Inage, S. Kusumoto, T. Shiba, and N. Kasai, Immunobiological activities of synthetic lipid A analogs and related compounds as compared with those of bacterial lipopolysaccharide, Re-glycolipid, lipid A, and muramyl dipeptide, Infect. Immun. 41:758 (1983).

30. S. Kotani, H. Takada, M. Tsujimoto, T. Ogawa, K. Harada, Y. Mori, A. Kawasaki, A. Tanaka, S. Nagao, O. Tanaka, T. Shiba, S. Kusumoto, M. Imoto, H. Yoshimura, M. Yamamoto, and T. Shimamoto, Immunologically active lipid A analogs synthesized according to revised structural model of natural lipid A, Infect. Immun. 45:293 (1984).

31. Y. Kumazawa, M. Matsuura, Y. Nakatsuru-Watanabe, M. Fukumoto, C. Nischimura, J. Y. Homma, M. Inage, S. Kusumoto, and T. Shiba, Mitogenic and polyclonal B cell activation activities of synthetic lipid A analogues, Eur. J. Immunol. 14:109 (1984).

32. Y. Kumazawa, M. Matsuura, J. Y. Homma, Y. Nakatsuru, M. Kiso, and A. Hasegawa, B cell activation and adjuvant activities of chemically synthesized analogues of nonreducing sugar moiety of lipid A, Eur. J. Immunol. 15:199 (1985).

33. S. Kusumoto, M. Yamamoto, and T. Shiba, Chemical synthesis of lipid X and lipid Y, acyl glucosamine-1-phosphates isolated from Escherichia coli mutants, Tetrahedron Lett. 25:3727 (1984).

34. V. Lehmann, Isolation, purification and properties of an intermediate in 3-deoxy-D-manno-octulosonic acid – lipid A biosynthesis, Eur. J. Biochem. 75:257 (1977).

35. V. Lehmann, E. Rupprecht, and M. J. Osborn, Isolation of mutants conditionally blocked in the biosynthesis of the 3-deoxy-D-manno-octulosonic acid – lipid A part of lipopolysaccharides derived from Salmonella typhimurium Eur. J. Biochem. 76:41 (1977).

36. O. Lüderitz, C. Galanos, V. Lehmann, H. Mayer, E.Th. Rietschel, and J. Weckesser, Chemical structure and biological activities of lipid A's from various bacterial families, Naturwissenschaften 65:575 (1978).

37. O. Lüderitz, K. Tanamoto, C. Galanos, O. Westphal, U. Zahringer, E.Th. Rietschel, S. Kusumoto, and T. Shiba, Structural principles of lipopolysaccharides and biological properties of synthetic partial structures, American Chemical Society Symposium Series 231:3 (1983).

38. O. Lüderitz, O. Westphal, A. M. Staub, and L. LeMinor, Preparation and immunological properties of an artificial antigen with colitose (3-deoxyl-L-fucose) as determinant group, Nature 188:556 (1960).

39. M. Matsuura, Y. Kojima, J. Y. Homma, Y. Kumazawa, Y. Kubota, T. Shiba, and S. Kusumoto, Biological activities of synthetic lipid A analogues, in: "Bacterial Endotoxins: Chemical, Biological, and Clinical Aspects," J. Y. Homma, S. Kanegasaki, O. Lüderitz, T. Shiba, and O. Westphal, eds., Verlag Chemie, Weinheim (1984).

40. M. Matsuura, Y. Kojima, J. Y. Homma, Y. Kubota, N. Shibukawa, M. Shibata, M. Inage, S. Kusumoto, and T. Shiba, Interferon-inducing, pyrogenic and proclotting enzyme of horseshoe crab activities of chemically synthesized lipid A analogues, Eur. J. Biochem. 137:639 (1983).

41. M. Matsuura, Y. Kojima, Y. Homma, Y. Kubota, A. Yamamoto, M. Kiso, and A. Hasegawa, Biological activities of chemically synthesized analogues of the nonreducing sugar moiety of lipid A, FEBS Lett. 167:226 (1984).

42. H. Mayer, P. V. Salimath, O. Holst, and J. Weckesser, Unusual lipid A types in phototrophic bacteria and related species, Rev. Infect. Dis. 6:542 (1984).

43. H. Mayer and J. Weckesser, Unusual lipid A's: structures, taxonomic relevance and potential value for endotoxin research, in: "Handbook of Endotoxins, Volume 1: Chemistry of Endotoxin," E.Th. Rietschel, ed., Elsevier, Amsterdam, New York, and Oxford (1984).

44. W. T. J. Morgan and S. M. Partridge, Studies on immunochemistry. 4. The fractionation and nature of antigenic material isolated from bacterium dysenteriae (Shiga), J. Biochem. 34:169 (1940).

45. C. A. Mulford and M. J. Osborn, An intermediate step in translocation of lipopolysaccharide to outer membrane of Salmonella typhimurium, Proc. Natl. Acad. Sci. USA 80:1159 (1983).

46. R. S. Munson, N. S. Rasmussen, and M. J. Osborn, Biosynthesis of Lipid A. Enzymatic incorporation of 3-deoxy-D-mannooctulosonate into a precursor of lipid A in Salmonella typhimurium, J. Biol. Chem. 253:1503 (1978).

47. M. N. Nashed and L. Anderson, A convergent and flexible approach to the synthesis of enterobacterial lipid A. Fully substituted disaccharides having palmitoyl as fatty acyl moiety, Carboh. Res. 92:C5 (1981).

48. A. Nowotny, Chemical structure of a phosphomucolipid and its occurrence in some strains of Salmonella, J. Chem. Soc. 83:501 (1961).

49. A. Nowotny, Heterogeneity of endotoxins, in: "Handbook of Endotoxins, Volume 1: Chemistry of Endotoxin," E.Th. Rietschel, ed., Elsevier, Amsterdam, New York and Oxford (1984).

50. M. J. Osborn, Biosynthesis and assembly of lipopolysaccharide of the outer membrane, in: Bacterial Outer Membranes, Biogenesis, and Functions," M. Inouye, ed., John Wiley and Sons, New York (1979).

51. M. Schindler, M. J. Osborn, and D. E. Koppel, Lateral diffusion of LPS in the outer membrane of S. typhimurium, Nature 285:261 (1980).

52. N. Quereshi, K. Takayama, D. Heller, and C. Fenselau, Position of ester groups in the lipid A backbone of lipopolysaccharides obtained from Salmonella typhimurium, J. Biol. Chem. 258:12947 (1983).

53. Ch.R. H. Raetz, The enzymatic synthesis of lipid A: molecular structure and biological function of monosaccharide precursors, Rev. Infect. Dis. 6:463 (1984).

54. Ch.R. H. Raetz, Escherichia coli mutants that allow elucidation of the precursors and biosynthesis of lipid A, in: "Handbook of Endotoxins, Volume 1: Chemistry of Endotoxin," E.Th. Rietschel, ed., Elsevier, Amsterdam, New York and Oxford (1984).

55. B. L. Ray, G. Painter, and Ch.R. H. Raetz, The biosynthesis of gram-negative endotoxin. Formation of lipid A disaccharides from monosac-charide precursors in extracts of Escherichia coli, J. Biol. Chem. 259:4852 (1984).

56. E.Th. Rietschel, H. Gottert, O. Lüderitz, and O. Westphal, Nature and linkages of the fatty acids present in the lipid A component of Salmonella lipopolysaccharides, Eur. J. Biochem. 28:166 (1972).

57. E.Th. Rietschel, Z. Sidorczyk, U. Zähringer, H.-W. Wollenweber, and O. Luderitz, Analysis of the primary structure of lipid A, in: "Bacterial Lipopolysaccharides, Structure, Synthesis, and Biological Activities, American Chemical Society Symposium Series 231:195 (1983).

58. E.Th. Rietschel, H.-W. Wollenweber, H. Brade, U. Zähringer, B. Lindner, U. Seydel, H. Bradaczek, G. Barnickel, H. Labischinski, and P. Giesbrecht, Structure and conformation of the lipid A component of lipopolysaccharides, in: "Handbook of Endotoxins, Volume 1: Chemistry of Endotoxin," E.Th. Rietschel, ed., Elsevier, Amsterdam, New York and Oxford (1984).

59. E.Th. Rietschel, H.-W. Wollenweber, R. Russa, H. Brade, and U. Zähringer, Concepts of the chemical structure of lipid A, Rev. Infect. Dis. 6:432 (1984).

60. M. R. Rosner, H. G. Khorana, and A. C. Satterthwait, The structure of lipopolysaccharide from a heptoseless mutant of Escherichia coli K-12. II. The application of ^{31}P NMR spectroscopy, J. Biol. Chem. 254:5918 (1979).

61. U. Seydel, B. Lindner, H.-W. Wollenweber, and E.Th. Rietschel, Structural studies on the lipid A component of enterobacterial

lipopolysaccharides by laser desorption mass spectrometry. Location of acyl groups at the lipid A backbone, Eur. J. Biochem. 145:505 (1984).

62. M. J. Shear and F. C. Turner, Chemical treatment of tumors. V. Isolation of the hemorrhage producing fraction from Serratia Marcescens (Bacillus prodigiosus) culture filtrate, J. Natl. Cancer Inst. 4:81 (1943).

63. T. Shiba, S. Kusumoto, M. Inage, H. Chaki, M. Imoto, and T. Shimamoto, How can chemical synthesis approach the entity of lipid A, in: "Bacterial Endotoxins: Chemical, Biological, and Clinical Aspects," J. Y. Homma, S. Kanegasaki, O. Lüderitz, T. Shiba, and O. Westphal, eds., Verlag Chemie, Weinheim (1984).

64. Z. Sidorczyk, U. Zähringer, and E.Th. Rietschel, Chemical structure of the lipid A component of the lipopolysaccharide from a Proteus miralilis re-mutant, Eur. J. Biochem. 137:15 (1983).

65. S. M. Strain, S. W. Fesik, and I. M. Armitage, Characterization of lipopolysaccharide from a heptoseless mutant of Escherichia coli by carbon 13 nuclear magnetic resonance, J. Biol. Chem. 258:2906 (1983).

66. K. Takayama, N. Qureshi, and P. Mascagni, Complete structure of lipid A obtained from the lipopolysaccharides of the heptoseless mutant of Salmonella typhimurium J. Biol. Chem. 258:12801 (1983).

67. K. Takayama, N. Qureshi, P. Mascagni, M. A. Nashed, L. Anderson, and Ch.R. H. Raetz, Fatty acyl derivatives of glucosamine 1-phosphate in Escherichia coli and their relation to lipid A. Complete structure of a diacyl GlcN-1-P found in a phosphatidylglycerol-deficient mutant, J. Biol. Chem. 258:7379 (1983).

68. K. Tanamoto, C. Galanos, O. Lüderitz, S. Kusumoto, and T. Shiba, Mitogenic activities of synthetic lipid A analogus and suppression of mitogenicity of lipid A, Infect. Immun. 44:427 (1984).

69. K. Tanamoto, U. Zähringer, G. R. McKenzie, C. Galanos, E.Th. Rietschel, O. Lüderitz, S. Kusumoto, and T. Shiba, Biological activities of synthetic lipid A analogs: Pyrogenicity, lethal toxicity, anticomplement activity, and induction of gelation of Limulus amoebocyte lysate, Infect. Immun. 44:421 (1984).

70. S. N. Vogel, G. S. Madonna, L. M. Wahl, and P. D. Rich, In vitro stimulation of C3H/HeJ spleen cells and macrophages by a precursor molecule derived from Salmonella typhimurium, J. Immunol. 132:347 (1983).

71. P. Waldstätten, R. Christian, G. Schulz, F. M. Unger, P. Kosma, C. Kratky, and H. Paulsen, Synthesis of oligosaccharides containing 3-deoxy-D-manno-2-octulopyranosylono (KDO) residues, American Chemical Society Symposium Series, 231:121 (1983).

72. J. Weckesser, G. Drews, and H. Mayer, Lipopolysaccharides of photosynthetic prokaryotes, Ann. Rev. Microbiol. 33:215 (1979).

73. O. Westphal and O. Lüderitz, Chemische Erforschung von Lipopolysacchariden gramnegativer Bakterien, Angew. Chemie. 66:407 (1954).

74. O. Westphal, O. Lüderitz, and F. Bister, Über die Extraktion von Bakterien mit phenol/wasser, Z. Naturf. 7b:148 (1952).

75. H.-W. Wollenweber, K. W. Broady, O. Lüderitz, and E.Th. Rietschel, The chemical structure of lipid A: demonstration of amide-linked 3-acyloxyacyl residues in Salmonella minnesota re lipopolysaccharides, Eur. J. Biochem. 124:191 (1982).

76. H.-W. Wollenweber, U. Seydel, B. Lindner, O. Lüderitz, and E.Th. Rietschel, Nature and location of amide-bound (R)-3-acyloxyacyl groups in lipid A of lipopolysaccharides from various gram-negative bacteria, Eur. J. Biochem. 145:265 (1984).

77. T. Yasuda, Sh. Kanegasaki, T. Tsumita, T. Tadakuma, J. Y. Homma, M. Inage, S. Kusumoto, and T. Shiba, Biological activity of chemically synthesized lipid A analogues. Demonstration of adjuvant effect in haptensensitized liposomal system, Eur. J. Biochem. 124:405 (1982).

78. T. Yasuda, S. Kanegasaki, T. Tsumita, T. Tadakuma, N. Ikewaki, J. Y. Homma, M. Inage, S. Kusumoto, and T. Shiba, Further study of biological activities of chemically synthesized analogues of lipid A in artificial membrane vesicles, Eur. J. Biochem. 140:245 (1984).

PRODUCTION OF HUMAN MONOCLONAL ANTIBODIES RECOGNIZING CROSS-REACTIVE DETERMINANTS ON LIPOPOLYSACCHARIDES

James W. Larrick, Mark Jahnsen, George Senyk,
Stefan Weiss, and Karen Watson

Cetus Immune Research Labs
Palo Alto, California

INTRODUCTION

Gram-negative bacterial infections are a major cause of morbidity and mortality in modern hospitals and account for up to one percent of admissions and over 20,000 deaths per year (12). Although the administration of potent antibiotics and aggressive support techniques have reduced the frequency of lethal infections, deaths from gram-negative bacteremia continue to occur with a high frequency due to the inability of antibiotics to neutralize the toxic effects of endotoxins. Polyclonal antisera raised against the lipid-containing inner core structure of lipopolysaccharides (LPS), presented in exposed form in outer membranes of rough mutant strains, exhibit cross-reactive anti-endotoxin and protective activities (19,2,3,4).

To establish the validity and immunochemical basis for these observations and to provide potential therapeutic agents for use in gram-negative sepsis, we have made human monoclonal antibodies that react with LPS core-determinants. Peripheral blood lymphocytes were Epstein-Barr virus (EBV) transformed. Core-specific, antibody secreting cells were fused with a mouse/human myeloma cell line and cloned for stable antibody production. Sixteen monoclonal antibodies from cloned hybridomas were further characterized; all were of the IgM class and most demonstrated binding to a wide range of purified lipid A's or rough LPS's. The antibodies demonstrated binding to various smooth LPS's and to a range of clinical bacterial isolates by ELISA. These studies demonstrate the feasibility of preparing human monoclonal antibodies that react with cross reactive determinants of the LPS core region, including those related to the biologically reactive lipid A moiety. One or more of these antibodies may prove useful as a novel therapeutic agent in gram-negative sepsis.

METHODS

Cell Lines

All cell lines were maintained in Iscove's DME medium supplemented with 10% fetal bovine serum (FBS) 2 mM glutamine and 5×10^{-5} M 2-mercaptoethanol. The cell lines were checked routinely for the presence of mycoplasma. For large scale production of human monoclonal antibodies, cell lines were adapted to serum-free growth in HL-1 (Ventrex Labs, Portland, Maine) and in HB104 (Hana Biologics, Berkeley, California).

Fusion Protocol

Cells were combined and fused as previously described (16). Briefly, cells were fused with 40% polyethylene glycol 4000 (BDH) and 10% dimethylsulfoxide in Hank's balanced salt solution without calcium and with magnesium (2 mM). Cells were cultured in Iscove's DME with 15% FBS for 24 hrs prior to addition of hypoxanthine (100 μm), azaserine (2 μg/ml) and ouabain (1 μM). Fusions were plated at a density of 10^5 cells/microtiter well for hybrid selection. Cultures were subsequently fed every three days.

EBV Transformation

Epstein-Barr virus transformations were performed as previously described (5) except that cells were usually cultured at 10^4/well in 96-well culture plates.

LPS Specific ELISA

Immulon I (Dynatech) flat-bottom microtiter plates were coated overnight with 50 μl of LPS (Ribi Immunochem, Hamilton, Montana) in pH 9.6 NaHCO$_3$ buffer (50 mM). Plates were washed with PBS++ (containing calcium, magnesium) and 0.05% Tween 20 (Sigma). Subsequently, 100 μl of 1% BSA plus 0.05% Tween 20 in PBS++, followed by 100 μl of the antibody containing supernatants, were added to each well. Supernatants were incubated for 30 minutes at 22°C. After another wash, peroxidase-conjugated goat antihuman immunoglobulin developing reagent (Tago) was added. After 30 minutes at 22°C and another wash, ABTS substrate plus H_2O_2 (0.03%) were added. Plates were read at OD$_{405}$ after 30 minutes at 37°C.

Bacterial Binding ELISA

Immulon I flat-bottom microtiter plates were coated with 50 μl of 1.0% glutaraldehyde (Sigma) in deionized water and incubated four hours at room

Table 1. Generation of Human Anti-Core LPS Antibodies

Day 1
    A. Draw 50 ml blood from preselected volunteers with high LPS antibody titers.
    B. Rosette cells with AET-treated SRBC (11).
    C. EBV transform with B958 supernatant.
    D. Plate at 1 x 10^4/well in 96-well plates.

Day 20
    Select positive wells by specific LPS ELISA and expand or subculture highest titer wells.

Day 27
    Retest subcultures and fuse to mouse/human cell line F3B6.

Days 40-47
    Select positive wells by specific LPS ELISA and clone in soft agar or by limiting dilution.

Day 61
    Replica plaque positive hybrids and expand positives (see 8).

Day 75
    Begin serum-free adaptation of hybridomas

Table 2A. Core Lipopolysaccharides

	E. coli J5	S. minn. R595	S. typhi Re	E. coli J5 Lipid A	S. minn. R595 Lipid A	S. typhi Lipid A
D253	0*	1	1	0	1	1
D234	7	7	4	7	1	9
D267	3	6	1	3	1	7
D250	+					
D244	+	+				
L116	8	9	0	8	9	8
L118	8	+	2	9	9	+
L119	9	+	6	+	9	+
L121	9	8	9	9	8	+
L123	8	9	8	9	9	9
L124	0	3	0	1	3	1
L126	0	7	0	3	8	0
WI-3	8	+	3	+	9	9
WI-4	3	+	0	+	9	+
WI-5	4	9	1	8	8	7
WI-6	6	+	0	9	+	+
WI-7	7	8	0	+	+	+

*Magnitude of numbers indicates degree of monoclonal binding. Zero is negative, + is off scale with plate reader set at 2.0 absorbance full scale.

temperature. Wells were aspirated with an eight-channel manifold. Sixty µl of 0.25% bacterial suspension were added to each well and incubated overnight. Plates were assayed as described for the LPS assay.

IgM Specific ELISA

Immulon II flat-bottom microtiter plates were coated with Tago goat anti-human IgM (1:100) in 50 mM bicarbonate buffer (pH 9.6). After 90 minutes at 37°C, plates were washed as above and 100 µl of PBS++, 1% BSA, 0.05% Tween 20, 0.01% thimerosal, followed by antibody containing supernatants was added to each well. Plates were incubated for 30 minutes at 22°C, washed and developer and substrate were added as described above.

Glucose Levels

Glucose consumption by hybridoma cell lines was measured by Yellow Spring Instrument's glucose analyzer model 23A.

RESULTS AND DISCUSSION

Recently we reported on the generation of human monoclonals using EBV transformation combined with cell fusion (8). Several observations prompted this approach: 1) instability of immunoglobulin secretion by the EBV transformed cells; 2) increased immunoglobulin secretion by these hybridomas when compared with EBV lymphoblastoid cell lines; and, 3) higher frequency of rescue of the antigen-specific B cell populations (1,9).

Table 2B. Rough Mutant - Bacteria

	SL3770[§]	SL3749	SL3750	SL3748	E. coli J5	SL3789	S. minn. R595
D253	0*	0	0	0	0	0	0
D234	7	+	+	+	+	+	+
D267	1	4	5	7	6	3	7
D250	ND	ND	ND	ND	9	ND	0
244	0	1	0	3	5	1	8
L116	0	0	0	5	0	0	7
L118	0	3	5	9	4	4	+
L119	0	0	0	+	2	0	+
L121	0	0	1	+	+	0	+
L123	0	1	+	+	9	1	+
L124	0	0	0	1	0	0	1
L126	0	0	0	2	0	1	5
WI-3	2	5	7	8	8	4	1
WI-4	1	2	2	3	2	1	1
WI-5	1	1	2	3	2	1	1
WI-6	1	1	2	2	2	1	1

[§]See (10); SL3770 is smooth; SL3749 and SL3750 are superficial rough (Ra,Rb); SL 3748 is Rc and SL3789 is Rc. *See legend Table 2A. ND = not determined.

Table 2C. Clinical Isolates

	K. pneumoniae SM-1	P. aeruginosa Type I	P. aeruginosa Type III	E. coli SM-1	E. coli SM-2	E. coli SM-5	E. aerogenes
253	0*	ND	0	0	ND	0	0
D234	2	1	0	2	0	3	3
D267	0	0	1	0	0	0	0
D250	0	0	0	4	0	1	0
D244	0	0	ND	0	0	0	0
L116	0	0	0	0	0	0	0
L118	7	0	0	8	3	8	8
L119	0	0	0	0	0	0	0
L121	2	0	0	5	0	1	5
L123	1	0	0	3	0	2	3
L124	0	0	0	0	0	0	0
L126	2	0	0	0	0	0	0
WI-3	7	7	1	3	4	5	6
WI-4	4	7	1	4	3	4	5
WI-5	3	5	0	3	1	2	4
WI-6	+	9	1	9	7	8	9
WI-7	4	5	ND	3	1	3	3

*See legend Table 2A

Table 1 demonstrates the basic method of EBV transformation and subsequent cell fusion for the generation of human monoclonal anti-lipopolysaccharide antibodies. To begin we chose individuals with high LPS titers. From these individuals we generated a total of 17 human monoclonal antibodies, all of the IgM class. All have been cloned and stably produce greater than 10 μg of antibody per ml of spent culture media. Table 2A presents the ELISA results of these anti-LPS hybridomas binding to various core lipopolysaccharides. D253 is a control human monoclonal antibody which does not bind LPS. Table 2B demonstrates binding of these monoclonal antibodies to various rough mutant bacteria. Table 2C demonstrates binding of these same antibodies to a variety of bacteremic gram-negative clinical isolates. This series of 16 antibodies shows a number of different binding patterns:

1. Some of the antibodies recognize only core antigenic determinants;

2. Some recognize core antigenic determinants as well as determinants found on certain smooth lipopolysaccharide molecules; and

3. Some antibodies demonstrate broad cross-reactivity to not only core antigenic determinants, but to these determinants found on whole bacteria.

Antibody D250 binds only to J5 E. coli (Rc) core determinants with minimal binding to other core lipopolysaccharides and rough mutant bacteria. It gives spotty binding to clinical isolates of E. coli. Antibody D244 binds to E. coli J5 and Salmonella minnesota R595 LPS and bacteria, but not

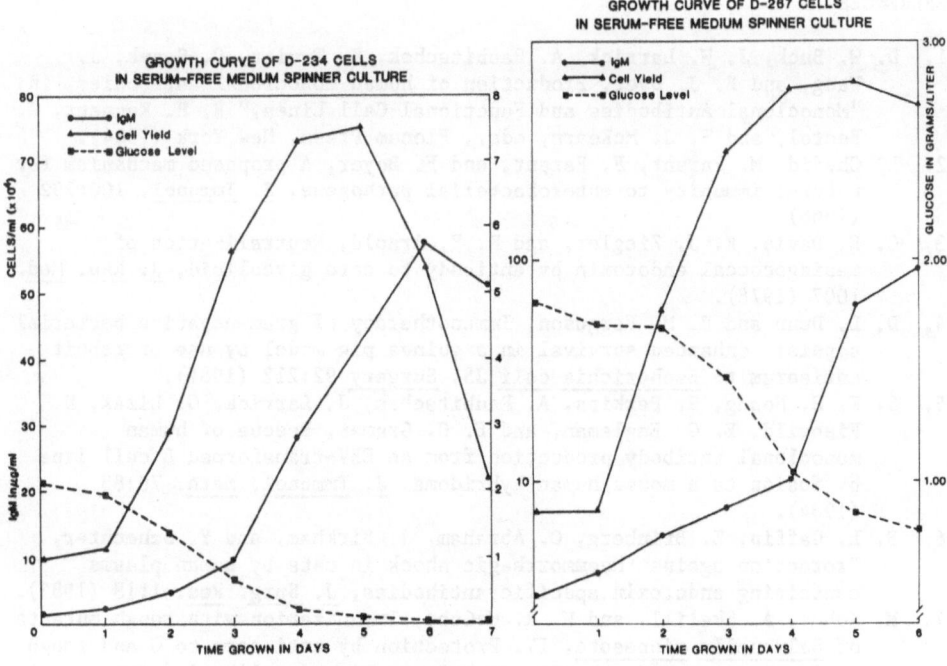

Figure 1. Growth of human hybridoma cell lines D234 in serum-free media HL-1 (A) and D267 in serum-free media HB104 (B). Cloned antibody producing cell lines were gradually adapted by growth in decreasing concentrations of fetal bovine serum prior to substitution of Iscove's DME with HB104 or HL-1 media.

to any other bacteria or lipopolysaccharides. Antibodies D234 and D267 show considerable binding to rough lipopolysaccharides and lipid A's with spotty binding on clinical isolates. The "L" series and "W" series of antibodies show high binding on rough lipopolysaccharides with a few "holes" in the binding patterns. Antibodies within the "L" series bind to more clinical isolates with a higher binding affinity. In general, those monoclonals showing the highest amount of core LPS or rough bacterial binding demonstrate the most cross-reactivity on clinical isolates.

Two of the hybridomas have been adapted to growth in serum-free medium for large scale spinner culture production of monoclonal antibodies (see Figures 1a and b). Under these conditions, as much as 50-75 µg/ml of specific human monoclonal antibody were produced. Evaluation of these antibodies in selected in vitro and in vivo models of endotoxin toxicity are in progress.

Previous work by Nelles and Niswander (15) and Mutharia et al. (14) demonstrated that mouse monoclonal antibodies reactive with the Rc-J5 mutant of E. coli exhibited extensive cross-reactivity with other gram-negative bacterial species. Our studies demonstrate the feasibility of generating human monoclonal antibodies that react with cross-reactive determinants of the LPS core region, including those related to the biologically reactive lipid A moiety. One or more of these antibodies may be useful as a novel therapeutic agent in gram-negative sepsis. Studies are in progress to see if monoclonals can have the anti-LPS protective capacity previously demonstrated for antisera by a large number of investigators (6,2,3,4,7,13,18,19).

REFERENCES

1. D. W. Buck, J. W. Larrick, A. Raubitschek, K. Truitt, G. Senyk, J. Wang, and B. J. Dyer, Production of human monoclonal antibodies, in: "Monoclonal Antibodies and Functional Cell Lines," R. H. Kennett, K. Bectol, and T. J. McKearn, eds., Plenum Press, New York (1984).
2. L. Chedid, M. Parant, F. Parant, and F. Boyer, A proposed mechanism for natural immunity to enterobacterial pathogens, J. Immunol. 100:292 (1968).
3. C. E. Davis, E. J. Ziegler, and K. F. Arnold, Neutralization of meningococcal endotoxin by antibody to core glycolipid, J. Exp. Med. 1007 (1978).
4. D. L. Dunn and R. M. Ferguson, Immunotherapy of gram-negative bacterial sepsis: enhanced survival in a guinea pig model by use of rabbit antiserum to Escherichia coli J5, Surgery 92:212 (1982).
5. S. K. H. Foung, S. Perkins, A. Raubitschek, J. Larrick, G. Lizak, D. Fishwild, E. G. Engleman, and F. C. Grumet, Rescue of human monoclonal antibody production from an EBV-transformed B cell line by fusion to a mouse/human hybridoma, J. Immunol. Meth. 70:83 (1984).
6. S. L. Gaffin, Z. Brinberg, C. Abraham, J. Birkham, and Y. Schechter, Protection against haemmorrhagic shock in cats by human plasma containing endotoxin specific antibodies, J. Surg. Res. 1:18 (1983).
7. M. Johns, A. Skelill, and W. R. McCabe, Immunization with rough mutants of Salmonella minnesota. IV. Protection by anti-sera to O and rough antigens against endotoxin, J. Infect. Diseases 147:57 (1983).
8. J. W. Larrick, B. Dyer, G. S. Senyk, S. Hart, R. Moss, D. Lippman, M. Jahnsen, J. Wang, H. Weintraub, and A. Raubitschek, In vitro expansion of human B cells for the production of human monoclonal antibodies, in: "Human Hybridomas and Monoclonal Antibodies," E. Engleman, S. Foung, J. Larrick, and A. Raubitschek, eds., Plenum Press, in press (1985).

9. J. W. Larrick and D. W. Buck, Practical aspects of human monoclonal antibody production, Biotechniques 2:6 (1984).

10. M. B. Lyman, J. P. Steward, and R. J. Roantree, Characterization of the virulence and antigenic structure of Salmonella typhimurium strains with lipopolysaccharide core defects, Infection and Immunity 13:1539 (1976).

11. M. Madsen and H. E. Johnson, A methodological study of E-rosette formation using AET-treated sheep red blood cells, J. Immunol. Meth. 27:61 (1979).

12. D. G. Maisi, Nosocomial bacteremia: an epidemiological overview, Am. J. Med. 70:719 (1981).

13. N. A. Mullan, P. M. Newsome, P. G. Cunnington, G. H. Palmer, and M. E. Wilson, Protection against gram-negative infections with antiserum to lipid A from Salmonella minnesota R595, Infect. Immunity 10:1195 (1974).

14. L. M. Mutharia, G. Crockford, W. C. Bogard, Jr., and R. E. W. Hancock, Monoclonal antibodies specific for Escherichia coli J5 lipopolysaccharide: cross-reaction with other gram-negative bacterial species, Infect. Immunity 45:631 (1984).

15. M. J. Nelles and C. A. Niswander, Mouse monoclonal antibodies reactive with J5 lipopolysaccharide exhibit extensive serological cross-reactivity with a variety of gram-negative bacteria, Infect. Immunity 46:677 (1984).

16. K. E. Truitt, J. W. Larrick, and A. Raubitschek, Fusion of non-adherent human cell lines, in: " Monoclonal Antibodies and Functional Cell Lines," R. H. Kennett, K. Bectol, and T. J. McKean, eds., Plenum Press, New York (1984).

17. L. S. Young, P. Stevens, and J. Ingram, Functional role of antibody against "Core" glycolipid of Enterobacteriaceae, J. Clin. Invest. 56:850 (1975).

18. E. J. Ziegler, H. Douglas, J. E. Sherman, C. E. Davis, and A. I. Braude, Treatment of E. coli and Klebsiella bacteremia in agranulocytic animals with antiserum to a UDP-GAL epimerase-deficient mutant, J. Immunol. 111:433 (1973).

19. E. J. Ziegler, J. A. McCutchan, J. Fierer, M. P. Glauser, J. C. Sadoff, H. Douglas, and A. I. Braude, Treatment of gram-negative bacteremia and shock with human antiserum to a mutant Escherichia coli, N. Eng. J. Med. 307:1225 (1982).

PHYSICAL PROPERTIES OF SHORT-CHAIN AND

LONG-CHAIN FRACTIONS OF LIPOPOLYSACCHARIDE

Arnold A. Peterson and Estelle J. McGroarty

Departments of Biophysics and Biochemistry
Michigan State University
East Lansing, Michigan

INTRODUCTION

The toxicity of lipopolysaccharide (LPS) isolated from different organisms varies greatly, as does the toxicity of different salt forms of a
single LPS isolate (4). Thus, it seems probable that the "endotoxic effects"
of LPS are due to the presence of particular chemical groups, and to the
accessibility of these groups. The accessibility may, in turn, depend on
the aggregate structure.

LPS helps to form a permeability barrier within the outer membrane of
gram-negative bacteria, and to prevent entry of hydrophobic antibiotics and
detergents into the cell (3,6). This barrier function has also been found
to be dependent on the length of the polysaccharide attached to lipid A
(7).

Many gram-negative bacteria produce LPS which are heterogenous in the
number of O-antigen repeat units (2,5,8,10). In this study we separated
the LPS of Escherichia coli 0111:B4 into two fractions based on O-antigen
length. The long O-antigen fraction (with an average of 18 repeat units)
and the short O-antigen fraction (with an average of one repeat unit) were
then analyzed for their fluid properties and cation binding affinities. We
found that of the two fractions, the long-chain LPS was more fluid in the
head group region than the short-chain LPS (suggesting a "masking" of anionic
sites in the LPS core-lipid A region).

METHODS

LPS from Escherichia coli 0111:B4 (phenol extracted) was purchased
from Sigma Chemical Company. Short- and long-chain LPS fractions were separated on a Sephadex G-200 column in a buffer containing NaEDTA and deoxycholate (10). Samples were extensively dialysed in buffer without deoxycholate to remove the detergent, followed by dialysis against $MgCl_2$, and then
distilled water to form the magnesium salt of LPS (MgLPS).

Electron spin resonance (ESR) spectroscopy was carried out with a Varian
X-band spectrometer. For measurements of LPS head group mobility, the probe
4-dodecyl dimethyl ammonium-1-oxyl-2,2,6,6-tetramethyl piperidine bromide
(CAT_{12}) was used, and LPS acyl chain mobility was detected with the probe

5-doxyl stearate (5DS). Spectra were analyzed for $2T_{11}$, the hyperfine splitting (a measure of probe mobility), S, the order parameter, and Ao, the isotropic hyperfine coupling constant (a measure of polarity). Titrations of LPS were performed by successive additions of cations to sample containing CAT_{12}, in 50 mM HEPES, pH 7.0, at 37°C. The partitioning of probe between free aqueous and LPS bound was measured using the parameter Ψ_1 (1).

RESULTS

Using the ESR lipid probe 5DS, the mobility of the acyl chain regions of the short-chain, long-chain, and unseparated Mg-LPS samples from E. coli 0111:B4 were shown to be very similar (Figure 1). The short-chain and unseparated fractions were slightly more fluid than the smooth fraction but this may be a reflection of probe penetrating deeper into the hydrophobic region of short-chain and unseparated samples. Measurement of the polarity of the spin probe environment (Ao) indicated that the probe sensed an increasing hydrophobicity in the order, long-chain < short-chain < unseparated fractions (Figure 2). These variations in polarity between the fractions indicate differences in LPS packing which result either from differences in the depth to which the probe partitions, or differences in hydration of the lipid region. The presence of long O-antigen may hinder close apposition of LPS head groups such that water can penetrate deeper into the acyl chain region of the long-chain LPS fraction than into the short-chain fraction, increasing the measured polarity.

In the single phospholipid system, the gel-to-liquid crystal phase transition was marked by a large, rapid increase in hydrophobicity (Ao) as the probe 5DS partitioned into the bilayer (Figure 2). However, the polarity of LPS measured by this probe increased relatively slowly with temperature, indicating a broad phase transition from approximately 20°C to above 50°C.

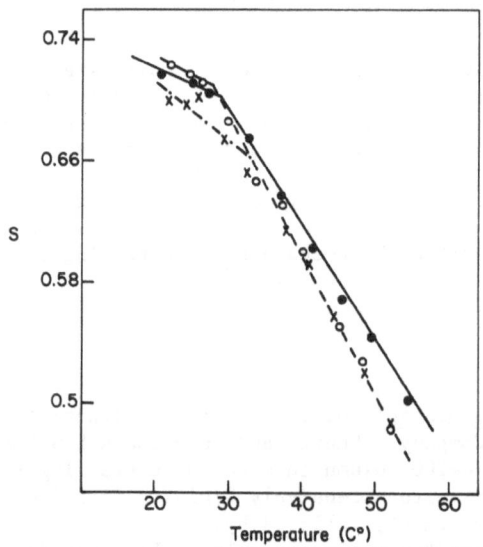

Figure 1. Mobility of the lipid probe in unseparated, O;
long-chain, ●; and short-chain, X, LPS. Mobility
was measured by the order parameter S, as a
function of temperature.

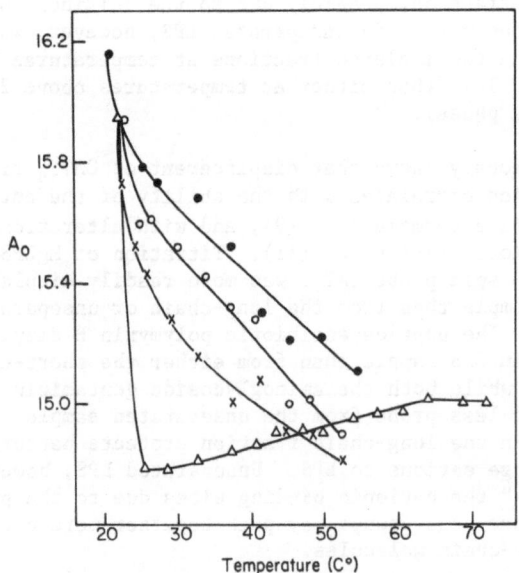

Figure 2. Polarity or hydration of the LPS lipid region as a
function of temperature, measured by 5DS, in unseparated, X; long-chain, ●; and short-chain, O, LPS and
in 1-palmitoyl-2-oleoyl phosphatidylethanol-amine, Δ.

The mobility ($2T_{11}$) of the cationic probe CAT_{12} was greater in the
long-chain fraction than in the short-chain fraction (Figure 3). This difference in $2T_{11}$ indicates that the LPS head group region in the short-chain
fraction was more rigid or more tightly packed than the head group region

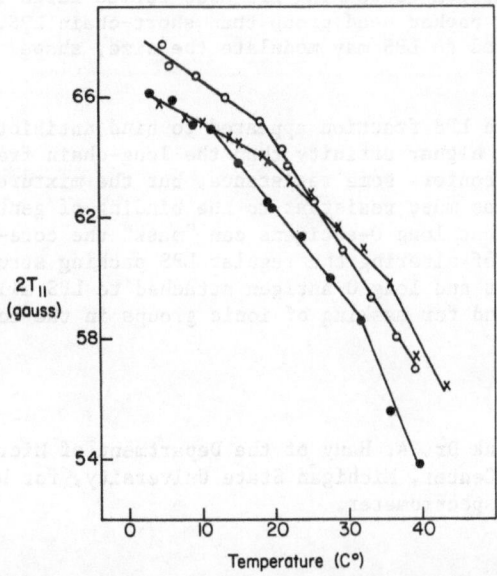

Figure 3. The ridigity of MgLPS head groups. Motion ($2T_{11}$) was
measured as a function of temperature in unseparated,
X; long-chain, ●; and short-chain, O, samples of LPS.

in the long-chain fraction, possibly due to the bulkiness of longer O-antigen. The head group motion in unseparated LPS, however, was intermediate between that of the two isolated fractions at temperatures below 25°C (in the gel phase) and less than either at temperatures above 25°C (in the gel-liquid crystalline phase).

We have previously shown that displacement of CAT_{12} from LPS by polycationic antibiotics correlates with the ability of the antibiotic to increase outer membrane permeability (9), and with alterations in LPS structure that confer antibiotic resistance (11). Titration of MgLPS with polycations indicated that the spin probe CAT_{12} was more readily displaced from the short-chain LPS sample than from the long-chain or unseparated LPS samples (data not shown). The peptide antibiotic polymyxin B displaced less probe from the long-chain LPS sample than from either the short-chain or unseparated LPS samples, while both the aminoglycoside gentamicin and the polyamine spermine displaced less probe from the unseparated sample. Possibly, the longer O-antigen in the long-chain fraction protects bacteria by hindering the binding of large cations to LPS. Unseparated LPS, however, may be able to partially "mask" the cationic binding sites due to the presence of long-chain LPS, while the head groups may pack together more closely due to the interspersed short-chain molecules.

CONCLUSIONS

The packing arrangement or aggregate structure of isolated long-chain and short-chain fractions of LPS appears to differ. The long-chain LPS isolate was more mobile in the head group region, and more polar in the acyl chain region compared to the short-chain isolate. Our results suggest that, above 25°C, aggregates of separated long-chain LPS have a greater surface curvature than those of the short-chain LPS fraction. Whereas short-chain LPS may form large lamellar sheets, long-chain LPS may tend to form small tubular micelles or ribbon-shaped aggregates. The lipid polarity and head group mobility of unseparated LPS indicate that a mixture of short- and long-chain O-antigen-containing LPS also formed large lamellar sheets, with a more tightly packed head group than short-chain LPS. Thus, the length of O-antigen attached to LPS may modulate the size, shape, and fluidity of the aggregates.

The short-chain LPS fraction appeared to bind antibiotics, such as polymyxin B, with a higher affinity than the long-chain fraction. Perhaps the long O-antigen confers some resistance, but the mixture of different O-antigen lengths was most resistant to the binding of gentamicin and spermine. We propose that long O-antigens can "mask" the core-lipid A region, but at the expense of altering the regular LPS packing structure, and that the mixture of short and long O-antigen attached to LPS allows for a stable packing structure and for masking of ionic groups in the core-lipid A.

ACKNOWLEDGMENTS

We wish to thank Dr. A. Haug of the Department of Microbiology and Pesticide Research Center, Michigan State University, for his advice and the use of the ESR spectrometer.

REFERENCES

1. R. T. Coughlin, C. R. Caldwell, A. Haug, and E. J. McGroarty, A cationic electron spin resonance probe used to analyze cation interactions with lipopolysaccharide, Biochem. Biophys. Res. Comm. 100:1137 (1981).

2. R. C. Goldman and L. Leive, Heterogeneity of antigenic-side-chain
    length in lipopolysaccharide for Escherichia coli 0111 and
    Salmonella typhimurium LT2, Eur. J. Biochem. 107:145 (1980).
3. L. Leive, The barrier function of gram negative envelope, Ann. NY
    Acad. Sci. 235:109 (1974).
4. O. Luderitz, M. A. Freudenberg, C. Galanos, V. Lehmann, E.Th.
    Rietschel, and D. H. Shaw, Lipopolysaccharides of gram-negative
    bacteria, Curr. Top. Membr. Transp. 17:79 (1982).
5. R. S. Munford, C. L. Hall, and P. D. Rick, Size heterogeneity of
    Salmonella typhimurium lipopolysaccharide in outer membranes and
    culture supernatant membrane fragments, J. Bacteriol. 144:630
    (1980).
6. H. Nikaido, Outer membranes of Salmonella typhimurium, transmembrane
    diffusion of some hydrophobic substances, Biochim. Biophys. Acta.
    433:118 (1976).
7. H. Nikaido and M. Vaara, Molecular basis of bacterial outer membrane
    permeability, Microbiol. Rev. 49:1 (1985).
8. E. T. Pava and P. H. Makela, Lipopolysaccharide heterogeneity in
    Salmonella typhimurium analyzed by sodium dodecyl
    sulfate/polyacrylamide gel electrophoresis, Eur. J. Biochem.
    107:137 (1980).
9. A. A. Peterson, R. E. W. Hancock, and E. J. McGroarty, Binding of
    cationic antibiotics to lipopolysaccharides of Pseudomonas
    aeruginosa, in preparation (1985).
10. A. A. Peterson and E. J. McGroarty, High molecular weight components
    in lipopolysaccharides of Salmonella typhimurium, Salmonella
    minnesota and Escherichia coli, J. Bact. 162:in press (1985).
11. A. A. Peterson and E. J. McGroarty, Decreased binding of antibiotics
    to lipopolysaccharides from polymyxin-resistant strains of
    Escherichia coli and Salmonella typhimurium, in preparation (1985).

ANALYSIS AND CHARACTERIZATION OF BACTERIAL ENDOTOXINS (LIPOPOLYSACCHARIDES)

BY SDS-POLYACRYLAMIDE GEL ELECTROPHORESIS FOLLOWED BY SILVER STAIN

Chao-Ming Tsai and Carl E. Frasch

Office of Biologics Research and Review
Food and Drug Administation
Bethesda, Maryland

INTRODUCTION

Lipopolysaccharide (LPS) is a surface component of gram-negative bacteria, and possesses both antigenic and pharmacological properties. The term endotoxin is used synonymously with LPS in emphasizing its toxic properties. Enterobacterial LPS consists of three regions, i.e., lipid A, core and O-side-chain, the latter which may not be present in certain bacteria. The antigenic specificities of LPSs reside primarily in the O-side-chains but also in the core regions, while the biological properties reside in the lipid A moiety (1).

Many LPS preparations contain a heterogenous population of LPS molecules. A great degree of heterogeneity in the LPS from Serratia marcescens has been demonstrated using ion-exchange column chromatography (9). Jann et al. (5) used SDS-PAGE as a simple technique to examine the heterogeneity of enterobacterial LPSs and showed that a Citrobactor LPS had four different size components on SDS-PAGE by dye-staining of LPS. However, extreme heterogeneity of the LPSs from E. coli 0111 (2) and Salmonella typhimurium (11) have been demonstrated by radioautography, a very sensitive detection method, following SDS-PAGE of the LPSs. More than 40 components could be detected in these LPSs.

We have recently developed a very sensitive silver stain for detecting LPS in polyacrylamide gels (14). All of the components in the LPSs from S. typhimurium and E. coli 0111 which were revealed by the radioautography were detected by the silver stain. In this communication, SDS-PAGE followed by silver stain was used to characterize and compare 30 LPSs from different bacterial species. We found considerable heterogeneity among the LPS preparations from the different species, many of the nonenteric bacterial species producing only low molecular weight LPS. The sensitivity of the silver stain also permitted as little as 0.01% LPS impurity in purified meningococcal polysaccharide vaccines to be detected.

MATERIALS AND METHODS

Endotoxins (LPSs)

Most of the enterobacterial LPSs prepared by the phenol-water procedure (17) were obtained either from Sigma Chemical Company, St. Louis, Missouri, or from List Biological Laboratories, Campbell, California. Meningococcal LPSs were prepared from N. meningitidis grown in tryptic soy broth as described (15). The semi-rough LPS of S. typhimurium was kindly provided by P. H. Makela of the Central Public Health Laboratory, Helsinki, Finland. The LPSs of N. gonorrhoeae and H. influenzae were obtained from P. Z. Allen and P. Anderson of the University of Rochester, Rochester, New York. The LPSs of B. pertussis and P. multocida were obtained from M. Naghdi of the Office of Biologics, Bethesda, Maryland and from P. A. Rebers of National Animal Research Lab, Ames, Iowa.

SDS-PAGE

The SDS-PAGE system of Laemmli (7) was used. 4M urea was incorporated into a 14% separating gel to improve separation of the LPS components and the stacking gel was increased to 5%. EDTA was added in a concentration of 2mM to the sample digestion buffer. Electrophoresis was carried out at 140 V for about 3 hrs until the bromophenol blue tracking dye reached the end of the gel.

Silver Stain

The silver stain for LPS (14) has been modified slightly and is outlined as follows: 1) fix LPS in polyacrylamide gel in 40% ethanol – 5% acetic acid overnight; 2) oxidize LPS for 10 min with 0.7% periodic acid in 40%

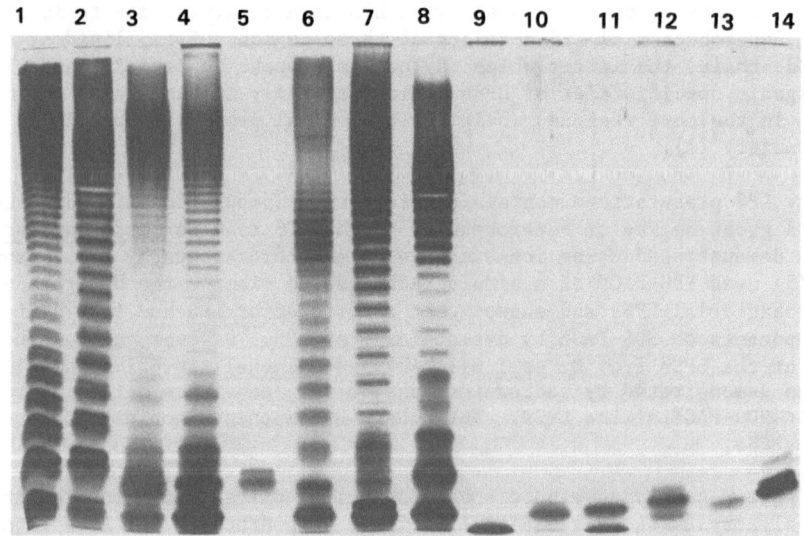

Figure 1. Characterization of bacterial lipopolysaccharides (LPS by SDS-PAGE and silver stain. The bacterial LPSs analyzed are as follows: 1) Salmonella abortus equi; 2) S. typhosa; 3) S. minnesota; 4) S. typhimurium; 5) S. typhimurium semi-rough; 6) Shigella flexneri; 7) Serratia marcescens; 8) Escherichia coli O111; 9) E. coli J5; 10) E. coli K12; 11) Neisseria meningtidis; 12) N. gonorrhorea; 13) Bordetella pertussis; and 14) Haemophilus influenzae. The sample load for each LPS was 5 μg for enterobacterial smooth LPSs and 0.2 μg for enterobacterial rough or semi-rough LPSs, and non-enteric bacterial LPSs. The faint lines across the upper part of the gel arose fron contaminants in the sample digestion buffer.

ethanol – 5% acetic acid; 3) perform three 20 min washes with deionized water; 4) form silver-LPS complex in fresh 0.2N NH$_4$OH-0.02N NaOH solution containing 0.8% silver nitrate for 10 min; 5) perform three 10 min washes with deionized water; 6) visualize silver-LPS complex in 0.02% formaldehyde – 0.005% citric acid for 2-5 min; 7) wash the gel as in step 3 and keep the finished gel in water; the stained LPS bands will slowly intensify for 10-20 min. After step 6, stain-developing can be stopped by placing the gel in 1% acetic acid for 2-3 min prior to the final water wash.

RESULTS AND DISCUSSION

In the presence of SDS, the high molecular weight (MW) aggregates of LPS will dissociate into monomers complexed with SDS due to hydrophobic binding of SDS to the lipid A of LPS (5,10). Upon SDS-PAGE, the negatively charged LPS-SDS complex will migrate in the polyacrylamide gel toward the anode, and the relative mobility of an LPS is inversely related to its molecular size (5,11,12). The LPS of 30 different bacteria were analyzed by SDS-PAGE, and the LPS components visualized by the silver strain. Figure 1 shows the SDS-PAGE profiles of seven smooth, one semi-rough and two rough enterobacterial LPSs and also the profiles of four nonenteric bacterial LPSs. All of the enterobacterial smooth LPSs were extremely heterogenous and each LPS had 30-40 orderly spaced bands which may appear as doublets in certain LPSs such as S. abortus equi and others. In contrast, the two enterobacterial rough LPSs had one band and the semi-rough LPS contained one major doublet band. By comparison, most of the nonenteric bacterial LPSs have 2-3 components with molecular sizes equivalent to that of the rough LPSs of enteric bacteria.

The extreme heterogeneity of enteric smooth LPSs including S. typhimurium and E. coli 0111 shown in Figure 1 are very similar to those reported by Palva and Makela (11) and Goldman and Leive (2) using radioautography as a detection method. They have shown that the 30-40 components of the LPSs

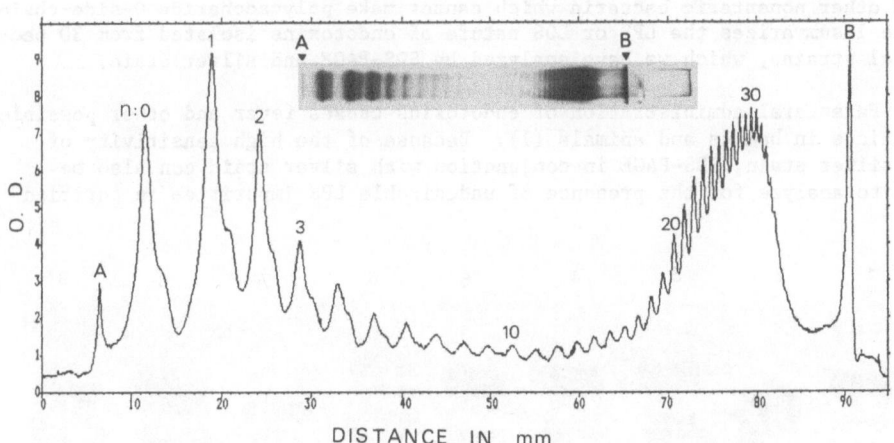

Figure 2. Densitometric scanning of S. typhosa lipopolysaccharide analyzed by SDS-PAGE and silver stain. The photo of the sample get scanned is shown at the top of the scanned profile. The gel was scanned with an LKB Ultrascan Laser Desitometer Model 2202. A and B are the stain marks of the ion front generated by electrophoresis and of the interface between the separating and stacking gels. N (0,1,2,3 etc.) is the number of the repeating unit in O-side chain.

from S. typhimurium and E. coli 0111 represent LPS molecules having an increasing number of repeating units in their O-side chains and that the fastest migrating major component is an LPS molecule having a complete core
only (i.e. a rough type LPS), the next fastest one with one repeating unit
in the O-side-chain and so forth.

Distribution of the different molecular species in a smooth LPS can be
analyzed by densitometric scanning of the silver-stained gel as illustrated
in Figure 2 for S. typhosa. Among all LPS components in S. typhosa LPS,
the component with only one repeating unit is the highest in quantity. As
the number of the repeating unit in the O-side-chain increases the relative
quantity of each LPS component progressively decreases to a minimum at about
10 units and then slowly increase back to reach a cluster of LPS components
having long O-side-chains centered at around 30 units. The shoulders seen
on the first six peaks reflect the doublet characteristic of these six bands.
The preferentially produced long O-side-chains of an LPS, which can be seen
as a cluster of bands on the gel, vary from strain to strain (see Figure
1). Some LPSs may have a minor second cluster of components at a higher MW
such as seen in E. coli 0111 LPS. All of the smooth LPSs analyzed contained
a rough LPS component as either the major component or at least one of the
major components as in the case of S. typhosa LPS.

A number of recent studies indicate that the LPSs of many nonenteric
bacteria contain a heterogenous population of low MW LPS molecules (4,15).
The heterogeneity of meningococcal LPSs as examined by SDS-PAGE is shown in
Figure 3 for eight serologically distinct LPSs from different meningococcal
strains (8). All of the meningococcal LPSs have two or three predominant
bands seen on the gel. The major bands often appeared as barely separated
doublets which are probably the same molecule having some charge differences
as with the doublet bands observed in E. coli LPS (2). The molecular sizes
of these meningococcal LPS components are equivalent to those of enterobacterial rough LPSs (Figure 1). The carbohydrate moiety in the meningococcal
LPS is not a long polysaccharide but is an oligosaccharide of about ten monosaccharide residues (6,15). Therefore, the term lipooligosaccharide (LOS)
may be more suitable to describe the endotoxins produced by meningococci and
some other nonenteric bacteria which cannot make polysaccharide O-side-chain.
Table 1 summarizes the LPS or LOS nature of endotoxins isolated from 30 bacterial strains, which we have analyzed by SDS-PAGE and silver stain.

Parenteral administration of endotoxins causes fever and other possible
reactions in humans and animals (1). Because of the high sensitivity of
the silver stain, SDS-PAGE in conjunction with silver stain can also be
used to analyze for the presence of undesirable LPS impurities in purified

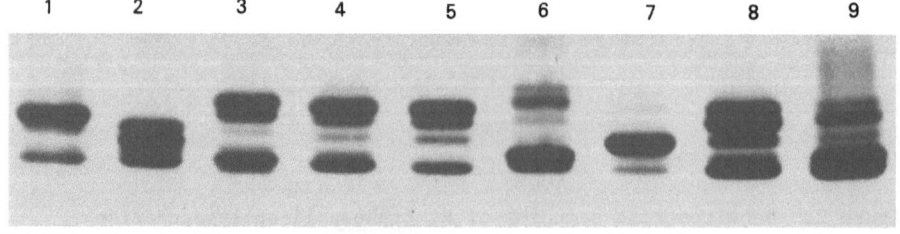

Figure 3. Heterogeneity of meningococcal lipopolysaccharides.
Meningcoccal LPSs were analyzed by SDS-PAGE in 15% gel and
stained with silver. Lane 1 is LPS from strain M986 and
Lanes 2 through 9 are 8 serologically different LPSs, L1
through L8, as described (8). The sample load for each
LPS was 0.5 μg.

Table 1. Summary of SDS-PAGE Pattern of Bacterial Endotoxins Analyzed
by SDS-PAGE and Visualized by Silver Stain

Source of Endotoxins			Smooth (S) and Rough (R) LPS or LOS by SDS-PAGE pattern#
Genera	Species	Strains	
Escherichia	coli	smooth strains: 026, 055, 0111, 0113, 0127	S-LPS
		rough strains: K12, CL29, J5	R-LPS
Salmonella	typhimurium	wild type	S-LPS
		semi-rough mutant	SR-LPS*
Salmonella	minnesota	wild type	S-LPS
		rough mutant Re595	R-LPS
Salmonella	abortus equi		S-LPS
Salmonella	typhosa		S-LPS
Shigella	flexneri		S-LPS
Serratia	marcescens		S-LPS
Neisseria	meningitidis	8 LPS types (L1-L8)	LOS**
Neisseria	gonorrhoeae		LOS**
Neisseria	lactamicae		LOS
Bordetella	pertussis		LOS
Haemophilus	influenzae		LOS
Pasteurella	multocidae		LOS
Vibrio	cholerae		"S" – LPS+

LPS = lipopolysaccharide; LOS = Lipooligosaccharide
* SR (semi-rough) LPS; a LPS mutant that produces only one unit of
  O-side-chain repeating unit in LPS
**Rough type LOS mutants have been isolated lacking the oligosaccharide
  region beyond heptose.
+ "S", atypical smooth LPS which contains five components with the smallest
  one equivalent to rough LPS.

gram-negative bacterial cell components such as capsular polysaccharide
vaccines. The LPS impurities in two meningococcal polysaccharides were
separated from the bulk of the polysaccharides by SDS-PAGE due to the large
differences in molecular size. During the process of silver stain the polysaccharide is leached out of the gel due to its high solubility and is therefore not detected. The amount of LPS in the polysaccharides was semi-quanti-
tated by comparing the stain intensity of the LPS in a polysaccharide with
that of an LPS standard (Figure 4). The PS-I in lane four contained about

20 ng LPS in 5 μg of polysaccharide (i.e. 0.4% LPS) and the PS-II in lane six contained about 30 ng in 100 μg of polysaccharide (i.e. ~0.03%). Proteinase K digestion (3) of the samples had no effect upon the stain intensity of LPS bands indicating that no proteins which are also stained with much less sensitivity by silver, comigrate with LPS. The silver-stain method can detect five nanograms of meningococcal LPS (14) therefore as low as 0.01% LPS impurity in a meningococcal polysaccharide can be detected using 50 μg of polysaccharide, one human vaccine dose, on the gel.

Silver stain was originally used to stain proteins (13). The silver stain modified for LPS preferentially stains LPS but also stains bovine

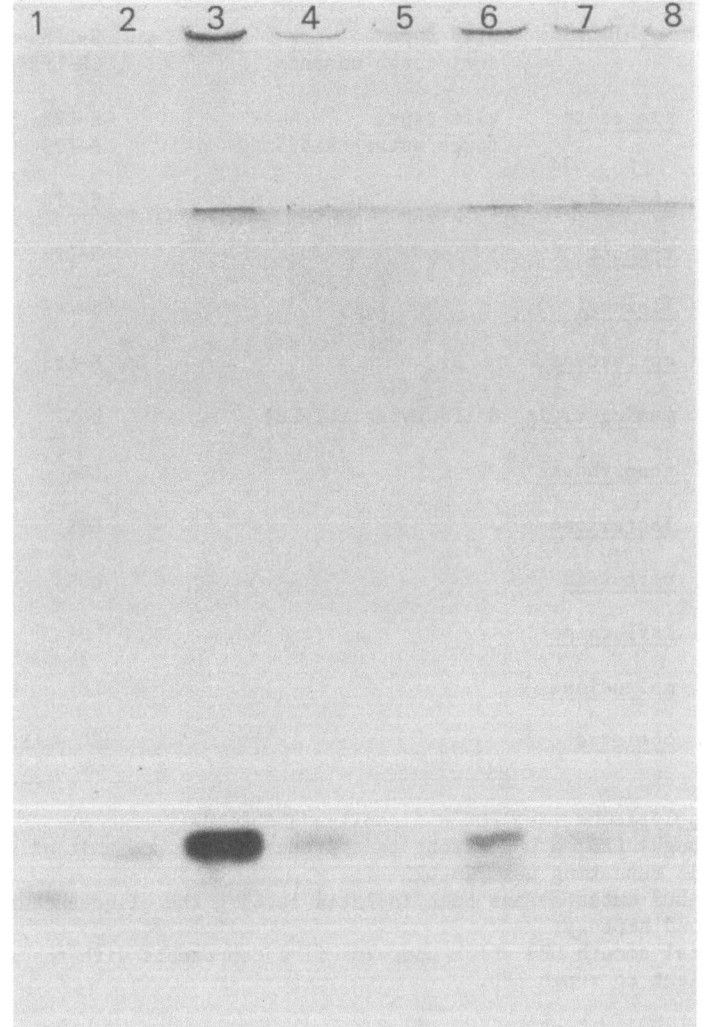

Figure 4. Analysis of endotoxin (i.e., LPS) impurity in meningcococcal capsular polysaccharides. Endotoxin was first separated from the bulk of the polysaccharide by SDS-PAGE and then stained with silver. Lanes 1 and 2 contained 10 nanograms of meningococcal LPSs from group A and C strains, respectively. Lanes 3-5 were loaded with 50, 5 and 0.5 μg of meningococcal group C polysaccharide lot PS-I. Lanes 6-8 were loaded with a total of 100, 50, and 25 μg of combined group A and C polysaccharides lots PS-II.

serum albumin with one third the sensitivity. The carbohydrate portion of an LPS is the reactive component in the silver stain because silver has been used to stain sugars on paper chromatograms (16) and the lipid A isolated from meningococcal LPS was not stained. The stain can be made specific for LPS by pretreatment of SDS-digested LPS samples with proteinase K (3). Phenol-water extracted LPSs contain about 1% protein. we found that the proteinase K treatment of the phenol-water purified LPS had no effect on their SDS-PAGE banding patterns and stain intensity.

In conclusion, well characterized endotoxins must be used for structural analysis as well as for immunological and biological studies of endotoxins because of the great deal of heterogeneity present in bacterial endotoxins. SDS-PAGE analysis followed by silver stain offers a simple yet powerful method for the characterization of bacterial endotoxins. The high sensitivity of silver stain also permits the method to be used to detect endotoxin impurities in purified gram-negative bacterial cell components.

REFERENCES

1. C. Galanos, O. Luderitz, E. T. Rietschel, and O. Westphal, Newer aspects of the chemistry and biology of bacterial lipopolysaccharides, with special reference to their lipid A component, Int. Rev. Biochem. 14:239 (1977).
2. R. C. Goldman and I. Leive, Heterogeneity of antigen side chain length in lipopolysaccharide from Escherichia coli 0111 and Salmonella typhimurium LT2, Eur. J. Biochem. 107:145 (1980).
3. P. J. Hitchcock and T. M. Brown, Morphological heterogeneity among Salmonella lipopolysaccharide chemotypes in silver-stained polyacrylamide gels, J. Bacteriol. 154:269 (1983).
4. T. J. Inzana, Electrophoretic heterogeneity and inter-strain variation in the lipopolysaccharide of Haemophilus influenzae, J. Infect. Dis. 148:492 (1983).
5. B. Jann, K. Reske, and K. Jann, Heterogeneity of lipopolysaccharides. Analysis of polysaccharide chain length by sodium dodecylsulfate-polyacrylamide gel electrophoresis, Eur. J. Biochem. 60:239 (1975).
6. H. J. Jennings, K. G. Johnson, and L. Keene, The structure of an R-type oligosaccharide core obtained from some lipopolysaccharides of Neisseria meningitidis, Carbohydr. Res. 121:233 (1983).
7. U. K. Laemmli, Cleavage of structural proteins during the assembly of the head of bacteriophage T4, Nature (London)277:680 (1970).
8. R. E. Mandrell and W. D. Zollinger, Lipopolysaccharide serotyping of Neisseria meningitidis by hemagglutination inhibition, Infect. Immun. 16:471 (1977).
9. A. Nowotny, Chemical and biological heterogeneity of endotoxins, in: "Microbial Toxins," Volume IV, Academic Press, New York and London (1971).
10. A. L. Olins and R. C. Warner, Physiochemical studies on a lipopolysaccharide from the cell wall of Azotobacter vinelandii, J. Biol. Chem. 242:4994 (1967).
11. E. T. Palva and P. H. Makela, Lipopolysaccharide heterogeneity in Salmonella typhimurium analyzed by sodium dodecyl sulfate-polyacrylamide gel electrophoresis, Eur. J. Biochem. 107:137 (1980).
12. R. Russell and K. Johnson, SDS-polyacrylamide gel electrophoresis of lipopolysaccharides, Canad J. Microbiol. 21:2013 (1975).
13. R. C. Switzer, III, C. R. Merril, and S. Shifrin, A highly sensitive silver stain for detecting proteins and peptides in polyacrylamide gels, Anal. Biochem. 98:231 (1979).
14. C. M. Tsai and C. E. Frasch, A sensitive silver stain for detecting lipopolysaccharides in polyacrylamide gels, Anal. Biochem. 119:115 (1982).

15. C. M. Tsai and C. E. Frasch, Heterogeneity and variation among
    Neisseria meningitidis lipopolysaccharides, _J. Bacteriol._ 155:498
    (1983).
16. W. E. Trevelyan, D. Proctor and J. S. Harrison, Detection of sugars on
    paper chromatograms, _Nature_ (London) 166:444 (1950).
17. O. Westphal and K. Jann, Bacterial lipopolysaccharides. Extraction with
    phenol-water and further application of the procedure, _Methods
    Carbohydr. Chem._ 5:83 (1965).

SECTION II. PHYSIOLOGICAL AND PHARMACOLOGICAL EFFECTS OF ENDOTOXINS

Metabolic Effects of Endotoxin
John J. Spitzer, Gregory J. Bagby, and Charles H. Lang

Glucose Dyshomeostasis in Endotoxicosis: Direct Versus Monokine-Mediated
Mechanisms of Endotoxin Action
Michael R. Yelich, Linda Witek-Janusek, and James P. Filkins

In Vivo Effects on Cellular Metabolism and Calcium Dynamics
by Continuous Infusion of Endotoxin
Judy A. Spitzer

Endotoxin (LPS) Toxicity to Hepatocytes (H) in Vitro -
Modulation by Macrophages (M)
Patricia S. Latham and Susan B. Sepelak

Effects of Endotoxin on Mixed Function Oxidase Activity
Joseph F. Williams and Andor Szentivanyi

Myocardial Performance and Adrenergic Modulation of Cyclic AMP
Following Endotoxin Administration
Raymond E. Shepherd, Charles H. Lang, Brent A. Brumfield,
Norman W. Robie, Karen R. DuSapin, and Kathleen H. McDonough

Cyclic Nucleotides in the Immunopharmacology of
Lipopolysaccharide Endotoxins
John W. Hadden, Anny Galy, Elba M. Hadden, J. L. Touraine,
and Ronald G. Coffey

METABOLIC EFFECTS OF ENDOTOXIN

John J. Spitzer, Gregory J. Bagby and Charles H. Lang

Department of Physiology
Louisiana State University Medical Center
New Orleans, Louisiana

GENERAL CONSIDERATIONS

Endotoxins, derived from gram-negative organisms, elicit a variety of effects in the host. Notable among these are the immunologic and anti-tumor effects, disseminated intravascular coagulopathies, the release of a large array of leukocyte-derived mediators, metabolic and endocrine alterations, etc. Many of these effects have been discussed extensively elsewhere in this monograph e.g., immunologic, anti-tumor effects, mediators. The aim of this communication is to briefly summarize the metabolic alterations affecting carbohydrate homeostasis following endotoxin administration. We will focus primarily on experimental results obtained in our own laboratories, fully recognizing that major contributions by other laboratories have continuously modified our own thinking.

The bacterial origin of endotoxins does not appear to influence their major biologic effects. Furthermore, while marked species differences exist in respect to endotoxin sensitivity, it appears that many of the hormonal and metabolic alterations elicited by endotoxin are quite similar in a variety of animal species that have been investigated. Because of the differences in sensitivity to endotoxin, we feel that in metabolic investigations studying endotoxin effects, it is important to relate the observed alterations to the severity of the insult, as expressed by the degree of lethality within a given time period (e.g. LD_{10}, or LD_{50} within 24 hrs), rather than indexing the produced effects to the absolute dose administered.

Administration of endotoxin to animals has been employed for some time as a tool to advance our understanding of sepsis and septic shock. At times, the use of endotoxin has been criticized because the changes that take place in endotoxin-treated animals do not always mimic those observed in septic patients. Indeed, one has to be quite careful in extrapolating the results obtained in animals following endotoxin to man in gram-negative sepsis. Several factors must be considered in this connection: first, the species under study may respond to sepsis and endotoxin differently from man (at least in certain respects); second, most endotoxin studies are designed to investigate the shock phase of sepsis and employ large bolus injections. At best, these types of studies can relate to only the severe stages of septic shock. Thus the dose of endotoxin used in comparison to the amount that might be present during sepsis must be considered; third, one must consider the overall state of the endotoxemic animal at the time the

specific measurements are made; fourth, it must be recognized that it is
very difficult to administer endotoxin in a manner that simulates its presence during sepsis; and, finally, one must consider that the septic patient
is usually studied while given clinical and pharmacologic support, which is
designed to improve the clinical outcome but which may complicate the comparison to animal studies. There is little doubt that the clinical manifestations of sepsis are not due exclusively to a response to in vivo released endotoxins; however, it is also evident that many aspects of sepsis
can be mimicked by endotoxin administration. This has been demonstrated
repeatedly in endotoxic shock. More recently we have found that the metabolic alterations described in hypermetabolic, hyperdynamic sepsis (12) are
quite similar to those found following the administration of very low, non-
lethal doses of endotoxin in experimental animals (6,13). Most of the
sepsis-induced metabolic alterations are also reproduced when endotoxin is
delivered over a period of several hours, and even days, by the recently-
developed experimental technique of subcutaneously-implanted osmotic mini-
pump (24). These new approaches promise more precise assessments of the
importance of endotoxins in producing some of the manifestations of sepsis.

DISSOCIATION OF THE HEMODYNAMIC, METABOLIC AND HYPERTHERMIC

EFFECTS OF ENDOTOXIN

Numerous investigators have described the cardiovascular sequelae of
the administration of lethal doses of endotoxin, which include systemic
hypotension, decreased cardiac output and increased heart rate. These cardiovascular changes are accompanied by an initial hyperglycemia followed by
hypoglycemia (if the administered dose is high), hyperlactacidemia and increased glucose turnover and hepatic gluconeogenesis (26) all of which appear
to be dose-dependent. Preterminally, the elevated rate of gluconeogenesis
is replaced by a marked decrease in this variable (27). In most species,
the administration of lethal doses of endotoxin also evokes hypothermia.
In a recent investigation, we attempted to separate the hemodynamic, metabolic and thermal effects of E. coli endotoxin by evaluating changes that
take place in chronically-catheterized, conscious rats when administered a
wide range of endotoxin doses (13). Very low, nonlethal doses of endotoxin
caused alterations in carbohydrate metabolism, (e.g., increased glucose
turnover and arterial lactate concentration) but did not produce hemodynamic
changes, (i.e., a fall in blood pressure and cardiac output). These metabolic changes could be elicited with a dose of endotoxin that was one hun-
dredth of the amount needed to produce cardiovascular alterations, thus
making possible the separation of metabolic effects from hemodynamic alterations. In the same studies, different doses of endotoxin elicited either
an increase or a decrease in body temperature: at high doses, hypothermia
was the characteristic feature, while at very low, nonlethal doses hyper-
thermia was present. Thus, the thermal effects could also be separated
from both cardiovascular and metabolic responses. These findings are summarized in Figure 1. Although the original study from which these data
were reproduced included six doses of endotoxin, for the sake of an easy
overview this figure represents results following only two: the higher one
represents an LD_{50} (in 24 hrs) whereas the lower is three orders of magnitude
less than that require to produce an LD_{10} effect. With the LD_{50} dose the
metabolic responses seemed to parallel hemodynamic changes. However, the
lower dose, which failed to alter the hemodynamic variables, still induced
metabolic effects as evidenced by alterations in glucose turnover and arterial lactate concentration. Also, while the high dose decreased core
body temperature, the lower one produced a fever response in these animals.

A closer perusal of the results indicates that the metabolic effects
observed following the very low dose of endotoxin were delayed when compared

to the higher dose; furthermore, they tended to parallel changes in body temperature. We postulate that endotoxin elicits the release of a wide range of hormones and leukocyte-derived mediators. The relative importance of these endogenously-produced agents may vary with the dose of endotoxin administered. For example, at high doses of endotoxin the decrease in blood pressure would result in increased sympatho-adrenal activity which would lead to the hyperglycemic response. At lower doses of endotoxin the hypotensive response is lost and catecholamine output would be reduced, or absent. Under these conditions the influence of other mediators which affect carbohydrate metabolism in a more subtle way may predominate. Further studies on these responses should be quite profitable.

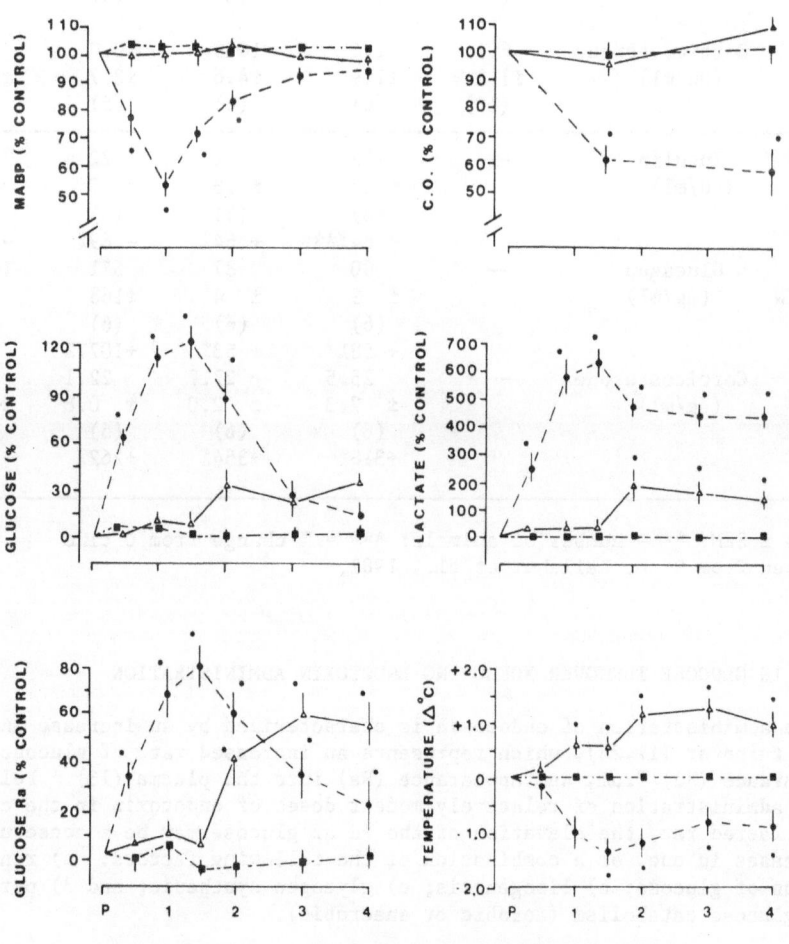

TIME POST-ENDOTOXIN

Figure 1. The effects of an approximately LD$_{50}$ (●--●) and LD$_{10}$ (△--△) of endotoxin on mean arterial blood pressure, cardiac output, arterial glucose concentration, arterial lactate concentration, glucose Ra and core body temperature in conscious rats. Time-matched control animals received saline (■--■). Data taken from (13). Values represent means ± SEM. Asterisks (*) indicate significant differences (p < 0.05) compared to time-matched controls.

Table 1. Changes in Insulin, Glucagon and Corticosterone
following Endotoxin Administration

			Hours after Saline or Endotoxin			
		0	1	2	5	8
CONTROL	Insulin (μU/ml)	69* ±13 (10)**	69 ± 14 (6)	60 ± 8 (6)	52 ± 3 (5)	50 ± 8 (5)
	Glucagon (μg/ml)	57 ± 5 (8)	35 ± 5 (6)	29 ± 6 (5)	30 ± 8 (5)	29 ± 8 (6)
	Corticosterone (μg/ml)	6.1 ±1.1 (10)	12.6 ±1.9 (6)	10.9 ±4.6 (6)	10.0 ±2.7 (5)	6.3 ± 1.4 (5)
ENDOTOXIN	Insulin (μU/ml)	--	113 ± 17 (6) + 64%***	114 ± 18 (5) + 64%	22 ± 7 (5) - 69%	24 ± 2 (5) - 65%
	Glucagon (μg/ml)	--	90 ± 9 (6) + 58%	87 ± 4 (6) + 53%	671 ±168 (6) +1077%	1000
	Corticosterone (μg/ml)	--	25.5 ± 2.3 (6) +318%	27.7 ± 2.0 (6) +354%	22.1 ± 0.8 (5) +262%	18.2 ± 1.5 (5) +198%

* = Mean ± SEM; ** = number of animals; *** = % change from 0 time
Data taken from D. L. Kelleher et al., 1982.

CHANGES IN GLUCOSE TURNOVER FOLLOWING ENDOTOXIN ADMINISTRATION

The administration of endotoxin is characterized by an increase in
glucose turnover (17,26), which represents an increased rate of glucose
disappearance (Rd) from, and appearance (Ra) into the plasma (13). Following the administration of relatively modest doses of endotoxin in the conscious, fasted rat, the elevation of the Rd of glucose may be a consequence
of increases in one, or a combination of the following factors: a) renal
excretion of glucose; b) lipogenesis; c) glycogen synthesis; and d) peripheral glucose catabolism (aerobic or anaerobic).

a) Renal glucose excretion is unlikely to be an important contributor
to the increased Rd, since following the administration of endotoxin no
significant glycosuria was found as compared to time-matched control animals
(19).

b) and c) Lipogenesis and glycogen synthesis are unlikely to be major
pathways for glucose carbons following glucose uptake by tissues under these
experimental conditions. The endocrine responses to endotoxin administration, consisting of increased plasma concentrations of glucagon, catecholamines and glucocorticoids, and the lack of sustained increase in insulin
levels, favor the mobilization rather than the deposition of lipids and

Table 2. Liver and Muscle Glycogen Changes in Fed Rats
Following Endotoxin Administration (μmole/g)

| | 0 | Hours after Saline or Endotoxin | | | |
		1	2	5	8
		Liver			
Control	212*	179	186	169	160
	± 24	± 37	± 23	± 23	± 21
	(13)**	(8)	(8)	(6)	(6)
Endotoxin		104	20	5	6
		± 26	± 7	± 1	± 3
		(6)	(6)	(5)	(5)
		- 51%***	- 93%	- 98%	- 97%
		Muscle			
Control	32	28	28	30	23
	± 2	± 3	± 2	± 2	± 1
	(12)	(8)	(8)	(6)	(6)
Endotoxin		25	22	18	31
		± 4	± 3	± 4	± 4
		(6)	(6)	(5)	(5)
		- 21%	- 32%	- 42%	- 3%

* = mean ± SEM; ** = number of animals; *** = % change from 0 time;
Data taken from D. L. Kelleher et al., 1982.

glycogen (10,13,25). The insulin response is generally biphasic (moderate
hyperinsulinism followed by hypoinsulinism), and leads to a marked decrease
of insulin-to-glucagon ratio several hours after endotoxin administration.
These hormonal changes are illustrated in Table 1. Table 2 indicates that
net glycogen synthesis is also not the fate of the peripherally-removed
glucose carbons since both hepatic and skeletal muscle glycogen contents
decreased following endotoxin administration.

    d) Thus, glycolysis and glucose oxidation are the most likely contrib-
uting factors to glucose removal under these conditions. Increased glycoly-
sis has been observed in tissues that do not oxidize glucose to CO_2, i.e.,
blood cells, macrophages, (2,8,15). In addition, elevated glycolysis in
skeletal muscle is also indicated by the increased lactate release from
this tissue and by the elevated rate of lactate turnover following endotoxin
(20,26). Skeletal muscle and adipose tissue seem to be the primary sites
of the elevated glucose utilization following endotoxin administration (5,-
18,20). Glucose uptake by skeletal muscle at 60, 120, 180 and 240 minutes
after endotoxin administration showed at 35, 21, 32, and 17% increase, respectively, as compared to the uptake prior to endotoxin (20). Skeletal
muscle glucose uptake was elevated in spite of a decrease in arterial blood
glucose concentration. Thus, when uptake is normalized to the prevailing
arterial concentration, the glucose clearance rate of skeletal muscle following endotoxin administration is significantly elevated as compared to time-
matched control values (20). Although it was calculated that the skeletal
muscle mass accounted for a substantial portion of the increased glucose
turnover following endotoxin administration, other tissues also appeared to
contribute to this increase. Adipose tissue is likely to be one of the

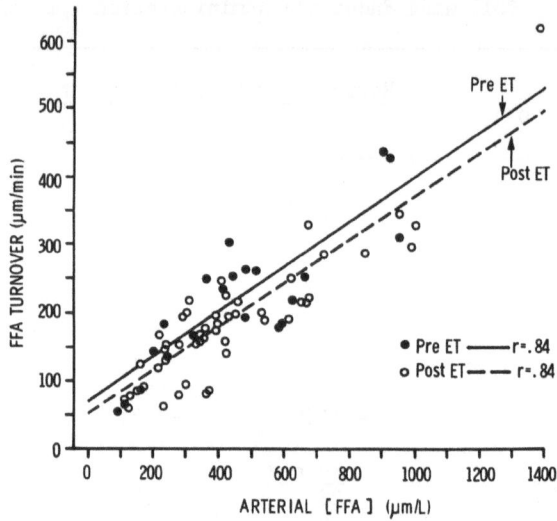

Figure 2. Linear correlation between arterial free fatty acid (FFA)
concentration and FFA turnover preceding and following endotoxin administration. The slopes of the lines represent
metabolic clearance rates. Figure reproduced from (19)
with permission.

other tissues (23). In fact, glucose uptake by adipose tissues was found
to be elevated following endotoxin under in vivo conditions (22), as was
the removal of glucose by isolated adipocytes taken from animals injected
with endotoxin (14). It is also of interest to consider that the elevated
glucose uptake may be present in the pancreas where it has been hypothesized
that the increased influx is responsible for the transient hyperinsulinemia
following endotoxin administration (4,7).

The increased glucose uptake by peripheral tissues elicited by endotoxin
appears to be a specific effect rather than a generalized increase in fuel
uptake. This statement is based on the findings that the existing linear

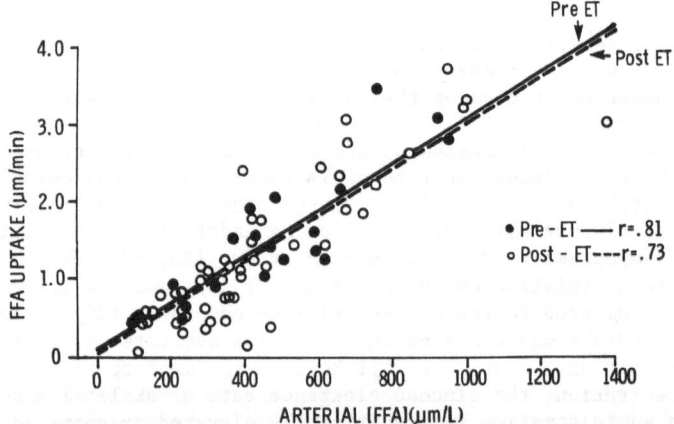

Figure 3. Linear correlation between arterial free fatty acid (FFA)
concentration and FFA uptake by skeletal muscle, preceding
and following endotoxin administration. Figure reproduced
from (19) with permission.

correlation between arterial free fatty acid (FFA) concentration and FFA
uptake by skeletal muscle, and also the correlation between arterial FFA
concentration and FFA turnover, did not change following endotoxin administration (19). These data are reproduced in Figures 2 and 3, where the
slope of the line indicates the metabolic clearance rate of FFA. It has
also been calculated from the data of Romanosky et al. (19) that the frac-
tional contribution of skeletal muscle FFA utilization was unchanged following the administration of endotoxin. Approximately 25% of the total FFA
turnover was accounted for by skeletal muscle FFA use. Similar studies
concerning the main ketone bodies (beta-hydroxybutyrate and acetoacetate)
have also indicated that the metabolic clearance rate of ketone bodies by
skeletal muscle or by the total body were unchanged following endotoxin
administration (21), as illustrated in Figures 4 and 5.

The changes discussed in the previous paragraph were observed in animals
following the administration of a relatively moderate dose of endotoxin.
Under these conditions, hepatic blood flow is quite well maintained. Hepatic
arterial blood flow is increased, while portal venous blood flow is either
unaltered or slightly decreased (3,16). However, when the endotoxin-produced
damage is very severe and lethality is high, hepatic glucose production can
decrease to below control rates (27). This is usually a preterminal event
and is due to both an impairment of hepatic function and the marked decrease
of hepatic blood flow resulting in a diminished delivery of oxygen and of
gluconeogenic precursors.

As indicated earlier, in addition to an increased rate of disappearance
of glucose, its rate of appearance (Ra) is also elevated following the administration of endotoxin (26). Three factors could potentially account for
such an increase: a) absorption of glucose from the gastrointestinal tract;
b) increased hepatic glycogenolysis; and c) elevated gluconeogenesis.

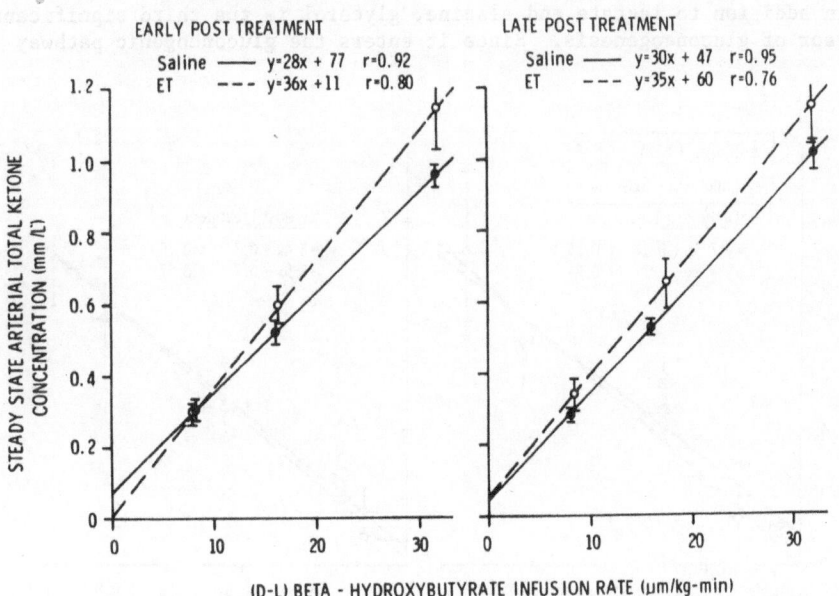

Figure 4. Linear correlation between ketone body turnover and dependence
of steady-state arterial total ketone body concentration during
DL-beta-hydroxybutyrate infusion in control and endotoxin-
treated dogs. The slopes of the lines represent metabolic
clearance ratees. Figure reproduced from (21) with permission.

a) The experiments showing the elevated Ra were performed in animals after a 12-18 hr fast. Glucose absorption from the intestine is not likely to have been very marked at that time and thus gastrointestinal absorption cannot be responsible for the elevated Ra following endotoxin administration.

b) During the initial period after endotoxin, hepatic glycogenolysis may provide a small portion of the increased glucose Ra. However, glycogen stores in the liver are low due to the fasting period prior to endotoxin administration. The remaining glycogen markedly decreased during the first hour after endotoxin and was virtually depleted by 2 hrs. Therefore, hepatic glycogenolysis is unable to support the elevated glucose production rate observed under these conditions (10).

c) Thus, increased gluconeogenesis is essential for the maintenance of glucose influx. Indeed, it has been repeatedly demonstrated that the turnover of the major gluconeogenic precursor, lactate, is markedly elevated following endotoxin administration. A substantial portion of this additional lactate turnover is converted to glucose (9,17,26).

Although the rate of gluconeogenesis from lactate is elevated following endotoxin administration, the efficiency of handling lactate by the liver is impaired. This is demonstrated by the fact that while arterial lactate concentrations are increased by approximately three-fold, hepatic lactate uptake does not change significantly (16). This results in a decreased metabolic clearance rate of lactate by the liver and is consistent with the postulated impairment of hepatic function under these conditions. Similarly, net gluconeogenesis from alanine was elevated following endotoxin (11). Arterial alanine concentration increased significantly, while hepatic removal of this precursor of gluconeogenesis remained unaltered, indicating an impairment of the metabolic clearance rate by the liver (16). These changes are indicated in Figure 6.

In addition to lactate and alanine, glycerol is the third significant precursor of gluconeogenesis. Since it enters the gluconeogenic pathway

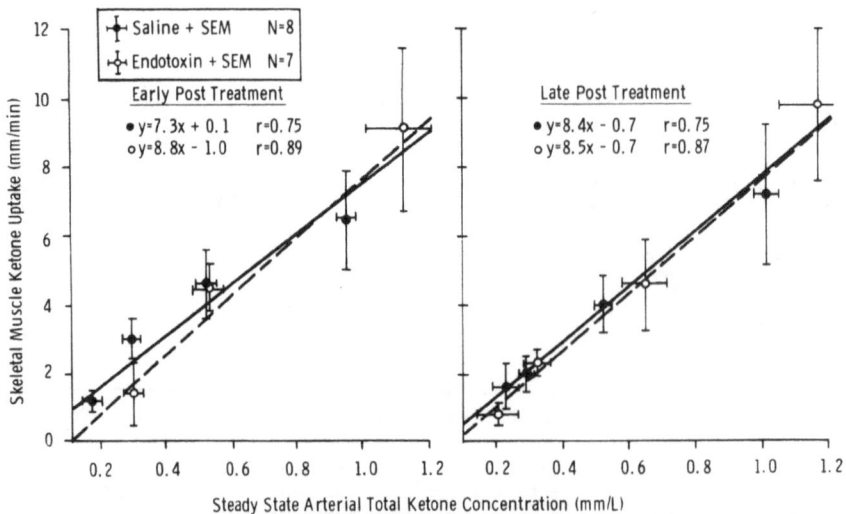

Figure 5. Dependence of skeletal muscle ketone uptake upon the prevailing arterial total ketone body concentrations in control and endotoxin-treated dogs. Figure reproduced from (21) with permission.

above phosphoenolpyruvate carboxykinase (PEPCK) and thereby bypasses a rate controlling step, it is of interest to consider the hepatic clearance of this precursor following endotoxin. As indicated in Figure 7, endotoxin treatment results in a significantly decreased hepatic glycerol clearance. This decrease was accompanied by an elevated hepatic glucose output and approximately 20-25% decrease in total hepatic blood flow (16).

Thus, it appears that although an elevated rate of hepatic gluconeogenesis is present following endotoxin administration, this is accompanied by a decreased hepatic clearance of all three major gluconeogenic precursors (lactate, alanine and glycerol) due to an impairment of hepatic function. However, the greatly increased arterial concentration of the three major precursors overcomes the hepatic impairment, resulting in a net elevation of hepatic glucose output.

APPARENT DISCREPANCY BETWEEN INCREASED GLUCONEOGENESIS AND

DECREASED PEPCK ACTIVITY

As indicated earlier, one of the hallmarks of the metabolic effects of low and moderate doses of endotoxin is an elevated rate of gluconeogenesis. It has also been known for many years that the activity of hepatic PEPCK is diminished following endotoxin administration (1). Since PEPCK is one of the rate-controlling enzymes of gluconeogenesis, the two findings (increased gluconeogenesis and decreased PEPCK activity) appear to be contradictory.

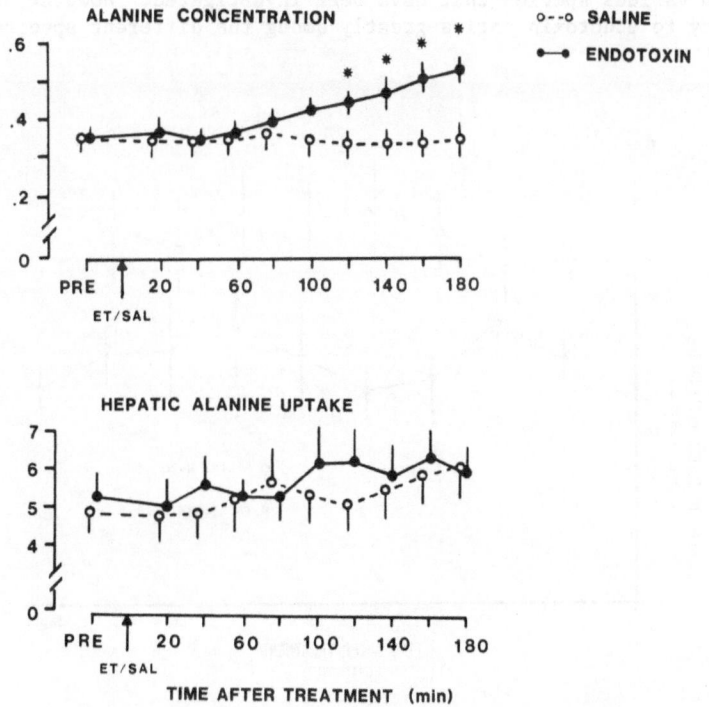

Figure 6. Arterial alanine concentration and hepatic alanine uptake following the administration of saline or E. coli endotoxin. Data are expressed as means ± SEM. Asterisks (*) indicate a significant difference (p < 0.05) between the time-matched saline and endotoxin-treated groups. (O. P. McGuiness and J. J. Spitzer, unpublished data.)

However, the apparent contradiction can be resolved easily by considering the available experimental data, and especially the information that has been obtained under in vivo conditions. As indicated in the previous section, the arterial concentration of the major gluconeogenic precursors is markedly elevated following endotoxin, while the total hepatic blood flow is only moderately decreased. This results in an increased deliverly of these substances to the liver, overcoming the inhibition in the activity of the rate-controlling enzyme and resulting in elevated glucose production. Furthermore, the hormonal milieu is such that the catabolic hormones (catecholamines, glucocorticoids, glucagon) predominate in the blood while the concentration of the major anabolic hormone, insulin, does not keep up with these changes. Therefore, the overall result of these changes is also an elevated rate of gluconeogenesis. The importance of gluconeogenic precursor supply and the hormonal environment has also been substantiated in in vitro studies. In these investigations hepatocytes isolated from endotoxin-treated rats produced glucose at slower rates than cells from control animals when the incubation media were identical. However, the rate of gluconeogenesis was higher by hepatocytes from endotoxin-treated rats when incubated with high precursor and glucagon concentrations (simulating the in vivo conditions after endotoxin), than by control cells incubated with lower precursor and glucagon concentrations which simulated the in vivo conditions in control rats (24).

SUMMARY AND CONCLUSIONS

1. The circulatory and metabolic effects of endotoxins appear to be similar in various species that have been investigated. However, the sensitivity to endotoxin varies greatly among the different species.

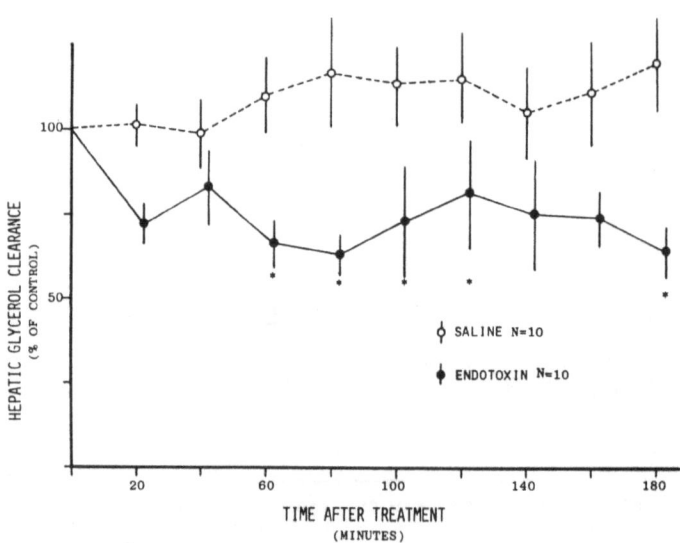

Figure 7. Hepatic glycerol clearance after administration of Escherichia coli endotoxin or saline. Data are expressed as percent of pretreatment values of 18 ± 2 and 15 ± 2 (SE) ml. kg^{-1}, min^{-1}, respectively. Asterisks (*) denote significant differences (p < 0.05) between endotoxin-injected and time-matched saline groups. Figure reproduced from (16) with permission.

2. Although the clinical manifestations of sepsis are not due exclusively to endotoxins, endotoxins of bacterial origin appear to play a role in producing the endocrine and metabolic alterations, which are quire similar in the experimental septic animal and in animals following the administration of endotoxin.

3. The metabolic and febrile effects of endotoxins can be dissociated, and can be separated from the hemodynamic alterations caused by these agents.

4. Endotoxin has pronounced effects on glucose metabolism, affecting both the disappearance and appearance rates. These effects are present even at doses several orders of magnitude below the LD_{50}.

5. It appears that the increased rate of disappearance of glucose is the primary response to endotoxin, and is followed by a compensatory elevation of Ra.

6. The increased rate of disappearance of glucose following endotoxin administration is due to elevated glucose utilization by skeletal muscle, adipose tissue, blood cells, etc.

7. Increased hepatic glycogenolysis and especially increased gluconeogenesis are responsible for the augmented rate of appearance of glucose under these conditions.

8. Although the gluconeogenic ability of the liver is impaired (as indicated by decreased PEPCK activity and decreased metabolic clearance rates of the major gluconeogenic precursors), the absolute rate of gluconeogenesis is elevated because of the altered hormonal environment (predominance of catabolic over anabolic hormones) and the increased rate of delivery of the gluconeogenic precursors to the liver.

ACKNOWLEDGMENTS

This work was supported by grant GM 32654.

REFERENCES

1. L. J. Berry, Metabolic effects of bacterial endotoxin, in: "Microbial Toxins," Volume V, S. Kadis, G. Weinbaum and S. J. Ajl, eds., Academic Press, New York (1971).
2. Z. A. Cohn and S. I. Morse, Functional and metabolic properties of polymorphonuclear leukocytes, J. Exp. Med. 3:689 (1960).
3. J. L. Ferguson, J. J. Spitzer, and H. I. Miller, Effects of endotoxin on regional blood flow in the unanesthetized guinea pig, J. Surg. Res. 25:236 (1978).
4. M. J. Fettman, M. S. Hand, L. G. Chandresena, J. L. Cleek, R. A. Mason, and R. W. Phillips, Hepatosplanchnic insulin kinetics in awake endotoxemic Yucatan minipigs: the "misinformed B-cell" hypothesis revisited, Circ. Shock 14:237 (1984).
5. J. P. Filkins and B. J. Buchanan, In vivo vs. in vitro effects of endotoxin in glucogenolysis, gluconeogenesis and glucose utilization, Proc. Soc. Exp. Biol. Med. 155:216 (1977).
6. R. E. Fish and J. A. Spitzer, Continuous infusion of endotoxin from an osmotic pump in the conscious, unrestrained rat: a unique model of chronic endotoxemia, Circ. Shock 12:135 (1984).
7. M. D. Hand, M. J. Fettman, L. G. Chandrasena, J. L. Cleek, R. A. Mason, and R. W. Phillips, Increased glucose uptake precedes

hyperinsulinemia in awake endotoxemic minipigs, Part II, <u>Circ.</u>
<u>Shock</u> 11:287 (1983).

8. L. B. Hinshaw, L. T. Archer, B. K. Beller, G. L. White, J. M.
    Schroeder, and D. D. Holmes, Glucose utilization and role of blood
    in endotoxin shock, <u>Am. J. Physiol.</u> 233:E71 (1977).

9. D. L. Kelleher, G. J. Bagby, B. C. Fong, and J. J. Spitzer, Glucose
    turnover five hours following endotoxin administration to normal
    and diabetic rats, <u>Circ. Shock</u> 9:47 (1982).

10. D. L. Kelleher, B. C. Fong, G. J. Bagby, and J. J. Spitzer, Metabolic
    and hormonal changes following endotoxin administration to diabetic
    rats, <u>Am. J. Physiol.</u> 243:R77 (1982).

11. R. E. Kuttner and J. J. Spitzer, Gluconeogenesis from alanine in
    endotoxin-treated dogs, <u>J. Surg. Res.</u> 25:166 (1978).

12. C. H. Lang, G. J. Bagby and J. J. Spitzer, Carbohydrate dynamics in
    the hypermetabolic septic rat, <u>Metabolism</u> 33:959 (1984).

13. C. H. Lang, G. J. Bagby and J. J. Spitzer, Glucose kinetics and body
    temperature following lethal and nonlethal doses of endotoxin, <u>Am.</u>
    <u>J. Physiol.</u> 248:R471 (1985).

14. G. J. Leach and J. A. Spitzer, Endotoxin induced alterations in
    glucose transport in isolated adipocytes, <u>Biochim Biophys. Acta.</u>
    648:71 (1981).

15. S. P. Martin, G. R. McKinney, and R. Green, The metabolism of human
    polymorphonuclear leukocytes, <u>Ann. N.Y. Acad. Sci.</u> 59:996 (1955).

16. O. P. McGuiness and J. J. Spitzer, Hepatic glycerol flux after <u>E. coli</u>
    endotoxin administration, <u>Am. J. Physiol.</u> 247:R687 (1984).

17. G. E. Merrill and J. J. Spitzer, Glucose and lactate kinetics in
    guinea pigs following <u>Escherichia coli</u> endotoxin administration,
    <u>Circ. Shock</u> 5:11 (1978).

18. R. M. Raymond, J. M. Harkema, and T. E. Emerson, Jr., Mechanism of
    increased glucose uptake by skeletal muscle during <u>E. coli</u>
    endotoxin shock in the dog, <u>Circ. Shock</u> 8:77 (1981).

19. A. J. Romanosky, G. J. Bagby, E. L. Bockman, and J. J. Spitzer, Free
    fatty acid utilization by skeletal muscle after endotoxin
    administration, <u>Am. J. Physiol.</u> 239:E391 (1980).

20. A. J. Romanosky, G. J. Bagby, E. L. Bockman, and J. J. Spitzer,
    Increased muscle glucose uptake and lactate release after endotoxin
    administration, <u>Am. J. Physiol.</u> 239:E311 (1980).

21. A. J. Romanosky, O. P. McGuiness, and J. J. Spitzer, Metabolic
    clearance rate of ketone bodies in dogs following Escherichia coli
    endotoxin administration, <u>Circ. Shock</u> 11:311 (1983).

22. J. A. Spitzer, A. G. B. Kovach, P. Sandor, J. J. Spitzer, and R.
    Storck, Adipose tissue and endotoxin shock, <u>Acta. Physiol. Acad.</u>
    <u>Sci. Hung.</u> 44:183 (1973).

23. J. A. Spitzer and J. J. Spitzer, Effect of LPS on carbohydrate and
    lipid metabolism, <u>in</u>: "Beneficial Effects of Endotoxins," A.
    Nowotny, ed., Plenum Publishing, New York (1983).

24. J. A. Spitzer, K. M. Nelson, and R. E. Fish, Time course of changes in
    gluconeogenesis from various precursors in chronically endotoxemic
    rats, <u>Metabolism</u>, April (1985).

25. J. J. Spitzer, J. L. Ferguson, H. J. Hirsch, S. Loo, and K. H. Gabbay,
    Effects of <u>E. coli</u> endotoxin on pancreatic hormones and blood flow,
    <u>Circ. Shock</u> 7:353 (1980).

26. R. R. Wolfe, D. Elahi, and J. J. Spitzer, Glucose and lactate kinetics
    after endotoxin administration in dogs, <u>Am. J. Physiol.</u> 232:E180
    (1977).

27. R. R. Wolfe, D. Elahi, and J. J. Spitzer, Glucose kinetics in dogs
    following a lethal dose of endotoxin, <u>Metabolism</u> 26:847 (1977).

GLUCOSE DYSHOMEOSTASIS IN ENDOTOXICOSIS:

DIRECT VERSUS MONOKINE-MEDIATED MECHANISMS OF ENDOTOXIN ACTION

Michael R. Yelich, Linda Witek-Janusek, and
James P. Filkins

Department of Physiology
Loyola University of Chicago
Stritch School of Medicine
Maywood, Illinois

"It is the information carried by the bacteria that we
cannot abide. The gram-negative bacteria are the best
examples of this. They display endotoxin in their walls,
and the macromolecules are read by our tissues as the
very worst of bad news. When we sense lipopolysaccharide,
we are likely to turn on every defense at our disposal;
we will bomb, defoliate, seal off, and destroy all the
tissues in the area. ... It is a shambles. ... All of
this seems unnecessary, panic-driven. There is nothing
intrinsically poisonous about endotoxin, but it must
look awful, or feel awful, when sensed by cells. ...
The self-disintegration of the whole animal that follows
a systemic injection (of endotoxin) can be interpreted
as a well-intentioned, but lethal error."

Lewis Thomas, The Lives of a Cell, 1974

A PERSPECTIVE ON CELL METABOLISM AND ENDOTOXIN

Among the many pathophysiologic disturbances which are elicited by
gram-negative sepsis and concomitant endotoxicosis are global disruptions
in the control mechanisms of normal metabolism. These mechanisms include
the regulation and integration of the storage, mobilization, transport, and
utilization of major cell nutrients such as fatty acids, ketone bodies,
amino acids, lactic acid and especially glucose. Since these logistical
considerations of physiologic nutrients constitute "caloric homeostasis"
(37), there is heuristic value in viewing the pathophysiologic metabolic
syndrome of sepsis and endotoxicosis as a paradigm of "caloric dyshomeostasis"--that is, a metastable steady state of life-threatening derangements
in the cellular mechanisms, neuroendocrine regulation and systemic integration of carbohydrate, lipid and protein metabolism.

One pivotal aspect of caloric dyshomeostasis in endotoxicosis deals
with the logistics of glucose metabolism--its synthesis, storage and ultimate
peripheral utilization, for either oxidative energetics (i.e., conversion
of glucose to carbon dioxide and water, as in the brain) or non-oxidative

energetics (i.e., conversion of glucose to lactate, as in white muscle). Among the earliest metabolic alteration observed in endotoxicosis is a transient hyperglycemia which deteriorates into a profound, progressive, relentless and refractory hypoglycemia. This pattern of endotoxicosis is in contrast to a typical stress-induced hyperglycemia which stabilizes to euglycemia when the stress is removed. Hypoglycemia may be either the sole factor in the convulsive demise of the endotoxic organism due to central nervous system glucopenia, or a major contributory factor in the initial cardiovascular system failure. This circulatory-metabolic disintegration generally progresses to the syndrome of multiple systems organ failure and cachexia which characterize fulminating sepsis.

As is unfortunately true for too many of the pathophysiologic constructs concerning endotoxicosis, the question remains as to the fundamental events underlying the pathogenesis of endotoxin-induced glucose dyshomeostasis. While it is generally accepted that the root cause of this problem is an imbalance between hepatic provision of glucose (via glycogenolysis and gluconeogenesis) and peripheral utilization of glucose, no consensus exists as to the precise cellular mechanisms responsible for this imbalance, at either the hepatic or the peripheral tissue levels. This brief review will focus on carbohydrate dyshomeostasis and the evidence which exists, or indeed does not exist, for direct effects of endotoxin as opposed to the effects of endotoxin mediated by macrophage-derived monokines. Within the framework of the overall problem of glucose dyshomeostasis during endotoxicosis, this evidence will relate primarily to the particular expertise of the authors: 1) the role of endotoxin-induced changes in peripheral glucose utilization; and 2) the role of endotoxin-induced alterations in the endocrine secretion of insulin.

In reference to the quotation from Lewis Thomas which introduces this chapter (46), our objective is to review just how the unabidable information residing in endotoxin is read by our tissues--it is a direct process by which lipopolysaccharide (endotoxin) commands the attention of the cell membranes and organelles, or is it dependent on the macrophages as a translation system which generates as a second level of command signals via the synthesis and release of regulatory monokines?

THE RETICULOENDOTHELIAL SYSTEM AND THE PATHOPHYSIOLOGY OF CIRCULATORY SHOCK

Abundant experimental evidence exists to substantiate the concept that the reticuloendothelial system (RES) plays a central role in the resistance to circulatory shock, particularly with reference to septic shock (1,26). The generally accepted mechanisms for the RES-shock relationship relates mainly to the phagocytic (or more precisely endocytic) functions of the RES. In septic shock, the standard prevailing explanation for the RES-shock relation is that the macrophage system is a phagocytic sink for the uptake and detoxification of endotoxin and/or a variety of potentially nocuous products generated during endotoxicosis.

While the endocytic and detoxification functions of the RES and especially the hepatic sinusoidal macrophages are of undoubted significance in the sequelae of endotoxic and septic shock, it is now recognized that the RES is also an important secretory (exocytic) system (38,47). Thus, in contrast to the endocytic explanation of the RES-septic shock relationship, there is the growing realization that mediators produced by and released from macrophages play a central role in shock pathogenesis. While most studies of the macrophage mediator hypothesis of septic shock have focused on the immunological and cardiovascular mediator-target interactions, recent work has suggested that macrophage secretory products also impact on carbohydrate, lipid, and protein metabolism in sepsis.

This section will focus on the metabolic aspects of shock relating to the disturbances which occur in carbohydrate homeostasis when endotoxin is injected into an organism. It is appropriate to provide a more thorough overview of glucoregulation during endotoxic and septic shock before considering evidence regarding the direct or mediated effects of endotoxin on peripheral glucose utilization.

Over the past fifteen years much effort has been expended to determine the pathophysiological manifestations of circulatory shock as they relate to carbohydrate metabolism. Most of this effort has been directed toward studying the effects of endotoxin on glucose dyshomeostasis during shock. While in the past there has been controversy as to whether endotoxin plays a significant role in the pathogenesis of clinical sepsis and septic shock, investigators recently have successfully employed anti-endotoxin serum to treat humans in septic shock (31,58). These findings strongly support the role of endotoxin in septic shock and justify the use of the endotoxic animal as a valid model of sepsis and septic shock.

Sepsis and endotoxicosis are characterized by marked abnormalities in the biochemical processes and inter-organ regulation of fuel substate dynamics. Sepsis and septic shock are now appreciated fundamentally as pathophysiological metabolic states which eventuate in host-defense, cardiovascular and total multiple organ system failure (3,23,42,44). Foremost among the metabolic abnormalities during endotoxicosis are profound alterations in carbohydrate homeostasis which are intimately linked to the ensuing circulatory syndrome and the insidious progression from hemodynamically reversible (high cardiac output) states to hemodynamically irreversible (low cardiac output) states. The early responses to endotoxicosis and sepsis are marked by hyperglycemia--a metabolic indicator of compensated carbohydrate homeostasis. If endotoxicosis is severe or if sepsis is overwhelming, then there occurs a lethal transition to overt shock, accompanied by the pathogenesis of profound, progressive hypoglycemia--the metabolic hallmark of decompensated carbohydrate homeostasis (13).

While the early hyperglycemic response to endotoxin is generally considered adaptive and beneficial to the stability of host metabolism, it breeds a disproportionate hyperinsulinemia. This in turn sets the metabolic stage for the progression to glucose dyshomeostasis and hypoglycemia.

Under conditions of low-dose endotoxemia or mild sepsis in the fed state, the initial metabolic response of hyperglycemia is due to accelerated glycogenolysis; in the fasted state it is due to increased glyconeogenesis. The functional status of these two hepatic processes are important determinants of circulating glucose levels during endotoxicosis. Hyperglycemia also occurs in fasted animals during endotoxicosis or sepsis due to increased substrate delivery (secondary to elevated catecholamine levels) or a favorable hormonal environment (elevated glucagon levels) (51). In either the fed or the fasted state, late (severe) endotoxic shock or sepsis is characterized by a relative depression of gluconeogenesis (51). While the effects of endotoxin on glycogenolysis and gluconeogenesis are well documented experimentally (13), it is not known whether endotoxin acts directly on these hepatic processes, or whether the effects of endotoxin are mediated in vivo.

In constrast to the early, mild effects of low-dose endotoxemia, the effects of late, high-dose endotoxemia are characterized by hypoglycemia (in both the fed and the fasted states), which signals the transition from compensated to decompensated carbohydrate homeostasis. The major determinant of circulating glucose levels is the status of peripheral glucose metabolism.

Early (mild) endotoxemia or sepsis is associated with increased glucose utilization in a variety of tissues. Especially prominant users of glucose are muscle, adipose tissue and liver (2,18,30,35,36,40,48). The transition to late (severe) endotoxicosis or sepsis is characterized by the continuance of exaggerated peripheral glucose utilization and impaired gluconeogenesis. The phases of endotoxin-induced glucose dyshomeostasis are shown diagramatically in Figure 1.

Several investigators have demonstrated that endotoxin acts directly to elevate peripheral utilization of glucose in a variety of tissues (i.e., endotoxin has insulin-like activity, ILA). However, there is now additional evidence which is consistent with the hypothesis that endotoxin induces a mediator in vivo which has ILA. The direct and mediated roles of endotoxin in glucose dyshomeostasis will be discussed in subsequent sections.

In summary, the prominent glucoregulatory alterations of late or severe endotoxemia and sepsis are a lethal combination of exaggerated peripheral glucose utilization and incompetent hepatic glucogenesis--a double-edged sword compatible with a functional hyperinsulinism.

DIRECT ACTION OF ENDOTOXIN ON GLUCOSE UTILIZATION

Although bacterial endotoxin is normally cleared from the blood by the endocytic process of the RES, it is likely that substantial amounts of endotoxin may escape RES processing. Such RES escape may occur in the immuno-incompetent host, e.g., the very young, the aged, and the chronically ill, as well as in the immunocompetent host whose RES system is overwhelmed by a fulminating bacteremia. This "unprocessed" endotoxin then unleashes its nocuous potential on host cells and organs by directly affecting cellular metabolism. Due to the lipid nature of the endotoxin molecule, the most likely target of its direct action is the cell membrane.

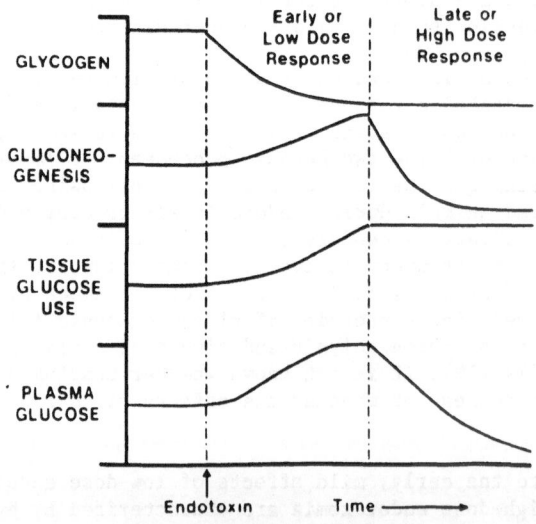

Phases of glucoregulation in endotoxicosis.

Figure 1. [From Filkins, 1978 (13).]

Table 1. Adipose Tissue Glucose Oxidation in Response to
Various Concentrations of Endotoxin*

Endotoxin (µg/ml)	N	Glucose Oxidation[a] (DPM/gm/180 min x 10^3)
0	37	170.4 ± 5.7
1	7	172.7 ± 15.8
10	10	225.7 ± 22.7**
50	7	265.0 ± 27.5**
100	6	327.6 ± 19.6**
500	7	367.0 ± 33.7**

* Glucose oxidation was measured for 180 min in paired epididymal fat pads in the presence of various concentrations of
endotoxin.
[a] Values represent the mean ± SEM.
**$P < 0.05$ compared to the 0 endotoxin group as determined by
the unpaired Student's t test. From (49).

The direct action of endotoxin on cellular glucose metabolism was first
reported by Woods and associates (52,53,54), who demonstrated that endotoxin
stimulated cellular glycolysis in both malignant and non-malignant cells
via a mechanism similar to that for the action of insulin. This insulin-like

Table 2. Adipose Tissue Glucose Oxidation in Response to Various
Preincubation Times with Endotoxin*

Preincubation Time (min)	Treatment	Glucose Oxidation DPM/gm/180 min x 10^3)	N
5	– ETX	108.5 ± 6.7	11
	+ ETX	143.1 ± 11.1**	11
15	– ETX	143.8 ± 9.3	14
	+ ETX	271.4 ± 24.6**	14
30	– ETX	129.5 ± 8.4	16
	+ ETX	241.8 ± 19.1**	16
60	– ETX	123.7 ± 9.7	10
	+ ETX	199.2 ± 11.0**	10
120	– ETX	167.9 ± 4.8	84
	+ ETX	267.9 ± 7.4**	84

* Paired epididymal fat pads were preincubated with and without S. enteriditis endotoxin (ETX) 100 µg/ml in Krebs-Ringer bicarbonate plus glucose.
Tissues were then rinsed and glucose oxidation was measured and is expressed as the mean ± SEM.
**$P < 0.01$ from the respective (-ETX) control as determined by the unpaired
Student's t test. From (49).

action of endotoxin was not observed in noninsulin-responsive cells and was antagonized by the synthetic glucocorticoid dexamethasone. Since insulin blunted the stimulatory action of endotoxin, a common site of action was suggested (53). Subsequently, endotoxin was reported to stimulate the direct uptake of glucose in the in situ perfused dog gracilis muscle (40) as well as to stimulate glucose oxidation by human leukocytes (19), rat adipocytes (45), and dog myocytes (32). Although Filkins and Buchanan (17) initially reported an inability to demonstrate a direct action of endotoxin on glucose oxidation by the rat epididymal fat pad, Witek-Janusek and Filkins (49) subsequently succeeded in demonstrating a consistent endotoxin-induced stimulation of glucose oxidation by the rat fat pad. The avoidance of cold shock (that is, placing the tissue in ice cold saline) prior to or between incubations was of primary importance for the demonstration of the insulin-like action of endotoxin. This temperature dependent effect explained the discrepancy between the reports of Filkins and Buchanan (17) and Witek-Janusek and Filkins (49). In further characterizing the direct actions of endotoxin on glucose oxidation by the rat fat pad, the latter investigators reported that the lowest concentration of endotoxin which produced a significant increment in glucose oxidation was 10 micrograms/ml (Table 1) and that a brief five minute exposure of the tissue to endotoxin was sufficient to produce this effect (Table 2). In this model the stimulatory action of 100 micrograms/ml of S. enteritidis endotoxin was reported to be equivalent to that of 1 ng/ml of insulin. Furthermore, this effect was found to be dependent on calcium, since it was reduced when tissue was incubated in a calcium-free medium (49). Using rat adipocytes, Holley and Spitzer (28) found that insulin binding was not appreciably altered by in vitro exposure to endotoxin and thus a post receptor site of synergy was suggested.

It is well documented that glucocorticoids protect against the lethal effects of both endotoxic and septic shock (27,41); however, the precise

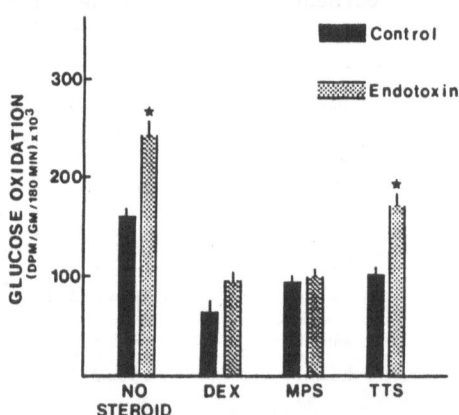

Figure 2. Paired epididymal fat pads were preincubated (60 min) with either 100 µg/ml dexamethasone sodium phosphate (DEX), 500 µg/ml methylprednisolone sodium succinate (MPS), or 500 µg/ml testosterone proprionate (TTS). Tissues were rinsed and glucose oxidation was measured with and without S. enteridi tis endotoxin (100 µg/ml). Bars represent the mean with the SEM above (N = 10/group). *P < 0.01 compared to the respective control group as determined by factorial analysis of variance. [From Witek-Janusek and Filkins, 1981 (49).]

mechanism for this protection is equivocal. The ability of glucocorticoids
to ameliorate the glucose dyshomeostasis of experimental endotoxic and septic
shock has been documented (3,4,5,11) and most likely represents an important
mechanism of their protective role. In the isolated fat pad assay of glucose
oxidation, both dexamethasone and methylprednisolone have been shown to
directly antagonize the insulinomimetic action of endotoxin (Figure 2).
This antagonism was evident when the tissue was treated simultaneously with
glucocorticoids and endotoxin, but was not evident when the steroid hormone
was added after pretreatment with endotoxin. These results correlated with
in vivo data, which emphasized that glucocorticoids must be administered
early during the course of endotoxicosis or sepsis in order to be beneficial
(27). Thus glucocorticoids have a direct anti-endotoxin action which may
be similar to their anti-insulin actions. This may be significant with
respect to the mechanism by which glucocorticoids prevent the hypoglycemia
of endotoxic shock. More importantly, studies of the interaction of endotoxin and hormones may reveal important information regarding the mechanism
of action of endotoxin.

Numerous studies have shown that non-steroidal anti-inflammatory agents
can alter many of the pathophysiological sequelae of shock (8,9,39,43,49).
Increased survival has been achieved in experimental septic shock models by
treatment with the cyclo-oxygenase inhibitors indomethacin or ibuprofen
(22,43). Indomethacin also has been shown to blunt the hypoglycemia of
endotoxin shock (39). With respect to the direct effects of endotoxin,
Witek-Janusek and Filkins (49) have documented that indomethacin directly
antagonizes the stimulatory action of in vitro endotoxin on glucose oxidation
by the rat fat pad (Figure 3). In addition, the direct insulin-like action
of endotoxin could not be demonstrated when using epididymal fat pads obtained from essential fatty acid deficient rats which are resistant to the
lethal effects of endotoxin (9). This latter observation along with the
similar antiendotoxin action of glucocorticoids and indomethacin suggests

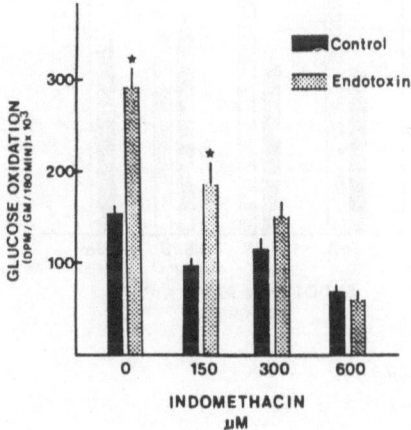

Figure 3. Paired epididymal fat pads were preincubated (60 min) with
either 0, 150, 300, or 600 μM indomethacin. Tissues were
rinsed and glucose oxidation was measured with and without
S. enteriditis endotoxin (100 μg/ml). Bars represent mean
with SEM above (N = 13/group). *P < 0.01 compared to the
respective control group as determined by factorial analysis
of variance. [From Witek-Janusek and Filkins, 1981 (49).]

that the direct insulin-like action of endotoxin may be related to altered
phospholipid metabolism and the generation of a glucoregulatory eicosanoid.

    Witek-Janusek and Filkins (50) also have evaluated the relation of
endotoxin structure to its direct insulin-like action on epididymal fat pad
glucose oxidation. They compared the direct insulin-like effects of a
variety of endotoxin preparations that contained various combinations of
the primary constituents of the endotoxin molecule, these being lipid A,
lipid associated protein and polysaccharide. Figure 4 shows the direct
effect of three different endotoxin preparations and lipid A on glucose
oxidation by the rat epididymal fat pad. The Salmonella enteritidis Boivin
preparation of endotoxin (SEB) consists of bacterial cell wall lipopolysaccharide (LPS) and lipid associated protein; whereas the Salmonella enteritidis Westphal preparation of endotoxin (SEW) consists of a protein-free
LPS (34). Salmonella minnesota Re595 glycolipid (SM) is a lipid A-rich
endotoxin preparation obtained from a mutant bacterial strain. It contains
the lipid A moiety of complete endotoxin and only a small amount of polysaccharide (16%). Pure SM lipid A, in contrast, contains less than 0.02%
polysaccharide (29). All of the endotoxin preparations used (SEB, SEW, and
SM-glycolipid) significantly increased glucose oxidation in the fat pads
relative to the control tissue oxidation; however, the increment was greatest

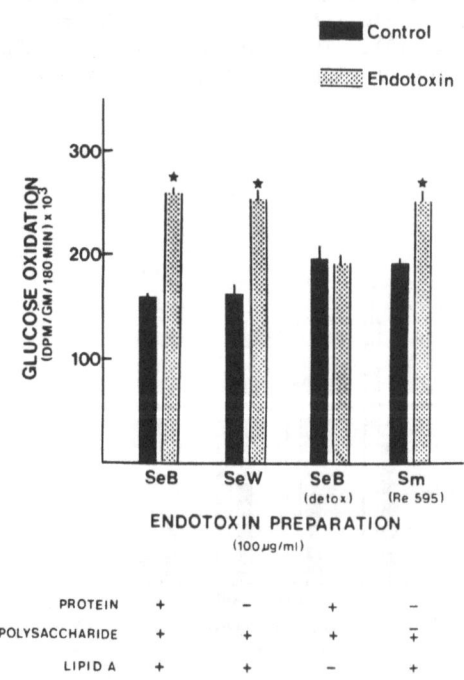

Figure 4.   Paired epididymal fat pads were placed in metabolic flasks
             containing ^{14}C-glucose (0.5 µCi/ml) Krebs-Ringer-Bicarbonate
             with 5.55 mM unlabeled D-glucose ± 100 µg/ml of S. enteritidis
             Boivin (SEB), S. enteritidis Westphal (SEW), alkaline hydro-
             lyzed (Detox) SEB, or S. minnesota (SM) glycolipid. Glucose
             oxidation was measured over a period of 180 min. Values
             represent the mean ± SEM. *P < .05. [From Witek-Janusek
             and Filkins, 1983 (50).]

Table 3. Effect of Lipid A on Glucose Oxidations in Epididymal Fat Pad[#]

Endotoxin Preparation	Glucose Oxidation DPM/g/180 min) x 10^3			% Change from Control
	N	Control	Test	
SM-glycolipid	10	189 ± 10	250 ± 17	+ 32*
SM-lipid A (sonicates)	10	142 ± 20	138 ± 18	- 3
SM-lipid A (TEA salt)	16	117 ± 8	131 ± 13	+ 12
SM-lipid A (BSA complex)	41	184 ± 10	204 ± 13	+ 11

[#]Paired epididymal fat pads were incubated ± 25 μg/ml of S. minnesota (SM) glycolipid, sonicates of SM-lipid A, triethylamine (TEA) salts of lipid A, or lipid A complexed to bovine serum albumin (BSA). Glucose oxidation was measured over a period of 180 min. Values represent the mean ± SEM. *$P < .05$ comparing test to control oxidation. From (50).

relative to the control tissue oxidation; however, the increment was greatest (60%) using SEB (i.e., the complete preparation of endotoxin) and the least (40%) using SM-glycolipid (lipid A + 16% polysaccharide). Conversely, mild alkaline hydrolysis of SEB, which removes the ester-linked fatty acids from lipid A, blocked the insulin-like action of SEB (Figure 4). Similarly, pretreatment of SEB with polymixin B, which binds to and blocks lipid A-mediated effects (33), also eliminated the insulin-like actions of endotoxin (50). Thus, these studies suggested that lipid A was the mediator of the insulin-like action of endotoxin; however, the essentially polysaccharide-free SM lipid A had no effect on fat pad glucose oxidation (Table 3). Since lipid A is insoluble in an aqueous medium, various attempts were made to increase its solubility. Hence, sonicates, triethylamine salts and bovine serum albumin complexes of SM lipid A were used in the fat pad assay. None of these complexes produced an increase of glucose oxidation in the in vitro assay system (Table 3). Interestingly, recombining the detoxified alkaline

Table 4. Effects of Combinations of Lipid A Plus Alkaline Hydrolysates of Endotoxins on Glucose Oxidation in Epididymal Fat Pads[#]

Endotoxin Preparation	Glucose Oxidation DPM/g/180 min) x 10^3			% Change from Control
	N	Control	Test	
Lipid A-BSA	41	184 ± 10	204 ± 13	+ 10
SEB	15	132 ± 6	178 ± 15	+ 35*
Alkaline hydrolysate of SEB	6	159 ± 14	155 ± 14	- 0.3*
Lipid A + alkaline hydro-lysate of SEB	18	177 ± 11	224 ± 20	+ 27*
Lipid A + alkaline hydro-lysate of SM-Glycolipid	35	202 ± 12	258 ± 19	+ 28*

[#]Paired epididymal fat pads were incubated ± 25 μg/ml of lipid A complexed to bovine serum albumin (BSA), the alkaline hydrolysate of S. enteritidis Boivin (SEB), lipid A plus alkaline hydrolysate of SEB or lipid A plus the alkaline hydrolysate of S. minnesota (SM) glycolipid. Glucose oxidation was measured over a period of 180 min. Values represent mean ± SEM. *$P < .05$ comparing test to control oxidation. From (50).

Table 5. Summary: Comparison of Glucoregulatory Alterations in Endotoxic and Non-endotoxic Perturbation of the RES

| Glucoregulatory Parameter | Index | RES Perturbations | | |
		Endotoxicosis	Carbon Blockade	Lead Salts
Glucose Tolerance	iv GTT	I	I	I
Insulin Tolerance	-Resistance to seizure deaths	D	D	D
	-Resistance to hypoglycemia	D	D	D
Glycogen Synthesis	-Hepatic depletion	D	D	D
Gluconeogenesis	-In vivo % conversion of ^{14}C-alanine to ^{14}C-glucose	D	D	D
	-In vitro assessment of isolated hepatocyte conversion of alanine or lactate to glucose	D	D	D
Glucose Oxidation	-In vivo whole body oxidation of ^{14}C-glucose to ^{14}CO$_2$	I	I	I
	-In vitro oxidation of ^{14}C-glucose to ^{14}CO$_2$	I	I	I

I = increase in parameter as compared to control groups.
D = decrease in parameter as compared to control groups. From (19).

hydrolysate with the SM lipid A restored the insulin-like activity (Table 4). It is possible that pure lipid A, due to its insolubility, cannot effectively reach target cells and that the addition of polysaccharide facilitates its interaction with the cell membrane.

In summary, the information provided above demonstrates evidence that bacterial endotoxin does have the ability to alter cellular glucose utilization in a direct manner. The importance of this action in the glucose dyshomeostasis of endotoxicosis and sepsis is unknown. However, in vitro models of endotoxin action and interaction with glucoregulatory hormones (such as insulin and glucocorticoids) have the potential of providing important insights into the mechanisms of the membrane perturbing effects of endotoxin.

MEDIATED ACTION OF ENDOTOXIN ON PERIPHERAL GLUCOSE UTILIZATION

As a part of the effort to understand the precise manner in which endotoxin leads to glucose dyshomeostasis, it has become clear that the reticuloendothelial system (RES) is at least partially responsible for many of the effects of endotoxin (21). For example, many of the glucoregulatory disturbances which had been noted to be secondary to endotoxin are also noted after affecting experimental animals with "RES blockade" techniques (2,19;

Table 5). These techniques basically entail injecting microparticles (such as colloidal carbon or lead acetate) into the circulation of experimental animals. The macrophages of the RES remove the particulate matter from the circulation by phagocytosis and the rate of removal can be measured. Upon a second challenge with particulate matter the rate of removal decreases, therefore indicating that the RES was functionally blocked. Since "RES blockade" could be shown to mimic the glucoregulatory disturbances during endotoxicosis (19) and since endotoxin is removed from the circulation primarily by the macrophages of the RES, it was logical to test the hypothesis that "RES blockade" could alter peripheral glucose utilization in a manner similar to that induced by endotoxemia.

Recent laboratory investigations have established that the macrophages of the RES are not merely phagocytic cells, but also secretory cells. It is now well appreciated that macrophages produce and secrete a large number of mediators often linked to shock (16). Among these mediators are a family of regulatory proteins termed monokines--because they are derived primarily from monocytic cell lines. It was thus necessary to test the hypothesis that some of the effects of in vivo endotoxin were mediated by the RES through monokines.

Macrophage-conditioned medium was prepared using peritoneal macrophages derived from male, Holtzman rats (14). Briefly, this procedure required that the rats receive an intraperitoneal injection with denatured sodium caseinate. The caseinate injection results in an inflammatory reaction and within 96 hours macrophages are harvested from the peritoneal fluid. The macrophages are then washed and quantified. The harvested macrophages were then used for immediate co-incubation with epididymal fat pads in the glucose oxidation assay. Alternately, glass-adhered macrophages were cultured for 24 hours with and without 100 micrograms/ml endotoxin. Macrophages were then removed by centrifugation and the resulting supernatant (i.e., the macrophage-conditioned medium) was then used as the incubation medium for the assay of glucose oxidation by the rat fat pad.

Table 6. Effect of Endotoxin on MILA Production When Using Coincubations of Peritoneal Macrophages with Epididymal Fat Pads

Experimental Group	Number of Preparations	ILA Assay[a] (Adipose Tissue Glucose Oxidation) ($^{14}CO_2$/g/hr)
A Epididymal Fat Pads	12	48,652 ± 4,251
B Epididymal Fat Pads + Peritoneal Macrophages (5 x 10^6/ml)	12	79,321 ± 5,656
C Epididymal Fat Pads + Endotoxin (100 µg/ml)	6	51,451 ± 3,759
D Epididymal Fat Pads + Endotoxin (100 µg/ml) + Macrophages (5 x 10^6/ml)	12	121,152 ± 8,491

[a]Groups B and D were corrected for macrophage glucose oxidation which was less than 6% of total. P values < 0.05 (Group A compared with Group B, B with C, and B with D). From (14).

Table 7. Effect of Endotoxin on MILA Production when Using Glucose
Oxidation in Epididymal Fat Pads Incubated with Supernatants
from Peritoneal Macrophage Cultures in DMEM

Experimental Group (Assay Incubation Media)	Number of Preparations	ILA Assay[a] (Adipose Tissue Glucose Oxidation) ($^{14}CO_2$/g/hr)
A Media Incubated at 37°C	12	51,667 ± 4,897
B Media Incubated at 37°C with Endotoxin (100 g/ml)	8	48,667 ± 4,110
C Supernatants from Peritoneal Macrophages Incubations (2.5 x 10^7/ml)	12	86,118 ± 6,256
D Supernatants from Peritoneal Macrophage Incubations (2.5 x 10^7/ml) with Endotoxins (100 µg/ml)	12	164,887 ± 12,112

[a]P values < 0.05 (Group A compared with Group C and C with D). (From 14.)

Glucose oxidation by fat pads co-incubated with peritoneal macrophages
was 63% greater than that for fat pads incubated in control medium alone.
Similarly, fat pads which were co-incubated with macrophages and endotoxin
(100 micrograms/ml) oxidized 135% more glucose than those incubated only
with endotoxin. The direct effects of endotoxin on fat pad glucose oxidation were minimized by using fat pads which had been previously chilled in
ice-cold saline. These results are shown in Table 6.

A similar experiment was then conducted, except that the macrophages
were not co-incubated with the fat pads. Conditioned-medium obtained from
cultured macrophages was added to the fat pad assay system. Fat pads which
were incubated with macrophage conditioned-medium oxidized 67% more glucose
than those incubated in control medium. Fat pads which were incubated with
conditioned-medium obtained from endotoxin-treated cultures demonstrated
91% more glucose oxidation than fat pads incubated only with conditioned-
medium and 239% more glucose oxidation than fat pads incubated only with
endotoxin. Again the direct effects of endotoxin on fat pad glucose oxidation were minimized by pre-chilling the fat pads in ice-cold saline. The
results of these experiments are shown in Table 7.

The results indicated that a product derived from macrophages demonstrated insulin-like activity. This product, a putative monokine, was thus
named MILA for Macrophage Insulin-Like Activity. In addition the data indicated that co-incubation of the macrophages with endotoxin increased the
insulin-like activity of the macrophage-conditioned medium. This result
underscores the relationship between the RES and the effects of endotoxin.
Thus it appears that endotoxicosis leads to the increased production of
monokines (e.g., MILA) from the macrophages of the RES; the MILA can in
turn contribute to the increased levels of glucose utilization which are
characteristic of endotoxicosis.

In summary, it is clear that in addition to the fact that endotoxin
can directly alter peripheral utilization of glucose, endotoxin can have

significant effects on glucose utilization which are mediated by a monokine produced by the mononuclear phagocyte system. Since plasma glucose levels are dependent on a precise balance of glucose production and glucose utilization within the organism, both the direct and RES-mediated effects of endotoxin can combine to elevate glucose use in excess of glucose production--the very conditions which lead to glucose dyshomeostasis and hypoglycemia. The following sections will consider how endotoxin is involved in the alteration of endocrine mechanisms which can severely impact on the overall problem of endotoxin-induced glucose dyshomeostasis.

ACTIONS OF ENDOTOXIN ON INSULIN SECRETION:

DIRECT VERSUS MONOKINE-MEDIATED MECHANISMS

The previous sections have outlined the important metabolic considerations regarding glucose dyshomeostasis which occurs during endotoxicosis and which eventually leads to a lethal hypoglycemia. As described previously, hepatic glycogenolysis and gluconeogenesis are the primary means of supporting blood glucose levels during the early stages of endotoxicosis. As such the endocrine hormonal mechanisms which facilitate these hepatic compensatory processes comprise changing levels of glucagon, catecholamines and glucocorticoids. These hormones are usually considered to be "anti-insulin" in nature as they increase hepatic glucose output. Endotoxin, however, leads to a state of increased tissue use of glucose by direct (49,50) and mediated (14,20) mechanisms as discussed earlier. Thus the so-called insulin-like activity of endotoxin acts in direct opposition to the hepatic compensatory mechanisms which support blood glucose levels. Elevated tissue use of glucose in excess of hepatic compensation results in progressive hypoglycemia.

Among the many physiological actions of insulin are its ability to inhibit hepatic glycogenolysis and gluconeogenesis, as well as its actions leading to the increased use of glucose by the peripheral tissues. The results of these actions are similar to metabolic alterations which are observed during late endotoxicosis. While endotoxin per se can have ILA (49,50) many experimental studies have been conducted to examine whether insulin itself plays an important role in the glucose dyshomeostasis of endotoxicosis.

There are several experimental observations which suggest that insulin plays a key role during endotoxicosis (6,7). For example, fasted rats are less susceptible to the hypoglycemic effects of endotoxin than are non-fasted rats; the stimulation of insulin secretion by endogenous secretagogues being greater in the non-fasted rats than in the fasted rats. In addition, the ability of rats to tolerate parenteral injections of small amounts of insulin is greatly diminished during endotoxicosis (6). Commensurate with these latter observations, if rats are injected with substances which stimulate insulin secretion, such as arginine, leucine or tolbutamide, the resulting insulin secretion and subsequent increase in insulin levels potentiates the lethal, hypoglycemic actions of endotoxin (6,24,56). Conversely, when endogenous insulin secretion is inhibited with mannoheptulose (6), the hypoglycemia of endotoxicosis is prevented. Collectively these observations concerning the interactions of insulin with the metabolic disturbances which occur during endotoxicosis (depressed gluconeogenesis and increased peripheral tissue utilization of glucose) are consistent with the hypothesis that the endotoxic animal demonstrates a functional state of hyperinsulinism.

Several studies have examined plasma insulin levels during controlled states of endotoxicosis, and hyperinsulinemia is usually demonstrated early during the course of endotoxemia (7,55,56). Buchanan and Filkins (7) and

Yelich and Filkins (55) have found in vivo evidence of insulin hypersecretion in endotoxic rats subsequent to a glucose challenge. Yelich and Filkins (55) have suggested that the early hyperglycemia of endotoxicosis is the endogenous stimulus for the insulin hypersecretion, but studies by other investigators have documented insulin secretion in the absence of hyperglycemia in the early stages of endotoxicosis (25). These latter studies have led to the suggestion that B cells of the pancreas may be "misinformed" and secrete insulin "inappropriately" in response to non-stimulatory levels of glucose. Amino acids, especially arginine (56) and leucine (24), also have been suggested as possible endogenous secretagogues for insulin secretion during endotoxicosis. The possibility that amino acids are responsible for endotoxin-induced hyperinsulinemia is intriguing, but there is only indirect evidence that amino acids are actually involved in the hyperinsulinemia of endotoxicosis (24,56).

While the in vivo evidence mentioned above strongly suggested that the endotoxic pancreases secreted inappropriate amounts of insulin, it was imperative to test the hypothesis in vitro in order to eliminate confounding factors which could not be controlled in vivo. Yelich and Filkins demonstrated, under clearly defined in vitro conditions, that pancreases obtained from endotoxic rats secreted quantitatively more insulin for a given glucose stimulus (300 mg/dl glucose) than pancreases obtained from control rats (55; Figure 5). It thus appeared that hyperinsulinemia per se can have a major effect on the glucose dyshomeostasis during endotoxic shock, and the endotoxic pancreas was capable of hypersecreting insulin in response to the

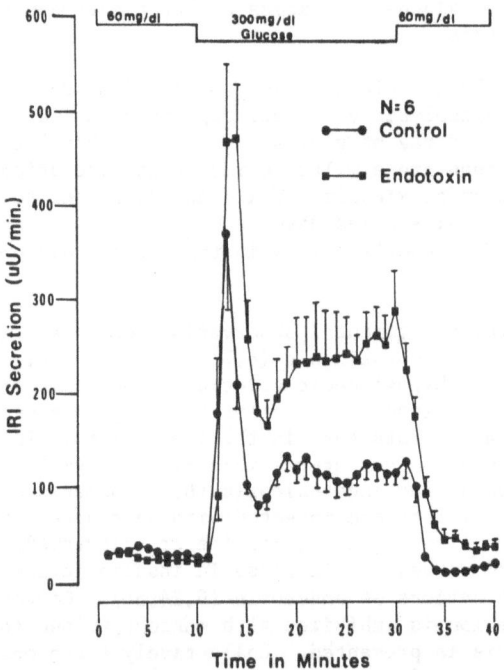

Figure 5. Effect of endotoxin on insulin secretion in isolated perfused rat pancreas. Pancreases were perfused for a total of 40 min. Glucose concentration of medium was increased from 60 to 300 mg/dl at 10 min period and returned to 60 mg/dl at 30 min period. Animals received either saline or 15 mg/kg endotoxin 3 hr prior to isolation and perfusion of pancreas. Very small standard errors are not shown. [From Yelich and Filkins, 1980 (55).]

appropriate stimulus. Therefore, during the course of endotoxicosis, hyperinsulinemia was deleterious whether it occurred as a result of the normal
stimulation of the B cells of the Islets of Langerhans by various physiological means or as a result of pathophysiological mechanisms. However it
appears that by whatever mechanism insulin secretion is provoked, the B
cells of the endotoxic pancreas respond by secreting more insulin than normal
B cells. This alteration of the B cell from a normal insulin secretory
state to an insulin hypersecretory state has been confirmed using a variety
of physiological and pharmacological secretagogues such as glucose, tolbutamide, arginine and 3-isobutyl-1-methylxanthine (20,55,56,57). The following
discussion will describe the approach which was used to identify the mechanism by which normal B cells become hypersecretory in response to
endotoxin.

THE DIRECT EFFECT OF ENDOTOXIN ON INSULIN SECRETION

In experiments conducted in a manner parallel to those examining the
direct effects of endotoxin on peripheral utilization of glucose by epididymal fat pads, two types of experiments were conducted to test the hypothesis that endotoxin can directly affect the pancreatic secretion of insulin
(57). Pancreases were perfused with medium containing glucose below the
threshold for insulin secretion (60 mg/dl). Subsequently endotoxin (200
micrograms/ml was perfused into the in vitro pancreas preparation in order
to determine whether it could act as a secretagogue. Under these conditions
no insulin secretion was observed (Figure 6). A second type of experiment
sought to determine whether endotoxin could alter the normal secretory

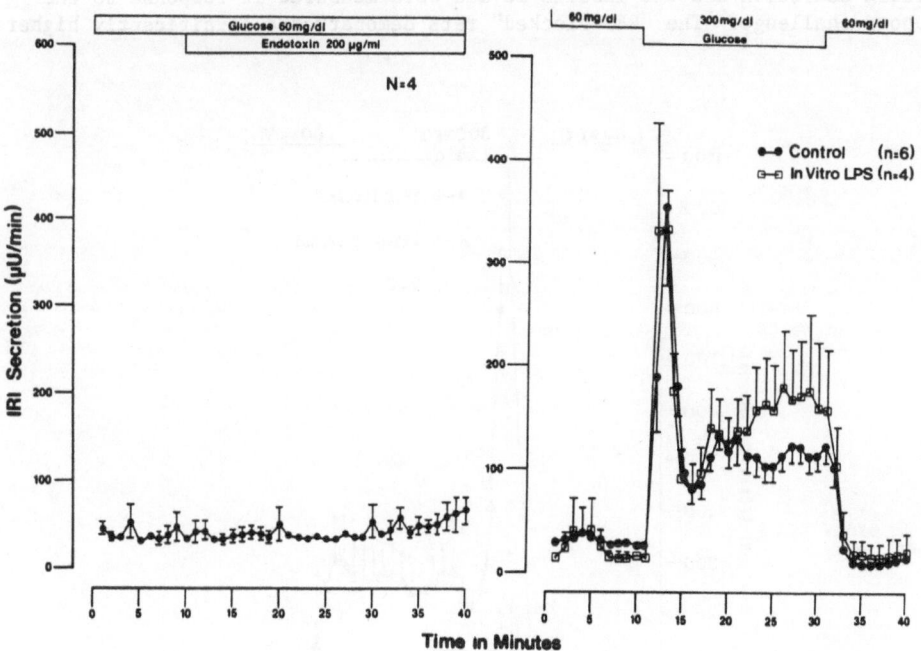

Figure 6.   Left panel: Effect of in vitro endotoxin on basal insulin
            secretion from perfused pancreas preparations obtained from
            control rats. Right panel: effect of in vitro endotoxin
            (200 μg/ml) on glucose-induced insulin secretion from perfused
            pancreas preparations obtained from control rats. [From
            Yelich and Filkins, 1984 (57).]

dynamics of insulin. Pancreases from normal rats were stimulated to secrete insulin using a medium containing 300 mg/dl glucose as the insulin secretagogue, with or without endotoxin in the medium. Endotoxin did not alter the normal quantitative or qualitative nature of glucose-induced insulin secretion in the endocrine pancreas (Figure 6).

Under the specific conditions of the experimental protocols (57), there was no evidence to suggest that endotoxin could directly stimulate insulin secretion or affect the manner in which normal insulin secretion occurs.

THE MONOKINE-MEDIATED EFFECTS OF ENDOTOXIN ON INSULIN SECRETION

It has been mentioned previously that the RES is at least partially responsible for many of the actions of endotoxin (21). Of special note is the fact that many of the glucoregulatory disturbances which characterize endotoxemia also occur subsequent to "RES blockade" (19). Since RES blockade mimics the glucoregulatory disturbances of endotoxicosis, and since endotoxin is removed from the circulation primarily by the macrophage of the RES, it was logical to test the hypothesis that "RES blockade" could induce an insulin hypersecretory state of the endocrine pancreas that was similar to that induced by endotoxin.

To pursue the potential mediating role of the RES on endotoxin-induced hyperinsulinemia and pancreatic insulin hypersecretion, initial studies were performed in vivo to determine whether "RES blockade" could induce insulin hypersecretion by the endocrine pancreas. Control rats and rats treated with colloidal carbon were given injections of glucose to stimulate insulin secretion and the insulin levels were measured in response to the glucose challenge. The "RES blocked" rats demonstrated significantly higher

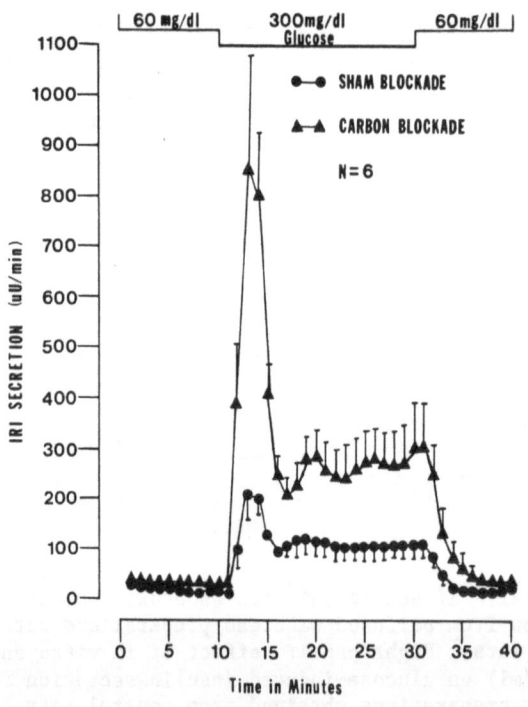

Figure 7. [From Filkins and Yelich, 1982 (20).]

plasma insulin levels than normal rats (20). These studies suggested that pancreases from "RES blocked" rats secreted more insulin in response to a given glucose stimulus than those from control rats. It was then necessary to test the pancreatic insulin secretion directly in vitro (20). Pancreas perfusion studies were thus performed on normal pancreases and those obtained from rats which had received "RES blockade"--either colloidal carbon or lead acetate. Glucose-induced insulin secretion by the pancreases from "RES blockade" rats was significantly greater than that by pancreases from normal rats or rats treated with the vehicle of the RES test substance (Figures 7 and 8). Since endotoxicosis is actually a form of "RES blockade", these experiments suggested that the endotoxin-induced hypersecretory state of the B cells of the pancreas was indirectly the result of "perturbation" of the RES by endotoxin (20).

Since it is now well appreciated that macrophages produce and secrete a variety of mediators, including monokines, which are often linked to shock (16), the question was asked whether monokines were the RES link between endotoxin and the insulin hypersecretory state of the B cell. To answer this question pancreas perfusions were performed wherein insulin secretion was measured from pancreases obtained from animals treated with macrophage-conditioned medium. The conditioned medium was obtained, as described previously for MILA, by culturing peritoneal exudate macrophages in control medium for 24 hours. The conditioned medium, when injected into rats, was shown to induce a state of pancreatic insulin hypersecretion which was qualitatively and quantitatively similar to that induced by endotoxin (20; Figure 9). Since the medium presumably contained a mediator derived from macrophages which potentiated insulin release from the endocrine pancreas, the functional name of MIRA (Macrophage Insulin Release-potentiating Activity) was arbitrarily assigned to the mediator.

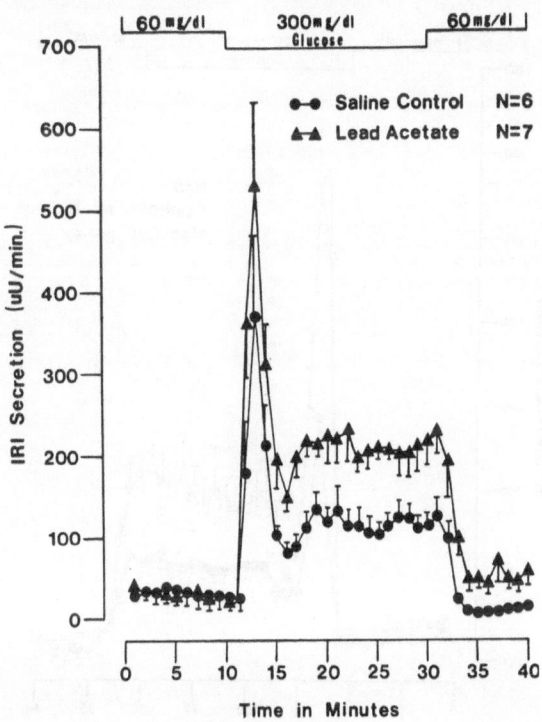

Figure 8. [From Filkins 1985 (16).]

In summary, endotoxicosis can induce alterations in the insulin secretory state of the endocrine pancreas, leading to hyperinsulinemia and altered glucose homeostasis. The effects of endotoxin on the endocrine pancreas appear to be indirect, most likely via the effects of monokines. No direct effects of endotoxin on the secretion of insulin from the endocrine pancreas were noted. These observations provide firm evidence to link the RES to the glucose and insulin dyshomeostasis of endotoxicosis. Tentatively this link is thought to be the regulatory protein products secreted by the macrophages--the monokines (13,15,16).

OVERVIEW

This chapter has discussed glucose dyshomeostasis during endotoxicosis and sepsis from a unique perspective. Figure 10 is a diagramatic representation of the preceding discussions. At the top of the diagram are the primary events of sepsis and endotoxemia. The bottom of the diagram shows just one endpoint of the various metabolic consequences which occur when the processes of sepsis or endotoxicosis proceeds to a pathophysiological conclusion--that is, the development of a profound and progressive hypoglycemia. Playing a central role in the entire process from endotoxemia to hypoglycemia is the reticuloendothelial system (RES). Of particular importance are the hepatic sinusoidal macrophages which phagocytize and detoxify endotoxin. In association with the mission of the macrophages to detoxify endotoxin, there are at least two other fates of endotoxin which contribute adversely to the state of the endotoxic organism. Endotoxin which is processed by the RES can trigger the production and release of mediators from the macrophages, especially regulatory proteins called monokines, which affect primary metabolic processes which control glucose homeostasis.

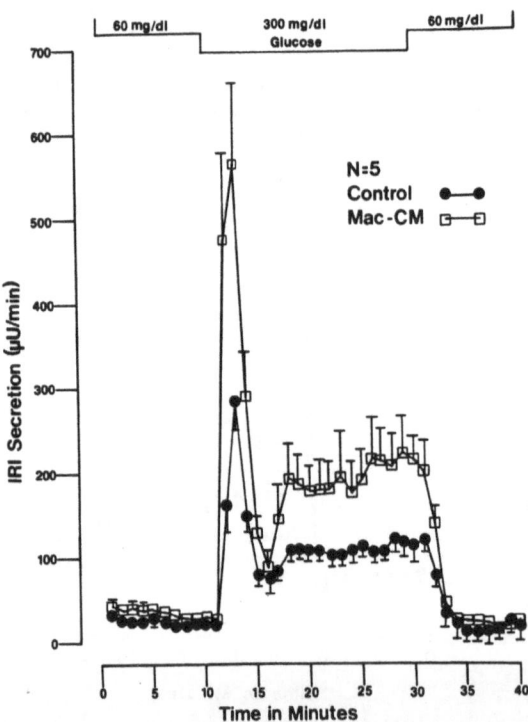

Figure 9. [From Filkins and Yelich 1982 (20).]

128

Endotoxin which escapes processing by the RES can directly affect the metabolic processes of peripheral tissues which are of primary importance for glucose homeostasis.

As discussed in this review, the interaction of endotoxin and macrophages results in the production of MILA (Macrophage Insulin-Like Activity) and MIRA (Macrophage Insulin Release-potentiating Activity). These functional monokines are mediators of the actions of endotoxin. They contribute to the development of a functional hyperinsulinism in the endotoxic animal which ultimately results in glucose dyshomeostasis and profound hypoglycemia. MILA and MIRA appear to act in distinctly different ways. MILA acts directly on periperal tissues to increase the utilization of glucose by those tissues. In contrast, MIRA acts on the β cells of the endocrine pancreas to alter its insulin secretory state, resulting in the hypersecretion of immunoreactive insulin and subsequently systemic hyperinsulinemia. Insulin then acts directly on peripheral tissue to increase glucose utilization and on the liver to inhibit the hepatic processes which produce glucose. Endotoxin which escapes processing by the RES can act directly (mimicking insulin) to increase glucose utilization by peripheral tissues.

Endotoxin thus acts both directly and by way of macrophage-derived mediators to contribute to a functional hyperinsulinism in the endotoxic organism. The primary mechanism by which endotoxin acts to produce glucose dyshomeostasis is by increasing insulin-like activity (immunoreactive insulin, MILS, insulin-like activity of endotoxin) which promotes utilization of glucose by tissues in excess of hepatic production of glucose and thus results in hypoglycemia. Insulin plays an important role in the overall process because not only does it stimulate increased glucose use, but is also inhibits glucose production.

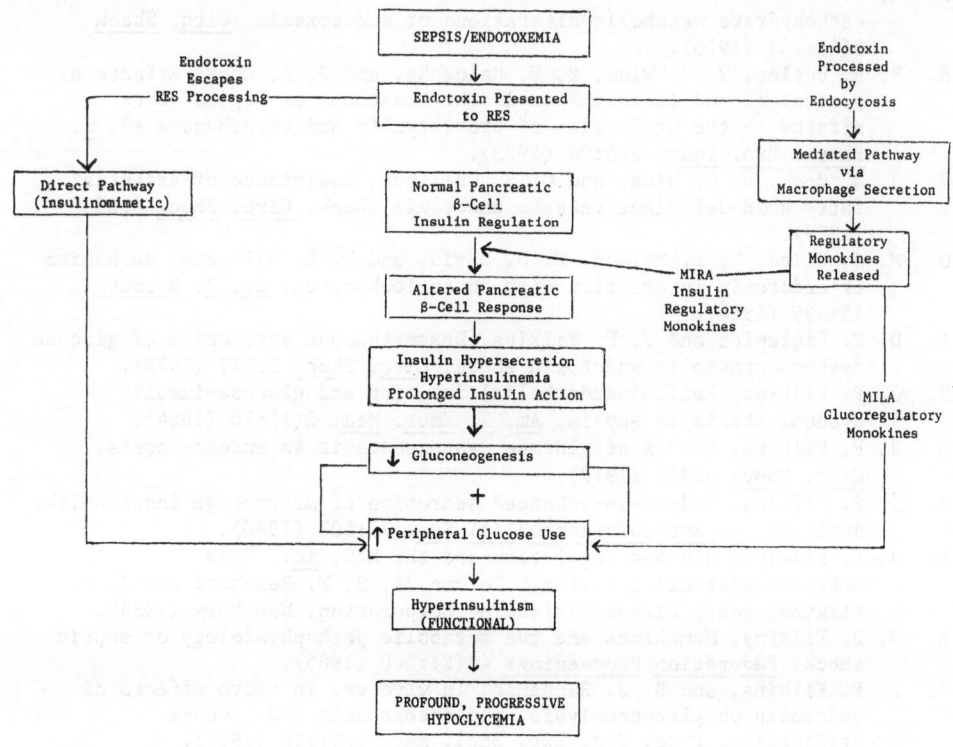

Figure 10.

As elaborated by Lewis Thomas (46), endotoxin is "read by our tissues as the very worst of bad news..." The unabidable information residing in endotoxin is read by our tissues by both a direct process commanding the attention of the cells and by a process which is dependent on the macrophage-derived regulatory proteins--the monokines. The response of the organism to endotoxin does appear to be "a well intentioned, but lethal error" (46).

ACKNOWLEDGMENT

This work was supported in part by National Institutes of Health grants GM 29619, AM 36044, and HL 31163.

REFERENCES

1. B. M. Altura, Reticuloendothelial cells and host defense, Adv. Microcirc. 9:252 (1980).
2. K. V. Anderson, K. L. Bleyl, and W. R. Drucker, Peripheral uptake of glucose during sepsis, Surg. Forum 30:40 (1979).
3. W. R. Beisel and R. W. Wannemacher, Gluconeogenesis, ureagenesis and ketogenesis during sepsis, J. Parent. and Enter. Nutr. 4:277 (1980).
4. L. J. Berry, Metabolic effects of bacterial endotoxins, in: "Microbial Toxins," Vol. 5, S. Kadis, G. Weinbaum, and S. J. Ajl, eds., Acadmic Press, New York (1971).
5. L. J. Berry, D. S. Smith, and L. G. Young, Effects of bacterial endotoxins on metabolism. I. Carbohydrate depletion and the protective role of cortisone, J. Exp. Med. 110:389 (1959).
6. B. J. Buchanan and J. P. Filkins, Insulin secretion and sensitization to endotoxin shock, Circ. Shock 3(3):233 (1976).
7. B. J. Buchanan and J. P. Filkins, Insulin secretion and the carbohydrate metabolic alterations of endotoxemia, Circ. Shock 3(4):267 (1976).
8. R. R. Butler, W. C. Wise, P. V. Halushka, and J. A. Cook, Effects of gentamicin and indomethacin in the treatment of septic shock: effects on the production of prostacyclin and thromboxane A2, J. Pharm. Exp. Ther. 225:94 (1983).
9. J. A. Cook, W. C. Wise, and C. S. Callihan, Resistance of essential fatty acid-deficient rats to endotoxic shock, Circ. Shock 6:333 (1979).
10. M. J. Cline, K. L. Melmon, W. C. Davis, and H. E. Williams, Mechanism of endotoxin interaction with human leukocytes, Br. J. Haematol. 15:539 (1968).
11. D. P. Figlewicz and J. P. Filkins, Dexamethasone antagonism of glucose dyshomeostasis in endotoxin shock, Circ. Shock 5:317 (1978).
12. J. P. Filkins, Reticuloendothelial function and glucose-insulin dyshomeostasis in sepsis, Am. J. Emer. Med. 2(1):70 (1984).
13. J. P. Filkins, Phases of glucose dyshomeostasis in endotoxicosis, Circ. Shock 5:347 (1978).
14. J. P. Filkins, Endotoxin-enhanced secretion of macrophage insulin-like activity, J. Reticuloendothelial Soc. 27:507 (1980).
15. J. P. Filkins, Glucose regulation and the RES, in: "The Reticuloendothelial System," Volume 7A, S. M. Reichard and J. P. Filkins, eds., Plenum Publishing Corporation, New York (1984).
16. J. P. Filkins, Monokines and the metabolic pathophysiology of septic shock, Federation Proceedings 44(2):300 (1985).
17. J. P. Filkins, and B. J. Buchanan, In vivo vs. in vitro effects of endotoxin on glycogenolysis, gluconeogenesis and glucose utilization, Proc. Soc. Exp. Biol. Med. 155:216 (1977).

18. J. P. Filkins and D. P. Figlewicz, Increased insulin responsiveness in endotoxicosis, Circ. Shock 6:1 (1979).

19. J. P. Filkins, L. W. Janusek and M. R. Yelich, Role of insulin and insulin-like activity in the hypoglycemic response to endotoxin, in: "Bacterial Endotoxin and Host Response," M. K. Agarwal, ed., Elsevier-North Holland Biomedical Press, Amsterdam (1980).

20. J. P. Filkins and M. R. Yelich, Mechanism of hyperinsulinemia after reticuloendothelial system phagocytosis, Am. J. Physiol. 242:E115 (1982).

21. J. P. Filkins and M. R. Yelich, RES function and glucoregulation in endotoxicosis, in: "The Reticuloendothelial System and the Pathogenesis of Liver Disease," H. Liehr and M. Grun, eds., Elsevier-North Holland Biomedical Press, Amsterdam (1980).

22. J. R. Fletcher and P. W. Ramwell, Prostaglandins in shock: to give or to block, Adv. Shock Res. 3:257 (1980).

23. D. E. Fry, L. Pearstein, A. Fulton, and H. C. Polk, Multiple system organ failure, Arch. Surg. 115:136 (1980).

24. R. D. Goldfarb, P. Weber and J. E. Eisenman, The effect of leucine on plasma insulin following endotoxin shock in the rat, Circ. Shock 8:343 (1981).

25. M. S. Hand, M. J. Fettman, L. G. Chandrasena, J. L. Cleek, R. A. Mason, and R. W. Phillips, Increased glucose uptake precedes hyperinsulinemia in awake endotoxic minipig, Circ. Shock 11 (4):287 (1983).

26. S. G. Hershey, The reticuloendothelial system: relationship to shock and host defense, in: "Microcirculation," Vol. III, G. Kaley and B. Altura, eds., University Press, Baltimore (1980).

27. L. Hinshaw, L. Archer, B. Beller-Todd, B. Benjamin, D. Flournoy and R. Passey, Survival of primates in lethal septic shock following delayed treatment with steroid, Circ. Shock 8:291 (1981).

28. D. C. Holley and J. A. Spitzer, Insulin action and binding in adipocytes exposed to endotoxin in vitro and in vivo, Circ. Shock 7:3 (1980).

29. N. Kasai and A. Nowotny, Endotoxic glycolipid from a heptose-less mutant of Salmonella minnnesota, J. Bacteriol. 94:1824 (1967).

30. D. L. Kelleher, P. A. Puinno, B. C. Fond, and J. A. Spitzer, Glucose and lactate kinetics in septic rats, Metabolism 31:252 (1982).

31. E. Lachman, S. B. Pitsoe, and S. Gaffin, Anti-lipopolysaccharide immunotherapy in management of septic shock of obstetric and gynegological origin, Lancet May:981 (1984).

32. M. S. Liu, W. M. Long and J. J. Spitzer, Influence of Escherichia coli endotoxin on palmitate, glucose and lactate utilization by isolated dog heart myocytes, in: "Pathophysiological Effects of Endotoxins at the Cellular Level," J. A. Majde and R. J. Person, eds., Alan R. Liss Publishing, New York (1981).

33. D. C. Morrison and D. M. Jacobs, Binding of polymixin B to the lipid A portion of bacterial lipopolysaccharides, Immunochemistry 13:813 (1976).

34. D. C. Morrison and R. J. Ulevitch, The effects of bacterial endotoxins on host mediation system, Am. J. Pathol. 93:527 (1978).

35. J. M. Naylor and D. S. Kronefeld, In vivo studies of hypoglycemia and lactic acidosis in endotoxic shock, Am. J. Physiol. 248:E309 (1985).

36. K. M. Nelson and J. A. Spitzer, Alteration of adipocyte calcium homeostasis by Escherichia coli endotoxin, Am. J. Physiol. 248:R331 (1985).

37. E. A. Newsholme and C. Start, "Regulation in Metabolism," John Wiley and Sons, New York (1973).

38. R. C. Page, P. Davies, and A. C. Allison, The macrophage as a secretory cell, Int. Rev. Cytol. 52:119 (1978).

39. J. R. Parratt and R. M. Sturgess, The effect of indomethacin on the cardiovascular and metabolic responses to E. coli endotoxin in the cat, Br. J. Pharmacol. 50:177 (1974).

40. R. M. Raymond, J. M. Harkema and T. E. Emerson, Mechanism of increased glucose uptake by skeletal muscle during E. coli endotoxin shock in the dog, Circ. Shock 8:77 (1981).

41. W. Schumer, Steroids in the treatment of clinical septic shock, Ann. Surg. 184:333 (1976).

42. W. Schumer, Cellular metabolism in shock, Circ. Shock 2:109 (1975).

43. B. L. Short, M. Gardiner, R. I. Walker, S. R. Jones, and J. R. Fletcher, Indomethacin improves survival in gram-negative sepsis, Adv. Shock Res. 6:27 (1981).

44. J. H. Siegel, Relations between circulatory and metabolic changes in sepsis, Ann. Rev. Med. 32:175 (1981).

45. J. A. Spitzer and D. C. Holley, Alterations in insulin action by endotoxin in vitro, Adv. Shock Res. 2:129 (1979).

46. L. Thomas, "The Lives of a Cell: Notes of a Biology Watcher," Viking Press, New York (1974).

47. E. R. Unanue, Secretory function of mononuclear phagocytes--a review, Am. J. Pathol. 83:396 (1976).

48. K. A. Wichterman, I. H. Chudry, and A. E. Baue, Studies of peripheral glucose uptake during sepsis, Arch. Surg. 114:740 (1979).

49. L. Witek-Janusek and J. P. Filkins, Insulin-like action of endotoxin: antagonism by steroidal and non-steroidal anti-inflammatory agents, Circ. Shock 8:573 (1981).

50. L. Witek-Janusek and J. P. Filkins, Relation of endotoxin structure to hypoglycemic and insulin-like actions, Circ. Shock 11:23 (1983).

51. R. R. Wolfe and J. H. F. Shaw, Glucose and FFA kinetics in sepsis: role of glucagon and sympathetic nervous system activity, Am. J. Physiol. 248:E236 (1985).

52. M. W. Woods, D. Burk, I. Howard, and M. Landy, Insulin-like action of endotoxins in normal and leukemic leukocytes and other tissues, Proc. Amer. Assoc. Cancer Res. 3:279 (1961).

53. M. W. Woods, M. Landy, D. Burk, and T. Howard, Effects of endotoxin on cellular metabolism, in: "Bacterial Endotoxins," M. Landy and W. Braun, eds., Quinn and Boden Publishers, Rahway, New Jersey (1964).

54. M. W. Woods, M. Landy, J. L. Whitby and D. Burk, Metabolic effects of endotoxin on mammalian cells, Bacteriol. Rev. 25:447 (1961).

55. M. R. Yelich and J. P. Filkins, Mechanism of hyperinsulinemia in endotoxicosis, Am. J. Physiol. 239(2):E156 (1980).

56. M. R. Yelich and J. P. Filkins, Insulin secretion and the potentiation of endotoxin shock in the rat, Circ. Shock 9(6):589 (1982).

57. M. R. Yelich and J. P. Filkins, Role for calcium in the insulin hypersecretory state of the endotoxic rat pancreas, Circ. Shock 14:49 (1984).

58. E. J. Ziegler, J. A. McCutchan, J. Fierer, M. P. Glauser, J. C. Sadoff, H. Douglas, and A. I. Braude, Treatment of gram-negative bacteremia and shock with human antiserum to a mutant Escherichia coli, New England Journal of Med. 307(20):1225 (1982).

IN VIVO EFFECTS ON CELLULAR METABOLISM AND CALCIUM DYNAMICS

BY CONTINUOUS INFUSION OF ENDOTOXIN

Judy A. Spitzer

Department of Physiology
Louisiana State University
Medical Center
New Orleans, Louisiana

INTRODUCTION

The pathophysiologic role of endotoxin in clinical sepsis is still incompletely defined and a subject of controversy (50,54). While some of the hallmarks of gram-negative sepsis, e.g., fever, shock, disseminated intravascular coagulation, complement activation, changes in leukocyte count can be simulated by intravenous administration of a bolus of endotoxin, such experimental models fail to reproduce several prominent features of clinical sepsis, e.g., elevated cardiac output and O_2 consumption and prolonged hyperglycemia (50). These observations are consistent with the concept that endotoxin may be important in the pathogenesis of the shock state, but fail to establish a role for endotoxin in the hypermetabolic or "high flow" state of sepsis.

Several years ago, we started out with the premise that valuable new information could be gleaned from an experimental small animal model in which a non-lethal dose of endotoxin was infused continuously over a period of days. Such a design would mimic clinical realities with greater fidelity than does a bolus injection. Also, this approach should permit the expression of a wide range of homeostatic mechanisms which are overwhelmed in lethal models of endotoxic or septic shock. Availability of the Alzet osmotic pump provided the means for continuous intravenous delivery of endotoxin in the conscious, unrestrained rat.

CHARACTERIZATION OF THE MODEL

We have successfully used the 2 ML1 osmotic pump in developing a chronic endotoxemia model in conscious, unrestrained rats (13). This pump has a reservoir volume of 2 ml and delivers reservoir contents at a nominal rate of 10 µl/hour for seven days. The pump consists of a collapsible reservoir with a flexible impermeable wall surrounded by an osmotic agent, all of this being encapsulated by a semi-permeable membrane. When the filled pump is put into an aqueous environment, the osmotic agent imbibes water which generates hydrostatic pressure on the flexible lining of the reservoir, gradually compressing it and producing a constant flow of its contents through the delivery portal.

Male Sprague-Dawley rats weighing 340-400 g were subjected to aseptic surgery under ether anesthesia. An indwelling PE 50 catheter was placed in the left carotid artery to the level of the aortic arch and exteriorized through the skin dorsally. The catheter was flushed daily with sterile heparinized isotonic saline. Blood was sampled and mean arterial blood pressure and heart rate measured through the arterial catheter.

In order to provide for a postsurgical recovery period before the onset of endotoxin delivery, a 100 cm length of PE 60 tubing was fabricated into a coil which was filled with saline and one end of it connected to the osmotic pump, actually encircling it. The other end of the coil was connected to PE 10 tubing, placed in the right jugular vein to the level of the heart. The pump was implanted subcutaneously over the scapulae. The use of the coil delayed delivery of the pump contents into the venous blood by approximately 42 hours.

The model is characterized by a three to four day period of anorexia and morbidity, followed by gradual recovery, with reproducible changes in plasma lactate, leukocyte count and O_2 consumption, not attributable to reduced food intake (13). Changes in plasma lactate and O_2 consumption in the course of continuous endotoxin and NaCl infusion are illustrated in Figures 1 and 2. The cardiovascular status of the animals does not deteriorate below physiological limits, cardiac output is maintained (14), and the nadir of mean arterial pressure is ca. 87 mm Hg.

The experimental design accommodates a wide range of measurements of physiologic, metabolic, and immunologic parameters. The spectrum of observations allows for assessment of function at the cellular, organ and whole body level.

The rest of this chapter will describe the dynamic changes in metabolic function that occur at the level of hepatocytes and adipocytes, respectively.

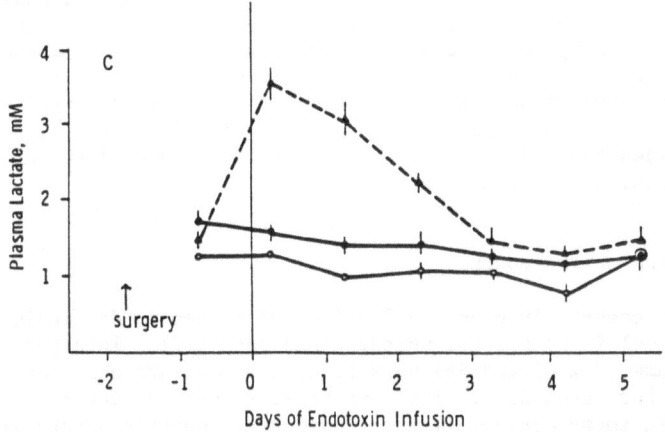

Figure 1. Arterial plasma lactate in fed control (CONTR), food deprived (FD) and endotoxemic (ET) rats; mean ± SEM. Data are presented in relation to onset of endotoxin infusion, which is approximately 42 hr after surgery. (Reproduced by permission of Alan R. Liss, Publisher.)

Hepatocytes

The character of changes in gluconeogenesis associated with endotoxemia is influenced by the point in the time course when the measurements are taken, the severity of the infectious insult and the nature of the techniques used to assess the process. Following endotoxin administration, there is enhanced availability of gluconeogenic precursors in the form of lactate and alanine (53,25); the circulating levels of the gluconeogenic hormones, glucagon and catecholamines are also increased (1,46). Kinetic studies have indeed demonstrated increased rates of glucose appearance in vivo in the initial phases after endotoxin administration (53). A significant decline in gluconeogenesis is apparent preterminally or with the use of highly lethal doses of endotoxin.

Previous in vitro studies of gluconeogenesis in endotoxicosis indicated depressed rates (26,11). However, such investigations usually employed overwhelming doses of endotoxin and/or made no attempt to evaluate the response in a milieu simulating in some respects the conditions likely to prevail in vivo. Exposing rats to a continuous infusion of non-lethal doses of endotoxin permits monitoring of changes in gluconeogenesis (and in other metabolic pathways) under circumstances that do not preclude the expression of adaptive mechanisms.

Rates of gluconeogenesis from lactate and precursors that enter the pathway at the triosephosphate level were measured in hepatocytes isolated from rats after six hours (day 2 postsurgery) and 30 hours (day 3 postsurgery) of continuous E. coli endotoxin (0111:B4 Lipopolysaccharide B; Difco ET) or 0.9% saline (NaCl) infusion from an implanted osmotic pump. The dosage of ET was 0.6 mg/100 g body weight/day (45).

Table 1 illustrates the internal environment with respect to circulating levels of glucose and lactate at these times. After six hours of continuous ET infusion, the plasma glucose and lactate concentrations were significantly higher than corresponding values in animals receiving continuous infusions of physiological saline. Twenty-four hours later, the difference in plasma

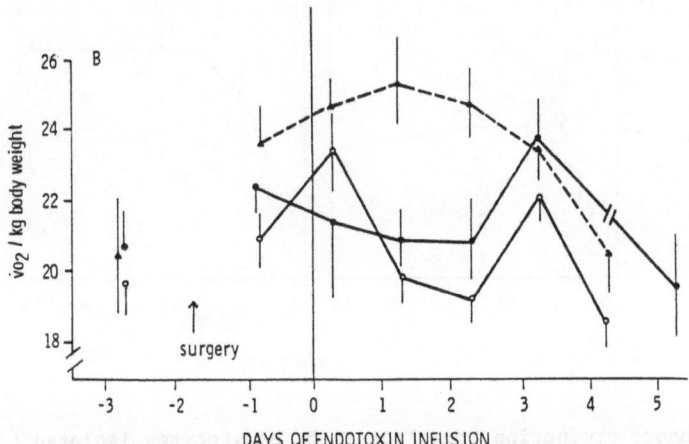

Figure 2. Oxygen consumption VO_2 per kg body weight of fed control (CONTROL), food deprived (FD) and endotoxemic (ET) rats; mean ± SEM. (Reproduced by permission of Alan R. Liss, Pub.)

Table 1. Plasma Glucose and Lactate Concentrations

Glucose ————— mM ————— Lactate

	NaCl	ET	NaCl	ET
Day 2 post-surgery	8.64 ± 0.13 (4)	$9.43^{*} \pm 0.34$ (4)	2.31 ± 0.17 (6)	$3.84^{*} \pm 0.44$ (5)
Day 3	$7.60^{\S} \pm 0.30$ (4)	$7.82^{\S} \pm 0.40$ (6)	2.22 ± 0.29 (4)	$4.29** \pm 0.51$ (6)

$^{*}p < 0.05$; $**p < 0.005$ vs. NaCl of equivalent time point
$^{\S}p < 0.01$ vs. its Day 2 value

glucose concentrations disappeared, while the elevated plasma lactate in endotoxemic rats was maintained. The early hyperglycemia obtained with this model is especially important in terms of the model's relevance to human disease, since human septic patients are generally (28), but not invariably (51), hyperglycemic due to an increased rate of glucose production and severe hypoglycemia is rarely present.

Glucose synthesis from lactate as a function of substrate concentration from NaCl and ET-infused rats after six and 30 hours of continuous infusion is shown in Figure 3. After six hours of continuous infusion, the rate of glucose production by ET cells was 26.4% higher than that by NaCl cells (two-way ANOVA, p < 0.05). At the conclusion of an additional 24 hours of

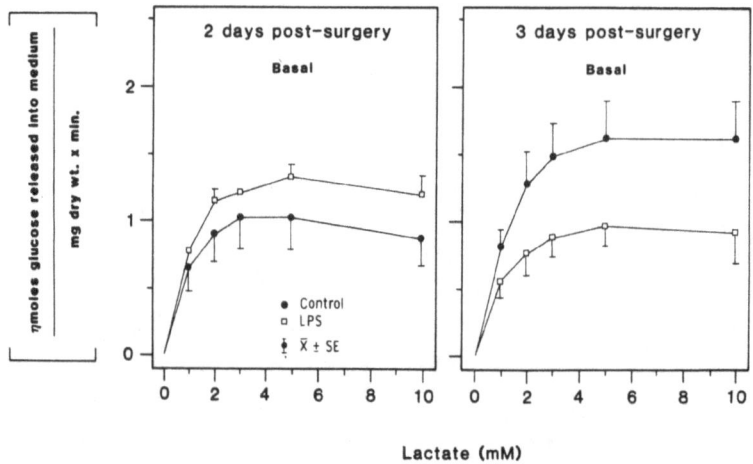

Figure 3. Glucose production from lactate by hepatocytes isolated 6 hrs (2 days postsurgery) and 30 hrs (3 days postsurgery) after onset of continuous endotoxin (LPS) or NaCl (Control) infusion. Each point represents mean ± SEM of 5 experiments on day 2, 6 experiments for control and 7 for endotoxic (LPS) cell preparations on day 3.

continuous infusion, the gluconeogenic rate of ET hepatocytes was depressed (39.6%, p < 0.001). It is important to note, however, that glucose production by ET cells at 4 mM lactate concentration (which corresponds to the circulating plasma level, as shown on Table 1), is not different from that by control cells at 2 mM lactate concentration, which corresponds to the plasma level of NaCl-infused animals at that time. Thus, by incorporating into the medium some of the hormonal and substrate components of the prevailing in vivo environment from which the hepatocytes had been isolated, hepatic glucose production from lactate by ET cells is found to be not depressed below the basal level of control cells.

Effect of Glucagon (0.5 μM) on Glucose Synthesis from Lactate

On day 2 there was no difference in the absolute values of glucose production by the two cell populations; however by day 3 postsurgery ET cells released significantly less (-40.8%, p < 0.001) glucose into the medium than did control cells (Figure 4). The extent of the glucagon-induced increase above basal gluconeogenesis from lactate in ET cells was depressed by 6 h of continuous ET infusion and also at 30 h (Figure 2).

Effect of Norepinephrine (5 μM) on Glucose Production from Lactate

Similar to glucagon stimulation, after six hours of continuous endotoxin infusion there was no significant difference in the quantities of glucose synthesized by NaCl and ET cells responding to norepinephrine stimulation. However, by 30 hours of continuous infusion, the ET cells produced significantly less glucose than cells of NaCl-infused controls (p < 0.001 by two-way ANOVA, Figure 5).

The percentage increase above basal value due to norepinephrine stimulation was significantly reduced in ET cells on day 2. On day 3 there was no further reduction in the responsiveness of the ET cells. However, at this time, the response of the NaCl control cells was also blunted compared with their previous day values, and thus the difference in response between the two types of cells disappeared (Table 2).

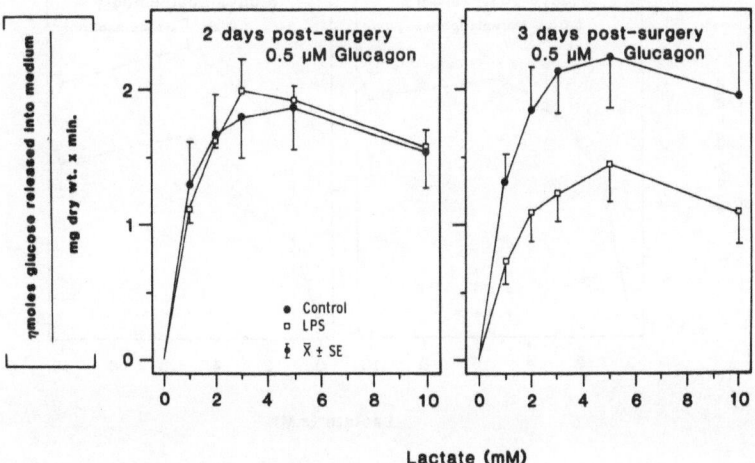

Figure 4. Glucagon-stimulated gluconeogenesis from lactate. Conditions as described for Figure 3.

	Glucagon		Norepinephrine	
	% increase above basal value			
	NaCl	ET	NaCl	ET
Day 2 post-surgery	97.8 ± 17.0	48.7* ± 3.8	71.5 ± 16.6	32.8* ± 2.2
	(5)		(5)	
Day 3 post-surgery	44.0 ± 4.5**	38.8 ± 6.4	26.7 ± 3.6**	32.9 ± 3.5
	(6)	(7)	(6)	(7)

1 – 0.5 µM, 2 – 5.0 µM
Each value represents the mean ± SEM of the number of experiments indicated in parentheses.
* Significantly different from NaCl; $p < 0.025$
**Significantly different from corresponding Day 2 values; $p < 0.025$
The values are derived from the actual rates of gluconeogenesis presented in Figures 3 – 5.

Glucose Production from Triosephosphate Precursors

After six hours of continuous infusion, both types of cell preparations produced glucose at comparable basal rates (Table 3). ET cells responded to glucagon stimulation with significantly more glucose production only with fructose as precursor. After 30 hours of continuous infusion, gluconeogenesis by ET cells was significantly depressed as compared to their previous day values (Table 4).

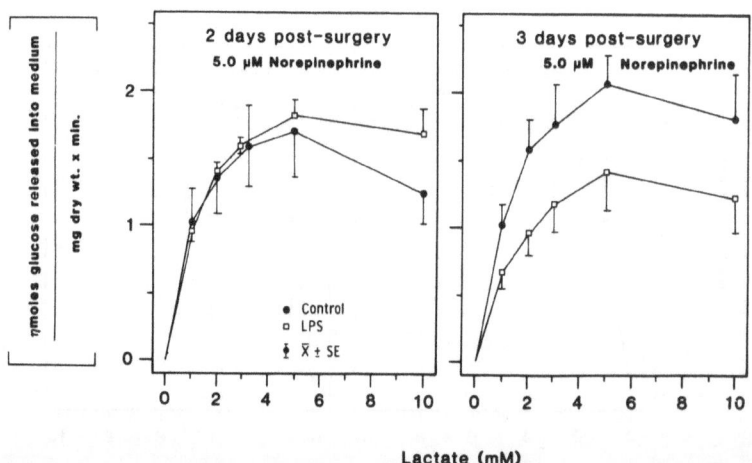

Figure 5. Norepinephrine-stimulated gluconeogenesis from lactate. Conditions as described for Figure 3.

Table 3. Glycerol (G) and Free Fatty Acid (FFA) Release by Adipocytes Isolated from Control and Endotoxemic Rats[a]

	Postsurgery					
	G	FFA	FFA:G	G	FFA	FFA:G
Control						
Basal	0.21 ±.04	0.42 ±.08	2.32 ±.68	0.34 ±.05	0.55 ±.12	1.61 ±.18
+ NE	1.05 ±.16	2.98 ±.44	2.89 ±.19	1.38 ±.10	3.25 ±.42	2.34 ±.28
+NE + I	0.37 ±.07	1.02 ±.14	2.92 ±.31	0.60 ±.06	1.15 ±.13	1.98 ±.20
Endotoxin						
Basal	0.18 ±.04	0.46 ±.09	2.79 ±.53	0.38[c] ±.08	0.54 ±.05	1.75 ±.28
+ NE	1.87 ±.53	5.25 ±1.02	3.06 ±.24	0.82[b,c] ±.14	1.31[b,c] ±.13	1.79 ±.20
+NE + I	0.43 ±.04	1.22 ±.14	2.86 ±.37	0.63 ±.11	0.84[d] ±.08	1.62 ±.25

[a] Data are μmoles FFA or G released/10^6 cells/h (mean ± SEM) under basal conditions and in the presence of norepinephrine (NE; 1.2 μM) and insulin (I; 5 ng/ml).
[b] Endotoxin vs. control; $P < .01$
[c] Day 3 vs. day 2; $P < .05$
[d] Endotoxin vs. control; $P < .06$

By demonstrating perturbations in the gluconeogenic pattern from these precursors, the cytoplasmic compartment of hepatocytes is also implicated as a site of metabolic injury induced by the continuous infusion of endotoxin.

Adipocytes

Lipid in the form of triglyceride is the major storage form of energy in the human body and is stored mainly in adipocytes. When calories are adequate or in excess and the individual is resting, fat accumulates in adipose tissue. On the other hand, periods of fasting, anxiety, or physical exertion are associated with mobilization of free fatty acids (FFA) in considerable quantities from their adipose storage. The released FFA, bound physically to plasma albumin, by which they are transported to various tissues (heart, skeletal muscle, liver, etc.) provide a source of metabolic energy or are incorporated into esterified lipids.

Although some degree of starvation is usually present in sepsis, the hormonal responses and availability of substrates are different during sepsis from those that characteristically occur during simple starvation. While

Table 4. Gluconeogenesis from Precursors Entering the
Pathway at the Triosephosphate Level

		Basal	
		NaCl	ET
Day 2 post-surgery	F	1.60 ± 0.15	1.54 ± 0.22
	S	1.37 ± 0.22	1.25 ± 0.17
	D	1.22 ± 0.16	1.47 ± 0.20
	G	0.58 ± 0.07	0.82 ± 0.12
		(4)	
Day 3 post-surgery	F	1.22 ± 0.29	0.71^b ± 0.24
	S	1.04 ± 0.23	0.64^b ± 0.18
	D	1.03 ± 0.25	0.54^b ± 0.18
	G	0.41 ± 0.14	0.34^b ± 0.08
		(4)	(6)

(nmoles glucose produced/mg. dry wt. min)

The values represent means ± SEM. NaCl and ET
refer to the contents of the pumps. Numbers in
parentheses indicate the number of animals.
b = significantly different from its day 2 value.
b ET basal: F – $p < 0.025$; S – $p < 0.025$;
D – $p < 0.005$; G – $p < 0.01$

serum FFA have been reported to increase during some gram-negative infections
in man (16,2), this lipid component is usually unaltered or decreased during
severe sepsis in other bacterial and viral infections as well as in endotoxemia in man and experimental animals (12,27,35). Alterations in serum
FFA concentrations may reflect changes in the rate of lipolysis and release
from adipocytes and/or utilization by tissues, such as the heart, skeletal
muscle and liver.

Infection-related alterations in lipolysis and tissue utilization of
FFA are variable and are affected by the severity and time course of illness. A bolus injection of endotoxin results in marked alterations in lipid
metabolism, including impaired free fatty acid mobilization from adipose
tissue (41a). Adipocytes isolated from E. coli endotoxin-injected rats,
however, exhibit an increased lipolytic response to norepinephrine, as well
as an increased sensitivity to the antilipolytic action of insulin (43).
The effects of sepsis and endotoxin treatment on adipocyte metabolism of
several species have been extensively investigated in our laboratory
(18-20,41,42).

Evaluation of the lipolytic responses of adipocytes at multiple time
points in the course of a sustained low dose endotoxin infusion affords a
more dynamic view of cellular metabolic events than static measurements
taken usually at one time point following a single high dose of endotoxin.

Table 5. Gluconeogenesis from Precursors Entering the Pathway at the Triosephosphate Level

| | | + Glucagon (0.5 µM) | |
		NaCl	ET
Day 2 post-surgery	F	$2.44^a \pm 0.27$	$2.54^a \pm 0.42$
	S	1.63 ± 0.22	1.57 ± 0.21
	D	$1.85^a \pm 0.22$	1.68 ± 0.11
	G	0.61 ± 0.07	0.80 ± 0.07
		(4)	
Day 3 post-surgery	F	1.97 ± 0.51	$0.82^b \pm 0.24$
	S	1.39 ± 0.36	$0.78^b \pm 0.22$
	D	1.47 ± 0.40	$0.68^b \pm 0.19$
	G	0.40 ± 0.14	$0.36^b \pm 0.11$
		(4)	(6)

(ordinate label: nmoles glucose produced/mg. dry wt. min)

The values represent means ± SEM. NaCl and ET refer
to the contents of the pumps. Numbers in parentheses
indicate the number of animals.
a = significantly different from its corresponding
   basal value
b = significantly different from its day 2 value
a NaCl: F – p < .025; D – p < .05; ET: F – p < 0.05
b ET glucagon: F – p < 0.005; S – p < 0.025; D –
   p < 0.005; G – p < 0.005

Norepinephrine-stimulated lipolysis and the antilipolytic effect of
insulin were studied in adipocytes isolated six hours and 30 hours after
continuous infusion of endotoxin and NaCl respectively. At six hours, stimulation by 1.2 µM norepinephrine resulted in a much more accentuated release
of glycerol and FFA in ET cells than in NaCl cells (12-fold increase over
basal in ET vs. ca. six-fold in NaCl cells). At this time, the endotoxemic
rats were outwardly indistinguishable from controls. At 30 hours, day 3
postsurgery, norepinephrine-stimulated lipolysis was similar to the previous
day's values in control cells, while continuous infusion of ET resulted in
a significant reduction in glycerol and FFA release with respect to both
control cells at the same time point and ET cells the day before (Table 5).

After 30 hours of continuous infusion, the relatively lesser effect of
insulin in diminishing norepinephrine-stimulated lipolysis in ET adipocytes
was reflected in a significantly higher percent residual lipolysis (p <
0.01) as shown in Figure 6. We define residual lipolysis as the amount of
glycerol and FFA released in the simultaneous presence of norepinephrine
and insulin, as compared to the amount released in the presence of the catecholamine alone, and consider this an index of the antilipolytic effect of
insulin on adipocytes. As can be observed from data presented in Table 5

and Figure 6, both the lipolytic effects of norepinephrine and the responsiveness to the antilipolytic effect of insulin were obtunded.

These adipocyte metabolic alterations in the course of continuous,
slow infusion of endotoxin are compatible with observations regarding the
metabolic profile of patients with sepsis (3), for example:

1. Metabolic changes can be documented within a few hours of the
onset of endotoxin infusion, in face of the absence of any clinical signs
of illness;
2. The metabolic responses exhibit biphasic patterns of sequential
change; there is a flip-flopping of hormonal responses.

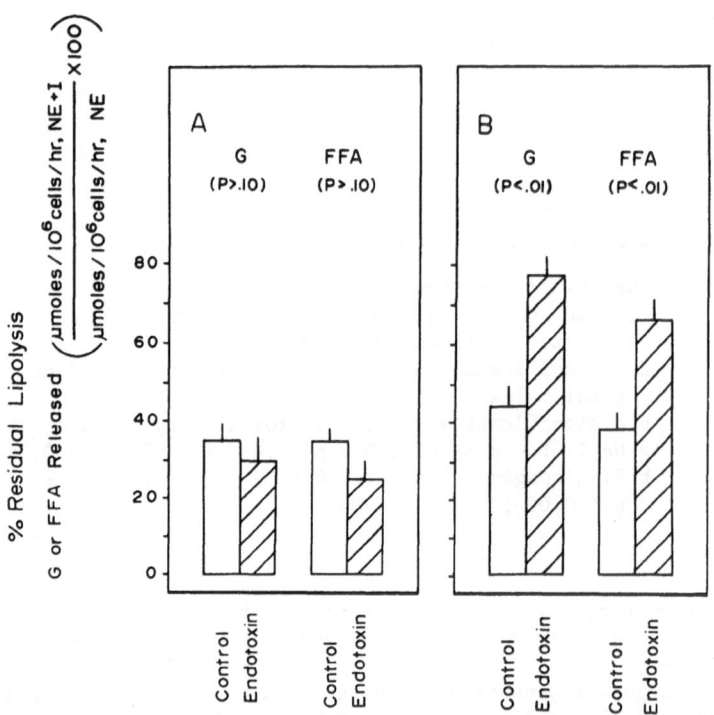

Figure 6. Modulation of the antilipolytic effect of insulin. __Panel A__
Adipocytes isolated from control, i.e., saline-infused and
endotoxin-infused rats, after 6 hrs of continuous infusion
(day 2 postsurgery) were incubated in the simultaneous presence of 1.2 µM norepinephrine (NE) and 5 mg/ml insulin (I),
as well as with NE alone. The lipolysis opposing effect of I
is expressed as % residual lipolysis, i.e., µmoles of
glycerol or FFA released

$$\frac{\text{in the presence of NE + I}}{\text{in the presence of NE}} \times 100$$

(n = 5 experiments for both control and ET-infused rats).
__Panel B__ Same as Panel A, but after 30 hrs of continuous infusion (day 3 postsurgery). (n = 7 for control, 8 for ET-
infused rats.)

The mechanism of enhanced norepinephrine responsiveness at six hours and its suppression at 30 hours is not entirely clear at this time. Circulating levels of glucagon, corticosteroids, ACTH, catecholamines and somatotropin are increased in endotoxemia (4,21), and each or all may elevate the activity and/or synthesis of hormon sensitive lipase.

The decreased norepinephrine responsiveness at 30 hours heralds an in vitro refractory state present at a time when in vivo demand for endogenous fuels was increased, i.e., in an anorectic animal with elevated whole body O_2 consumption (13). This state is also similar to that seen following experimental peritonitis induced in the rat by cecal ligation and puncture (43). The reduced norepinephrine responsiveness at 30 hours of continuous endotoxin infusion is likely to involve desensitization of β-adrenergic receptors induced by the continued presence of agonist (48).

The pattern of changes exhibited by adipocytes shares some important characteristics with those observed in gluconeogenesis by hepatocytes isolated at the same time points in the course of sustained endotoxemia. Adipocyte metabolic responses are manifest within hours of continuous endotoxin infusion, preceding overt clinical signs of illness. Furthermore, the hormonal responses mediating lipolytic adjustments follow a biphasic pattern of sequential change.

Summary

1. Simulation of some clinically observed metabolic events with continuous infusion of non-lethal doses of endotoxin is evidenced by:

a) early hyperglycemia and later absence of hyperglycemia;

b) metabolic responses begin within hours of continuous infusion, well before the onset of any clinical signs of illness; and

c) metabolic responses exhibit biphasic patterns of sequential change.

2. Efficiency of endotoxic hepatocytes to convert lactate to glucose is compromised: when glucose synthesis from lactate by isolated control and endotoxic hepatocytes after 30 hours of continuous infusion is assessed in context of their respective substrate and hormonal environments, comparable rates of glucose production by the two types of cell preparations are revealed.

3. The pattern of changes in NE-stimulated lipolysis in isolated adipocytes is likely to be a reflection of the levels of catecholamines and other stress hormones prevailing in vivo.

CALCIUM HOMEOSTASIS

Several lines of evidence implicate Ca^2 translocations in the cellular mechanisms of action of endotoxin (17,19,20,32,33,36,40). In addition, in recent years it has become increasingly evident that changes in intracellular free Ca^{2+} play a vital role in the mediation of chemical events linking hormonal or other external stimuli to alterations of intracellular enzyme activities (6,10,24,37,38). Specifically, alterations in cellular $[Ca^{2+}]$ have been proposed as essential factors in insulin and catecholamine action in adipocytes (39,29-31,49) and in the regulation by catecholamines and other hormones of hepatic metabolic pathways (5,8,15,22). In our previous work, we related changes in cellular calcium homeostasis to prevailing derangements in the hormonal regulation of some metabolic functions in acute endotoxicosis (34). To explore further whether perturbations in cellular

Table 6. The Effect of Continuous Endotoxin Administration on the Rate of Ca^{2+} Uptake and Ca Content in Isolated Rat Liver Mitochondria

Experimental Conditions	Ca^{2+} Uptake (nmol min^{-1} mg $prot^{-1}$)	Ca Content (nmol mg $prot^{-1}$)
Food restricted, 6 h	196.3 ± 15.8	7.05 ± 0.59
Osmotic pump, SAL, 6 h	184.4 ± 14.0	6.07 ± 0.27
Osmotic pump, ET, 6 h	159.8 ± 25.4	6.32 ± 1.09
Food restricted 30 h	221.5 ± 12.4	6.53 ± 0.13
Osmotic pump, SAL, 30 h	159.3 ± 12.9[a]	6.05 ± 0.31
Osmotic pump, ET, 30 h	164.2 ± 20.4[a]	6.00 ± 0.23
Food restricted, 102 h	139.7 ± 14.6	5.70 ± 0.45
Osmotic pump, SAL, 102 h	74.7 ± 7.0[a]	4.90 ± 0.52
Osmotic pump, ET, 102 h	78.2 ± 12.8[a]	4.83 ± 0.51

The values are means ± SEM for 4 - 8 animals in each group.
a = $p < 0.05$ vs. corresponding food restricted groups.

Ca^{2+} distribution may possibly contribute to the genesis of pathophysiologic patterns associated with endotoxemia and sepsis, we are currently investigating subcellular mechanisms instrumental in the regulation of cellular calcium metabolism. So far, we have focused on Ca^{2+} uptake and content of isolated liver mitochondria and microsomes, and on respiratory activity of mitochondria in endotoxemia and sepsis (44).

In these chronic infusion experiments, male Sprague-Dawley rats received 0.1 mg ET (026:B6)/100 g body weight/day. Mitochondrial and microsomal fractions were prepared 6, 30, and 102 hours after the onset of endotoxin infusion.

We found no difference in the amount of Ca^{2+} taken up by mitochondria of NaCl and ET-infused rats up to 102 hours of continuous infusion, although both groups take up significantly less Ca^{2+} after 30 and 102 hours of infusion, respectively, than their non-operated controls subjected to food restriction to match the voluntary intake of the ET animals at each point (Table 6).

The respiratory control index (RCI), i.e., the ratio of State 3/State 4 respiration, was assessed with succinate and with malate + glutamate. The results of these measurements are presented in Table 7.

The RCI with succinate exhibited an interesting dynamic change. Whereas at six hours of continuous infusion the RCI of ET cells was significantly lower than the corresponding value in mitochondria of NaCl-infused rats, at the 30 hours time point the difference was in the opposite direction, i.e., the RCI of ET cells was significantly higher than that of NaCl cells (Figure 7). At this time, the animals' whole body O_2 consumption was also significantly elevated (13).

In addition to mitochondria, endoplasmic reticulum represents another intracellular site with Ca^{2+}-sequestering and releasing abilities. We observed a significant depression in the Ca^{2+} transport activity of ET microsomes after 102 hours of ET infusion, as compared to microsomes of NaCl-

Table 7. The Effect of Continuous Endotoxin Administration on the Respiratory Control Index (RCI) of Isolated Rat Liver Mitochondria

	RCI with	
Experimental Conditions	Succinate	Mal + Glu
Food restricted, 6 h	3.90 ± 0.25	4.02 ± 0.23
Osmotic pump, SAL, 6 h	4.52 ± 0.40	3.55 ± 0.25
Osmotic pump, ET, 6 h	3.07 ± 0.20	3.97 ± 0.30
Food restricted, 30 h	5.23 ± 0.41	5.02 ± 0.33
Osmotic pump, SAL, 30 h	2.76 ± 0.13^{b}	3.03 ± 0.29^{b}
Osmotic pump, ET, 30h	$3.84 \pm 0.28^{a,b}$	3.43 ± 0.55^{b}
Food restricted, 102 h	4.13 ± 0.26	4.46 ± 0.41
Osmotic pump, SAL, 102 h	4.02 ± 0.31	4.19 ± 0.39
Osmotic pump, ET, 102 h	3.67 ± 0.36	3.53 ± 0.32

The values are means ± SEM for 4 - 9 animals in each group.
a = $p < 0.05$ vs. Osmotic pump, SAL; b = $p < 0.05$ vs. Food restricted
SAL = saline; ET = endotoxin

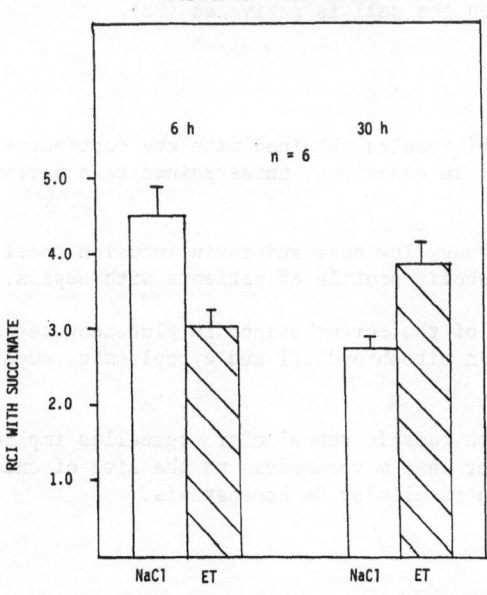

Figure 7. Respiratory control index (RCI) of isolated mitochondria. Mitochondria were prepared from livers of control (NaCl-infused) and endotoxic (ET-infused) rats after 6 and 30 hrs of continuous infusion. Respiration was measured in the presence of 7 mM succinate, in the absence and in the presence of ADP (after 3 separate additions, 0.12 mM each time).

Table 8. The Effect of Continuous Endotoxin Administration on the Activity
of Ca^{2+}-Activated, Mg^{2+}-Dependent ATPase in Rat Liver Microsomes

Experimental Conditions	ATPase Activity (nmol. mg prot.$^{-1}$)
Food restricted, 6 h	87.0 ± 4.8
Osmotic pump, SAL, 6 h	75.5 ± 2.8
Osmotic pump, ET, 6 h	74.2 ± 1.6
Food restricted, 30 h	85.1 ± 1.7
Osmotic pump, SAL, 30 h	82.3 ± 2.9
Osmotic pump, ET, 30 h	76.5 ± 4.0
Food restricted, 102 h	72.0 ± 6.9
Osmotic pump, SAL, 102 h	85.4 ± 10.9
Osmotic pump, ET, 102 h	84.3 ± 8.7

The values are means ± SEM for 4 - 8 animals in each group.

infused rats, and the change occurred in the absence of any concurrent
alteration in the activity of the Ca^{2+}-activated, Mg^{2+}-dependent ATPase
(Table 8). This finding suggests the endoplasmic reticulum, rather than
mitochondria, as the target of endotoxin action in possibly modifying intracellular Ca^{2+} metabolism. It is noteworthy in this context that the
endoplasmic reticulum pool is believed to be the source of Ca^{2+} released
into the cytosol when the cell is activated (38).

CONCLUSIONS

Our experimental results obtained with the continuous, low dose endotoxin infusion model in conscious, unrestrained rats warrant the following
conclusions:

1. The continuous, low dose endotoxin infusion model mimics some key
features of the metabolic profile of patients with sepsis.

2. The nature of the perturbations in gluconeogenesis implicates metabolic lesions both in mitochrondrial and cytoplasmic compartments of isolated
hepatocytes.

3. Results with hepatic subcellular organelles implicate the endoplasmic reticulum, rather than mitochondria as the site of endotoxin-induced
interference with intracellular Ca homeostasis.

REFERENCES

1. M. K. Agarwal and G. Lazar, Metabolic basis of endotoxicosis,
    Microbios. 20:183 (1977).
2. W. R. Beisel and R. H. Fiser, Jr., Lipid metabolism during infectious
    illness, Am. J. Clin. Nutr. 23:1069 (1970).
3. W. R. Beisel, Metabolic response to infection, Ann. Rev. Med. 26: 9-21
    (1975).

4. W. R. Beisel, Alterations in hormone production and utilization during infection, in: "The Physiologic and Metabolic Responses of the Host," M. C. Powanda and P. G. Camonico, eds., Elsevier/North Holland Biomedical Press (1981).

5. P. F. Blackmore, F. Assimacopoulos-Jeannet, T. M. Chan, and J. H. Exton, Studies on α-adrenergic activation of hepatic glucose output: insulin inhibition of α-adrenergic and glucagon actions in normal and calcium-depleted hepatocytes, J. Biol. Chem. 254:2828 (1979).

6. F. L. Bygrave, Mitochondria and the control of intracellular calcium, Biol. Rev. Camb. Philos. Soc. 53:43 (1978).

7. E. Carafoli and M. Crompton, The regulation of intracellular calcium, Curr. Top. Membr. Trans. 10:151 (1979).

8. J. J. Chen, D. F. Babcock, and H. A. Lardy, Norepinephrine, vasopressin, glucagon and A23187 induce efflux of calcium from an exchangeable pool in isolated rat heptocytes, Proc. Natl. Acad. Sci. USA 75:2234 (1978).

9. J. Connar, J. Fine, K. Kusano, M. J. McCrea, I. Parnas, and C. L. Prosser, Potentiation by endotoxin of responses associated with increases in calcium conductance, Proc. Natl. Acad. Sci. 70:3301 (1973).

10. J. H. Exton, Molecular mechanisms involved in α-adrenergic responses, Mol. Cell. Endocrinol. 23:233 (1981).

11. J. P. Filkins and R. P. Cornell, Depression of hepatic gluconeogenesis and the hypoglycemia of endotoxin shock, Am. J. Physiol. 227:778 (1974).

12. R. H. Fiser, J. C. Denniston, and W. R. Beisel, Infection with Diplococcus pneumoniae and Salmonella typhimurium in monkeys: changes in plasma lipids and lipoproteins, J. Infect. Dis. 125:54 (1972).

13. R. E. Fish and J. A. Spitzer, Continuous infusion of endotoxin from an osmotic pump in the conscious, unrestrained rat: a unique model of chronic endotoxemia, Circ. Shock 12:135 (1984).

14. R. E. Fish, A. H. Burns, C. H. Lang, and J. A. Spitzer, Myocardial dysfunction in a nonlethal, non-shock model of chronic endotoxemia, Circ. Shock 14:in press (1985).

15. S. Foden and P. J. Randle, Calcium metabolism in rat hepatocytes, Biochem. J. 170:615 (1978).

16. J. I. Gallin, D. Kaye, and W. M. O'Leary, Serum lipids in infection, New Eng. J. Med. 281:1081 (1969).

17. A. G. Garcia and S. M. Kirpekar, Release of noradrenaline from slices of cat spleen by pre-treatment with calcium, strontium and barium, J. Physiol. 235:693 (1973).

18. I. Kikawyj-Yevich and J. A. Spitzer, Endotoxin influence on lipolysis in isolated human and primate adipocytes, J. Surg. Res. 23:106 (1977).

19. I. Hikawyj-Yevich and J. A. Spitzer, The role of adrenergic receptors and Ca^{2+} in the action of endotoxin on human fat cells, J. Surg. Res. 23:233 (1977).

20. D. C. Holley and J. A. Spitzer, Insulin action and binding kinetics in adipocytes exposed to endotoxin in vitro and in vivo, Circ. Shock 7:3 (1980).

21. D. L. Kelleher, B. C. Fong, G. J. Bagby, and J. J. Spitzer, Metabolic and hormonal changes following endotoxin administration to diabetic rats, Am. J. Physiol. 243:R77 (1982).

22. S. Keppens, J. R. Vandenheede, and H. deWulf, On the role of calcium as second messenger in liver for the hormonally induced activation of glycogen phosphorylase, Biochem. Biophys. Acta. 496:448 (1977).

23. S. M. Kirpekar and Y. Misu, Release of noradrenaline by splenic nerve stimulation and its dependence on calcium, J. Physiol. 188:219 (1967).

24. R. H. Kretsinger, Evolution and function of calcium binding proteins, Int. Rev. Cytol. 46:323 (1976).

25. R. Kuttner and J. J. Spitzer, Gluconeogenesis from alanine in endotoxin-treated dogs, J. Surg. Res. 25:166 (1978).

26. K. LaNoue, A. D. Mason, Jr., and J. Daniels, The impairment of glucoenogenesis by gram negative infection, Metabolism 17:606 (1968).

27. R. S. Lees, R. H. Fiser, S. R. Beisel, and P. J. Bartelloni, Effects of an experimental viral infection on plasma lipid and lipoprotein metabolism, Metabolism 21:825 (1972).

28. C. L. Long, J. M. Kinney, and J. W. Geiger, Nonsuppressability of gluconeogenesis by glucose in septic patients, Metabolism 25:193 (1976).

29. J. M. McDonald, D. E. Bruns, and L. Jarett, Ability of insulin to increase calcium binding by adipocyte plasma membrane, Proc. Natl. Acad. Sci. 73:1542 (1976).

30. J. M. McDonald, D. E. Bruns, and L. Jarett, Ability of insulin to increase calcium uptake by adipocyte endoplasmic reticulum, J. Biol. Chem. 253:3504 (1978).

31. J. M. McDonald and L. Jarett, The effect of epinephrine on calcium handling by adipocyte plasma membranes, endoplasmic reticulum and mitochrondria, Endocrinology 107:1105 (1980).

32. S. C. Moreau and R. C. Skarnes, Host resistance to bacterial endotoxemia: mechanisms in endotoxin-tolerant animals, J. Infect. Dis. 128:S122 (1973).

33. D. C. Morrison, Z. G. Oades, and D. DiPietro, Endotoxin-initiated membrane changes in rabbit platelets, in: "Pathophysiological Effects of Endotoxins at the Cellular Level," J. A. Majde and R. J. Person, eds., Alan R. Liss, Inc., New York (1981).

34. K. M. Nelson and J. A. Spitzer, Alteration of adipocyte calcium homeostasis by E. coli endotoxin, Am. J. Physiol. 248:R331 (1985).

35. H. A. Neufeld, J. A. Pace, and F. E. White, The effect of bacterial infections on ketone concentrations in rat liver and blood and on free fatty acid concentrations in rat blood, Metabolism 25:877 (1976).

36. R. M. Person, Endotoxin alters spontaneous transmitter release at the frog neuromuscular junction, J. Neurosci. Res. 3:63 (1977).

37. H. Rasmussen and D. B. P. Goodman, Relationships between calcium and cyclic nucleotides in cell activation, Physiol. Rev. 57:421 (1977).

38. H. Rasmussen and P. Q. Barrett, Calcium messenger system: an integrated view, Physiol. Rev. 64:938 (1984).

39. R. J. Schimmel, The role of calcium ion in epinephrine activation of lipolysis, Horm. Metab. Res. 8:195 (1975).

40. M. E. Soulsby, F. D. Bruin, T. J. Looney, and M. L. Hess, Influence of endotoxin on myocardial calcium transport and the effect of augmented venous return, Circ. Shock 5:23 (1978).

41. J. A. Spitzer, Endotoxin induced alterations in isolated fat cells: effect on norepinephrine stimulated lipolysis and cyclic 3',5'-adenosine monophosphate accumulation, Proc. Soc. Exp. Biol. and Med. 145:186 (1974).

42. J. A. Spitzer, A. G. B. Kovach, P. Sandor, J. J. Spitzer, and R. Storck, Adipose tissue and endotoxin shock, Acta. Physiol. Acad. Sci. Hung. 44:183 (1973).

43. J. A. Spitzer and D. C. Holley, Alterations in insulin action by endotoxin in vitro, in: "Advances in Shock Research," W. Schumer, J. J. Spitzer, and B. E. Marshall, eds., Alan R. Liss, Inc., New York (1979).

44. J. A. Spitzer, G. J. Leach, and J. A. Reeves, Metabolic and endocrine alterations induced by endotoxin and sepsis at the cellular level, in: "Pathophysiological Effects of Endotoxin at the Cellular Level," J. A. Majde and R. J. Person, eds. (Vol. 62 in "Progress in

Clinical and Biological Research"), Alan R. Liss, Inc., New York (1981).

45. J. A. Spitzer, K. M. Nelson, and R. E. Fish, Time course of changes in gluconeogenesis from various precursors in chronically endotoxemic rats, Metabolism, April (1985).

46. J. J. Spitzer, J. L. Ferguson, H. J. Hirsch, S. Loo, and K. H. Gabbay, Effect of E. coli endotoxin on pancreatic hormones and blood flow, Circ. Shock 7:353 (1980).

47. J. J. Spitzer and J. A. Spitzer, Alterations in carbohydrate and lipid metabolism following administration of endotoxin, Klin. Wochenschrift 60:717 (1982).

48. M. L. Toews and J. P. Perkins, Agonist-induced changes in alpha-adrenergic receptors on intact cells, J. Biol. Chem. 259:2227 (1984).

49. N. Vydelingum, A. H. Kissebah, and W. Wyn, The role of calcium in insulin action, Horm. Metab. Res. 10:38 (1978).

50. K. A. Wichterman, A. E. Baue, and J. H. Chaundry, Sepsis and septic shock: a review of laboratory models and a proposal, J. Surg. Res. 29: 189 (1980).

51. D. W. Wilmore, A. D. Mason, Jr., and B. A. Pruitt, Jr., Impaired glucose flow in burned patients with gram negative sepsis, Surg. Gynecol. Obstet. 143:720 (1976).

52. R. R. Wolfe, D. Elahi, and J. J. Spitzer, Glucose kinetics in dogs following a lethal dose of endotoxin, Metabolism 26:847 (1977).

53. R. R. Wolfe, D. Elahi, and J. J. Spitzer, Glucose and lactate kinetics after endotoxin administration in dogs, Am. J. Physiol. 232:E180 (1977).

54. L. S. Young, P. Stevens, and B. Kayser, Gram-negative pathogens in septicemic infections, Scand. J. Infect. Dis. (Suppl.)31:78 (1982).

ENDOTOXIN (LPS) TOXICITY TO HEPATOCYTES (H) IN VITRO –

MODULATION BY MACROPHAGES (M)

Patricia S. Latham and Susan B. Sepelak

Departments of Medicine and Pathology
University of Maryland Hospital and the
Baltimore Veterans Administration Hospital
Baltimore, Maryland

INTRODUCTION

The liver plays a key role in the systemic response of a body to endotoxin exposure. The liver is not only a important site of metabolism and
metabolic regulation, but it is composed of a large capillary bed rich in
macrophages, the Kupffer cells. This sinusoidal network may contain at any
time up to twenty percent of the circulating blood volume. The liver is
responsible for clearing greater than ninety percent of endotoxin presented
to it as an exogenous bolus (6,10,14), and it most likely also responsible
for intermittently filtering endotoxin arising from the bacterial-infested
gut into the portal vein under various pathological conditions (6,14). It
is demonstrated that blockade of the liver's reticuloendothelial system of
macrophages can potentiate the systemic injury of endotoxin (6). On the
other hand, endotoxin is also known to have a number of effects on macrophages directly, which are capable of producing secondary and possible injurious effects themselves (1,4,11,15). A variety of possible endotoxin-cellu-
lar interactions, particularly those with macrophages, might tip the balance
toward the ultimate benefit or detriment of the organ in which they occur
and the organism at large. The current study explores the endotoxin-macrophage interaction in the context of the liver.

One of the effects of endotoxin on the liver in vivo is known to be
hepatocellular necrosis (2,8,23). It is not known, however, whether this
injury is the result of a direct effect of the endotoxin, or whether it
requires the mediation in some way of the sinusoidal/circulatory cells.
The current study analyzes rat hepatocytes to address this question.

METHODS

Hepatocytes were isolated from nembutal anesthetized, 250-350 g male
SD rats by a standard double-perfusion technique through the portal vein as
described in full by Seglen (21). The first four minute perfusion consisted
of oxygenated Ca++ free Hank's Solution (HBSS) at pH 7.4 to which 0.5 mM
EGTA had been added. The second seven minute perfusion consisted of oxygenated Williams E media also at pH 7.4 which contained calcium and 0.035%
collagenase. The hepatocytes were further washed and separated by differential centrifugation or centrifugal elutriation to result in a population of

cells after plating which was greater than 98% homogeneous and greater than
90% viable. Other cell types were identified on reviewing light and electron
microscopy, and phagocytic cells were identified after incubation with latex
particles and/or iron. The hepatocytes were allowed to incubate in Waymouths
or Williams E media containing 10% FCS/2 μg/ml insulin and containing a
standard antibiotic and fungicidal commercial additive (of 100 μ/ml pen:
100 μg/ml strep:25 μg/ml fungizone). The petri dishes were collagen-coated
with Vitrogen 100 (Collagen, Inc.) and hepatocytes were plated at a density
of 0.5×10^6/ml.

Macrophages were harvested three days after i.p. injection of 3% thioglycollate broth. The resulting cells were greater than 98% viable, greater
than 95% were esterase-positive after 24 hours in culture, and greater than
95% were phagocytic of latex particles. These cells were plated in the
same media described above at a density of 0.5 to 2×10^6/ml, in petri dishes
which did not contain collagen; or the macrophages were co-cultured with
previously plated hepatocytes at a ratio of 2-4:1.

The endotoxin used was commercial E. coli 0128:B8 (phe) obtained from
Sigma and used without further purification. It was added to the 24 hr old
cultures of hepatocytes and macrophages in doses of 0 to 100 μg/ml in the
media previously described, but in the absence of serum.

Hepatocellular injury was sequentially assessed by studying the cellular
morphology on transmission (TEM) and scanning electron microscopy (SEM),
and by serial assessments of hepatocellular death measured by the activity
of lactate dehydrogenase released into the supernatant, ^{51}Cr released from
preloaded hepatocytes, and by uptake of trypan blue.

Lactate dehydrogenase enzyme activity (LDH) was measured spectrophoto-
metrically (Sigma kit). The amount of LDH was measured in the supernatant
and in the cells to obtain a total for each individual sample. The percentage of LDH release was then obtained as a ratio of the amount in the supernatant to the total for that sample. The use of LDH and trypan blue in
co-cultures, however, required some modification for the interpretation of
the results to account for two cell types which, when dead, could not be
adequately separated in the context of the experiment. The hepatocellular-
derived LDH in the supernatant was estimated by subtracting the total amount
of LDH contained in simultaneous macrophage cultures plated at the same
density of cells. This correction attempted to compensate for the inability
to separate the LDH cell-types by assuming that all macrophages had died
and released their LDH. This modification of the calculated LDH-release
should over-estimate the number of dead hepatocytes.

Trypan blue uptake by dead cells was also utilized as a measure of
cell death after exposure of the cells to endotoxin. The cells were exposed
to 0.4% trypan blue with subsequent analysis of 100 cells in triplicate to
obtain the number of dead cells unable to exclude the dye as a percentage
of the total. The cell number was obtained by determining the protein content for each cell culture (protein content was found to have a linear correlation with cell number). In co-cultures, the protein content of the hepatocytes to the cell culture was obtained by subtracting the protein measured
in a simultaneous culture of macrophages at an identical density.

^{51}Cr release from hepatocytes exposed to endotoxin was measured by
pre-loading the hepatocytes for two hours with 2 to 5 μCi ^{51}Cr. The media
was then changed and the macrophages added to those cells which were to be
co-cultured. It was found that the hepatocytes retained ^{51}Cr very well
(150,000 cpms/culture at 24 hrs). Macrophages, on the other hand, accumu-
lated relatively little ^{51}Cr (< 1500 cpm/culture at 24 hrs). The amount of
^{51}Cr released from the hepatocytes was expressed as a percentage of the

total obtained by combining that measured in the supernatant to that measured in the cells.

RESULTS

Hepatocytes did not replicate under the conditions of this experiment, and instead, showed a consistent loss of viability in vitro, accelerated by the absence of serum. In each of the assays used to measure hepatocellular death, the loss of viability in control cultures approximated 15% per 24 hrs. Under the conditions of this experiment total LDH was not seen to significantly increase over time in either hepatocytes or macrophages, whether or not endotoxin was present.

LDH release after exposure of hepatocytes to endotoxin in vitro is expressed in Figure 1. The results demonstrate a dose-dependent decrease in hepatocellular survival with increasing doses of endotoxin. The difference between the treated and control cultures is statistically significant at 10 µg/ml endotoxin and is apparent at 24 hrs.

Thioglycollate-elicited peritoneal macrophages showed a similar loss of viability to endotoxin exposure (Figure 1). The degree of macrophage cell death was similar to that previously reported by other investigators (5,9,17).

The presence of macrophages in co-culture with hepatocytes appeared to inhibit the lethal effect of endotoxin upon the hepatocytes as measured by LDH release (Figure 2); although the results are not significantly different from LPS effect on hepatocytes alone. The lethal cellular effect of endo-

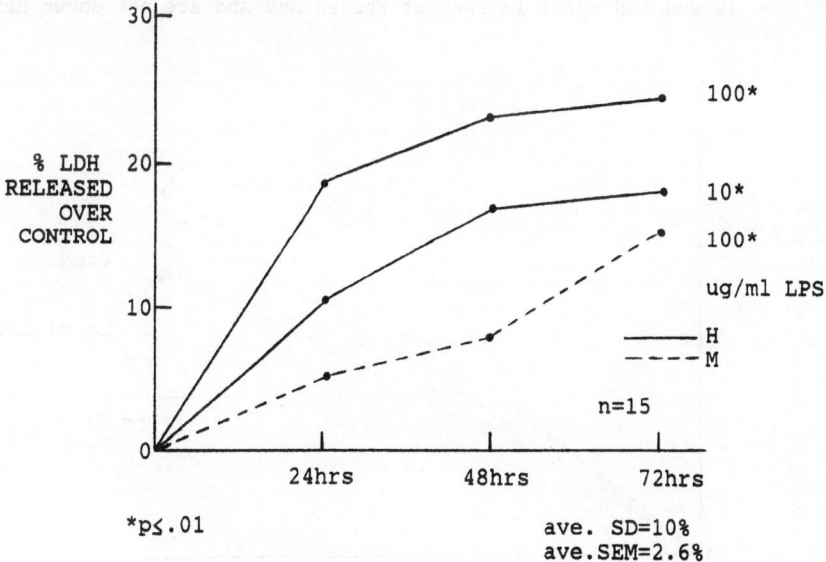

Figure 1. In vitro endotoxin (LPS) effects to rat hepatocytes (H) and macrophages (M). Graph indicates the release of LDH from hepatocytes or macrophages exposed to endotoxin. The data are generated from 6 experiments done in duplicate or triplicate (N=15). The control value obtained in appropriate cell cultures not exposed to endotoxin has been substracted. Note in these experiments that the amount of total LDH contained in hepatocytes is approximately 10-fold that found in macrophages.

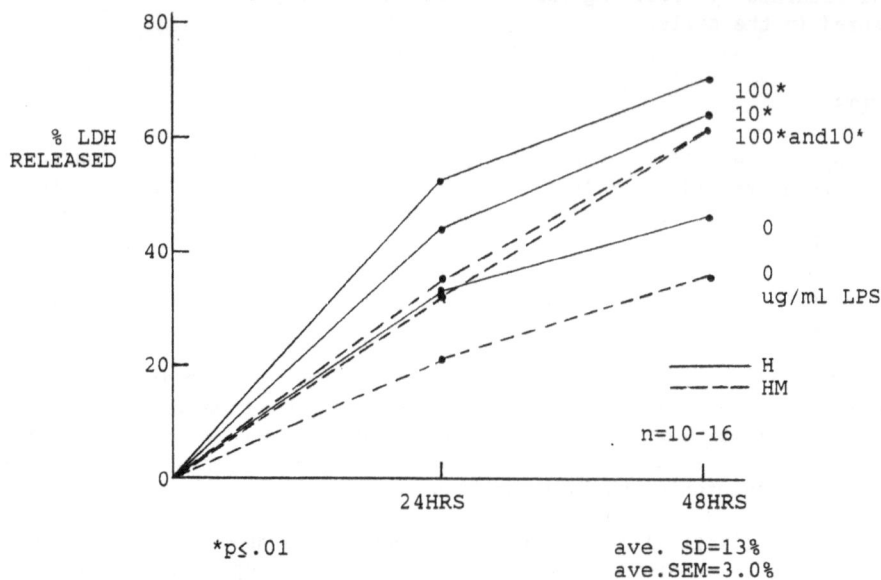

Figure 2. In vitro endotoxin (LPS) effects to rat hepatocytes (H) and
macrophages (M). Shows the LDH release from hepatocytes exposed to endotoxin, includes the results of hepatocyte:macro-
phage co-cultures at a ratio of 2-4:1 macrophages to hepatocytes. Control value is not subtracted. Results represent 6
separate experiments with a total of 16 samples at 0 and 100
µg/ml LPS, and 10 samples at 10 and 50 µg/ml LPS. The results
obtained with 50 µg/ml LPS are intermediate to those between
10 and 100 µg/ml in each of the assays and are not shown here.

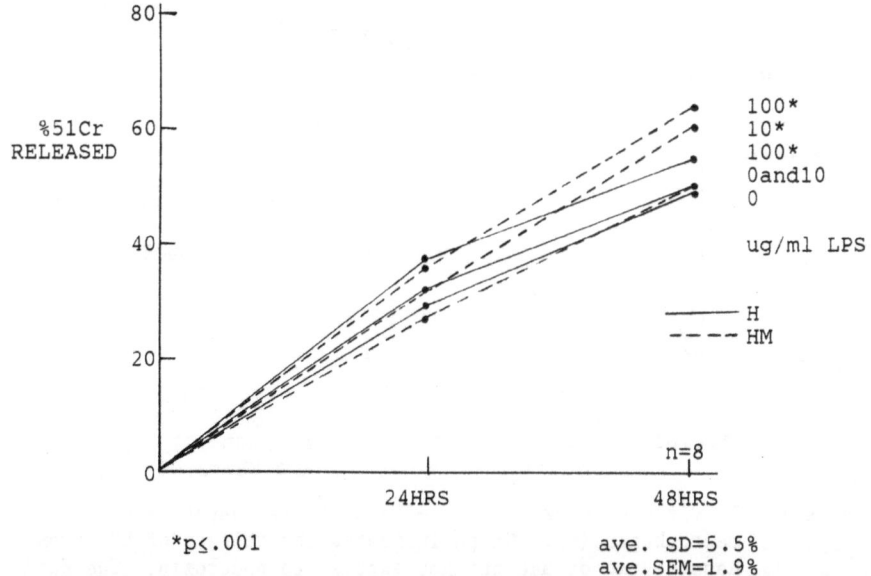

Figure 3. In vitro endotoxin (LPS) effects to rat hepatocytes (H) and
macrophages (M). Indicates release of ^{51}Cr from hepatocytes
and hepatocyte-macrophage co-cultures exposed to endotoxin.
Results are pooled from 2 experiments in quadruplicate total-
ing 8 samples.

toxin as measured by trypan blue uptake analysis (data not shown) appeared
to support the results seen with the LDH assay, as previously described.

When ^{51}Cr was used as the measure of hepatocellular death in response
to endotoxin, the lethal effect of endotoxin on hepatocytes appeared to be
enhanced in the presence of macrophages by a mild, but statistically significant amount (Figure 3).

The morphology of hepatocytes exposed to endotoxin revealed some findings which were not apparent in simultaneous cultures of unexposed control
cells. The findings with endotoxin were similar whether the hepatocytes
were treated alone or in co-culture. The most striking alteration of hepatocytes exposed to endotoxin was a blebbing of the plasma membrane apparent
after 24 hrs (Figures 4 and 5). On TEM these blebs were seen as outpouchings
of the plasma membrane containing few organelles.

TEM revealed only a progressive increase in signs of hepatocellular
distress and then degeneration. These findings were not unique to the endotoxin-treated hepatocytes and also occurred in the control cells; however,
they appeared in a greater proportion of those cells exposed to endotoxin.
A condensation of the mitochondrial matrix and dilatation of the endoplasmic
reticulum were noted among the earliest findings.

DISCUSSION

All of the data obtained in these studies support the conclusion that
endotoxin is hepatotoxic without the need for additional cells such as macro-

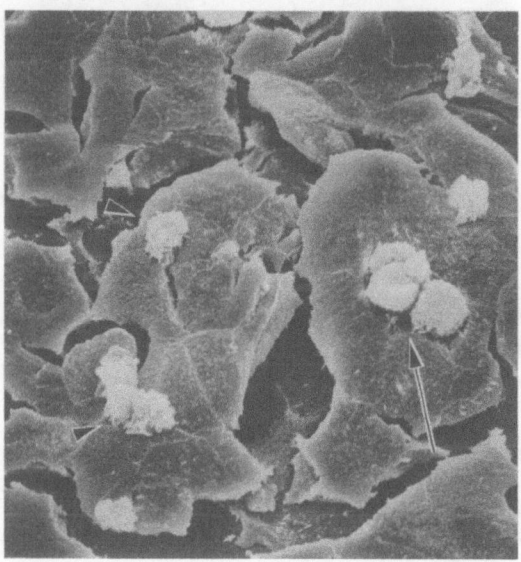

Figure 4.   Representative scanning electronmicrograph of hepatocyte-
macrophage co-cultures at 72 hrs of culture. These cells
serve as the control for endotoxin-treated cells seen in
Figure 5. The hepatocytes are seen as cell plates over the
ground of culture dish. Note the fine microvilli over the
surface of hepatocytes and the paucity of cell debris. A
pair of rounded dead hepatocytes are identified by the
arrow. Macrophages are identified by the arrowheads and are
seen to be rounded with ruffled membranes.

phages to mediate an injurious outcome. In contrast to these results, previous studies of hepatocytes exposed to endotoxin have suggested that no toxic injury occurs (12,13). Several factors may explain why the current experiments demonstrate a toxic effect. The previous studies were not specifically designed to determine the lethal effect of endotoxin upon hepatocytes; therefore, the time of endotoxin exposure in these studies [up to 2 hrs in McGivney's study (12), and without data in Mela's study (13)] was less than the 24 hrs we have found to be necessary to demonstrate a statistically significant lethal effect. In addition, only one type of endotoxin was evaluated in each study (12,13). We and others have found that endotoxins are extremely variable in their actions and potency on cultured cells (5,17,22). Several endotoxins were studied in our laboratory before selecting E. coli 0128:B8 (phe, Sigma) which gave the most demonstrable and reproducible results under the conditions of our experiment.

The findings with endotoxin-treated co-cultures in the current study suggest that macrophages are not necessary to the response of hepatocytes toward endotoxin, but they do appear to modify that response. The nature of the effect of macrophages on the hepatocyte-endotoxin interaction appears to be dependent on the parameter used to analyze the results. The data obtained with ^{51}Cr release is particularly convincing because the hepatocytes were labeled and studied independently of the macrophages, even though the cells were present together in culture. The interpretation of the LDH results using co-cultured cells, however, may be misleading because of the LDH released from hepatocytes must be estimated (refer to methodology). In support of the LDH data were the results obtained by measurement of trypan blue uptake; although the calculations in the trypan blue assay also included an

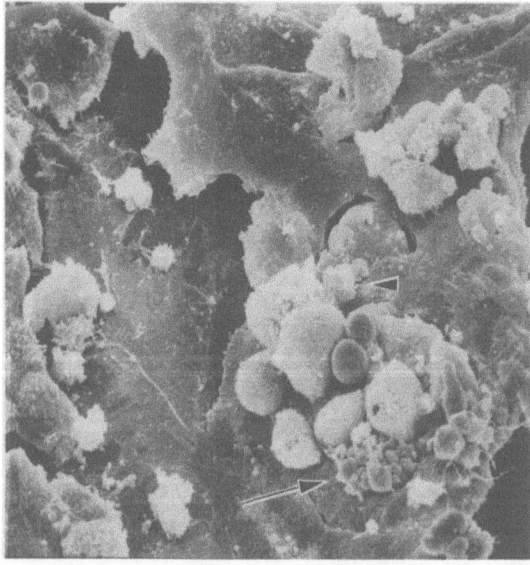

Figure 5. Representative electronmicrograph of a hepatocyte-macrophage co-culture under the same culture conditions as cells seen in Figure 4 (72 hrs of culture) but treated with endotoxin (100-500 µg/ml) for the last 48 hrs of culture. Note increase in dead heaptocytes and cellular debris over control culture. Arrow identifies numerous smooth-membraned vesicles having the same texture as plasma membrane of hepatocytes. These are hepatocullular blebs. Arrowhead identifies the smooth, ruffled membrane of a macrophage.

estimate in calculating the total viable and nonviable cell number. The explanation for the discrepancy in the results is not clear at this time, but may possibly be explained by the technical difficulties described in the analyses by LDH and trypan blue.

The sublethal injury of hepatocytes by endotoxin includes a condensation of the mitochondrial matrix as one of the early signs of toxic injury, and a striking increase in the formation of plasma membrane blebs. Other investigators have also noted an early cellular effect of endotoxin upon mitochondrial morphology and function (7,12). It has been suggested that the mitochondrion may be the initial target site of cellular injury due to endotoxin (3,7). Mitochondrial condensation, however, is also one of the earlier features of normal cellular degeneration of hepatocytes in primary culture; therefore, it is difficult in this context to attach much significance to the observation. On the other hand, the plasma membrane blebbing which was seen in hepatocytes was much more extensive than that seen at any time in control cultures. Plasma membrane blebbing is itself a nonspecific response of hepatocytes to a variety of toxins. In some cases, blebbing may be the result of injury to the cytoskeleton as occurs with cytochalasin B, or it may be an effort of the cell to rid itself of intracellular toxin or damaged organelles by localizing them to an area of the cell which can then be exteriorized and discarded, so-called "focal cytoplasmic degeneration" (18). Finally, blebbing may be an expression of injury to the plasma membrane itself. Endotoxin has been demonstrated to bind to hepatocyte plasma membranes (16,19); and it has also been described to cause injury to isolated membranes (20). Further studies are necessary to determine the significance of blebbing in the context of endotoxin injury to the hepatocyte.

REFERENCES

1. M. K. Agarwal and G. Lazar, Metabolic basis of endotoxicosis, Microbios. 20:183 (1977).

2. R. K. Boler and A. J. Bibighaus, Ultrastructural alterations of dog livers during endotoxin shock, Lab Inv. 17:537 (1967).

3. S. G. Bradley, Cellular and molecular mechanisms of action of bacterial endotoxins, Ann. Rev. Microbiol. 33:67 (1979).

4. C. G. Crafton, and N. R. DiLuzio, Relationship of reticuloendothelial functional activity to endotoxin lethality, Am. J. Physiol. 217:736 (1969).

5. L. M. Glode, A. Jacque, S. E. Mergenhagen, and D. L. Rosenstreich, Resistance of macrophages from C3H/HeJ mice to the in vitro cytotoxic effects of endotoxin, J. Immunol. 119:162 (1977).

6. M. Grun, H. Liehr, and U. Rosenack, Significance of endotoxemia in experimental galactosamine hepatitis in the rat, Acta. Hepatogastroenterol. 23:64 (1976).

7. L. Kilpatrick, M. Smith, and I. A. Silver, Early cellular responses in vitro to endotoxin administration, Circ. Shock 8:585 (1981).

8. D. W. Lee, C. S. Kim, Y. B. Lee, and D. S. Kim, A study of hepatic injury induced by endotoxin in rats, Yonsei Med. J. 19:19 (1973).

9. R. V. Maier and R. J. Ulevitch, The response of isolated hepatic macrophages (H-Mo) to lipopolysaccharide (LPS), Circulatory Shock 8:165 (1981).

10. R. V. Maier, J. C. Mathieson, and R. J. Ulevitch, Interactions of bacterial lipopolysaccharides with tissue macrophages and plasma lipoproteins, in: "Pathophysiological Effects of Endotoxins at the Cellular Level," J. A. Majde and R. J. Person, eds., Alan R. Liss, Inc., New York (1981).

11. R. E. McCallum, Hepatocyte-Kupffer cell interactions in the inhibition of hepatic gluconeogenesis by bacterial endotoxins, in: "Pathophysiological Effects of Endotoxins at the Cellular Level," J.

A. Majde and R. J. Person, eds., Alan R. Liss, Inc., New York (1981).

12. A. McGivney and S. G. Bradley, Susceptibility of mitochondria from endotoxin-resistant mice to lipopolysaccharide, Proc. Soc. Exp. Bio. Med. 163:56 (1980).

13. L. Mela, L. V. Bacalzo, and L. D. Miller, Defective oxidative metabolism of rat liver mitochondria in hemorrhagic and endotoxin shock, Am. J. Physiol. 220:571 (1971).

14. J. P. Nolan and D. S. Camara, Endotoxin, sinusoidal cells, and liver injury, in: "Progress in Liver Diseases," H. Popper and F. Schaffner, eds., Grune and Stratton, New York (1982).

15. M. J. Pabst and R. B. Johnston, Increased production of superoxide anion by macrophages exposed in vitro to muramyl dipeptide or lipopolysaccharide, J. Exp. Med. 151:101 (1980).

16. R. Pagani, M. T. Portoles, and A. M. Municio, The binding of Escherichia coli endotoxin to isolated hepatocytes, FEBS Letters 131:103 (1981).

17. D. L. Peavy, R. E. Baugn, and D. L. Musher, Strain dependent cytotoxic effects of endotoxin for mouse peritoneal macrophages, Infect. Immun. 21:310 (1978).

18. M. J. Phillips and P. S. Latham, Electron microscopy of human liver disease, in: "Diseases of Liver," 5th Edition, L. Schiff and E. R. Schiff, eds., Lippincott, Philadelphia (1982).

19. G. Ramadori, U. Hopf, S. Meyer, and K. H. Buschenfelde, Binding sites for endotoxic lipopolysaccharide on the plasma membrane of isolated hepatocytes, Acta. Hepatogastroenterol. 26:368 (1979).

20. B. G. Schuster, R. F. Palmer, and R. S. Aronson, The effect of endotoxin on thin bi-layer membranes, J. Membrane Biol. 3:67 (1970).

21. P. O. Seglen, Preparation of isolated rat liver cells, in: "Methods in Cell Biology," D. M. Prescott, eds., Academic Press, New York (1976).

22. I. A. Silver, Some effects of Escherichia coli endotoxin on cells in culture, in: "Pathophysiological effects of endotoxins at the cellular level," J. A. Majde and R. J. Person, eds., Alan R. Liss, Inc., New York (1981).

23. R. R. White, L. Mela, L. V. Bacalzo, K. Olofsson, and L. D. Miller, Hepatic ultrastructure in endotoxemia, hemorrhage, and hypoxia: emphasis on mitochondrial changes, Surgery 73:525 (1973).

EFFECTS OF ENDOTOXIN ON MIXED FUNCTION OXIDASE ACTIVITY

Joseph F. Williams and Andor Szentivanyi

Department of Pharmacology and Therapeutics
University of South Florida College of Medicine
Tampa, Florida

Over the last several years there has been considerable effort channeled toward development of immunotherapeutic agents. Great strides have been made in the understanding of the immune system, and the elaboration of a multitude of mediators elicited by stimulation of the various cellular components in response to various immunomodulators. However, it is also appreciated that these immunomodulators or the mediators elicited by them may possess activity on cellular systems other than those of the immune network. Understanding these non-immune effects is of major importance for the future clinical application of these agents.

One of the effects that many of the immunomodulators share in common is to decrease the activity of the hepatic microsomal mixed-function oxidase system. This enzyme system is responsible for the biotransformation of many drugs, other xenobiotics, and certain endogenous substances including steroids, fatty acids and prostaglandins. The activity of this drug-metabolizing enzyme system is important for the inactivation of many diverse therapeutic agents as well as for the activation of pro-drugs, such as cyclophosphamide. Therefore, it is obvious that modification of this enzyme system by immunoactivators could result in significant changes in the toxicity or therapeutic efficacy of pharmacological agents used concurrently during immunotherapy.

Of the various immunomodulators shown to alter hepatic mixed-function oxidase activity, endotoxin is one of the most potent. In keeping with the orientation of this book, discussion will be limited primarily to our work and that of other investigators concerning the alteration of drug-metabolizing activity by endotoxin. However, it should be pointed out that many of the results to be presented with respect to the effect of endotoxin are similar to those obtained with other immunomodulating substances. This apparent similarity may indicate a common mechanism that is yet to be established.

A brief overview of the salient features of the hepatic mixed-function oxidase system may be appropriate for those who may be unfamiliar with this research area.

Figure 1 illustrates the key components of the hepatic mixed function oxidase system. NADPH-cytochrome P-450 reductase (previously referred to as NADPH-cytochrome c reductase) functions in the reduction of oxidized

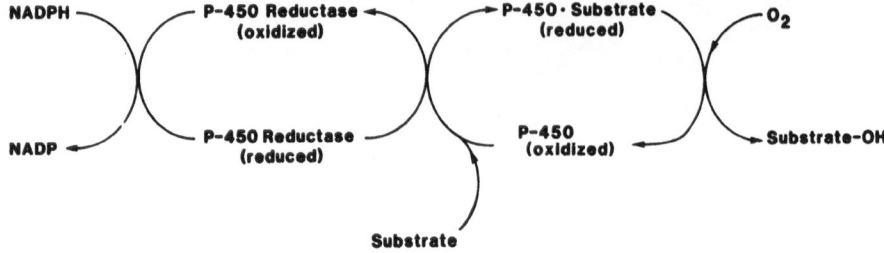

Figure 1. A brief schematic representation of the hepatic microsomal
mixed-function oxidase system.

cytochrome P-450, the terminal oxidase. Substrate binding occurs with the
oxidized P-450 and upon reduction the P-450 drug complex combines with
molecular oxygen. Subsequent steps result in the insertion of one atom of
oxygen into the drug-product, the other oxygen atom appears in water, and
the regeneration of oxidized P-450. The term cytochrome P-450 refers collectively to a family of hemoproteins which in their reduced form will combine with carbon monoxide eliciting an absorption spectra whose peak occurs
at about 450 nm. The currently recognized multiple isozymic forms of P-450
offer some partial explanation for the wide variety of substrates shown to
be biotransformed by this system. It should be mentioned that comparable
mixed-function oxidase systems have been shown to be present in tissues other
than the liver. Although the activity of the systems shown to be present in
the intestine, skin, lung, brain, and lymphocytes, just to name a few, is
generally considerably lower than that of the liver, they are, nonetheless,
of interest because of their possible involvement in the regulation of local
drug concentrations (6).

Figure 2 is a brief schematic presentation of our current understanding
of the regulation of the synthesis and degradation of P-450. The apoprotein
is synthesized on the endoplasmic reticulum, and combines with heme to form
the holo-enzyme. The heme may exist in a preformed heme pool or be provided
by increased activity of δ-aminolevulinic acid (ALA) synthetase, the rate
limiting step in heme biosynthesis. Many compounds have been shown to induce
the synthesis of cytochrome P-450 and the associated drug-metabolizing enzyme
activities (20). Both an increase of RNA and protein synthesis are stimulated by the inducing agents and induction of δ-ALA synthetase is also

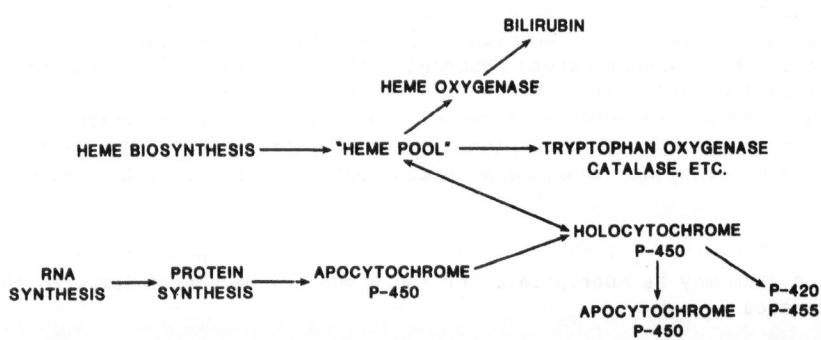

Figure 2. A model of the regulation of synthesis and degradation
of liver cytochrome P-450.

160

Table 1. Effect of Heated and Unheated B. pertussis vaccine on
Hepatic Mixed-Function Oxidase Activity

Treatment	Cytochrome P-450	Ethylmorphine N-Demethylase	Aniline Hydroxylase Activity
	(nmole/mg)	(μmole HCHO/hr/ mg protein)	(nmole pAP/hr/ mg protein)
Control	1.11 ± 0.18	0.555 ± 0.062	50.25 ± 3.22
Unheated Vaccine			
Day 1	0.67 ± 0.08	0.224 ± 0.034*	36.07 ± 2.23*
Day 5	0.60 ± 0.08	0.338 ± 0.027*	30.41 ± 2.21*
80°-Heated Vaccine			
Day 1	0.69 ± 0.01	0.241 ± 0.018*	36.69 ± 1.17*
Day 5	1.02 ± 0.07	0.530 ± 0.058	46.91 ± 3.66

*Significantly diffrent ($P < 0.05$) from control value. Each value repre-
sents mean ± S.E. for three to six experiments with pooled livers from
animals per group.

frequently seen. Less is known about the turnover of cytochrome P-450.
Estimates of half-life based on the heme moiety have been determined and
values of 36-48 hours have been reported. However, it is not known whether
there is a coordinated turnover of both the heme and apoprotein. The observation of Gasser et al. (12) would suggest that biologically there is a coordinated loss of both moieties. On the other hand, some agents apparently
can cause the dissociation of heme from the holocytochrome, leading to an
increase in the free heme pool, without a loss of the apoprotein. Apparent
consequences of the increase in the free heme pool concentration are (a) a
repression of δ-ALA synthetase activity; (b) an increased heme saturation of
other heme-requiring enzymes and; (c) an induction of heme oxygenase activity, the rate limiting step of heme catabolism to bilirubin. In addition,
other agents have been shown to cause a suicidal destruction of P-450 during
metabolism with the formation of spectrally altered inactive forms of the
cytochrome.

Our interest in the effect of endotoxin on cytochrome P-450 dependent
drug oxidation reactions grew out of the observation that animals inoculated
with Bordetella pertussis vaccine had markedly reduced drug-metabolizing
activity (22,23). Subsequently, it has been found that (Table 1) the effect
of B. pertussis was apparently due to two bacterial components, one that was
labile while the other was stable to heat of 80° for 0.5 hr. Table 2 shows
that injection of mice with endotoxin caused a loss of mixed-function oxidase
activity to a similar extent and duration as seen with the heated vaccine
(29,31). Other investigators (3,11,13,21) had previously observed that administration of endotoxin to mice and rats significantly decreased hepatic
drug-metabolizing activity and the level of cytochrome P-450.

Bissell and Hammaker (3,4,5) showed that concomitant with the effect of
endotoxin on mixed-function oxidase activity there was an increase in the
heme saturation of tryptophan oxygenase activity, an induction of heme oxygenase activity, and a decrease in δ-ALA synthetase activity. We have observed similar effects (25) as well as an effect of endotoxin to apparently

Table 2. Effect of E. coli 026:B6 Endotoxin on Hepatic Mixed Function Oxidase Activity

Treatment	Cytochrome P-450	Ethylmorphine N-Demethylase Activity	Aniline Hydroxylase Activity
	(nmole/mg)	(μmole HCHO/hr/mg protein)	(nmole pAP/hr/mg protein)
Control	1.14 ± 0.05	0.579 ± 0.057	53.18 ± 1.41
Endotoxin (10 μg)			
Day 1	0.73 ± 0.04*	0.249 ± 0.041*	43.00 ± 1.78*
Day 5	1.21 ± 0.08	0.583 ± 0.055	62.76 ± 4.98

*Significantly different (P < 0.05) from control value. Values represent mean ± S.E. of four experiments with pooled livers from five animals per group.

inhibit the induction of cytochrome P-450 by phenobarbital (30). These results are consistent with the interpretation that administration of endotoxin causes a dissociation of heme from cytochrome P-450, thereby reducing the detectable levels of the hemoprotein. The released heme is available for utilization by tryptophan oxygenase or by increasing the "free heme pool" to induce heme oxygenase activity. In addition, endotoxin may either block the synthesis of apoprotein caused by phenobarbital or affect the ability of the produced apoprotein to combine with heme.

The C3H/HeJ mouse strain has been shown to be genetically resistant to many of the biological effects of endotoxin and has been used extensively to explore the mediator aspects of endotoxin actions. It was therefore of interest to investigate the effect of endotoxin on the hepatic mixed-function oxidase system of this endotoxin-resistant mouse strain (32). Results of

Table 3. Mixed Function Oxidase Activity in C3H/HeJ and C3H/HeN Mice Injected with E. coli 026:B6 Endotoxin (10 μg)

	Cytochrome P-450	Ethylmorphine N-Demethylase Activity	Aniline Hydroxylase Activity
	(nmole/mg)	(μmole HCHO/hr/mg protein)	(nmole pAP/hr/mg protein)
C3H/HeN			
− endotoxin	0.81 ± 0.05	0.254 ± 0.021	84.31 ± 1.02
+ endotoxin	0.52 ± 0.02*	0.100 ± 0.012	50.45 ± 3.04
C3H/HeJ			
− endotoxin	0.76 ± 0.04	0.227 ± 0.020	87.90 ± 3.88
+ endotoxin	0.53 ± 0.03*	0.149 ± 0.021*	73.05 ± 6.01

*Significantly different (P < 0.05) from control value. Each value represents mean ± S.E. of five experiments with pooled livers from three animals per group.

Table 4. Effect of Acute Challenge Doses of LPS on Mixed-Function
Oxidase System of Control and LPS-Tolerant Rats

	LPS Dose	Aniline Hydroxylase Activity	Ethylmorphine N-Demethylase Activity	Cytochrome P-450
	(mg/kg)	(nmole pAP/hr/ mg protein)	(μmole HCHO/hr/mg protein)	(n mole/ mg protein)
Control	0	35.24 ± 2.20	0.393 ± 0.024	0.86 ± 0.03
	2	10.00 ± 0.92*	0.113 ± 0.010*	0.52 ± 0.02*
	5	14.83 ± 1.74	0.167 ± 0.020*	0.54 ± 0.02*
	10	Lethal		
	20	Lethal		
LPS- Tolerant	0	33.80 ± 3.32	0.224 ± 0.041*	0.74 ± 0.05*
	2	29.00 ± 2.68	0.214 ± 0.028	0.68 ± 0.03
	5	30.41 ± 2.10	0.309 ± 0.040	0.68 ± 0.03
	10	24.57 ± 1.99*	0.303 ± 0.036	0.63 ± 0.02*
	20	12.43 ± 1.68*	0.107 ± 0.025*	0.59 ± 0.02*

*Significantly different ($P < 0.05$) from appropriate control value. Each
value represents the mean ± S.E. for four animals.

this study (Table 3) showed that the C3H/HeJ mouse strain was almost as sensitive to endotoxin as the endotoxin sensitive C3H/HeN strain. Some attenua-
tion, particularly with respect to aniline hydroxylase activity, of the
effect of endotoxin did occur in the C3H/HeJ mice, but the effects of endotoxin in the two mouse strains on ethylmorphine N-demethylase activity and
cytochrome P-450 levels were approximately equivalent. At the time this was
somewhat surprising and disappointing because we had hoped to be able to use
the C3H/HeJ animals to investigate the role of possible mediators in the
mechanism of LPS action.

Since many physiological and immunological effects of LPS have been
attributed to the release of endogenous substances our attention was directed
to examine whether pharmacological antagonists of adrenergic amines and prostaglandins might ameliorate these effects of LPS (24,25). However, pretreatment of animals with alpha or beta adrenergic antagonists or indomethacin
did not alter the ability of LPS to decrease microsomal drug metabolizing
activity nor to induce heme oxygenase. The effect of LPS was also not prevented by dexamethasone. Metyrapone and SKF-525A, inhibitor of P-450 activity, have been shown to protect against the loss of mixed-function oxidase
activity caused by other agents (7,16). However, LPS was still effective in
reducing P-450 levels and enzyme activities in animals pretreated with these
inhibitors (unpublished observations).

Abernathy et al. (1) reported that mice made tolerant to LPS had a
longer hexobarbital sleep time and zoxazolamine paralysis time than control
animals. However, when LPS-tolerant mice were challenged with LPS there was
no further change in these in vivo correlates of microsomal mixed-function
oxidase activity, suggesting tolerance to the effect of LPS had been established. Recently we have examined the microsomal enzymatic activity of mice
and rats made tolerant to LPS and subsequently challenged with LPS (26,28).
Rats were injected with 0.01 mg, 0.1 mg, and 1.0 mg LPS on three consecutive

Table 5. Effect of Injection of LPS (2 mg/kg) on Heme Oxygenase
          Activity of Control and LPS-Tolerant Rats

	Heme Oxygenase Activity
	(nmole bilirubin/mg protein/10 min.)
Control	0.13 ± 0.02
Acute LPS	0.70 ± 0.08*
LPS-Tolerant	0.32 ± 0.09
LPS-Tolerant and Acute LPS	0.26 ± 0.07

*Significantly different (P < 0.05) from control value. Each
value represents the mean ± S.E. for nine animals.

days as described by Nolan and Ali (15). Two days after the last dose, animals received either isotonic, nonpyrogenic saline or a challenge dose of
LPS. Nontolerant animals received saline or LPS. Enzyme activities were
determined 24 hours later (Table 4). Tolerant animals had significantly
less ethylmorphine N-demethylase activity than control animals, but aniline
hydroxylase activity and cytochrome P-450 levels were approximately equivalent for these two groups. Administration of LPS to control animals caused
a marked decrease in mixed-function oxidase activity and was lethal at doses
of 10 mg/kg and 20mg/kg. In contrast, challenge doses of LPS to tolerant
animals caused no significant change in the enzyme activity until doses of

Table 6. Mixed-Function Oxidase Activity of Control and LPS-Tolerant
          Rats Injected with Serum Obtained from Rats Given LPS[+]

	Aniline Hydroxylase Activity (nmole pAP/hr/mg protein)		Cytochrome P-450 (nmole/mg protein)	
	Control	Tolerant	Control	Tolerant
Control Serum	38.85±1.56	35.56±1.51	0.91±0.06	0.67±0.05
2 hr LPS Serum	27.14±4.24*	36.30±2.02	0.64±0.01*	0.69±0.10
4 hr LPS Serum	31.99±1.77*	36.55±0.99	0.73±0.04*	0.57±0.08
8 hr LPS Serum	20.96±2.87*	36.56±1.59	0.75±0.03*	0.74±0.06
12 hr LPS Serum	29.58±2.59*	37.39±1.82	0.73±0.06*	0.72±0.07

[+]Blood collected via cardiac puncture at the indicated times after injection
of 5 mg/kg LPS to untreated rats. Control and LPS-tolerant rats were
injected i.p. with 1.0 ml of the post-LPS serum and sacrificed 24 hr later.
Tabular values are mean ± SE of four to five animals per group.
*Significantly different (P < 0.05) from value obtained after injection of
control serum.

Table 7. Mixed-Function Oxidase Activity of Control and LPS-Tolerant
Mice Injected with Serum Obtained from Mice Given LPS[+]

	Aniline Hydroxylase Activity (nmole pAP/hr/mg protein)		Cytochrome P-450 (nmole/mg protein)	
	Control	Tolerant	Control	Tolerant
Control Serum	65.06±3.44	63.11± 5.44	1.05±0.03	0.71±0.02
3 hr LPS Serum	47.79±3.99*	67.97±11.12	0.84±0.04*	0.60±0.06
6 hr LPS Serum	49.30±4.08*	67.29± 2.10	0.77±0.04*	0.73±0.08

[+]Control and LPS-tolerant mice injected with 0.6 ml of serum collected from
mice 3 hr or 6 hr after injection of 5 mg/kg LPS. Tabular values are mean
± SE of six animals/group.
*Significantly different (P < 0.05) from control serum values).

10 mg/kg were administered, doses lethal for control animals. These results
suggest that rats can be made at least partially tolerant to the effects of
LPS on mixed-function oxidase activity. Table 5 shows that the effect of
LPS to induce heme oxygenase activity is also attenuated by induction of
LPS-tolerance. Comparable results were also obtained in mice made tolerant
to LPS (data not shown).

A major question we wished to explore was whether the effects of LPS on
hepatic drug metabolizing activity was a direct effect, or like many other
hepatic responses to LPS, was mediated by other endogenously released
factors. As mentioned earlier, C3H/HeJ animals are not useful for this
purpose, but the results with tolerant animals afforded an intriguing
alternative. The possibility of an indirect effect of LPS was suggested by
experiments using isolated hepatic parenchymal cells incubated for up to
eight hours with 400 mg/ml of LPS. No significant loss of P-450 occurred
during this time period. These results are consistent with those of other
investigators who, using isolated hepatic parenchymal cells, have failed to
demonstrate direct effects of LPS on hepatic metabolism similar to those
seen after in vivo LPS administration (2,9,10).

Egawa et al. (8) reported that injection of mice with serum obtained
from LPS-treated animals caused a loss of cytochrome P-450. The serum factor
was reported to be heat stable. However, since these investigators used
untreated mice we considered the possibility that the effect they observed
might be due to residual, circulating serum LPS (26,27). We have used untreated and LPS-tolerant mice and rats to look for a serum factor from LPS-
treated animals (Tables 6 and 7). Mice were injected with LPS and serum was
collected at various times and injected intraperitoneally into control animals and animals rendered tolerant according to the previously described
protocol. Injection of serum from control animals had no significant effect
on hepatic mixed-function oxidase activity. Serum from LPS-treated animals
when injected in untreated animals caused a marked loss in drug-metabolizing
activity and in the level of cytochrome P-450. In contrast, injection of
the LPS-serum into tolerant animals had no significant effect. These
results suggest that the serum factor reported by Egawa et al. (8) may have
been LPS itself.

Table 8. Mixed-Function Oxidase Activity in Control and LPS-Tolerant Mice Injected with Acute Challenge Dose of LPS[+]

	Aniline Hydroxylase Activity (nmole pAP/hr/ mg protein)	Ethylmorphine N-Demethylase Activity (μmole HCHO/hr/ mg protein)	Cytochrome P-450 (nmole/ mg protein)
Saline	33.98 ± 4.05	0.181 ± 0.019	0.55 ± 0.03
Mφ Medium	35.71 ± 2.61	0.268 ± 0.025	0.57 ± 0.02
Mφ Medium + added LPS	31.69 ± 0.36	0.183 ± 0.020	0.57 ± 0.03
Mφ -LPS Medium	20.23 ± 4.40	0.114 ± 0.017*	0.40 ± 0.06*

[+]Tolerant animals were injected i.p. with 1.0 ml of culture medium from rat peritoneal macrophages incubated for 24 hr in the absence (Mφ Medium) or presence (Mφ -LPS Medium) of 100 μg/ml LPS. Mφ - Medium to which (100 μg/ml) was added just prior to injection was also tested. Tabular values are mean ± SE of four animals per group.
*Significantly different (P < 0.05) from the appropriate control value.

The macrophage is well-known as a target for endotoxin and releases a number of secretory products upon exposure to LPS. We have recently (26,27) shown that injection of LPS-tolerant animals with supernatant fluids from cultures of peritoneal macrophages incubated with LPS will depress hepatic microsomal mixed-function oxidase system (Tables 8 and 9). These results suggest that a macrophage-derived product and not LPS per se may be the ultimate effector. The implication for such a macrophage product is

Table 9. Mixed Function Oxidase Activity of LPS-Tolerant Mice Injected with Medium Obtained after 24 hr Culture of Mouse Peritoneal Macrophage with LPS[+]

	Aniline Hydroxylase Activity (nmole pAP/hr/ mg protein)	Ethylmorphine N-Demethylase Activity* (μmole HCHO/hr/ mg protein)	Cytochrome P-450 (nmole/ mg protein)
Saline	52.52 ± 6.71	0.341 ± 0.025	0.68 ± 0.07
Mφ Medium	53.92 ± 5.03	0.293 ± 0.025	0.59 ± 0.03
Mφ Medium + added LPS	44.80 ± 5.23	0.220 ± 0.010*	0.53 ± 0.02
Mφ -LPS Medium	31.87 ± 3.27*	0.143 ± 0.010*	0.38 ± 0.01*

[+]LPS-tolerant animals were injected i.p. with 0.5 ml of culture medium obtained from mouse peritoneal macrophages incubated for 24 hr in the absence (Mφ Medium) or presence (Mφ - Medium) of 100 μg/ml LPS. Mφ - Medium to which LPS (100 μg/ml) was added just prior to injection was also tested. Tabular values are mean ± SE for five animals per group.
*Significantly different (P < 0.05) from appropriate value.

certainly not unprecedented. Many investigators have obtained similar conclusions with respect to the interplay of macrophage-derived factors and various hepatic metabolic activity.

At present the identity of the factor(s) is unknown. Endotoxin shares with many immuno-active compounds the activity to depress hepatic mixed-function oxidase. Thus, besides B. pertussis and muramyl dipeptide that we have studied, other bacterial immunoadjuvants such as BCG and C. parvum, have been shown to depress drug-metabolizing enzyme activity. In addition, many interferon inducing agents, such as tilerone, poly I - poly C, pyran copolymer, as well as interferon itself, have been shown to cause similar depression of hepatic mixed-function oxidase system (for review see 28). Indeed, it has been suggested that the common denominator in the effect of all the above named agents, including LPS, may be their ability to cause induction of interferon (17). However, certain observations may moderate this hypothesis. First, LPS is not a very good interferon inducer and yet is one of the more potent with respect to causing the decrease in drug-metabolizing activity. Second, our macrophage culture supernatant fluids did not have significant interferon titers. Renton and Peterson (18) have recently reported that incubation of hepatic parenchymal cells with Kupffer cells actively phagocytizing dextran resulted in a decrease in hepatic drug-metabolizing activity. As in our studies, these investigators did not detect interferon to be present. Finally, incubation of isolated hepatic parenchymal cells with interferon resulted in an increase, not a decrease, in the drug-metabolizing activity (14,19).

Figure 3 is a schematic presentation suggesting the possible sequence of events involved in the effect of endotoxin, and probably other immuno-active agents, on the hepatic mixed-function oxidase system. Many steps are yet to be unraveled. For example, how does the interaction of the putative mediators with the hepatic parenchymal cell cause the dissociation of heme from the holocytochrome P-450? Of course, the identity of the mediator needs to be established. Is the loss of mixed-function oxidase activity secondary to other effects of endotoxin on hepatic function and does the perturbation of hepatic heme metabolism play an important regulatory role in the activity of other hepatic enzyme activities?

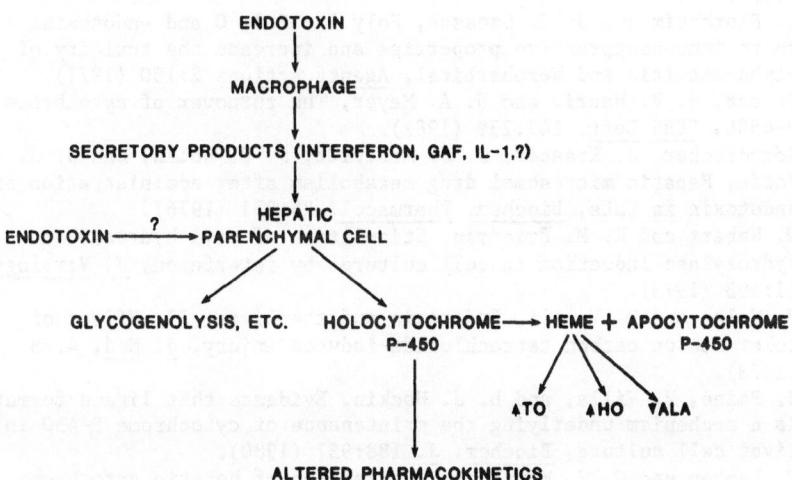

Figure 3. A tentative model of the possible effects of endotoxin in the alteration of the hepatic cytochrome P-450 system.

From a therapeutic point of view the possible alteration in hepatic drug-metabolizing activity caused by endotoxin and other immunoactive substances may be extremely important. Such an effect on this important pharmacokinetic parameter may influence the efficacy of these agents used as immunomodulators in conjunction with other therapeutic agents. In addition, the present observations appear to indicate a unique interaction between the immune system and hepatic metabolic activity not previously recognized.

REFERENCES

1. C. O. Abernathy, H. J. Zimmerman, and R. Utili, Effects of endotoxin tolerance on in vivo drug metabolism in mice, Res. Commun. Chem. Path. Pharmacol. 29:193 (1980).
2. L. J. Berry, Bacterial toxins, CRC Crit. Rev. Toxicol. 5:238 (1977).
3. D. M. Bissell and L. E. Hammaker, Cytochrome P-450 heme and the regulation of hepatic heme oxygenase activity, Arch. Biochem. Biophys. 176:91 (1976a).
4. D. M. Bissell and L. E. Hammaker, Cytochrome P-450 heme and the regulation of δ-aminolevulinic acid synthetase in the liver, Arch. Biochem. Biophys. 176:103 (1976b).
5. D. M. Bissell and L. E. Hammaker, Effect of endotoxin on tryptophan pyrrolase and delta-aminolevulinate synthetase: evidence for an endogenous regulatory haem fraction in rat liver, Biochem. J. 166:301 (1977).
6. M. D. Burke and S. Orrenius, Isolation and comparison of endoplasmic reticulum membranes and their mixed function oxidase activities from mammalian extrahepatic tissues, Pharmacol. Therap. 7:549 (1979).
7. G. Drummond, D. Rosenberg, and A. Kappas, Metal induction of haeme oxygenase without concurrent degradation of cytochrome P-450, Biochem. J. 202:59 (1982).
8. K. Egawa, M. Yoshida, and N. Kasai, An endotoxin-induced serum factor that depresses hepatic -aminoleuvulinic acid synthetase activity and cytochrome P-450 levels in mice, Microbiol. Immunol. 25:1091 (1981).
9. J. P. Filkins and B. J. Buchanan, In vivo vs. in vitro effects of endotoxin on glycogenolysis, gluconeogenesis, and glucose utilization, Proc. Soc. Exp. Biol. Med. 155:216 (1977).
10. J. P. Filkins and R. P. Cornell, Depression of hepatic gluconeogenesis and the hypoglycemia of endotoxin shock, Am. J. Physiol. 277:778 (1974).
11. G. L. Florsheim and J. J. Szeszak, Poly I - Poly C and endotoxins share immunosuppressive properties and increase the toxicity of alpha-amanitin and hexobarbital, Agents Actions 2:150 (1971).
12. R. Gasser, H. P. Hauri, and U. A. Meyer, The turnover of cytochrome P-450b, FEBS Lett. 147:239 (1982).
13. R. Gorodischer, J. Krasner, J. J. McDevitt, J. P. Nolan, and S. J. Yaffe, Hepatic microsomal drug metabolism after administration of endotoxin in rats, Biochem. Pharmacol. 25:351 (1976).
14. D. W. Nebert and R. M. Friedman, Stimulation of aryl hydrocarbon hydroxylase induction in cell cultures by interferon, J. Virology 11:193 (1973).
15. J. P. Nolan and M. V. Ali, Endotoxin and the liver. II. Effect of tolerance on carbon tetrachloride-induced injury, J. Med. 4:28 (1973).
16. A. J. Paine, P. Villa, and L. J. Hockin, Evidence that ligand formation is a mechanism underlying the maintenance of cytochrome P-450 in rat liver cell culture, Biochem. J. 188:937 (1980).
17. K. W. Renton and G. J. Mannering, Depression of hepatic cytochrome P-450 dependent monooxygenase systems with administered interferon inducing agents, Biochem. Biophys. Res. Commun. 73:343 (1976).

18. K. W. Renton and T. G. Peterson, Depression of cytochrome P-450-dependent drug biotransformation in hepatocytes after the activation of the reticuloendothelial system by dextran sulfate, J. Pharmacol. Exp. Therap. 229:299 (1984).

19. K. W. Renton, L. B. DeLoria, and G. J. Mannering, Effects of polyriboinosinic acid polyribocytidylic acid and a mouse interferon preparation on cytochrome P-450-dependent monooxygenase systems in cultures of primary mouse hepatocytes, Mol. Pharmacol. 14:672 (1978).

20. R. Snyder and H. Remmer, Classes of hepatic microsomal mixed function oxidase inducers, Pharmacol. Therap. 7:203 (1979).

21. H. Vainio, Defective drug metabolism in rat liver in endotoxin shock, Ann. Med. Exp. Biol. Fenn. 51:65 (1973).

22. J. F. Williams and A. Szentivanyi, Effect of Bordetella pertussis vaccine on the drug-metabolizing enzyme system of mouse liver, Fed. Proc. 34:1001 (1975).

23. J. F. Williams and A. Szentivanyi, Depression of hepatic drug-metabolizing enzyme activity by B. pertussis vaccination, Eur. J. Pharmacol. 43:281 (1977).

24. J. F. Williams and A. Szentivanyi, Possible involvement of alpha-adrenergic mechanisms and reticuloendothelial activation in the effects of bacterial lipopolysaccharide on hepatic enzyme activities, Fed. Proc. 41:1722 (1982).

25. J. F. Williams and A. Szentivanyi, Investigation of adrenergic and prostaglandin influences in the endotoxin alteration of hepatic heme oxygenase, microsomal mixed-function oxidase, and glucocortiocoid-induced tryptophan oxygenase activities, Immunopharmacol. 6:75 (1983a).

26. J. F. Williams and A. Szentivanyi, Effect of endotoxin in endotoxin-tolerant animals on mixed function oxidase and heme oxygenase activities, Pharmacologist 25:218 (1983b).

27. J. F. Williams and A. Szentivanyi, Induction of tolerance in mice and rats to the effect of endotoxin to decrease the hepatic microsomal mixed-function oxidase system. Evidence for a possible macrophage-derived factor in the endotoxin effect, Int. J. Immunopharmacol. (1985a).

28. J. F. Williams and A. Szentivanyi, Pharmacokinetic and pharmacodynamic parameters affected by RE cell activators, in: "The Reticuloendothelial System: A Comprehensive Treatise," Vol. VIII: "The Pharmacology of the Reticuloendothelial System," J. W. Hadden, and A. Szentivanyi, eds., Plenum Publishing, New York (1985b).

29. J. F. Williams, S. Lowitt, and A. Szentivanyi, Depression of hepatic drug-metabolizing activity by Bordetella pertussis, Ann. Allergy 38:376 (1977).

30. J. F. Williams, S. Lowitt, and A. Szentivanyi, Effect of endotoxin and phenobarbital on heme enzymes of rat liver, Pharmacologist 21:232 (1979).

31. J. F. Williams, S. Lowitt, and A. Szentivanyi, Involvement of a heat-stable and heat labile component of Bordetella pertussis in the depression of the murine hepatic mixed-function oxidase system, Biochem. Pharmacol. 29:1483 (1980a).

32. J. F. Williams, S. Lowitt, and A. Szentivanyi, Endotoxin depression of hepatic mixed-function oxidase system in C3H/HeJ and C3H/HeN mice, Immunopharmacol. 2:285 (1980b).

MYOCARDIAL PERFORMANCE AND ADRENERGIC MODULATION OF

CYCLIC AMP FOLLOWING ENDOTOXIN ADMINISTRATION

Raymond E. Shepherd[1], Charles H. Lang[1], Brent A. Brumfield[1],
Norman W. Robie[2], Karen R. DuSapin[2], Kathleen H. McDonough[1]

Departments of Physiology[1] and Pharmacology and Experimental
Therapeutics[2], Louisiana State University Medical Center
New Orleans, Louisiana

INTRODUCTION

Myocardial contractility may change in response to altered circulatory demands imposed by physiological stimuli. Many studies have explored the behavior of cardiac contraction during shock of various etiologies. In the final phase of circulatory shock, clinical signs of circulatory insufficiency are indicated by low peripheral perfusion, low mean arterial blood pressure, high heart rate, and central venous congestion. A point is reached in the progression of the syndrome where fatal cardiovascular collapse occurs.

Over the last few decades myocardial function has been investigated during hemorrhage, endotoxemia and sepsis. Impaired cardiac function, as indicated by lower left ventricular stroke work for a given central venous pressure, was found in septic patients (55). The dysfunction became more severe as sepsis progressed from the hyperdynamic to the hypodynamic state. In addition, right heart function was also decreased in hypodynamic septic patients (29,61). Winslow et al. (62), however, reported that septic patients failed to demonstrate any overt signs of myocardial dysfunction although a diminished cardiac reserve was seen. This was evidenced by the absence of a significant increase in either cardiac output or stroke work despite an increase in left ventricular filling pressure following volume loading.

An intrinsic defect in mechanical function, independent of peripheral alterations, has also been observed in experimentally induced hyperdynamic sepsis (40). Using an isolated perfused working rat heart preparation, myocardial performance was severely impaired as indicated by a relatively flat left atrial filling pressure versus myocardial work curve. The increased heart rate and the apparent capacity to maintain adequate cardiovascular status in vivo, coupled with the depressed function in vitro, are consistent with a greater reliance of the heart upon sympathetic stimulation in sepsis to maintain adequate cardiac output and arterial blood pressure. Although endotoxin administration is not synonymous with septic shock, endotoxicosis has been proposed as an animal model of septic shock (27), and recent reports have implicated a key role for endotoxin in the pathophysiology of septic shock in the human patient (27,37,54).

Bacterial endotoxins, possibly released upon lysis of gram-negative bacteria, produce severe hypotension and circulatory shock accompanied by impaired myocardial work capacity (15,46,60). Left ventricular dysfunction, present three hours following endotoxin and at the onset of reduced coronary perfusion pressure, was consistently associated with an elevated LVEDP, depressed dp/dt_{max}, and decreased myocardial power and efficiency. These changes suggested that both endotoxin and inadequate coronary perfusion were important factors in the etiology of cardiac dysfunction (4,24,25,26). A decrease in the rate of left ventricular relaxation (-dP/dt) was also observed in these studies. Other reports have also indicated the importance of the elapsed time in the development of the myocardial dysfunction. Endotoxin-treated dogs that eventually died had a depressed end-systolic pressure-diameter relationship as early as two hours, whereas, in survivors contractility was unchanged (17,18). These data implicate cardiac contractile function as an important contribution to cardiovascular collapse following endotoxin administration.

Myocardial functional deterioration is also present in isolated cardiac tissue following in vivo administration of endotoxin, as indicated by depressed isometric contractile tension (3,43,44) and a fall in dP/dt (21,22, 23). However, there appears to be no direct influence of endotoxin on myocardial performance (17,37,39,44,47). The in vitro isolated perfused working heart preparation has also been used to assess myocardial function. These studies were initiated three hours post-endotoxin and demonstrated a reduced performance of the isolated perfused heart from endotoxin-treated rats (LD_{50}-6hr) in response to increasing preloads from 5-30 cm H_2O (46).

Plasma catecholamine concentrations are known to increase following ET administration and during sepsis. Although this is an indirect measure of sympathetic stimulation, these circulating catecholamines can activate post-synaptic receptors and enhance the functional sympathetic discharge. The effect of catecholamines on the rate of myocardial tension development is associated with stimulation of beta-adrenergic receptors thereby generating cyclic AMP. Although the precise role of cyclic AMP in the inotropic response remains equivocal, the positive inotropic effect of catecholamines is preceded by an increase in cyclic AMP levels in the perfused heart (49). Cyclic clic AMP initiates a number of biochemical events at distinct intracellular sites within the myocardial cell resulting in changes in inotropic state. Most of the investigations concerning the effects of catecholamines on the myocardium has been performed in vivo, although, more recently, some studies have also utilized in vitro perfused hearts and subcellular membrane preparations. Few studies have been conducted using isolated myocardial cells which lend themselves uniquely to these types of investigations. The use of isolated myocytes provides a homogeneous preparation of heart muscle cells, which is quite important, since muscle cells comprise only approximately 15% of all the cells found in the heart (41,57). Also, this preparation eliminates the interfering influence of the central nervous system, peripheral circulation, blood volume, endocrine products, etc., and permits hormonal influences (e.g., catecholamines) to be studied independently of their effects upon work.

Catecholamines also can influence the responsiveness of their own target cells. Cells adapt to exposure to catecholamines with a progressive loss of responsiveness to subsequent catecholamine stimulation. This selective desensitization can occur rapidly and is not necessarily associated with a decrease in receptor density. Harden et al. (19) demonstrated that a 15 min incubation of astrocytoma cells with an agonist resulted in a rapid reduction in beta-adrenergic stimulated adenylate cyclase activity that paralleled a reduction in receptor affinity for agonists but was not correlated with a reduction in receptor density. Krall et al. (34) incubated human leukocytes with 0.01 μM isoproterenol, a concentration of agonist that would occupy only a small fraction of the total beta-receptor population, and found that the ability of isoproterenol to elevate cyclic AMP levels was impaired without any change in receptor density. Similarly, Feldman et al. (10) observed that infusion of isoproterenol into humans promoted a loss in beta-adrenergic mediated cyclic AMP accumulation in leukocytes, with no change in leukocyte beta-receptor density. These data suggest that a loss of beta-adrenergic responsiveness without change in beta-receptor density might occur in our endotoxin-treated animals in vivo at physiological concentrations of circulating catecholamines.

Beta-adrenergic adenylate cyclase systems appear to consist of at least three distinct components (51,58): the hormone receptor which recognizes and binds biologically active agonist hormones and inactive antagonists; the catalytic moiety of the adenylate cyclase, which converts ATP to cyclic AMP, and the guanine nucleotide regulatory protein, which binds and hydrolyzes GTP and which functionally couples the hormone receptor with the catalytic moiety of the enzyme (58). In the plasma membrane the guanine nucleotide binding protein couples receptors to enzyme activation through the action of guanine nucleotides. Since formation of a "high affinity complex", an agonist-receptor-guanine regulatory protein complex, appears to be required for agonist activation of adenylate cyclase, alterations in the ability to form high-affinity complexes could account for the observed changes that accompany densensitization as well as diminished receptor amounts. A reduction in the number of high affinity receptor-agonist complexes can be detected subsequent to chronic catecholamine exposure (33). Modulation of the formation of this crucial agonist-receptor-guanine nucleotide regulatory protein complex and, therefore, of the coupling of receptor with adenylate cyclase activation may represent a locus of pathophysiological alteration in sepsis and endotoxemia. The preponderance of evidence supports the contention that myocardial function is impaired prior to the development of circulatory shock, and that this dysfunction may occur despite the maintenance of coronary perfusion pressure. The exact mechanisms of the dysfunction remain to be elucidated, but may involve alterations in catecholamine sensitivity of the tissues and subsequent changes in cAMP generation, as well as alterations in cellular calcium homeostasis.

In the present study, many of the problems associated with in vivo determinations of cardiac function were circumvented through the use of an isolated working heart preparation. This methodology has been used to study myocardial function in hearts removed from animals in hyperdynamic sepsis (40), endotoxin shock (11,50) and burn shock (59). Ventricular function was assessed in these studies independently of peripheral defects, in the absence of blood born factors or products of vascular stasis, and in the presence of a constant supply of substrate and oxygen. These studies demonstrated an inability of the heart to pump fluid and generate pressure following the insult. The aims of the present studies were to determine whether the myocardial dysfunction observed subsequent to endotoxin administration was dependent on the dose of endotoxin administered and to ascertain the mechanism for the depressed myocardial function during endotoxemia. The present report indicates a graded increase in plasma catecholamine levels in response to the administration of increasing doses of endotoxin. These changes were associated with endotoxin dose-dependent decreases in myocardial function and cyclic AMP accumulation.

MATERIALS AND METHODS

Male Sprague-Dawley rats (350-400g, Harlan) were housed in a controlled environment, exposed to a 12/12hr light/dark cycle and provided standard rodent chow (Purina) and water ad libitum for a minimum of two weeks before initiation of experimental procedures.

Animal Preparation

On the day prior to the experiment animals were anesthetized with ether and chronic catheters placed in the arch of the aorta, via the left carotid artery, and the right jugular vein using aseptic surgical techniques. Following catheterization, animals were returned to individual cages without food but with water ad libitum. The following day lyophilized E. coli endotoxin (025:B6, Difco) was reconstituted with a 0.9% nonpyrogenic saline, filtered through a 0.45 μm filter (Millex-HA), and injected intravenously at doses of 1000, 100 and 10 μg/100 g body weight. The volume of endotoxin injected was 0.2 ml/100 g. Time-matched control animals received an equal volume of pyrogen-free saline instead of endotoxin. Four hours following the administration of endotoxin or saline, blood pressure and heart rate were measured through the arterial catheter. In addition, an arterial blood sample (3 ml) was rapidly (<15 sec) collected into chilled syringes containing EGTA and glutathione and the plasma stored at -80°C until assayed. Plasma catecholamine concentrations were determined by high pressure liquid chromatography as described by Hjemdahl et al. (28).

Working Heart Preparation

Following the designated in vivo experimental measurements and blood withdrawal, animals were anesthetized with sodium pentobarbital and hearts removed for mechanical studies in vitro. The intrinsic myocardial performance of hearts from endotoxin and time-matched control rats was assessed in vitro using the isolated perfused working heart preparation of Neely et al. (42). The heart was excised and mounted on the perfusion apparatus by means of a stainless steel cannula in the aorta. Retrograde perfusion of the coronary arteries was begun immediately to allow stabilization of cardiac rate and force, and to allow recovery of the heart from the period of anoxia associated with excision and cannulation. During retrograde perfusion the left atrial appendage and the pulmonary artery were cannulated. At this point the retrograde perfusion of the heart was terminated and antegrade perfusion initiated via the left atrial cannula at a left atrial filling height of 15 cm H_2O. Cardiac work was increased by elevating left atrial filling pressure at a constant aortic outflow resistance. Following ten minutes of antegrade perfusion, baseline data for pressure, flow and rate were recorded on a Narco Biosystems DMP 4A physiograph. Coronary flow, defined as pulmonary artery flow plus any fluid dripping off the heart (usually less than ten percent of the pulmonary artery flow), aortic flow, heart rate, peak systolic and diastolic pressure and cardiac output were then determined at 15, 20, and 25 cm H_2O.

Preparation of Myocytes

Myocytes were prepared according to previously described methodology (41,57). Briefly, hearts were excised and rinsed in calcium-free Joklik's Minimum Essential Medium (MEM) (pH 7.4, 290 mOsm, oxygenated with 100% O_2) supplemented with 8 mM glutamate and 1.2 mM $MgSO_4$ (medium A). The heart was retrogradely perfused (8 ml/min) with 20 ml of medium A to remove blood. Medium A was then replaced with medium B (medium A supplemented with 0.1% collagenase and 0.1% BSA) and perfused until the hearts were enlarged and very soft (about 30 minutes). The ventricular tissue was diced and then incubated with shaking in a Dubnoff metabolic shaker (90 cps, 37°C). After ten minutes, 10 ml of medium C (medium A plus one percent BSA) was added to each flask and the cell suspension filtered through a nylon mesh. The first filtrate was discarded since many damaged cells are typically found in the first dispersion. The remaining tissue fragments were reincubated as before. Suspended cells were filtered and collected in siliconized 40 ml centrifuge tubes. The suspended cells were centrifuged at low speed (30 x g, 2-3 min), the supernatant discarded, and the resulting cell pellet resuspended in medium C. After several purifications of the cells (by settling in the one percent BSA medium), the cells were suspended in gradually increasing concentrations of $CaCl_2$ until a final Ca^{2+} concentration of 1.5 mM. All subsequent studies with myocytes were performed using 1.5 mM calcium.

Cell viability was determined on the basis of morphology. A myocyte was considered viable if it was elongated and clearly striated and had no blebs or a grainy appearance.

Cyclic AMP Accumulation in Isolated Myocytes

Cyclic AMP accumulation in calcium-tolerant myocytes was probed by stimulating the beta-adrenergic receptor (isoproterenol) or by bypassing the receptor and stimulating the adenylate cyclase directly (forskolin). Cyclic AMP accumulation was assessed in the presence of isobutylmethylxanthine (MIX). We have shown that isoproterenol (10^{-9}M to 10^{-4}M) creates an asymptotic dose-response curve, indicating that cAMP accumulation is maximal under conditions of the assay.

Myocytes (1 mg protein per ml) were incubated in triplicate 0.5 ml samples in 17 x 100 mm polyethylene tubes at 37°C in a Dubnoff metabolic shaking water bath. Myocytes were added to tubes that contained the desired concentrations of isobutylmethylxanthine (10^{-4}M) and isoproterenol (10^{-9}M to 10^{-5}M) or forskolin (10^{-6}M) and the tubes were then placed in the shaking water bath. After exactly two minutes, the tubes were removed from the incubator, acidified to 0.2 N HCl, placed in a boiling water bath for one minute, then removed and subjected to at least one freeze-thaw procedure prior to neutralization with 2 N NaOH. Each tube was titrated to pH 6-8 using indicator paper. The tubes were then centrifuged at 3000 x g for 15 min at 4°C and cyclic AMP assayed on the neutralized supernatant fraction using a modification (6) of the Gilman (16) protein kinase binding procedure. The pellet of myocytes remaining in the tubes following centrifugation was washed twice in saline to remove albumin prior to determination of protein using the method described by Lowry et al. (38). Briefly, in the cyclic AMP assay, an aliquot of the supernatant fluid was incubated at 0-4°C for 60 min in the presence of a known amount of ^{3}H-cyclic AMP (10,000 cpm) and a known quantity of bovine adrenal protein kinase (concentration to give 33% binding of cAMP). The reaction was stopped by addition of 0.5 ml of 200 mM phosphate-buffered solution (pH 6.0) containing 5 mg/ml charcoal and 2.5 mg/ml albumin. The free cAMP was adsorbed by the charcoal leaving the cAMP bound to protein kinase in solution. The charcoal was pelleted by centrifugation at 300 x g for ten minutes at 4°C and the supernatant fluid was decanted into scintillation vials and the radioactivity determined by liquid scintillation spectrometry. The content of cAMP in the aliquot was calculated from a standard curve of unlabeled cAMP prepared in the same incubation buffer and treated in the same manner as the samples.

Statistical Analysis

Data from both the in vivo and in vitro studies are presented as mean ± standard error. Differences between treatment means for the in vivo experiments were analyzed by the Students' t-test, using a two tailed hypothesis. For the perfused heart studies, analysis of variance was used to determine differences between means. If the analysis of variance was significant, differences between the ET and control groups were tested with the Dunnett's test. A p < 0.05 was considered statistically significant.

Table 1. In Vivo Measurements 4 Hours After Endotoxin Administration

| | Blood Pressure (mm Hg) | Heart Rate (beats/min) | Plasma Catecholamines | |
			Norepinephrine (pg/ml)	Epinephrine (pg/ml)
Controls (n=12)	119 ± 4	348 ± 9	<150 (11/12)[a]	<150 (10/12)[a]
Endotoxin 10 µg (n=11)	119 ± 3	389 ± 11*	<150 (10/12)[a]	<150 (11/12)[a]
Endotoxin 100 µg (n=11)	123 ± 3	383 ± 11*	883 ± 91	2268 ± 390
Endotoxin 1000 µg (n=12)	120 ± 3	398 ± 16*	2430 ± 700	7680 ± 1280

[a]Numbers inside parenthesis represent the number of animals with catecholamine concentrations <150 pg/ml, compared to the total number of animals in that group.
Values are mean ± SE; (n) = number of animals per group
*p < 0.05, endotoxin compared to time-matched controls

RESULTS AND DISCUSSION

In Vivo Studies

Mean arterial blood pressure (in vivo) was not different between controls and experimental groups four hours following the injection of endotoxin (Table 1). However, a significant tachycardia was evident at this time for each of the treatment groups compared to the time-matched controls. These data demonstrate the stability of the endotoxemic animals at the time in vitro measurements were made, and is in contrast to previous studies where a hypotensive condition may have contributed to the depressed cardiac performance (50). McDonough and associates (40) have reported that gram-negative septicemia, in which mean arterial blood pressure was normal and cardiac output was elevated in vivo, produced intrinsic defects within the myocardium that were evident during in vitro perfusion studies. Lang et al. (36) reported that rats treated by the same regime as used in the present study exhibited dose-dependent changes in cardiac output. After injections of 1000 µg endotoxin per 100 g body weight, cardiac output was clearly depressed (about 50%) throughout the following four hours; 100 and 10 µg produced a milder depression at 30 min that returned toward normal after four hours (See J. J. Spitzer et al., this volume). These doses of endotoxin were associated with 50% (1000 µg), 90% (100 µg), and 100% survival (10µg) after 24 hours.

Plasma Catecholamines

Plasma concentrations of norepinephrine and epinephrine were determined in quiescent animals four hours after endotoxin administration (Table 1).

Table 2. Cardiac Output x Peak Systolic Pressure
at Varying Levels of Preload

| | Left Atrial Filling Pressure, cm H_2O | | |
	15	20	25
Control (7)	4059 ±328	5741 ±427	6703 ±541
Endotoxin, 10 µg (7)	4809 ±511	5628 ±445	6284 ±408
Endotoxin 100 µg (7)	4108 ±246	4605* ±308	5048 ±343
Endotoxin 1000 µg (6)	3491 ±324	4252* ±220	4892* ±257

Values are mean ± SE, (n), mmHg x ml/min; Endotoxin dose is
given as 1000, 100 or 10 µg/100 gram body weight.
*$p < 0.05$ endotoxin vs control at the equivalent left atrial
filling pressure.

These data indicate an increased sympathetic discharge in response to endotoxin administration that was dependent on the dose of endotoxin. The non-
lethal dose of endotoxin (10 µg) did not elevate plasma catecholamines above
control levels. Increasing the endotoxin dose to 100 and 1000 µg produced
a step-wise increment in both plasma norepinephrine and epinephrine. These
data are indicative of elevated sympathetic nervous system activity, albeit
indirect evidence to this effect, and suggest that both neural and adrenal
release of catecholamines are involved. While our data do not indicate a
time course of catecholamine release, marked elevations of plasma catecholamines have been reported as early as 30 min in the development of septic
shock in humans (5) and rats (30). Further, Jones and Romano (30) reported
that norepinephrine concentrations remained unchanged during six hours post-
endotoxin (200 µg/100 g) whereas epinephrine levels decreased to about 235%
of the value reported after 30 min endotoxin; increasing the severity of
the insult resulted in larger increases in plasma catecholamines (30).

Isolated Working Heart Studies

Assessment of the in vitro mechanical function of hearts from saline
and endotoxin-treated rats were performed fours hours after the in vivo
bolus injection of the prescribed dose of endotoxin. Heart rates in vitro
were not different, ranging from 252 ± 18 to 281 ± 22 for all groups. Hearts
removed from control and endotoxin-treated rats responded to increasing
left atrial filling pressure by increasing their mechanical performance
denoted as cardiac output X peak systolic pressure (CO X PSP) (Table 2).
The product of cardiac output and peak systolic pressure can be used as an
estimate of myocardial work and is increased as a consequence of increasing
preload, as predicted by the Frank-Starling mechanism. The product, CO X
PSP, was decreased in rats given 1000 µg endotoxin at 20 and 25 cm H_2O left

atrial filling pressure. Myocardial work also was reduced at 20 and 25 cm H_2O in the rats given 100 µg endotoxin. The dose of endotoxin that resulted in no lethality produced results similar to those from control saline-injected animals. Cardiac output showed a somewhat similar profile as CO X PSP except that 100 and 1000 µg of endotoxin produced approximately a 20% decrease in cardiac output at all three left atrial filling pressures and 10 µg endotoxin produced an 11% and 13% decrease at 20 and 25 cm H_2O, respectively. Thus, nonlethal doses of endotoxin had an effect on pump function, in particular, a ten percent reduction in the cardiac output. McDonough et al. (40) reported that hearts removed during the hyperdynamic phase of sepsis showed a myocardial dysfunction that involved a 40% decrease in cardiac output while pressure generation was only reduced 20%. In the present study, even at the highest dose of endotoxin, peak systolic pressure was not statistically different from the saline controls. Thus, at fairly low doses of endotoxin, a dysfunction that is similar to the dysfunction induced by hyperdynamic sepsis can be demonstrated. McDonough and associates (40) also observed that animals that had survived the septic episode showed normal work responses to changing left atrial filling pressure. Goldfarb et al. (17,18) has reported that dogs administered endotoxin that eventually died showed signs of reduced contractility, whereas in the survivors contractility was unchanged despite the fact that elevated catecholamines (probably present subsequent to endotoxin administration) should induce an increase in contractility.

Our data are consistent with an elevated sympathetic support of heart function in vivo. At the time of sacrifice, mean arterial blood pressure was normal and heart rate was elevated in the endotoxin-treated rats. Previous studies showed that cardiac output was still reduced at four hours post-endotoxin treatment in all three groups. In vitro, all three endotoxin doses resulted in decreased cardiac output. Cardiac output x peak systolic pressure, however, was reduced only in those endotoxemic animals that showed elevated plasma catecholamines. The apparent in vitro alteration was an inability to pump fluid, rather than a depression in pressure development, that may be attributed to an inability of the ventricle to relax during diastole. In this regard, Hess and coworkers have reported a reduced contractility, myofibrillar ATPase activity and calcium uptake by sarcoplasmic reticulum from mayocardium of endotoxemic animals (21,22,23). These results suggest that hearts from endotoxemic animals have a decreased ability to sequester calcium which might lead to a reduced rate of relaxation.

The effects of endotoxin appear to be seen only when administered to the whole animal. Lefer and Rovetto (37) demonstrated that in vitro administration of endotoxin (10 µg/ml) had no direct effect on the inotropic response of the isolated cap papillary muscle. Similar results were obtained by McCaig and associates (39) up to one hour after application of endotoxin. Endotoxin (10-1000 µg/ml) also did not alter the contractile tension or the maximal rate of tension development in isolated atrial strips from control guinea pigs (44). Furthermore, Raffa and Trunkey (47) were unable to demonstrate myocardial depression related to the infusion of either endotoxin or live gram-negative bacteria into an isolated perfused rabbit interventricular septum. Therefore, the mechanism for the cardiac depression seen during shock is not related to the direct effect of endotoxin.

Adrenergic Responsiveness

The mammalian heart adjusts its pumping capacity to physiological need by changing the tension developed by its fibers. Increases in pumping capacity may be modulated, at least in part, by catecholamines released from the adrenal medulla or sympathetic nerves in the myocardium. The major biochemical response to catecholamines is stimulation of membrane bound adenylate cyclase producing cyclic AMP. Cyclic AMP activates protein kinase(s) that can phosphorylate a large number of proteins including phosphorylase kinase, troponin, cardiac sarcoplasmic reticulum, and the sarcolemma (32). The phosphorylation of intracellular proteins may produce the positive inotropic response induced by an elevation in sympathetic nervous system activity. At the cellular level, catecholamines may combine with the beta-adrenergic receptors to activate the enzyme adenylate cyclase and may, therefore, result in an increase in contractility. If, however, an alteration occurs in the beta-adrenergic pathway, such that the number of receptors is decreased or there is a decrease in affinity of the catecholamine to bind to the receptor, then these changes can alter myocardial contractility. Additionally, the changes may involve the adenylate cyclase enzyme. Figure 1 portrays a general schematic of the beta-adrenergic pathway in cardiac muscle and the resultant processes modulated by changes in cyclic AMP accumulation. One can determine if the depressed myocardial function involves changes in the beta-adrenergic pathway by stimulation of the myocardium with a beta-adrenergic agonist to activate the adenylate cyclase through interaction with the hormone receptor, or alternatively, adenylate cyclase can be stimulated directly.

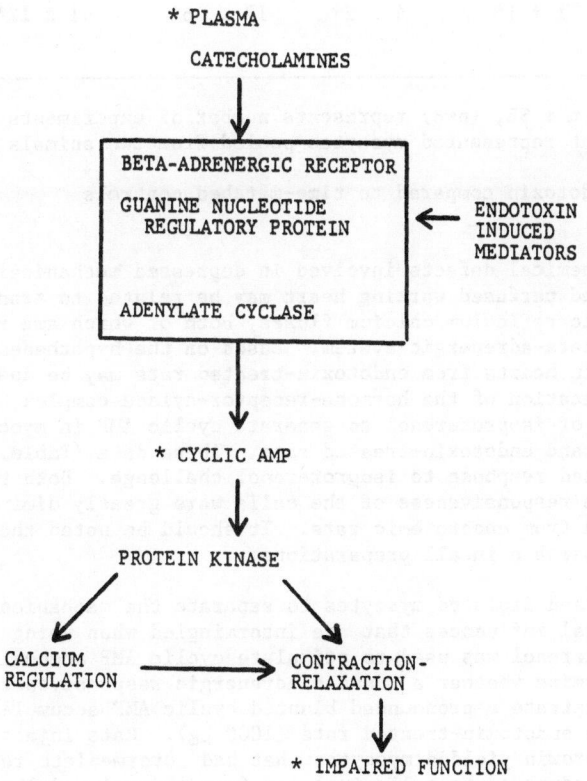

Figure 1. Proposed schematic for mechanism of endotoxin-induced myocardial dysfunction. Asterisks (*) represent measured alterations in endotoxin-treated animals compared to saline-treated controls. Our basic hypothesis implicates membrane alterations caused by endotoxin factors that result in reduced cyclic AMP levels that may lead to the imparied function.

Table 3. Cyclic AMP Accumulation in Myocytes Isolated from Rats Four Hours After Endotoxin Administration

Treatment	Basal	10^{-8}M	Isoproterenol 10^{-7}M	10^{-6}M	Forskolin 10^{-6}M
			pmols /mg protein /2 min		
Control (n=8)	5 ± 2	12 ± 3	46 ± 15	106 ± 19	115 ± 16
Endotoxin 10 µg (n=8)	2 ± 2	4 ± 2*	40 ± 10	87 ± 16	99 ± 16
Endotoxin 100 µg (n=8)	1 ± 1*	5 ± 1*	32 ± 10	87 ± 15	86 ± 13
Endotoxin 1000 µg (n=8)	1 ± 1*	4 ± 2*	17 ± 6*	41 ± 12*	58 ± 14*

Values are mean ± SE; (n=8) represents number of experiments per group. Each experiment represented myocytes pooled from two animals treated at the same time.
*$P < 0.05$, endotoxin compared to time-matched controls

The biochemical defects involved in depressed mechanical performance in the isolated perfused working heart may be related to transsarcolemmal or sarcoplasmic reticulum calcium fluxes, both of which are regulated, in part, by the beta-adrenergic system. Based on the hypotheses that functional derangements in hearts from endotoxin-treated rats may be due to faulty signal communication of the hormone-receptor-cylase complex, we investigated the influence of isoproterenol to generate cyclic AMP in myocytes isolated from vehicle- and endotoxin-treated rats. These data (Table 3) indicate a severely blunted response to isoproterenol challenge. Both the sensitivity (EC50) and the responsiveness of the cells were greatly diminished in myocytes prepared from endotoxemic rats. It should be noted that cell viabilities were comparable in all preparations.

We have used isolated myocytes to separate the mechanical influence from the humoral influences that are intermingled when using intact myocardium. Isoproterenol was used to stimulate cyclic AMP accumulation in myocytes to determine whether a reduced adrenergic responsiveness existed. Our data demonstrate a pronounced blunted cyclic AMP accumulation in myocytes harvested from endotoxin-treated rats (1000 µg). Rats injected with 100 µg and 10 µg endotoxin yielded myocytes that had intermediate responses to isoproterenol compared to cells isolated from control and 1000 µg endotoxin rats. These data indicate that a reduced sensitivity to catecholamines existed either by changes in the beta-adrenergic receptor or in the adenylate cyclase itself. To test whether a biochemical defect existed in the adenylate cyclase, we used forskolin to stimulate the cyclase directly, bypassing the beta-adrenergic receptor (Table 3). These data clearly indicate that, again, there was a dose-dependent decrease in cyclic AMP accumulation. These data suggested that the biochemical defect may reside within the adenylate cyclase enzyme itself.

Although a fully functioning beta-adrenergic system is not necessary for normal cardiac function in non-diseased hearts performing under non-stressed conditions (56), beta-adrenergic stimulation does contribute to the mechanical response to stress (20) and is important in supporting cardiac function during heart failure (8,14). It is tempting to speculate that an abnormality of the beta-adrenergic receptor pathway could indicate the transition from hyperdynamic to hypodynamic shock.

Beta-adrenergic receptor-mediated function is attenuated by persistent exposure to catecholamines (58). This process has been correlated with decreased beta-receptor mediated cyclic AMP production without changes in receptor density (34), and in some circumstances, with beta-receptor loss (58). A decrease in beta-receptor density has been demonstrated in vitro (12,31,53) and in vivo (1,2,7,9,10,13). A loss of responsiveness without reduction in receptor density suggests a functional uncoupling of the beta-adrenergic receptor adenylate cyclase system.

The possibility that myocytes from endotoxin-treated rats were in a desensitized state was tested by assessing receptor density and affinity in binding experiments using (^{3}H)-dihydroalprenolol. Our data indicate no difference in receptor density (35 vs 42 fmol/mg of membrane protein) or receptor affinity (0.4 vs 0.5 nM) as calculated from Scatchard analysis (52) of saline-treated and endotoxin-treated rats. Thus, our data indicate that the defect is not at the beta-adrenergic receptor level but at a regulatory step of adenylate cyclase, perhaps in the guanine nucleotide regulatory protein or within the adenylate cyclase itself. Further evidence for this is derived from analysis of adenylate cyclase activity in membranes prepared from hearts of endotoxin-treated rats. The ability of sodium fluoride to activate adenylate cyclase is impaired in the membranes from endotoxin-treated rats compared to membranes from control rats, 52 ± 5 vs. 78 ± 4 pmols cyclic AMP/mg protein/min, respectively (54).

In summary, we have used the isolated perfused working rat heart preparation to study intrinsic myocardial mechanical function during endotoxemia. Ventricular function was assessed in these studies independently of peripheral defects, in the absence of bloodborn factors or products of vascular stasis and in the presence of a constant supply of substrate and oxygen. Preload was varied, afterload resistance was maintained constant, and the effects of various doses of endotoxin on pressure development were separated from effects on cardiac output. The endotoxin-induced dysfunction seemed to involve the ability of the heart to pump fluid more than its ability to generate pressure. We have demonstrated that myocytes from endotoxin-treated rats are less responsive to isoproterenol challenge and stimulation by forskolin, indicating that the locus of this pathophysiological defect may be within the guanine nucleotide regulatory protein or within the adenylate cyclase itself. The increased plasma catecholamines may have produced agonist-induced desensitization of the beta-adrenergic receptor; we were unable to demonstrate differences in receptor density or affinity. Further evidence that the defect is a post-receptor event is suggested by the finding that sodium fluoride, a compound that stimulates adenylate cyclase directly, was unable to increase the activity of this enzyme in membranes from endotoxin-treated rats compared to control rats.

ACKNOWLEDGMENTS

The authors acknowledge the helpful assistance of Ms. Jane Henry and Ms. Cindy Wang. This project was supported by HL 32749 and a grant from AHA-LA, Incorporated.

REFERENCES

1. R. D. Aarons and P.B. Molinoff, Changes in the density of beta-adrenergic receptors in rat lymphocytes, heart, and lung after chronic treatment with propranolol, J. Pharmacol. Exptl. Ther. 221:439 (1981).

2. R. D. Aarons, A. S. Nies, J. G. Gerber, and P. B. Molinoff, Decreased beta-adrenegic receptor density on human lymphocytes after chronic treatment with agonists, J. Pharmacol. Exptl. Ther. 224:1 (1982).

3. H. R. Adams, C. R. Baxter, J. L. Parker, and N. B. Watts, Contractile function and rhythmicity of cardiac preparations from E. coli endotoxin-shocked guinea pigs, Circ. Shock 13:241 (1984).

4. L. T. Archer, B. A. Benjamin, B. K. Beller-Todd, D. J. Brackett, M. F. Wilson, and L. B. Hinshaw, Does LD_{100} E. coli shock cause myocardial failure? Circ. Shock 9:7 (1982).

5. C. R. Benedict and D. G. Grahame-Smith, Plasma noradrenaline and adrenaline concentrations and dopamine-beta-hydroxylase activity in patients with shock due to septicaemia, trauma and hemorrhage, Quart. J. Med. 185:1 (1978).

6. B. L. Brown, J. D. M. Albano, R. P. Ekins, A. M. Sgherzi, and A. L. Tampion, A simple and sensitive saturation assay method for the measurement of adenosine 3'5'-cyclic monophosphate, Biochem. J. 121:561 (1971).

7. W. S. Colucci, R. W. Alexander, G. H. Williams, R. E. Rude, B. L. Holman, M. A. Konstam, J. Wayne, G. H. Mudge, and E. Braunwald, Decreased lymphocyte beta-adrenergic receptor density in patients with heart failure and tolerance to the beta-adrenergic agonist, Pirbuterol. N. Eng. J. Med. 305 (1981).

8. S. E. Epstein and E. Braunwald, The effect of beta adrenergic blockade on patterns of urinary sodium excretion: studies in normal subjects and in patients with heart disease, Ann. Intern. Med. 65:20 (1966).

9. R. D. Feldman, L. E. Limbird, J. Nadeau, D. Robertson, A. J. J. Wood, Leukocyte beta-receptor alterations in hypertensive subjects, J. Clin. Invest. 73:648 (1984).

10. R. D. Feldman, L. E. Limbird, J. Nadeau, G. H. Fitzgerald, D. Robertson, and A. J. J. Wood, Dynamic regulation of leukocyte beta-adrenergic receptor-agonist interactions by physiological changes in circulating catecholamines, J. Clin. Invest. 72:164 (1983).

11. R. E. Fish, A. H. Burns, C. H. Lang, and J. A. Spitzer, Myocardial dysfunction in a non-lethal, non-shock model of chronic endotoxemia, Circ. Shock 16 (1985).

12. T. J. Franklin, W. P. Morris, and P. A. Twose, Densensitization of beta-adrenergic receptors in human fibroblasts in tissue culture, Mol. Pharmacol. 11:485 (1975).

13. J. Fraser, J. Nadeau, J. D. Robertson, and A. J. J. Wood, Regulation of human leukocyte beta receptors by endogenous catecholamines. Relationship of leukocyte beta receptor density to the cardiac sensitivity to isoproterenol, J. Clin. Invest. 67:1777 (1984).

14. T. E. Gaffney and E. Braunwald, Importance of the adrenergic nervous system in the support of circulatory function in patients with congestive heart failure, Am. J. Med. 34:320 (1963).

15. R. P. Gilbert, Mechanisms of the hemodynamic effects of endotoxin, Physiol. Rev. 40:245 (1960).

16. A. G. Gilman, A protein kinase binding assay for adenosine 3',5'-cyclic monophosphate, Proc. Natl. Acad. Sci. USA 67:305 (1970).

17. R. D. Goldfarb, Cardiac dynamics following shock: role of circulating cardiodepressant substances, Circ. Shock 9:317 (1982).

18. R. D. Goldfarb, W. Tambolini, S. M. Wiener, and P. B. Weber, Canine left ventricular performance during LD_{50} endotoxemia, Am. J. Physiol. 244:H370 (1983).

19. T. K. Harden, Y. F. Su, and J. P. Perkins, Catecholamine-induced desensitization involved in uncoupling beta-adrenergic receptors and adenylate cyclase, J. Cyclic Nucleotide Res. 5:99 (1979).

20. D. C. Harrison, Effects of beta-blockade on circulatory dynamics, in: "Beta-Adrenergic Blockade: A New Era in Cardiovascular Medicine," E. Braunwald, ed., Excerpta Medica, Amsterdam (1978).

21. M. L. Hess, Subcellular function in the acutely failing myocardium, Circ. Shock 6:119 (1979).

22. M. L. Hess and S. M. Krause, Contractile protein dysfunction as a determinant of depressed cardiac contractility during endotoxin shock, J. Mol. Cardiol. 12:715 (1981).

23. M. L. Hess, A. Hastillo, and L. J. Greenfield, Spectrum of cardiovascular function during gram-negative sepsis, Prog. Cardiovascular Dis. 23:279 (1981).

24. L. B. Hinshaw, Role of the heart in the pathogenesis of endotoxin shock: a review of the clinical findings and observations on animal species, J. Surg. Res. 17:134 (1974).

25. L. B. Hinshaw, L. T. Archer, J. J. Spitzer, M. R. Black, M. D. Peyson, and L. J. Greenfield, Effects of coronary hypotension and endotoxin on myocardial performance, Am. J. Physiol. 227:1051 (1974).

26. L. B. Hinshaw, B. Benjamin, L. T. Archer, and M. D. Payton, The heart and endotoxin shock, Tex. Rep. Biol. Med. 39:173 (1979).

27. L. B. Hinshaw, Overview of endotoxin shock, in: "Pathophysiology of Shock, Anoxia and Ischemia," R. A. Cowley and B. F. Trump, eds., Williams and Wilkins, Baltimore (1982).

28. P. Hjemdahl, M. Daleskog, and T. Kahan, Determination of plasma cate-cholamines by high-performance liquid chromatography with electro-chemical detection: Comparison with a radioenzymatic method, Life Sciences 25:131 (1979).

29. M. J. Hoffman, L. J. Greenfield, H. J. Sugerman, and J. L. Taturm, Unsuspected right ventricular dysfunction in shock and sepsis, Ann. Surg. 198:307 (1983).

30. S. B. Jones and F. D. Romano, Plasma catecholamine in the conscious rat during endotoxicosis, Circ. Shock 14:189 (1984).

31. G. L. Johnson, B. B. Wolfe, T. K. Harden, P. B. Molinoff, and J. P. Perkins, Role of beta-adrenergic receptors in catecholamine-induced desensitization of adenylate cyclase in human astrocytoma cells, J. Biol. Chem. 253:1472 (1978).

32. A. M. Katz, Role of the contractile proteins and sarcoplasmic reticulum in the response of the heart to catecholamine: a historic review, Adv. Cyclic Nucleotide Res. 11:303 (1979).

33. R. S. Kent, A. Delean, and R. J. Lefkowitz, A quantitative analysis of beta adrenergic receptor interactions: resolution of high and low affinity states of the receptor by computer modelling of ligand binding data, Mol. Pharmacol. 17:14 (1980).

34. J. F. Krall, M. Connelly, and M. L. Tuck, Acute regulation of beta adrenergic catecholamine sensitivity in human lymphocytes, J. Pharmacol. Exptl. Ther. 214:554 (1980).

35. E. Lachman, S. B. Pitsoe, and S. L. Goffin, Anti-lipopolysaccharide immunotherapy in management of septic shock of obstetric and gynaecological origin, Lancet 8384:981 (1984).

36. C. H. Lang, G. J. Bagby, and J. J. Spitzer, Glucose kinetics and body temperature after lethal and nonlethal doses of endotoxin, Am. J. Physiol. 248:R000 (1985).

37. A. M. Lefer and M. J. Rovetto, Influence of myocardial depressant factor on physiologic properties of cardiac muscle, Proc. Soc. Exp. Biol. Med. 134:269 (1970).

38. O. H. Lowry, N. J. Rosebrough, A. L. Farr, and R. J. Randall, Protein measurement with the folin phenol reagent, J. Biol. Chem. 193:265 (1951).

39. D. J. McCaig, K. A. Kane, G. Bailey, P. F. Millington, and J. R. Parratt, Myocardial function in feline endotoxin shock: a correlation between myocardial contractility, electrophysiology, and ultrastructure, Circ. Shock 6:201 (1979).

40. K. H. McDonough, C. H. Lang, and J. J. Spitzer, Depressed function of isolated hearts from hyperdynamic, septic rats, Circ. Shock 12:241 (1984).

41. J. Montini, G. J. Bagby, A. H. Burns, and J. J. Spitzer, Am. J. Physiol. 240:H659 (1981).

42. J. R. Neely, H. Liebermeister, E. J. Battersby, and H. E. Morgan, Effect of pressure development on oxygen consumption by isolated rat heart, Am. J. Physiol. 212:804 (1967).

43. J. L. Parker and H. R. Adams, Myocardial effects of endotoxin shock: characterization of an isolated heart muscle model, Adv. Shock Res. 2:163 (1979).

44. J. L. Parker and H. R. Adams, Contractile dysfunction of atrial myocardium from endotoxin-shocked guinea pigs, Am. J. Physiol. 240:H954 (1981).

45. M. Pollack, A. I. Huang, R. K. Prescott, L. S. Young, K. W. Hunter, D. F. Cruess, and C.-M. Tsai, Enhanced survival in Pseudomonas aeruginosa septicemia associated with high levels of circulating antibody to E. coli endotoxin core, J. Clin. Invest. 72:1874 (1983).

46. J. Postel and P. R. Schloerb, Cardiac depression in bacteremia, Ann. Surg. 186:74 (1977).

47. J. Raffa and D. D. Trunkey, Myocardial depression in sepsis, J. Trauma 18:617 (1978).

48. E. Remold-O'Donnell, Stimulation and desensitization of macrophage adenylate cyclase by prostaglandins and catecholamines, J. Biol. Chem. 249:3615 (1974).

49. G. A. Robison, R. W. Butcher, I. Oye, H. E. Morgan, and E. W. Sutherland, The effect of epinephrine on adenosine 3',5'-phosphate levels in isolated perfused rat heart, Mol. Pharmacol. 1:168 (1965).

50. A. J. Romanosky, A. H. Burns, and R. E. Shepherd, In vitro myocardial performance following in vivo administration of E. coli endotoxin, Fed. Proc. 42:608 (1983).

51. E. M. Ross and A. G. Gilman, Biochemical properties of hormone-sensitive adenylate cyclase, Ann. Rev. Biochem. 49:533 (1980).

52. G. Scatchard, The attractions of proteins for small molecules and ions, Ann. N.Y. Acad. Sci. 51:660 (1949).

53. M. Shear, P. Insel, K. L. Melmon, and P. Coffino, Agonist specific refractoriness induced by isoproterenol. Studies with mutant cells, J. Biol. Chem. 251:7572 (1976).

54. R. E. Shepherd, K. H. McDonough, and A. H. Burns, Mechanism of cardiac dysfunction in hearts from endotoxin-treated rats, Circ. Shock 13:95 (1984).

55. J. H. Siegel, F. B. Cerra, B. Coleman, I. Giocannini, M. Shetye, J. R. Border, and R. H. McMenamy, Physiological and metabolic correlation in human sepsis, Surgery 86:163 (1979).

56. J. F. Spann, E. H. Sonneblick, T. Cooper, C. A. Chidsey, V. L. William, and E. Braunwald, Cardiac norepinephrine stores and the contractile state of heart muscle, Circ. Res. 19:317 (1966).

57. J. J. Spitzer, Studies of substrate metabolism in isolated myocytes, in: "Myocardial Injury," J. J. Spitzer, ed., Plenum Publishing Corporation, New York (1983).

58. G. L. Stiles, M. G. Caron, and R. J. Lefkowitz, Beta-adrenergic receptors: biochemical mechanisms of physiological regulation, Physiol. Rev. 64:661 (1984).

59. T. E. Temples, A. H. Burns, F. C. Nance, and H. I. Miller, Effect of burn shock on myocardial function in guinea pigs, Circ. Shock 14:81 (1984).

60. C. S. Thomas, M. A. Melly, M. G. Koenig, and S. K. Brockman, The hemodynamic effects of viable gram-negative organisms, Surg. Gynecol. Obstet. 128:753 (1969).

61. J. P. Weisul, T. F. O'Donnell, M. A. Stone, and G. H. Clowes, Myocardial performance in clinical septic shock: effects of isoproterenol and glucose potassium insulin, J. Surg. Res. 18:357 (1975).

62. E. J. Winslow, H. S. Loeb, S. Kamath, and R. M. Gunnar, Hemodynamic studies and results of therapy in 50 patients with bacteremic shock, Am. J. Med. 54:421 (1973).

63. E. J. Ziegler, J. A. McCutchan, J. Furer, M. P. Glauser, J. C. Scadoff, H. Douglas, and A. I. Braude, Treatment of gram-negative bacteremia and shock with human antiserum to mutant Escherichia coli, N. Eng. J. Med. 307:1225 (1982).

CYCLIC NUCLEOTIDES IN THE IMMUNOPHARMACOLOGY OF

LIPOPOLYSACCHARIDE ENDOTOXINS

John W. Hadden, Ann Galy, Elba M. Hadden, J. L. Touraine,
and Ronald G. Coffey

Program of Immunopharmacology
Department of Internal Medicine
University of South Florida College of Medicine
Tampa, Florida

INTRODUCTION

Lipopolysaccharides (LPS) are responsible for the characteristic immunopharmacologic activities of the endotoxins of gram-negative bacteria. This
chapter will discuss their immunopharmacologic actions including some new
observations and the evidence that their actions are mediated in part by
cyclic nucleotides. The lipopolysaccharides are to be distinguished, for
the purposes of this discussion, from bacterially derived protein toxins,
like choleratoxin and E. coli enterotoxins, which have also been implicated
as immunomodulators acting via cyclic nucleotide pathways.

E. coli protein enterotoxins have been classified into two forms, a
heat labile enterotoxin acting to regulate adenylate cyclase and to increase
cellular levels of cyclic 3',5' adenosine monophosphate (cyclic AMP) in
several tissues (14) and a heat stable enterotoxin acting to regulate guanylate cyclase and to increase cyclic 3'5' guanosine monophosphate (cyclic
GMP) levels in gastrointestinal epithelial cells (16). In both cases, these
cyclic nucleotide related actions are thought to contribute to the mechanisms
by which these organisms induce diarrhea and intestinal cramps. Another
example is the Vibrio cholera protein toxin called choleratoxin which has
been shown to activate adenylate cyclase in lymphocytes, macrophages, neutrophils and mast cells and to increase cellular levels of cyclic AMP in association with the inhibition of secretory, motile, and proliferative functions
(see 6 for review). The role of these protein toxins in immune regulation
in gastrointestinal disease has not been ascertained and their relative
rarity in nature would indicate that they play little role in the healthy
individual.

In contrast to the protein toxins, the lipopolysaccharides are ubiquitously present in nature, and have many immunomodulatory activities, to the
extent that many immunologists consider them ever present nuisances bent on
fouling their experiments. Clearly in the context of sepsis and endotoxemia
the toxicities of endotoxins on the body's defense systems, including hyper-
pyrexia, intravascular coagulation, RES blockade, etc., can be considered
host destructive. On the other hand, in the absence of disease, low doses
of endotoxins may be considered benign or even positive immunoregulators
contributing to more effective host defense. It is even possible to envision

them as being a very part of the host defense mechanism. It may be that our nonpathogenic gastrointestinal flora provides us with a continuous low level of LPS molecules acting as hormonal signals to promote the development and enhance the function of the entire immune system. This latter comment may, for the traditionally trained immunologist, seem heretic, yet we hope in the course of this chapter to make clear several points which support the tenability of this notion. In brief, endotoxins regulate, either directly or indirectly, almost every phase of the development and function of the major natural and specific immune defense systems of the body. Their effects are positive, i.e., they generally promote growth and function of the cell populations involved. They do so at µg/ml concentrations and in some cases at ng/ml concentrations, i.e., at concentrations which may be periodically, even regularly, achieved locally and perhaps also systemically. Finally, where analyzed, they appear to act via the cyclic nucleotide pathways by which many of the body's hormones act and many of the molecules mediating immune function are also thought to act. They, therefore, qualify as specific messengers acting via receptors and hormone pathways in physiologically constructive ways. In order to allow the reader to evaluate this speculative notion in greater depth, a brief review of the immunopharmacology of endotoxins is essential.

IMMUNOPHARMACOLOGIC ACTIONS OF ENDOTOXIN

In Vivo Studies

Firstly, endotoxins by nature of their polysaccharide components are antigenic and through their diversity elicit a spectrum of antibodies capable of neutralizing their function and preventing the endotoxins, in general, from being immunotoxic and dysregulatory to the system. The lipid A moiety is generally considered to provide the basis of the immunoregulatory functions although there is not unanimity on this point for all aspects of the nonspecific action of LPS on the immune system and evidence for a role of the polysaccharide moiety exists (see Friedman and Nowotny in this volume). The many actions of LPS on the immune system have been reviewed extensively in this book and elsewhere (see 22 and 34). Endotoxins generally enhance host resistance to pathogen challenge particularly when administered just prior to challenge. They have been shown to enhance resistance to transplantable tumors and are thought to be part of tumor destructive processes initiated by bacterial immunotherapies such as BCG, C. Parvum and mixed bacterial vaccines (7). They are potent adjuvants for antibody production when administered with or just following antigen. They restore humoral immunity in aged mice, enhance the response to weak antigens, render tolerogenic doses of antigen immunogenic and prevent the induction of B-cell tolerance but not T-cell tolerance. They inhibit cellular immune responses under circumstances where they concomitantly enhance humoral immunity yet they enhance cellular immunity to unrelated antigens including delayed-type hypersensitivity, graft-versus-host reaction, and allograft rejection.
They expand the hematopoietic and reticuloendothelial systems and enhance nonspecific resistance mechanism. LPS also promotes complement (C) activation via the alternate pathway which leads to the production of activated C_3. B cells and macrophages bear C_3 receptors and C_3 activation is thought to play a role in stimulating these cells. LPS-induced production of chemotactic peptides C_3A and C_5A can also be envisioned to participate in granulocyte and macrophage chemotaxis and accumulation.

In Vitro Studies

The cellular targets and molecular mediators of these various LPS actions are many; the essential features are summarized in Figure 1.

188

B Lymphocyte. The best characterized action of LPS is that of a polyclonal B cell activator. In vitro studies initiated by Gery et al., (19)
and Melchers and Andersson (31) have clearly demonstrated that LPS nonspecifically activates resting B cells to proliferate and to produce various
classes of immunoglobulins. In the context of antigen, LPS stimulates clonal
expansion of antigen reactive B cells and accelerates their differentiation
into antibody secreting cells. LPS's effects on macrophages are thought to
contribute to B cell differentiation and activation through the elaboration
of interleukin I (IL-1), B cell differentiation factor (BDF) and B cell
activating factor (BAF) (26,27,56). Williamson and coworkers (54) have
suggested that the direct B cell effects of LPS are mediated by lipid A
while those involving the macrophage are mediated by the polysaccharide
moiety of LPS. In the context of a T dependent antigen challenge, endotoxin
effects on T helper cells contribute perhaps through involvement of antigen
specific T helper factors and/or B cell growth factor. Suppressor mechanisms
for LPS-induced B cell responses have been described and are apparently
mediated by macrophages and both T and B cells (see 55). LPS also induces
mature B cells to make colony stimulating factor (CSF) through a macrophage
dependent mechanism.

Besides the resting B cell, the immature B lymphocyte is a target of
LPS-induced macrophage and T cell produced growth and differentiation factors.
In addition the precursor of the B cell (pre B cell) is apparently a target
of direct LPS action (25,40,41). LPS induces pre B cell differentiation in
vitro in two hours as measured by induction of complement and other surface
B cell receptors. Thus, LPS appears to regulate directly or indirectly B
cell differentiation from the earliest known precursor to the immunoglobulin
producing plasma cell.

Macrophage. Extensive studies initiated by the works of Cohn (11) and
Strauss (45) demonstrate that many macrophage functions are very sensitive
to stimulation by LPS including phagocytosis, migration, lysosomal enzyme
content, metabolism, secretion, microbicidal activity and cytotoxicity (see
34 and 39 for review). Macrophage microbicidal activity is stimulated

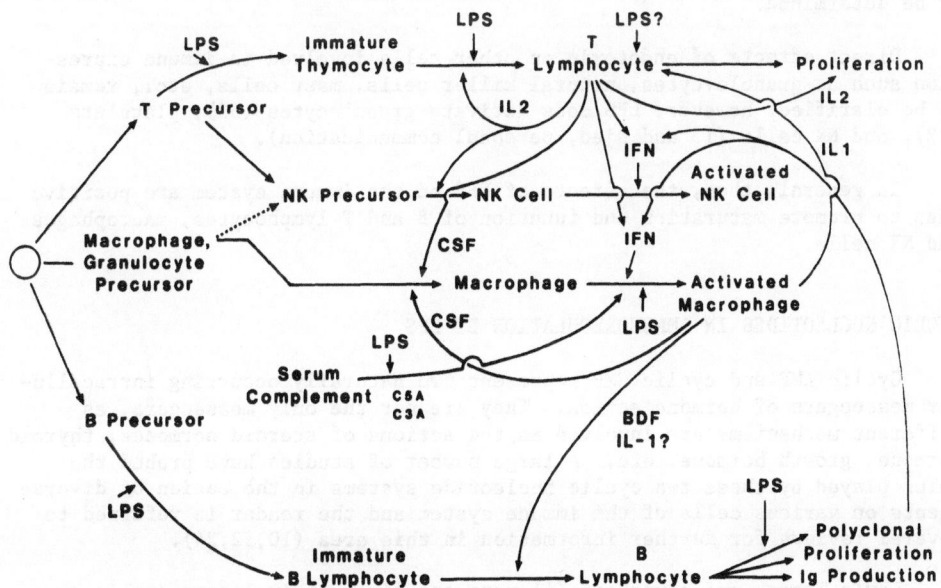

Figure 1. Actions of endotoxin (LPS) on various cell
targets in the immune system.

directly by LPS with macrophage activating lymphokines like γ interferon (IFN) (39). These macrophage activities are clearly important in the enhancement of nonspecific resistance mechanism mediated by the macrophage. Important in the indirect effects of LPS is the capacity to induce macrophages to secrete a number of important mediators regulating other cell populations. LPS is one of the most potent inducers of macrophage-derived: 1) IL-1, which regulates T cells by inducing Interleukin II (IL-2) which in turn induces T cell proliferation and differentiation; 2) colony stimulating factors (CSF) which induce clonal growth of macrophage and granulocyte precursors and also induce mature macrophages to proliferate; 3) endogenous pyrogen (through to be identical to IL-1) which provides the mechanism for LPS-induced fever; 4) B cell growth and differentiation factors (BDF; BAF) which regulate B cell maturation and function as discussed; 5) α interferon (IFN) which induces antiviral resistance; and 6) tumor necrosis factor.

T cell. Scheid et al. (40) showed that LPS induces prothymocytes to differentiate in culture into cells bearing the markers of intrathymic lymphocytes. We (Galy, Hadden and Hadden, in preparation, 1985) present in this chapter the first evidence that LPS induces immature thymocytes to mature into mitogen responsive T cells. Several investigators (17,37,48,49) have shown that, while LPS does not induce polyclonal activation of thymocytes, it does synergize with concanavalin A (Con A) to increase T cell proliferation particularly at low doses of Con A. Whether this latter effect is mediated by IL-1 produced by thymic macrophages is not clear since mixed thymocyte populations were used in these studies. Schmidke and Nagarian (42) showed that phytohemagglutin (PHA) and Con A responses of human peripheral blood lymphocytes, mainly T cells, is synergistically augmented by LPS, another action which may be mediated by monocyte-produced IL-1. Vogel et al. (47) have shown that LPS can stimulate growth of one T cell line and a small population (3%) of splenic T cells. These studies indicate that T cells at various stages of development are sensitive to LPS action. A major role of IL-1 is the induction of IL-2 by T cells; thus, LPS induces IL-2 indirectly via its effect on macrophages (See Nakano and Nitta in this volume). Whether LPS induces or potentiates the induction of other T cell-produced lymphokines such as macrophage activating factor (MAF), migration inhibitory factor (MIF), colony stimulating factor (CSF) and γIFN remains to be determined.

Direct effects of endotoxin on other cells involved in immune expression such as granuloycytes, natural killer cells, mast cells, etc., remain to be clarified; however, LPS does activate granulocytes (52), platelets (12), and NK cells (13 and Djeu, personal communication).

In general, then, the effects of LPS on the immune system are positive ones to promote maturation and function of B and T lymphocytes, macrophages and NK cells.

CYCLIC NUCLEOTIDES IN IMMUNOREGULATION BY LPS

Cyclic AMP and cyclic GMP represent two naturally occurring intracellular messengers of hormone action. They are not the only messengers, as different mechanisms are involved in the actions of steroid hormones, thyroid hormone, growth hormone, etc. A large number of studies have probed the roles played by these two cyclic nucleotide systems in the action of diverse agents on various cells of the immune system and the reader is referred to several reviews for further information in this area (10,22,24).

Lewis Thomas and coworkers (4) were the first to implicate cyclic nucleotides in LPS action and subsequently a number of papers have suggested that cyclic nucleotides are important mechanisms by which LPS acts in various immune cell populations.

Lymphocyte Proliferation

Watson (48-50) was the first to show that LPS induces early increases (three to five fold at 5-15 min) in cyclic GMP levels in mouse spleen lymphocytes without significant effects on cyclic AMP levels. He showed that cyclic GMP but not cyclic AMP induced proliferation in these cells (51) and that cyclic AMP inhibits the action of LPS to induce B cell proliferation, and the inhibition is reversed by cyclic GMP (49).

The effect of LPS to increase cyclic GMP levels in B lymphocytes have been confirmed by Shenker and Grey (43,44), and Bomboy and Graber (5). Freedman (18) extended the observation to show that, like T cell mitogens, LPS induces early calcium influx in murine splenocytes, apparent within six minutes and persisting up to 20 hours. He also showed that cyclic AMP inhibited LPS-induced calcium influx, that cyclic GMP enhanced the uptake, and that cyclic AMP antagonized the effect of cyclic GMP. Using fluorescein-labelled antisera to the cyclic nucleotides and their protein kinases, Largen and Votta (30) found that LPS decreases the number of cells staining for cyclic AMP at five minutes and increased the number of cells staining for cyclic GMP at 30-60 minutes. LPS had no effect on the cells staining for Type I or II cyclic AMP-dependent protein kinase but markedly increased for five to 60 minutes the number of cells staining for cyclic GMP-dependent protein kinase. Similarly, Ohara et al. (35) showed that microinjection of antibody to cyclic GMP suppressed and antibody to cyclic AMP enhanced LPS-induced B lymphocyte proliferation. These collected observations directly parallel those made in T lymphocytes activated by PHA or Con A and support a role for cyclic GMP and calcium in B lymphocyte activation by LPS.

B and T Precursor Cell Differentiation

Scheid et al. (40) showed that LPS induces prothymocyte maturation in the Komuro-Boyse assay and B cell maturation in a similar assay. The effect of LPS in both assays was apparent at two hours, was maximal at 20-50 μg/ml LPS, and was enhanced by theophylline which inhibits cyclic AMP phosphodiesterase and antagonized by imidazole which promotes cyclic AMP catabolism (Figure 2). In these studies (25,40,41) cyclic AMP and agents which increase cyclic AMP also induce both prothymocyte and pre B cell differentiation. Theophylline potentiates induction by agents like poly A:U, choleratoxin, and prostaglandin PGE_1; and imidazole antagonizes their induction.

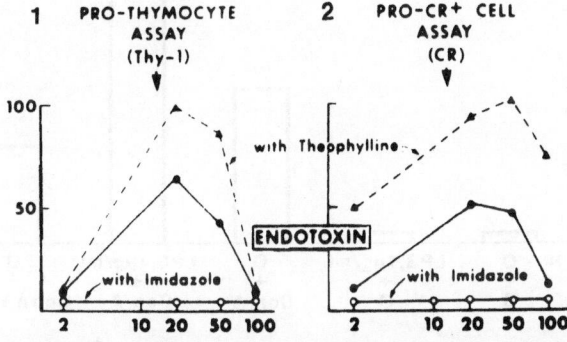

Figure 2. Induction by endotoxin of prothymocyte (1) and pro-complement receptor (CR) B cells (2). Differentiation with enhancement of induction by theophylline (10^{-3}M) and inhibition by imidazole (10^{-7}M). (Republished in part from reference (41) with permission of the author.)

These induction assays were performed on athymic nu/nu mouse spleen cell populations fractionated on discontinuous albumin gradients to enrich the precursor cells. A number of experiments were performed under similar conditions to measure cyclic AMP levels (Scheid, Hadden, Coffey unpublished observations). In four experiments LPS (25 µg/ml) significantly increased cyclic AMP levels two fold at five to 30 minutes of incubation. The increases occurred in precursor rich fractions (A and B) and not in mature cell fractions (C and D). The increases were augmented, not ablated, by depletion of mature B cells with anti LYB 2.1. These preliminary observations suggest that, in contrast to that on mature B cell proliferation, its action to induce T and B cell maturation LPS action is mediated by cyclic AMP.

Fairchild and Cohen (15) have also provided support for a parallelism of LPS and cyclic AMP induced B cell differentiation in mouse bone marrow cells.

Thymocyte Maturation

We recently observed that LPS induces PNA[+] immature cortical thymocytes to mature to a Con A responsive state characteristic of the PNA[-] mature thymocyte (Galy, Hadden and Hadden, in preparation, 1985). This action is shared by thymic epithelial cell supernatants and IL-2 but not by thymosin or IL-1. In these experiments PNA[+] thymocytes were prepared by double rosetting of mouse thymocytes to ensure purity of the immature population. We (8) confirmed previously that the surface marker profile of the these cells conforms to the phenotype of the immature thymocyte and these cells show little or no reactivity to Con A in contrast to PNA[-] mature T cells. Addition of LPS (.1 - 10 µg/ml) alone to these PNA[+] cells had no effect on thy-

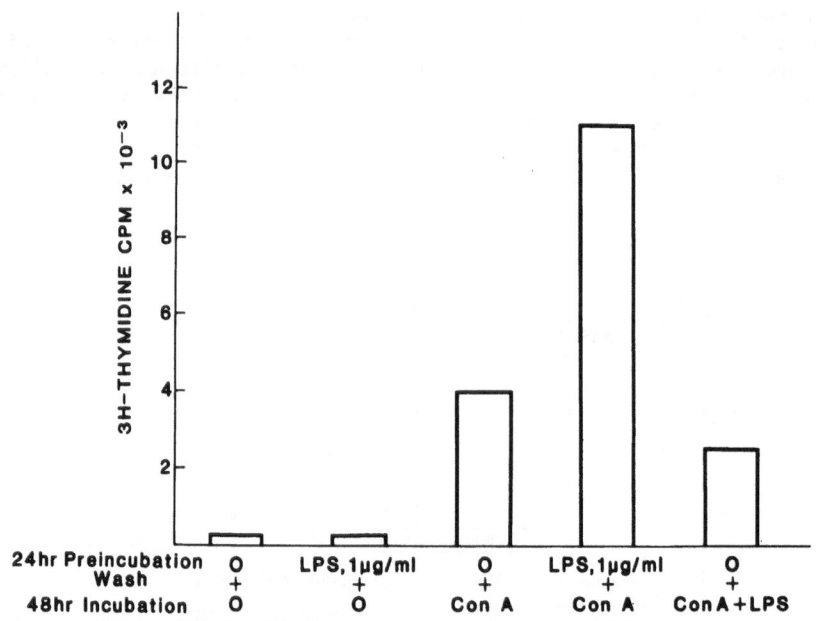

Figure 3. Effect of LPS on Con A responsiveness of PNA[+] immature thymocytes. Cells were prepared as described (8) and incubated for 24 hrs in the presence and absence of LPS 1 µg/ml (Difco endotoxin), then washed four times with media and incubated an additional 48 hrs in the presence of Con A (2.5 µg/ml). Tritiated thymidine incorporation was measured as described (20).

midine incorporation measured at 72 hours. Under similar circumstances
with PNA⁻ cells or unfractionated thymocytes endotoxin has a small effect
to augment thymidine incorporation, presumably the result of IL-1 production by macrophages. If endotoxin is added at the time 0 to PNA⁻ thymocytes
and Con A is added at 24 hours to the PNA⁻ thymocyte, a marked increase of
Con A-induced thymidine incorporation occurs at 72 hours. If LPS is added
at 24 hours at the same time as the Con A, an increase of thymidine incorporation occurs compared to Con A alone; however, the increase is significantly
less than that observed with preincubation with LPS. These results imply
that LPS acts to differentiate the PNA⁻ cell to make it Con A responsive
rather than an action to allow Con A to induce immature thymocyte proliferation by overcoming a deficiency in IL-1 and, therefore, IL-2 in the response.
To clarify the point PNA⁻ cells were incubated for 24 hours with endotoxin
and washed four times to remove all but trace endotoxin (< 1 ng) before Con
A was added. Under these circumstances a marked Con A response was induced
by LPS, absent in the control (Figure 3). These experiments confirm that
the action of LPS occurs prior to the action of Con A, and causes a Con A-
unresponsive population to become responsive. IL-1 kindly provided by
Charles Dinarello was inactive in the preincubation experiment indicating
the action of LPS is not mediated by macrophage-produced IL-1. Neither LPS
nor IL-1 showed IL-2 activity in the IL-2 dependent cell line (CTLL) assay.
That the effect is LPS⁻ and probably lipid A is specific was confirmed by
the relative inability of C_3H/HeJ mice to respond compared to C57B1/6 or
BALB/c mice and by the action of polymyxin B to prevent the response. Preliminary data indicate that LPS induces two fold increases of cyclic GMP
but no change of cyclic AMP levels in PNA⁻ thymocytes at five to 20 minutes.

T Lymphocyte Proliferation

The addition of LPS to unfractionated thymocytes results in an augmentation of basal thymidine incorporation and the preexisting Con A response of
mature thymocytes and this effect is comparable to the effect of exogenously
added IL-1 and is probably mediated by thymic macrophages. The effects of
LPS on thymus cyclic nucleotide levels have previously been studied by Naylor
et al. (32) and Tsien et al. (46). Both of these groups showed that LPS
increases cyclic AMP levels in unfractionated mouse thymocytes. The increases occurred with LPS concentrations above 1 μg/ml and ranged up to 12
fold. It is notable that LPS induces PG production in blood leukocyte populations and results in the production of cyclic AMP (36). The cyclic AMP
increases observed with LPS in unfractionated thymocytes may well result
from the induction of prostaglandin synthesis by thymic macrophages.

Naylor (1978 Ph.D. thesis at George Washington University Medical
School) also examined the effects of LPS on cyclic GMP levels of unfractionated thymocytes and observed that LPS (100 μg/ml) induces early increases
in cyclic GMP levels as well. The effect of LPS on cyclic nucleotide levels
of PNA⁻ mature thymocytes has only recently been examined by us and the
preliminary data indicate that LPS at doses of 1 μg/ml increases cyclic GMP
levels. It is notable that the effect of LPS to promote mature thymocyte
proliferation in response to Con A is mimicked by cyclic GMP but not cyclic
AMP (48) and the effects of LPS to induce T suppressor cell function has
been related to the cyclic AMP changes (55). Under these circumstances, it
seems likely that the cyclic GMP changes induced by LPS in mature T cells
are related to the enhancement of proliferation and the cyclic AMP changes
are related to suppressor influences. However, more experiments are needed
to clarify what populations of immature and mature thymocytes are responsive
to LPS and what the mechanisms are.

Macrophage

Macrophage regulation by LPS is a central phenomenon in LPS immunopharmacology. LPS at high concentrations induces macrophage prostaglandin

synthesis (29) and therefore, corresponding cyclic AMP increases in association with collagenase induction (3). We (24) previously examined the effects
of LPS on macrophage microbicidal activity under circumstances in which
macrophages show enhanced listericidal capacity following stimulation with
LPS (Figure 4). We observed that low concentrations (.05 and 0.5 μg/ml) of
LPS induce early increases in macrophage cyclic GMP levels (p's < .05; three
experiments) without significant effect on cyclic AMP levels (Figure 5).
It is notable that CSF, PMA, MDP and Tuftsin, all of which have similar
effects as LPS on macrophages, i.e., activation for cytotoxicity and monokine
production, also raise macrophage levels of cyclic GMP (9,20,21). Wrightman
and Raitz (53) have described an interesting effect of LPS to activate protein kinase C in a macrophage cell line. This action has also been described
for the phorbol myristate acetate (PMA) (33) which increases cyclic GMP in
both lymphocytes and macrophages (9,20). The relation of cyclic GMP generation and protein kinase C activation remains to be determined.

As discussed, LPS is known to induce production of IL-1, CSF, and αIFN
by macrophages, αIFN, IL-2 and CSF by lymphocytes. IL-1 has been reported
to induce late changes (20 hr) but not early changes (0-1 hr) in cyclic GMP
levels of lymphocytes (28). αIFN has been reported to induce cyclic GMP
increases in lymphoid cells (38). CSF$_1$ has been reported by us to increase
cyclic GMP levels in macrophages (20) and IL-2 has been shown to increase
cyclic GMP levels in PNA$^+$ thymocytes as well as mitogen-primed blood lymphocytes (Hadden, Hadden, and Coffey, unpublished observations). Thus each of
the mediators induced by LPS has been implicated in inducing cyclic GMP in
one or another cell or circumstance.

CONCLUSION

The foregoing, although so far incomplete, attests to important roles
played by cyclic nucleotides in the direct action of LPS and in the indirect
action of LPS-induced mediators on various cell populations. While enticing,

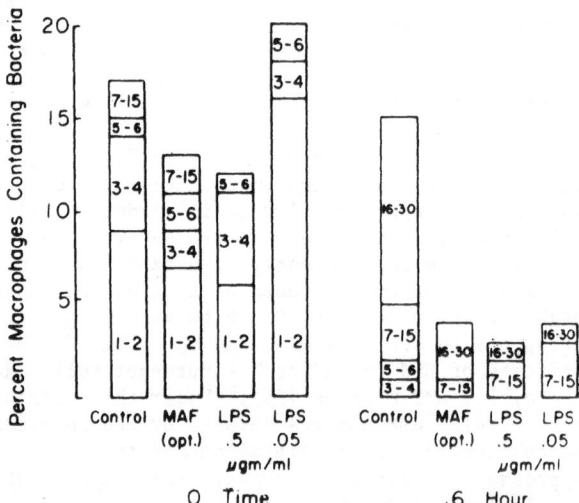

Figure 4. Effects of Macrophage activating factor (MAF) and endotoxin
           (LPS) on the capacity of guinea pig peritoneal exudate macro-
           phages to kill intracellular Listeria Monocytogenes. The
           experiments were performed as previously described (24) and are
           republished here with permission of the editor.

194

these studies have intrinsic defects which limit conclusions about the relationship of LPS action to the cyclic nucleotide change and of these changes to the subsequent biological response. A wide variety of LPS preparations have been analyzed; the roles of Lipid A and polysaccharide components and contaminating enterotoxins have not been assessed for some of the biological responses or any of the cyclic nucleotide responses. The cell populations employed are generally enriched but not purified and LPS-induced direct versus indirect actions have not been adequately assessed in purified cell populations. The levels of cyclic nucleotides have been measured in relatively haphazard ways without attention to the contribution of prostaglandin influence, often without purification of the cyclic nucleotide (in the case of cyclic GMP, an essential issue), and without phosphodiesterase inhibitors (to block cyclic nucleotide turnover and to show that the effect is on production and not catabolism). Importantly, with a single exception (30), the experiments are lacking to show the relation of the cyclic nucleotides change to metabolic events, like protein kinase activation, which would link the cyclic nucleotide to the effector response. The possible importance of endotoxin mechanisms in normal development and regulation of immune function make it important to correct these deficiencies.

Endotoxins are likely candidates for positive regulators of the development and differentiation of both classes of lymphocytes and of macrophages and various granulocyte populations. The evidence indicating nontoxic physiological roles is more compelling than that supporting toxicity and suggests a reevaluation of the beneficial effects of endotoxins on immune function. The cyclic nucleotides are likely to be involved in the action of endotoxin. The pharmacologic data are strong in indicating a role for cyclic AMP in actions of endotoxin to induce precursor cell differentiation and for cyclic GMP in actions of endotoxin to promote proliferative or secretory functions of mature lymphocytes and macrophages. In addition, endotoxin-induced mediators such as CSF, IFN, IL-1 and IL-2 appear to involve cyclic nucleotide

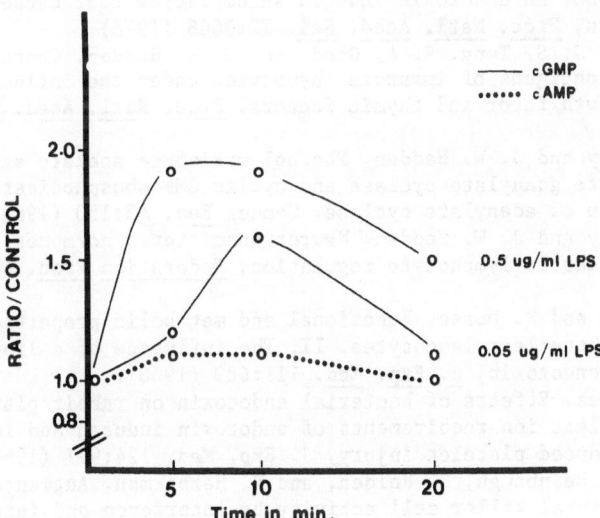

Figure 5. Effects of endotoxin on cyclic nucleotide levels of peritoneal exudate macrophages. Adherent monolayers were incubated for various times with LPS (Difco) at .05 and 0.5 μg/ml and cyclic AMP and cyclic GMP levels of macrophages were measured as previously described (24). The data are expressed as percent of control levels and the cyclic AMP data were pooled for the two concentrations of LPS.

mediation as part of their action. Further experimental development in
this area using defined endotoxins and components and purified cell populations is definitely warranted. The many positive effects of endotoxins
indicate that as the molecule can be detoxified without losing important
biological activities, as described by both Ribi and Nowotny in this volume,
powerful immunopharmacologic agents will emerge. The possible therapeutic
implications are great for use of detoxified endotoxin alone and in combination with other agents in cancer, immunodeficiency, autoimmunity and
infection.

REFERENCES

1. R. Apte, C. Hertogs, and D. Pluznik, Generation of colony-stimulating
    factor by purified macrophages and lymphocytes, J. Recituloen.
    26:491 (1979).

2. R. Apte, C. Hertogs, and D. Pluznik, Regulation of lipoplysaccharide-
    induced granulopoiesis and macrophage formation by spleen cells,
    J. Immunol. 124:1223 (1980).

3. R. Bhatnager, U. Schade, E. Rietschel, and K. Decker, Involvement of
    prostaglandin E and adenosine 3',5'-monophosphate in lipopolysac-
    charide-stimulated collagenase release by rat Kupffer cells, Eur.
    J. Biochem. 125:125 (1982).

4. M. W. Bitensky, R. E. Gorman, and L. Thomas, Selective stimulation of
    epinephrine-responsive adenyl cyclase in mice by endotoxin, Proc.
    Soc. Exp. Biol. Med. 138:773 (1971).

5. J. D. Bomboy, Jr., and S. E. Graber, Stimulation of cyclic 3',5'-guano-
    sine monophosphate levels in rat spleen cells by lipopolysaccharide
    preparations, J. Lab. Clin. Med. 95:654 (1980).

6. H. R. Bourne, M. Lichtenstein, K. L. Melmon, C. S. Henney, Y.
    Weinstein, and G. M. Shearer, Modulation of inflammation and
    immunity by cyclic AMP, Science 184:19 (1974).

7. E. A. Carswell, L. J. Old, R. L. Kassel, S. Green, N. Fiore, and B.
    Williamson, An endotoxin-induced serum factor that causes necrosis
    of tumors, Proc. Natl. Acad. Sci. 72:3666 (1975).

8. S. S. Chen, J. S. Tung, R. A. Good, and J. W. Hadden, Changes in
    surface antigens of immature thymocytes under the influence of T
    cell growth fctor and thymic factors, Proc. Natl. Acad. Sci. 80:5980
    (1983).

9. R. G. Coffey and J. W. Hadden, Phorbol myristate acetate stimulation of
    lymphocyte guanylate cyclase and cyclic GMP phosphodiesterase and
    reduction of adenylate cyclase, Cancer Res. 43:150 (1982).

10. R. G. Coffey and J. W. Hadden, Neurotransmitters, hormones and cyclic
    nucleotides in lymphocyte regulation, Federation Proc. 44:112
    (1985).

11. Z. A. Cohen and S. Morse, Functional and metabolic properties of
    polymorphonuclear leucocytes. II. The influence of a lipopolysac-
    charide endotoxin, J. Exp. Med. 111:689 (1960).

12. R. M. DesPrez, Effects of bacterial endotoxin on rabbit platelets. IV.
    The divalent ion requirements of endotoxin induced and immunologi-
    cally induced platelet injury, J. Exp. Med. 124:971 (1964).

13. J. Djeu, J. Heinbaugh, H. Holden, and R. Herberman, Augmentation of
    mouse natural killer cell activity by interferon and interferon
    inducers, J. Immunol. 122:175 (1979).

14. D. J. Evans, Jr., C. Chen, G. T. Curlin, and D. G. Evans, Stimulation
    of adenyl cyclase by Escherichia coli enterotoxin, Nature New Biol.
    236:137 (1972).

15. S. Fairchild and J. Cohen, B lymphocyte precursors, J. Immunol. 12:1227
    (1978).

16. M. Field, L. H. Graf, Jr., W. Laird, and P. L. Smith, Heat-stable
    enterotoxin of Escherichia coli: in vitro effects on guanylate

cyclase activity, cyclic GMP concentration, and ion transport in small intestine, Proc. Nat. Acad. Sci. 75:2800 (1978).

17. J. Forbes, Y. Nakao, and R. Smith, T mitogens triger LPS responsiveness in mouse thymus cells, J. Immunol. 114:1004 (1975).

18. M. H. Freedman, Early biochemical events in lymphocyte activation. I. Cell. Immunol. 44:290 (1979).

19. I. Gery, J. Kruger, and S. Spiesel, Stimulation of B lymphocytes by endotoxin: reactions of thymus-deprived mice and karyotypic analysis of dividing cells in mice bearing T_6-T_6 thymus grafts, J. Immunol. 108:1088 (1972).

20. E. M. Hadden, J. R. Sadlik, R. G. Coffey, and J. W. Hadden, Effects of phorbol myristate acetate (PMA) and lymphokine on cyclic GMP levels and proliferation of macrophages, Cancer Res. 42:3064 (1982).

21. J. W. Hadden, The action of immunopotentiators in vitro on lymphocytes and macrophage activation, in: "The Pharmacology of Immunoregulation," G. Werner and F. Floch'h, eds., Academic Press, London (1978).

22. J. W. Hadden and R. G. Coffey, Cyclic nucleotides in mitogen induced lymphocyte proliferation, Immunol. Today 3:299 (1982).

23. J. W. Hadden, R. G. Coffey, and F. Spreafico, "Immunopharmacology," Plenum Press, New York (1977).

24. J. W. Hadden, A. Englard, J. R. Sadlik, and E. M. Hadden, The comparative effects of isoprinosine, levamisole, muramyl dipeptide and SM1213 on lymphocyte and macrophage proliferation and activation in vitro, Int. J. Immunopharm. 1:17 (1979).

25. U. Hammerling, A. Chin, and M. Scheid, The ontogeny of murine B lymphocytes, J. Immunol. 115:1425 (1975).

26. M. Hoffmann, O. Weiss, S. Koenig, J. Hirst, and H. Oettgen, Suppression and enhancement of the T cell dependent production of antibody to SRBC in vitro by bacterial lipopolysaccharide, J. Immunol. 114:738 (1975).

27. M. Hoffmann, C. Galanos, S. Koenig, and H. Oettgen, B cell activation by lipopolysaccharide, J. Exp. Med. 146:1640 (1977).

28. S. Katz, F. Kierszenbaum, and B. Waksman, Mechanisms of action of lymphocyte-activating factor, J. Immunol. 121:2386 (1978).

29. J. Kurland and R. Bockman, Prostaglandin E production by human blood monocytes and mouse peritoneal macrophages, J. Exp. Med. 147:952 (1978).

30. M. T. Largen and B. Votta, Immunocytochemical evidence for 3',5'-cGMP and 3',5'-cGMP-dependent protein kinase involvement in lymphocyte proliferation, J. Cyclic. Nucleotide and Protein Phosphorylation Res. 9:231 (1983).

31. F. Melchers and J. Andersson, IgM in bone marrow-derived lymphocytes, Eur. J. Immunol. 4:181 (1974).

32. P. Naylor, C. Camp, A. Phillips, G. Thurman, and A. Goldstein, Effect of thymosin and lipopolysaccharide on murine lymphocyte cyclic AMP, J. Immunol. Meth. 20:143 (1978).

33. Y. Nishizuka, The role of protein kinase C in cell surface signal transduction and tumor promotion, Nature 308:693 (1984).

34. A. Nowotny, "Beneficial Effects of Endotoxin," Plenum Press, New York (1983).

35. J. Ohara and T. Watanabe, Microinjection of macromolecules into normal murine lymphocytes by cell fusion techniques, J. Immunol. 128:1090 (1982).

36. J. Oppenheim, W. Koopman, L. Wahl, and S. Dougherty, Prostaglandine E_2 rather than lymphocyte-activating factor produced by activated human mononuclear cells stimulates increases in murine thymocyte cAMP, Cell. Immunol. 49:64 (1980).

37. K. Ozato, W. Adler, and J. Ebert, Synergism of bacterial lipopolysaccharide and concanavalin A in the activation of thymic lymphocytes, Cell. Immunol. 17:532 (1975).

38. C. Rochette-Egly and M. G. Tovey, Interferon enhances guanylate cyclase activity in human lymphoma cells, Biochem. Biophysica. Res. Commun. 107:105 (1982).

39. S. W. Russell, Involvement of endotoxin in macorphage activation for antitumor immunity, in: "Proceedings of International Symposium on the Immunobiology and Immunopharmacology of Bacterial Endotoxins - Basic and Clinical Aspects," A. Szentivanyi, H. Friedman, and A. Nowotny, eds., Plenum Press, New York (1986).

40. M. Scheid, M. Hoffmann, K. Komuro, U. Hemmerlin, J. Abbott, E. Boyse, G. Cohen, J. Hooper, R. Schulof, and A. Goldstein, Differentiation of T cells induced by preparations from thymus and by nonthymic agents, J. Exp. Med. 138:1027 (1973).

41. M. P. Scheid, G. Goldstein, and E. A. Boyse, The generation and regulation of lymphocyte populations, J. Exp. Med. 147:1727 (1978).

42. J. Schmidtke and J. Najarian, Synergistic effects on DNA synthesis of phytohemagglutinin or concanavalin A and lipopolysaccharide in human peripheral blood lymphocytes, J. Immunol. 114:742 (1975).

43. B. J. Shenker and I. Gray, Cyclic nucleotide metabolism during lymphocyte transformation. I. Cell. Immunol. 43:11 (1979).

44. B. J. Shenker and I. Gray, Cyclic nucleotide metabolism during lymphocyte transformation. II. Cell. Immunol. 43:23 (1979).

45. B. S. Strauss and C. A. Stetson, Studies on the effect of certain macromolecular substances on the respiratory activity of the leucocytes of peripheral blood, J. Exp. Med. 112:653 (1960).

46. W.-H. Tsien, M. Sampson, and H. Sheppard, The elevation of mouse thymus cell cyclic adenosine monophosphate (cAMP) by lipopolysaccharide, Immunopharm. 3:253 (1981).

47. S. Vogel, M. Hilfiker, and M. Caulfield, Endotoxin-induced T-lymphocyte proliferation, J. Immunol. 130:1774 (1983).

48. J. Watson, The influence of intracellular levels of cyclic nucleotides on cell proliferation and the induction of antibody synthesis, J. Exp. Med. 141:97 (1975).

49. J. Watson, The involvement of cyclic nucleotide metabolism in the initiation of lymphocyte proliferation induced by mitogens, J. Immunol. 117:1656 (1976).

50. J. Watson, Involvement of cyclic nucleotides as intracellular mediators in the induction of antibody synthesis, in: "Comprehensive Immunology," R. A. Good and S. B. Day, eds., Plenum Press, New York (1977).

51. J. Watson, R. Epstein, and M. Cohn, Cyclic nucleotides as intracellular mediators of the expression of antigen-sensitive cells, Nature 246:404 (1973).

52. G. Weissman and L. Thomas, On a mechanism of tissue damage by bacterial endotoxins, in: "Bacterial Endotoxin," M. Landy and W. Braun, eds., Rutgers University Press, New Brunswick, New Jersey (1964).

53. P. Wrightman and C. Raetz, The activation of protein kinase C by biologically active lipid moieties of lipopolysaccharide, J. Biol. Chem. 259:10048 (1984).

54. S. Williamson, M. Wannemuehler, E. Jirillo, D. Pritchard, S. Michalek, J. McGhee, LPS regulation of the immune response: separate mechanisms for murine B cell activation by lipid A (direct) and polysaccharide (macrophage-dependent) derived from bacteroide LPS, J. Immunol. 133:2294 (1984).

55. R. A. Winchurch, C. Hilberg, W. Birmingham and A. Munster, Lipopolysaccharide-induced activation of suppressor cells: reversal by an agent which alters cyclic nucleotide metabolism, Immunol. 45:147 (1982).

56. D. Wood and P. Cameron, Stimulation of the release of a B cell activating factor from human monocytes, Cell. Immunol. 21:133 (1976).

SECTION III. IMMUNOGENICITY AND NON-SPECIFIC EFFECTS OF ENDOTOXIN

Nonspecific Effects of LPS on Bacterial Infections
L. Chedid, M. Parant, F. Parant, and G. Riveau

Bacteroides Species Increases Lipopolysaccharide Susceptibility
of Experimental Animals
A. C. Rodloff, P. Gadke, F. Lux, and H. Hahn

The Role of Post-Endotoxin Serum Components from BCG Infected Mice
in the Protection of Compromised Hosts
Renate Urbaschek, Daniela N. Mannel, Stephan E. Mergenhagen,
and Bernhard Urbaschek

Protection Against Lethal Haemophilus Pleuropneumoniae Infection
in Swine by Antibodies to LPS Core Antigens
B. W. Fenwick, J. S. Cullor, and H. J. Olander

Salmonella Antigens as Protective Immunogens in Salmonella Infection
Toby K. Eisenstein, Loran M. Killar, Barnet M. Sultzer,
and Marshall Phillips

Protective Function of Neutrophils During Experimental Endotoxic Shock
B. W. Fenwick, J. S. Cullor, and A. Kelly

Active Immunization with E. coli J5 and its Protective Effects from
Endotoxic Shock in Calves
J. S. Cullor, B. Fenwick, B. P. Smith, K. Pelzer, A. Kelly, and B. I. Osburn

Clinical Recognition and Treatment of Endotoxinemia
Joseph G. Sinkovics

Transient Suppression of Yeast Phagocytosis Induced by Legionella
Pneumophila in Cultures of Murine Resident Peritoneal Macrophages
Jeanne Becker, Robert J. Grasso, and Herman Friedman

The Effect of Endotoxin on Migration Inhibitory Factor and Interferon
Samuel B. Salvin and Pamela B. Renda

NONSPECIFIC EFFECTS OF LPS ON BACTERIAL INFECTIONS

L. Chedid, M. Parant, F. Parant, and G. Riveau

Institute Pasteur, Immunotherapie Experimentale
Paris, France

INTRODUCTION

It is obvious for an investigator in the field of resistance to bacterial infections that specific antibody formation has its limitations as a
defense mechanism. A primary immune response is the end product of a sequence of events and would occur too late to rescue the host in many septicemias, which is even more evident if we consider neonatal resistance.
Antibody formation is probably one of the highest forms of protein synthesis
in evolution (13,17). In contrast, other major defense mechanisms such as
phagocytosis or elevation of temperature have been shown to be more powerful
and more anciently established in many phylogenetic studies (17).

Almost twenty years ago, we presented data relating to a system by
which the host may be capable of coping rapidly with a spectrum of gram
negative pathogens. This hypothesis assumed that smooth virulent strains
could be modified by the host in such a manner that these organisms were
rendered sensitive to serum factors which could react with rough antigens
(24). For instance, a limited number, or conceivably a single type of antibody could deal with a great variety of strains which were clearly related
in terms of rough antigens although serologically distinct in smooth phenotype. Our experiments used an infection by Klebsiella pneumoniae as a
model, and also the common rough structure underlying various types of
lipopolysaccharides extracted from smooth strains of Salmonellae (5).

Previous experiments had shown that endotoxins could activate enzymes
that alter cell wall components, and are capable of detoxifying endotoxins
in vitro (4,36). In vivo these same lipopolysaccharides are capable of
enhancing resistance to a lethal challenge by endotoxin or by Klebsiella
organisms (24). It was first observed that mice treated with a minute amount
of endotoxin can survive for several days even if they have been inoculated
with one million lethal doses of an extremely virulent strain of K.
pneumoniae. During this prolonged period of survival, the number of bacteria
remains at a stationary level in infected mice. This equilibrium was not
related to a bacteriostatic effect but instead to a synchronism between the
rate of division of the microorganisms and the rate of their destruction by
the host as experiments with sulfonamide have shown (24). Rate of clearance
of ^{51}Cr-labeled bacteria, and determination of the number of organisms in
the RES argued against the possibility that this effect was mediated by
opsonins. Experiments in which bacteria recovered from the liver of

endotoxin-treated donors have been reinoculated into normal recipients, indicated also that a nonspecific mechanism capable of causing a modification of the phenotype of the microorganisms plays a significant role. The nature of this modification in endotoxin-treated hosts appears to be an alteration of endotoxic cell wall constituents of the bacteria. This alteration enhances the bacteria's susceptibility to phagocytosis but is not maintained during subsequent multiplication in vitro. This view was based on findings showing that serial passage of bacteria recovered from endotoxin-stimulated hosts and cultured before reinoculation in mice consistently fails to show any decrease in virulence such as would be associated with appreciable loss of smooth antigens (24). This phenotypic modification which could not be detected by a change of the colonial aspect, or by agglutination techniques was assumed to be related to a discrete exposure or production of a few antigenic surface sites capable of reacting with "rough" antibodies. Indeed, later findings showed that administration of high titer antiserum prepared in a horse immunized with a rough Salmonella typhimurium strain TV 119 protected mice infected with a highly virulent strain of K. pneumoniae. The protective effect of this antiserum was removed by absorption with rough organisms but not by smooth strains of S. typhimurium or S. typhi. Provided the Klebsiellae are preincubated in normal mouse serum, bactericidal activity can also be demonstrated in vitro. This activity does not require the presence of complement and can be removed either by the same Klebsiella strain or by rough strains of Salmonellae, but not by the smooth strain of Salmonella from which these mutants originate (5).

The bactericidal effect of normal mouse serum against a very limited number of smooth gram negative bacteria, generally ascribed to natural "O" antibodies, can be removed by absorption with rough microorganisms. Passive hemagglutination tests disclose the presence of thermolabile, non-agglutinating "R" antibodies in normal mouse serum; titers in germ-free animals are much lower (5). More recently, a complement-dependent bactericidal factor specific for the Ra chemotype was found in sera of mice. This factor was also present in all species tested by these investigators, be they of mammals, birds, reptiles, amphibians, and fish. This factor which has been conserved by vertebrates for more than 300 million years, was found to bind specifically N-acetyl-α-D-glucosaminyl and L-glycero-D-mannoheptosyl residues of rough core polysaccharide (15).

In any event such findings, together with previous observations concerning degradation of somatic antigen and phenotypic modification of Klebsiellae recovered from endotoxin-treated mice, have led us to the development of a different concept of natural immunity to Enterobacteriaceae. In this hypothesis, factors capable of attacking cell wall components unmask R antigenic sites common to many bacterial strains and species. It should be noted that serum contains factors capable of detoxificating the lipopolysaccharide molecule in the cell wall (21) as well as in its isolated form (4,20,36). Following this de-aggregation, a few types of "R" antibodies or of serum factors reacting with rough antigens should have the capability of coping, like masterkeys, with a very wide range of infections due to serologically unrelated organisms.

We will now describe some results which are more recent or are even unpublished, and which relate to other aspects of the possible mechanisms of enhancement of the nonspecific effect of LPS on bacterial infections. We will also report results concerning detoxified LPS preparations and synthetic immunomodulating molecules which copy part of the peptidoglycan structure. Finally, we will discuss the possible role of these agents on the production of fever, and of tumor necrotizing factor (TNF). It is indeed well established that fever plays an important role in nonspecific immunity, and we have previously reported that mouse serum containing TNF activity had the capacity of enhancing resistance to bacterial infections (29).

Table 1. Comparison of LPS, SPLPS and Muramyl Dipeptides

	MDP*	Effective Doses Toxicity[+]	Anti-Klebsiella activity
	(µg/kg)	(µg)	(µg)
LPS	0.0016	0.02	0.001
SPLPS	25	2,000	10
MDP	25	>10,000	10
Murabutide	>10,000	>10,000	10

*MDP = minimal pyrogenic dose in rabbit by the intravenous route
[+]LD_{50} by the intravenous route in adrenalectomized mice

Comparison of Toxicity and of Anti-Klebsiella Activity of LPS Preparations and of Muramyl Peptides Administered by Various Routes

Lipopolysaccharides (LPS) are very potent adjuvants and also very strong immunostimulating agents. Detoxification of LPS by phthalylation produces molecules (hereafter referred to as SPLPS) which retain its adjuvant activity (7,22). More recently, a synthetic glycopeptide, MDP (Nac-Mur-L-Ala-D-iGln) and several analogues have been shown to be adjuvant-active compounds (1,18). Contrary to LPS, muramyl peptides have no detectable toxicity in small rodents. However, like LPS, MDP can also produce fever and represents the minimal pyrogenic structure of bacterial peptidoglycan (19,32). In contrast, an MDP analogue named Murabutide (AcMur-L-Ala-D-Gln-OnBu) is devoid of pyrogenicity (9). In adrenalectomized mice, the compared toxicity of LPS, SPLPS and muramyl peptides are the following: 0.02 µg, 2 mg and >10 mg per mouse (Table 1).

Administered by the parenteral route, SPLPS has lost the LPS capacity of protecting mice against Klebsiella infection since 2 mg of SPLPS may still contain 2 ng of free toxic LPS (Table 1). MDP and Murabutide administered by the same route are able to enhance resistance to a consecutive Klebsiella infection although this activity is clearly weaker than in the case of LPS (9). However, MDP can protect mice by the oral route whereas LPS is ineffective (8). Also in contrast to LPS, MDP can protect newborn mice by parenteral or even by oral route (26).

Influence of Age on Nonspecific Resistance to Klebsiella Induced by LPS or MDP

During early postnatal life, mice have been found to be immunologically deficient and to be extremely susceptible to bacterial infections. Their poor immune status may be related to suppressor cells, and also to the lack of maturity of macrophages and lymphocytes. Little is known concerning the capacity of newborn mice to respond to nonspecific immunostimulants except that LPS is devoid of adjuvant activity (37). Previous studies in these animals have shown that endotoxin weakly stimulates nonspecific resistance to Klebsiella infection although it can increase the phagocytic activity (29). As depicted in Figure 1, mice challenged with Klebsiella could be protected by endotoxin only from 14 days of age. In contrast, MDP can protect seven and even one day old mice (26). In these experiments, both the treatment and the infectious challenge were administered by the subcutaneous route, and under these conditions LPS was completely devoid of protective effect. MDP can also protect young animals by the oral route.

Comparison of Anti-Klebsiella Activity of LPS and of Muramyl Peptides in Splenectomized Mice

Various forms of immunosuppressive treatment have been used to evaluate the potency of these immunostimulants. Most of these treatments render the host highly susceptible to bacterial infections, and treatment with immuno-stimulants may turn out to be of great importance. Irradiation, cyclophos-phamide or corticoid administration reduced natural resistance to a bacterial challenge, but like LPS (23) muramyl peptides greatly enhanced nonspecific immunity (8).

Patients, mainly children, who have undergone splenectomy are known to suffer often from severe infections with microorganisms possessing polysac-charide capsules. In our experimental model, mice were splenectomized at seven days of age and were challenged after five or six weeks. No mortality and a normal development were observed in these animals. They were however, highly susceptible to Klebsiella infection, the minimal lethal dose being ten organisms compared to 8×10^3 in sham-operated controls. LPS, MDP and

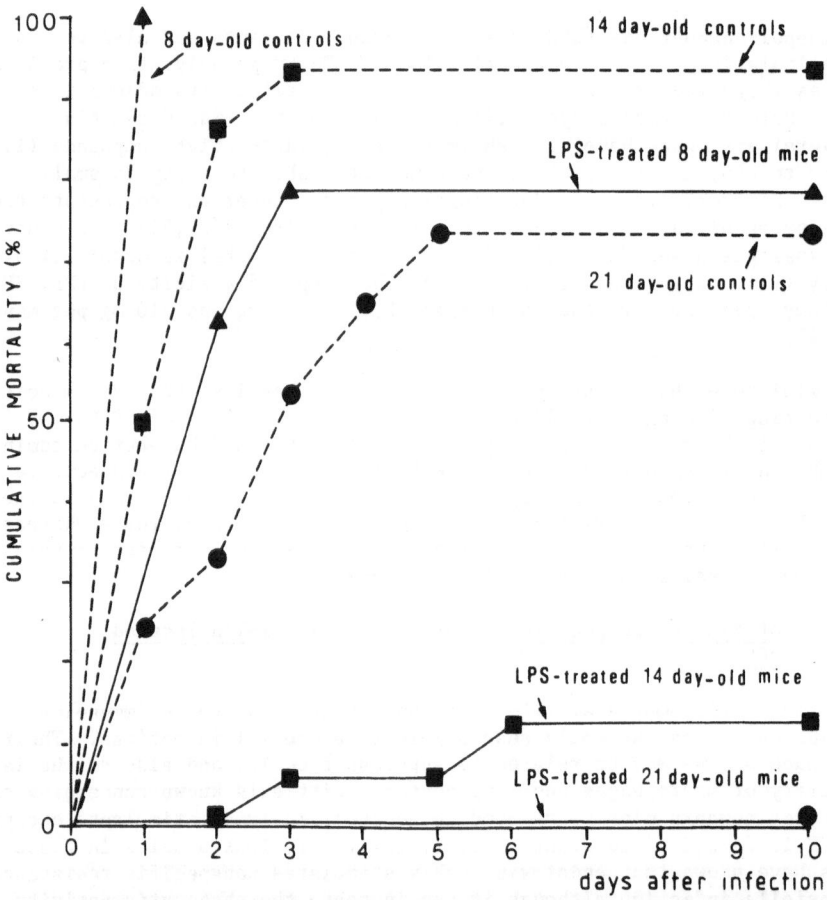

Figure 1. Influence of age on nonspecific resistance to Klebsiella pneumoniae infection induced by LPS. Mice were treated sub-cutaneously with 1 μg LPS one day before being infected intravenously with 10^3 organisms.

Table 2. Protective Effect of LPS, SPLPS, and Muramyl Dipeptide Against a _Klebsiella pneumoniae_ Infection in Neonatally Splenectomized Mice*

Treatment (day -1)	Mortality (dead/total)			$\underline{p}^{+}$
	day +3	day +6	day +9	
Saline	25/32	30/32	30/32	
LPS 1 µg	1/32	17/32	17/32	<0.01
SPLPS 1 µg	24/32	32/32	32/32	
MDP 100 µg	0/32	9/32	9/32	<0.01
Murabutide 100 µg	0/32	10/32	10/32	<0.01

*Adult DBA/2 mice that had been splenectomized at seven days of age. They were infected intravenously with ten organisms.
+By the adjusted chi-squared method.

Murabutide are efficient in stimulating the resistance of splenectomized mice whereas SPLPS was inactive (Table 2).

Influence of Mouse Strain on Nonspecific Resistance to Infection Induced by LPS or MDP

Whereas LPS is devoid of protective activity in the LPS low-responder C3H/HeJ mice, it protects effectively the high responder C3HeB/He strain and F1 hybrids (6). Even against a minimal lethal dose (about ten organisms in all these mice), LPS failed to protect the C3H/HeJ mice, whereas the high-responder strain was fully protected by the same treatment against 10^5 K. pneumoniae. However, the protection conferred by LPS in F1 hybrids was less marked as the challenge became more severe (23). Thus only 80% of the F1 hybrids were protected unless they were challenged by a smaller inoculum. A delay of several days in mortality was often observed in hybrids. Similarly, pretreatment with Corynebacterium could protect F1 hybrids when they were infected with a small number of Klebsiella organisms, but this stimulation was still less effective than LPS. The high-responder mouse strain was still protected when the number of viable organisms was increased (up to 3 x 10^4 K. pneumoniae), whereas no F1 hybrids survived. Therefore, results have shown that the inability of C3H/HeJ mice to respond to endotoxin does not depend on the virulence of the bacterial strain. The data obtained with Corynebacterium suggest, however, that unresponsiveness to LPS-enhanced resistance to infection may be related to a genetic defect in the capacity of C3H/He cells to respond to activation stimuli rather than to their unresponsiveness only to LPS as has been proposed by Ruco and Meltzer (35) and by Rosenstreich and Vogel (34). Thus, neither the previous results (25) nor the present ones permit a definite answer on the inheritance of LPS-induced resistance to infection. In such experiments studying resistance to a systemic challenge, both the responsiveness to LPS and the innate ability to clear and destroy invading bacteria are involved (23).

Transfer of cells from the histocompatible C3HeB/Fe mice into thymectomized and irradiated C3H/HeJ recipients showed that bone marrow cells completely restored the capacity of low-responder mice to be stimulated by LPS and fully protected against the Klebsiella challenge. Conversely, high-responder mice similarly reconstituted one month before with bone marrow cells from low-responder mice could not be definitively protected but retained their capacity to clear and destroy Klebsiella organisms during a few hours

205

after injection. Therefore, several types of cells could be incriminated and LPS unresponsiveness appears to be related to a defect of a radioresistant and long-lived cell type, and also to a radiosensitive cell (12).

In contrast to LPS and to <u>Corynebacterium</u>, MDP was found inactive in both low- and high-responder C3H strains.

<u>Induction of LAF and Endogenous Pyrogen Production by LPS, MDP and Murabutide</u>

In rabbits the minimal pyrogenic dose of LPS is 1.65 ng/kg and of MDP 25 µg/kg, whereas Murabutide is non-pyrogenic. During the febrile response, circulating endogenous pyrogen (EP) can be detected but not with Murabutide. Besides its capacity to elicit fever, EP can produce hypoferremia, hypercupremia and an increase in the synthesis of serum acute-phase proteins by the liver (14). EP is currently thought to be identical with or closely associated with Interleukin-1 (IL-1) (11). The pyrogenic adjuvant MDP is known to stimulate the monocyte/macrophage to produce pyrogenic IL-1. Nevertheless, by using certain MDP derivatives such as Murabutide it has been possible to selectively dissociate the production of EP and of "lymphocyte-activating factor" activity (LAF/IL-1) (10). MDP and its derivatives that generated circulating EP could elicit hypoferremia and hypercupremia in the rabbit. Moreover, upon injection of various macrophage supernatants, only those containing EP evoked the typical changes in metal concentrations, whereas non-pyrogenic IL-1 did not alter circulating metal levels. These results suggest that LAF production may be independent from MDP-induced hypoferremia and hypercupremia (33).

How do these findings and considerations relate to resistance to infection? Studies of Kluger (16) have shown that elevation of temperature can protect the host against infection. This was shown to be even true in lower vertebrates which can benefit through behavioral changes of environmental elevation of their temperature. This was also shown by the same author in rabbits infected with <u>Pasteurella</u> <u>multocida</u>. In vitro studies allowed him to demonstrate that concomitant increase of temperature and decrease of

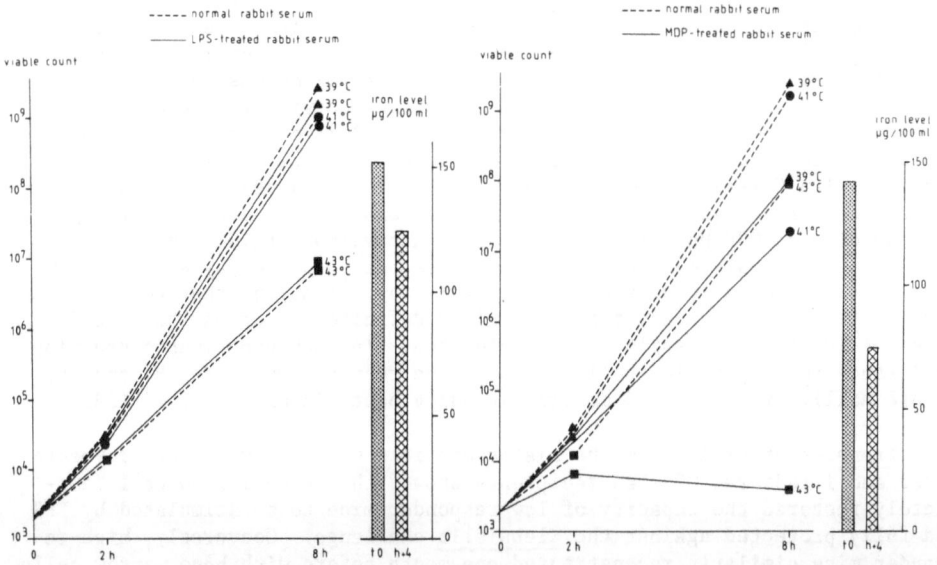

Figure 2. Bacterial growth of <u>Klebsiella</u> <u>pneumoniae</u> incubated in normal rabbit serum or in rabbit given LPS or MDP 4 hrs previously.

iron and zinc levels in the blood could check the growth without killing
the organisms. Thus, in a normal serum P. multocida proliferates in the
same fashion at 39°C, 40°C, 41°C or 42°C. When the metal levels are lowered,
the organisms still proliferate at 39°C (which is the normal temperature in
the rabbit) but their growth is inhibited completely at 42°C and partially
at 41°C. This inhibition is related to the influence of temperature on the
"iron pump" which by being blocked can no longer actively concentrate these
ions. Low internal iron level allows the organisms to survive but hinders
several of the syntheses required for production of virulence structures.

Transfer of Klebsiella Incubated at Various Temperatures in Plasma
Containing Low Concentrations of Iron

We have been able to obtain similar results using Klebsiella and the
serum of rabbits pretreated by either 1 mg/kg of MDP or by 10 µg/kg of LPS.
Both agents at these dosages produce very high fever. Yet four hours later,
the iron levels were lowered 47.2% after MDP and only 18.4% after LPS, because this dose is too high and toxic for the rabbit. Klebsiella organisms
(2×10^3) were incubated in the serum collected from the same donors either
before or four hours after the intravenous injection of MDP or LPS. Serum
samples were incubated at 39°C, 40°C, and 43°C for two or eight hours. As
can be seen on Figure 2, no effect was observed after LPS at all temperatures,
whereas the organisms proliferated much less when they were incubated in
serum samples obtained from rabbits pretreated with MDP (and which had a
50% reduction of its iron content) (Figure 2).

The morphology of these organisms is changed and so is their virulence
if they are administered to normal mice. Klebsiellae were incubated at
41°C for eight hours in the serum from a normal or from an MDP-treated

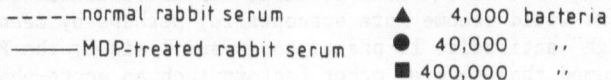

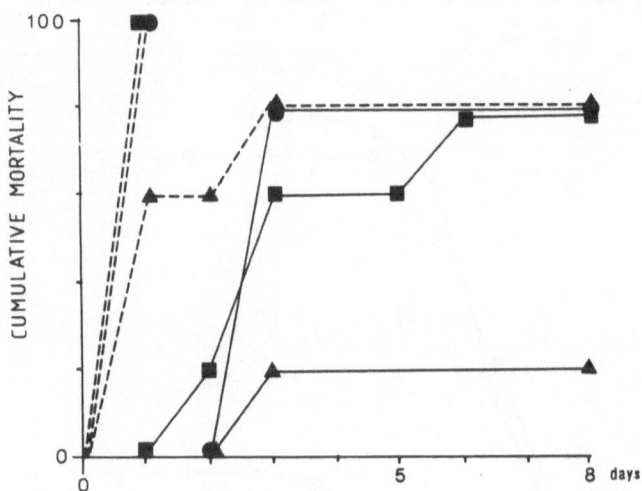

Figure 3. Influence of incubation at 41°C for 8 hrs in an iron-poor
serum (MDP-treated) rabbit) on virulence of Klebsiella
pneumoniae in mice. Controls receiving bacteria incubated
in broth medium (not reported in figure) died like those
receiving organisms incubated in normal rabbit serum.

Table 3. Lethal Effect of LPS and Muramyl Dipeptide Injected
Simultaneously into DBA/2 Mice

Treatment (dose per mouse)	Mortality
LPS 30 µg	0/12
LPS 30 µg + MDP 100 µg	8/12
LPS 30 µg + MDP 300 µg	8/12
LPS 30 µg + Murabutide 300 µg	0/12
LPS 50 µg	1/22
LPS 50 µg + MDP 30 µg	13/22
LPS 50 µg + MDP 100 µg	19/22
LPS 50 µg + Murabutide 100 µg	1/12

LPS was given by i.v. route and MDP by i.p. route.

rabbit. They were then inoculated at various amounts in normal mice. When
400,000 or 40,000 organisms had been incubated in normal serum, control
mice died in less than three days. In contrast, the same number of <u>Klebsi
ella pneumoniae</u> were less virulent after incubation in the serum from MDP-
treated rabbit, which had a low iron concentration (Figure 3).

At this stage, one could assume that the rapid production of EP by
activated macrophages is capable of modifying the virulent phenotype by the
combined action of fever and lowered metal concentration in the blood.
These organisms would become more susceptible, perhaps by means of pre-
existing "rough" antibodies to phagocytosis and lysis by the RES. It can
be easily assumed that several other factors such as acute-phase reactants,
must also be engaged in such a process. However, the above explanation
(temperature increase and iron decrease) does not answer two major questions:

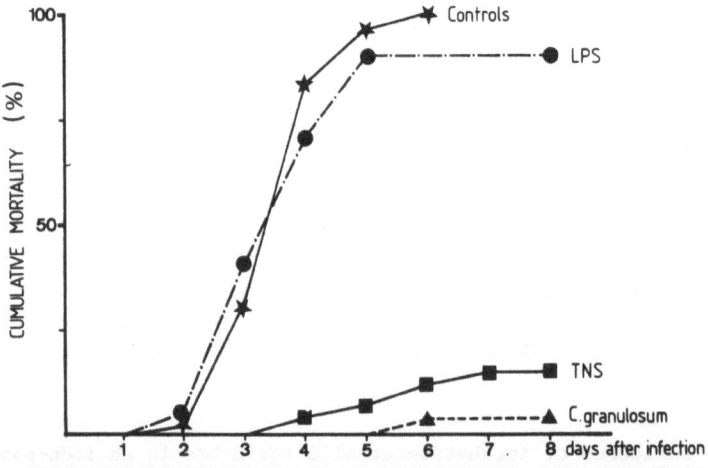

Figure 4. Protective activity of tumor necrotizing serum (TNS) in C3H/HeJ
mice infected by the intravenous route with 6×10^4 <u>Listeria
monocytogenes</u>. TNS (0.5 ml) was given i.v. 2 hrs before
challenge.

a) LPS and MDP, which are both pyrogenic in rabbits, can enhance the resistance of mice to <u>Klebsiella</u> although they do not elevate their body temperature;

b) Murabutide, which is not pyrogenic, does not produce hypoferremia or endogenous pyrogen in rabbits and can yet protect mice against <u>Klebsiella</u>.

There must, therefore, exist besides these mechanisms one or several other nonspecific factors. An interesting candidate could be tumor necrotizing factor (TNF). As mentioned, macrophage activation produces several monokines other than IL-1 such as CSF or TNF.

Synergism Between LPS and MDP

High levels of TNF are usually produced in mice inoculated with viable BCG and given two weeks later 25 µg of LPS which represents about 50 lethal doses in BCG-treated mice. Two hours later and just before they would die, they are bled and high levels of TNF are found in their serum (3).

For BCG, there also exists a toxic synergism between LPS and MDP which was first demonstrated by Ribi et al. in guinea pigs (31). We confirmed these results in mice, and also showed that in contrast to MDP, Murabutide does not enhance the toxicity of LPS under the same conditions (Table 3).

It is also possible to produce TNF by administering MDP and LPS simultaneously. In all experiments, the presence of TNF was detected by measuring the cytotoxicity of the serum on L 929α target cells in vitro. Care was taken to verify that TNF, in contrast to EP, is not destroyed by heating at 56°C.

Previous experiments have shown us that mouse serum containing TNF activity protected adult mice against two types of challenges with <u>Klebsiella pneumoniae</u> or with the intracellular parasite <u>Listeria monocytogenes</u>. Moreover, TNF activity was demonstrated in animals which are refractory to LPS and very susceptible to infections, such as adult C3H/HeJ mice (Figure 4) and seven day old mice (Figure 5). Protection passively transferred by TNF

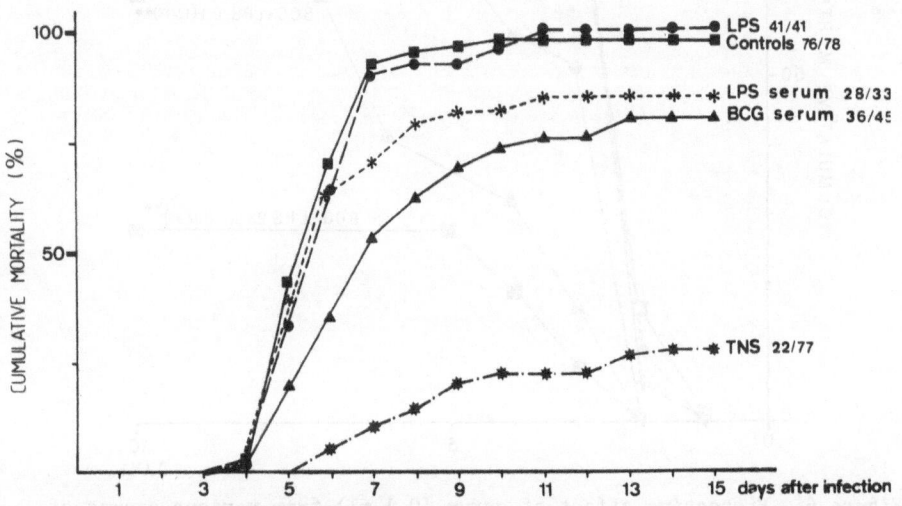

Figure 5. Protective activity of tumor necrotizing serum (TNS) in 7-day-old mice infected subcutaneously with 1.5 x 10^3 <u>Listeria monocytogenes</u>. TNS (0.1 ml) was given subcutaneously 24 hrs before challenge.

was not related to antibodies, since it was not decreased by absorption with homologous organisms (27).

In the following experiment, adult mice were treated by the intravenous route with MDP or BCG separately or in combination with LPS. These sera were injected subcutaneously to seven day old mice which were challenged 24 hours later with 10^3 K. pneumoniae. As can be seen, mice which received sera from MDP, LPS, or BCG plus MDP died as rapidly as PBS-treated controls. In contrast, sera from mice treated with LPS and BCG, or LPS and MDP were capable of increasing the resistance to infection (Figure 6). Further purification will be needed to ascribe the whole protective activity to TNF.

CONCLUSION

It was probably most unfortunate that aspirin, besides being an anti-pyretic agent, relieves discomfort and headache, leading people to believe that its use is more beneficial than it really is during infectious diseases. Nevertheless, and although the fortunes of fever have waned in recent years, a renewed interest in the field has lately developed (17).

It seems rather paradoxical that selection would have retained such a costly process as fever (a rise in body temperature of 3°C will result in approximately 25% of increased energy expenditure) to help the host to get rid of thermosensitive bacterial and viral agents. Yet in recent years it

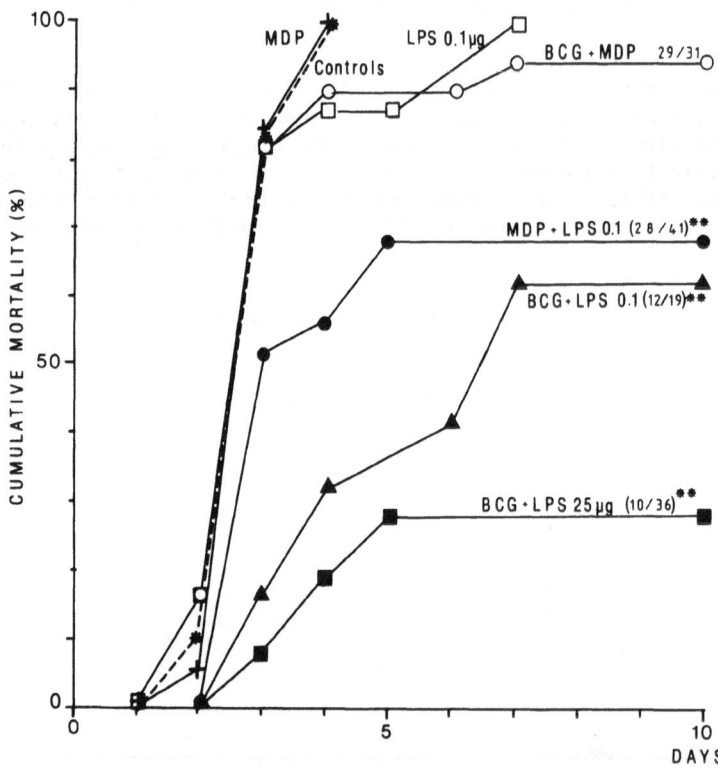

Figure 6. Protective effect of serum (0.1 ml) from various groups of adult mice in 7-day-old challenged subcutaneously one day later with 1 x 10^3 K. pneumoniae. Treatment of donors: BCG 1 mg on day −14; MDP 100 μg at h −2; LPS 25 μg or 0.1 μg at h −2 before bleeding.

has been well established that this unique response is very generalized in vertebrates. Thus, following inoculation of microbial organisms, ectotherms select a warmer environmental temperature and fever developed by this process in many nonmammalian vertebrates increased their resistance to infections (16). Similar experiments have been made in mammals. When rabbits were infected with Pasteurella multocida, a beneficial effect was observed if their temperature was increased about 2°C, whereas a statistically higher mortality rate was observed when it was lowered by administration of sodium salicylate (17). Theoretically, fever could enhance resistance by two possible mechanisms: 1) inhibition of proliferation of thermosensitive organisms; and 2) production by exogenous pyrogens of monokines and/or other mediators which stimulate more directly the immune system.

It has been postulated that a macrophage factor (IL-1) could be responsible for both fever production and for lymphocyte activation. A consensus developed stating that several monokines previously identified as being separate entities represented different facets of the same molecule (11). However, it seems that different activities attributed to IL-1 such as pyrogenicity or lymphocytic activation can be selectively produced (2,10). Interestingly, pyrogenic IL-1 can lower the concentration of plasma iron which in turn can influence bacterial proliferation during fever. Acute-phase proteins also have a role in nonspecific resistance as indicated by studies concerned with leukocytic endogenous mediator (LEM), a soluble mediator which may be identical to EP (see 14 for review). Phenotypic alteration of bacterial cell wall could render the organisms more sensitive to certain other serum factors.

Recently, Kawakami et al. (1984) found a bactericidal factor specific for the Ra chemotype strains of Salmonella in sera of normal mice and humans. This factor, called the Ra-reactive factor (RaRF), binds specifically to residues of rough core polysaccharide and presumably activates the classical pathway of complement after the binding to the determinant. The mouse RaRF is a protein with an approximate MW of 300,000 and is composed of polypeptides with MW of 28,000 and 70,000. Similar results were obtained with the human RaRF molecule. Its polypeptide composition, sensitivity to heat and reducing agents, and the requirement of Ca^{++} for its binding to the determinant indicate that it is not an immunoglobulin. By titration of complement activity remaining in the RaRF-deprived sera, it was demonstrated that the RaRF is not a complement component (15).

There likely exist other factors since resistance can be enhanced by agents, or under conditions, which do not produce temperature elevation or hypoferremia (28). One of these factors could be TNF which has been shown to enhance resistance to bacterial infections. Our experiments were performed with crude preparations. Since TNF has recently been cloned and identified (30), it would be most interesting to see whether this molecule is endowed with both anti-tumor and anti-bacterial activities, or whether another molecule is responsible for the latter effect.

REFERENCES

1. A. Adam, J. F. Petit, P. LeFrancier, and E. Lederer, Muramyl peptides. Chemical structure, biological activity and mechanism of action, Mol. Cell. Biochem. 41:27 (1981).
2. N. E. Byars, Two adjuvant-active muramyl dipeptide analogs induce differential production of lymphocyte-activating factor and a factor causing distress in guinea pigs, Infect. Immun. 44:344 (1984).
3. E. A. Carswell, L. J. Old, R. L. Kassel, S. Green, N. Fiore, and B. Williamson, An endotoxin-induced serum factor that causes necrosis of tumors, Proc. Natl. Acad. Sci. USA 72:3666 (1975).

4. L. Chedid, M. Parant, F. Boyer, and R. C. Skarnes, Non-specific host responses in tolerance to the lethal effect of endotoxin, in: "Bacterial Endotoxins," M. Landy and W. Brown, eds., Rutgers University Press, New Brunswick, NJ (1964).

5. L. Chedid, M. Parant, F. Parant, and F. Boyer, A proposed mechanism for natural immunity to enterobacterial pathogens, J. Immunol. 100:292 (1968).

6. L. Chedid, M. Parant, C. Damais, F. Parant, D. Juy, and A. Galelli, Failure of endotoxin to increase nonspecific resistance to infection of lipopolysaccharide low-responder mice, Infect. Immun. 13:722 (1976).

7. L. Chedid, F. Audibert, C. Bona, C. Damais, F. Parant and M. Parant, Biological activities of endotoxins detoxified by alkylation, Infect. Immun. 12:714 (1975).

8. L. Chedid, M. Parant, F. Parant, P. LeFrancier, J. Choay, and E. Lederer, Enhancement of nonspecific immunity to Klebsiella pneumoniae infection by a synthetic immunoadjuvant (N-acetylmuramyl-L-alanyl-D-isoglutamine) and several analogs, Proc. Natl. Acad. Sci. USA 74:2089 (1977).

9. L. Chedid, M. Parant, F. Audibert, G. Riveau, F. Parant, E. Lederer, J. Choay, and P. LeFrancier, Biological activity of a new synthetic muramyl peptide adjuvant devoid of pyrogenicity, Infect. Immun. 35:417 (1982).

10. C. Damais, G. Riveau, M. Parant, J. Gerota, and L. Chedid, Production of lymphocyte-activating factor in the absence of endogenous pyrogen by rabbit or human leukocytes stimulated by a muramyl dipeptide derivative, Int. J. Immunopharmac. 4:451 (1982).

11. C. A. Dinarello, Interleukin 1, Rev. Infect. Dis. 6:51 (1984).

12. A. Galelli, Y. LeGarrec, and L. Chedid, Transfer by bone marrow cells of increased natural resistance to Klebsiella pneumoniae induced by lipopolysaccharide in genetically deficient C3H/HeJ mice, Infect. Immun. 23:232 (1979).

13. R. A. Good and B. W. Papermaster, Ontogeny and phylogeny of adaptive immunity, Adv. Immunol. 4:1 (1964).

14. R. F. Kampschmidt, Leukocytic endogenous mediator/endogenous pyrogen, in: "Infection, the Physiologic and Metabolic Responses of the Host," M. C. Powanda and P. G. Canonico, eds., Elsevier/North Holland Biomedical Press (1980).

15. M. Kawakami, I. Ihara, S. Ihara, A. Susuki, and K. Fukui, A group of bactericidal factors conserved by vertebrates for more than 300 million years, J. Immunol. 132:2578 (1984).

16. M. J. Kluger, "Fever, Its Biology, Evolution, and Function," Princeton University Press, Princeton, NJ (1979).

17. M. J. Kluger, Historical aspects of fever and its role in disease, in: "Thermoregulatory Mechanisms and Their Therapeutic Implications," S. Karger, Basel (1980).

18. S. Kotani, H. Takada, M. Tsujimoto, T. Kubo, T. Ogawa, J. Azuma, H. Ogawa, K. Matsumoto, W. A. Siddiqui, A. Tanaka, S. Nagao, O. Kohashi, S. Kanoh, T. Shiba, and S. Kusumoto, Nonspecific and antigen-specific stimulation of host defense mechanisms by lipophilic derivatives of muramyl dipeptides, in: "Bacteria and Cancer," J. Jeljaszewicz, G. Pulverer, and W. Roszkowski, eds., Academic Press (1982).

19. S. Kotani, Y. Watanabe, T. Shimono, K. Harada, T. Shiba, S. Kusumoto, K. Yokogawa, and M. Taniguchi, Correlation between the immunoadjuvant activities and pyrogenicities of synthetic N-acetylmuramyl peptides or -amino acids, Biken J. 19:9 (1976).

20. M. Landy, R. C. Skarnes, F. S. Rosen, R. J. Trapani, and M. J. Shear, On activation of biologically active ("endotoxic") polysaccharides by fresh human serum, Proc. Soc. Exp. Biol. Med. 96:744 (1957).

21. M. Landy, R. Trapani, and F. S. Rosen, Inactivation of endoxotin by a humoral component. VI. Two separate systems required for viable and killed Salmonella typhosa, J. Clin. Invest. 39:352 (1960).

22. F. C. McIntire, M. P. Hargie, J. R. Schenck, R. A. Finley, H. W. Sievert, E. T. Rietschel, and D. L. Rosenstreich, Biologic properties of non-toxic derivatives of a lipopolysaccharide from Escherichia coli K235, J. Immunol. 117:674 (1976).

23. M. Parant, Effect of LPS on nonspecific resistance to bacterial infections, in: "Beneficial Effects of Endotoxins," A. Nowotny, ed., Plenum Press, New York and London (1983).

24. M. Parant, F. Parant, L. Chedid, and F. Boyer, On the nature of some nonspecific host responses in endotoxin-induced resistance to infection, in: "The Reticuloendothelial System and Atherosclerosis," Plenum Press, New York and London (1967).

25. M. Parant, F. Parant, and L. Chedid, Inheritance of lipopolysaccharide-enhanced nonspecific resistance to infection and of susceptibility to endotoxic shock in lipopolysaccharide low-responder mice, Infect. Immun. 16:432 (1977).

26. M. Parant, F. Parant, and L. Chedid, Enhancement of the neonate's nonspecific immunity to Klebsiella infection by muramyl dipeptide, a synthetic immunoadjuvant, Proc. Natl. Acad. Sci. USA 75:3395 (1978).

27. M. Parant, F. Parant, and L. Chedid, Enhancement of resistance to infections by endotoxin-induced serum factor from Mycobacterium bovis BCG-infected mice, Infect. Immun. 28:654 (1980).

28. M. Parant, G. Riveau, F. Parant, C. A. Dinarello, S. M. Wolff, and L. Chedid, Effect of Indomethacin on increased resistance to bacterial infection and on febrile responses induced by muramyl dipeptide, J. Infect. Dis. 142:708 (1980).

29. M. Parant and E. Sacquet, Augmentation de la resistance a l'infection du souriceau conventionnel ou axenique apres une injection d'endotoxine, C.R. Acad. Sc. Paris 262:1914 (1966).

30. D. Pennica, G. E. Nedwin, J. S. Hayflick, P. H. Seeburg, R. Derynck, M. A. Palladino, W. J. Kohr, B. B. Aggarwal, and D. V. Goeddel, Human tumor necrosis factor: precursor structure, expression and homology to lymphotoxin, Nature, 312:724 (1984).

31. E. E. Ribi, J. L. Cantrell, K. B. Von Eschen, and S. M. Schwartzman, Enhancement of endotoxic shock by N-acetyl-muramyl-L-alanyl-(L-seryl)-D-isoglutamine (muramyl dipeptide), Cancer Res. 39:4756 (1979).

32. G. Riveau, K. Masek, M. Parant, and L. Chedid, Central pyrogenic activity of muramyl dipeptide, J. Exp. Med. 152:869 (1980).

33. G. Riveau, M. Parant, C. Damais, F. Parant, and L. Chedid, Dissociation between muramyl peptide-induced fever and changes in plasma metal levels, in press (1985).

34. D. L. Rosenstreich and S. N. Vogel, Central role of macrophages in the host response to endotoxin, in: "Microbiology," D. Schlesinger, ed., American Society of Microbiology, Washington, (1980).

35. L. P. Ruco and M. S. Meltzer, Defective tumoricidal capacity of macrophages from C3H/HeJ mice, J. Immunol. 120:329 (1978).

36. R. C. Skarnes and L. Chedid, Biological degradation and inactivation of endotoxin (chromate-labeled), in: "Bacterial Endotoxins," M. Landy and W. Braun, eds., Rutgers University Press, New Brunswick, NJ (1964).

37. J. Sterzl, M. Holub, and I. Miler, Effect of endotoxin on antibody response and resistance to infection in newborn animals, Folia Microbiol. (Prague) 6:289 (1961).

BACTEROIDES SPECIES INCREASES LIPOPOLYSACCHARIDE

SUSCEPTIBILITY OF EXPERIMENTAL ANIMALS

A. C. Rodloff, P. Gädke, F. Lux, and H. Hahn

Institute for Medical Microbiology
Free University of Berlin
Berlin, Federal Republic of Germany

INTRODUCTION

Using different animal models, several authors have reported on synergistic infections with enterobacteria and Bacteroides species (1,6,7,8,10,-11). Synergism was extensively described for abscess formation and lethality of animals. A number of different hypotheses have been offered to provide explanations for this type of synergistic action. Lowered tissue oxygen tension (2,8) changing electrode potential (6), production of soluble growth factors (2,7) and phagocytosis inhibition were discussed as possible mechanisms (for review see 3).

With our study, we have evaluated the potential of different Bacteroides species to enhance the lethality of intravenously induced infections with E. coli in mice. In addition, the effect of these anaerobes on experimental intoxication with E. coli-derived lipopolysaccharide was studied.

MATERIALS AND METHODS

Animals

Female (C57B1/6xDBA/2)FL mice, approximately ten weeks of age weighing about 22 g were used.

Bacteria

E. coli ATTC 25922, B. fragilis ATTC 25285 (encapsulated), B. thetaiotaomicron ATTC 29741, B. distasonis B 24 (gift of Dr. C. Krasemann, Bayer AG, Wuppertal, Germany), B. asaccharolyticus ATTC 25260.

LPS

A phenol extracted LPS of E. coli (0 111:B 4) was obtained commercially (Sigma, St. Louis) and used after being dissolved (PBS).

Inoculum

Bacteroides species were grown in Schaedler broth, E. coli was grown in glucose broth. Optical density was determined by nephelometric measure-

Table 1. Lethality of Mice Injected with Single
Bacterial Strains or with LPS Only

Inoculum	Number of mice injected	Survivors 9 Days Later
1×10^7 E. coli	20	18
100 µg LPS	20	18
1×10^8 B. fragilis	20	20
1×10^8 B. thetaiotaomicron	20	20
1×10^8 B. distasonis	20	20
1×10^8 B. asaccharolyticus	20	20

ments and if necessary adjusted by adding fresh medium. Bacterial concentrations were confirmed by plating on suitable media.

Injections

Mice were inoculated with broth cultures of E. coli and of Bacteroides species, with LPS dissolved in PBS and with sterile broth via separate intravenous injections.

Lethality

After the inoculations, mice were monitored for nine days and lethality was recorded.

Table 2. Lethality of Mice Injected with E. coli in
Combination with Different Bacteroides Species

Inoculum	Number of Mice Injected	Survivors 9 Days Later
1×10^7 E. coli plus 1×10^8 B. fragilis	20	9
1×10^7 E. coli plus 1×10^8 B. thetaiotaomicron	20	4
1×10^7 E. coli plus 1×10^8 B. distasonis	20	14
1×10^7 E. coli plus 1×10^8 B. asaccharolyticus	20	16

Table 3. Lethality of Mice Injected with LPS and Different Bacteroides Species

Bacterial Inoculum	Time of LPS injection (100 μg each)	Number of Mice injected	Survivors 9 Days Later
1 x 10^8 B. fragilis	immediately	20	19
	3 hrs later	20	18
	6 hrs later	20	9
	12 hrs later	20	13
	24 hrs later	20	16
	48 hrs later	20	20
1 x 10^8 B. thetaiotaomicron	6 hrs later	20	3
1 x 10^8 B. distasonis	6 hrs later	20	12
1 x 10^8 B. asaccharolyticus	6 hrs later	20	5

RESULTS

Monoinfections

Lethalities of mice infected with a single bacterial strain or injected with LPS only are summarized in Table 1. Injection of 1 x 10^7 E. coli killed ten percent of the animals challenged. The same lethality was established by injecting 100 μg of LPS. Injections of 1 x 10^8 of any of the Bacteroides species tested did not result in animal lethality.

Mixed Infections

Injections of both E. coli and anyone of the Bacteroides species, resulted in enhanced lethality of animals challenged (Table 2). While for B. fragilis and for B. thetaiotaomicron this effect was significant (2p < 0.01, chi-square-test), it was not for B. distasonis and B. asaccharolyticus.

LPS Injections

If LPS was used instead of E. coli and injected together with B. fragilis, no synergistic lethality was observed. However, mice pretreated with B. fragilis and subsequently challenged with LPS exhibited increased lethality. This effect was clearly time dependent and maximal death rates were observed if B. fragilis was injected six hours prior to LPS challenge (Table 3). Pretreatment with other Bacteroides species was capable of inducing synergistic lethality as well.

DISCUSSION

It is generally accepted that LPS plays an important role in the pathogenesis of infections with gram-negative bacteria. Furthermore, it has been shown that it is possible to modify the host response to bacterial endotoxin. Watson and Kim (13) explained the age-dependent development of hypersensitivity to LPS in animals by contacts with intestinal flora. The results presented here demonstrate that different Bacteroides species are capable of enhancing susceptibility to E. coli infections or to E. coli-derived LPS. Heat-killed B. fragilis was less effective than viable organisms (data not shown). Furthermore, pretreatment with Bacteroides species was effective only for a limited period of time, however, since maximum effects of Bacteroides species were seen after six hours, this could be of importance in clinical situations in which usually both aerobic and anaerobic bacteria infect the host at the same time. In addition, it has been shown (6,8,10) that in animals infected with E. coli and B. fragilis, the anaerobes may survive in animal organs or in abscesses for long periods of time. Hence, the effect on the LPS susceptibility of the host might be prolonged. The parallelism of death rates of animals infected with E. coli plus Bacteroides species and of mice pretreated with corresponding Bacteroides species and challenged with LPS suggests that a major mechanism of synergistic lethality of mixed aerobic/anaerobic infections might be the increased host susceptibility to bacterial LPS. In contrast to other pospossible mechanisms discussed earlier (9), this increase in LPS susceptibility is operative in vivo.

ACKNOWLEDGMENTS

The authors would like to thank Miss S. Thiemann for skillful technical assistance and Mrs. A. Hausler for secretarial aid.

REFERENCES

1. I. Brook, V. Hunter, and R. I. Walker, Synergistic effect of bacteroides, clostridium, fusobacterium, anaerobic cocci, and aerobic bacteria on mortality and induction of subcutaneous abscesses in mice, J. Infect. Dis. 149:924 (1984).
2. J. C. Hagen, W. S. Wood, and T. Hashimoto, In vitro stimulation of Bacteroides fragilis growth by E. coli, Eur. J. Clin. Microbiol. 1:338 (1982).
3. T. Hofstad, Pathogenicity of Anaerobic gram-negative rods: possible mechanisms, Rev. Infect. Dis. 6:189 (1984).
4. H. R. Ingham, P. R. Sisson, D. Tharagonnet, J. B. Selkon, and A. A. Codd, Inhibition of phagocytosis in vitro by obligate anaerobes, Lancet 1252 (1977).
5. G. R. Jones and C. G. Gemmell, Impairment by bacteroides species of opsonisation and phagocytosis of enterobacteria, J. Med. Microbiol. 15:351 (1982).
6. M. J. Kelly, The quantative and histological demonstration of pathogenic synergy between E. coli and Bacteroides fragilis in guinea-pig wounds, J. Med. Microbiol. 11:513 (1978).
7. D. Mayrand and B. C. McBride, Ecological relationships of bacteria involved in a simple mixed anaerobic infection, Infect. Immun. 27:44 (1980).
8. A. B. Onderdonk, J. G. Bartlett, T. Louie, N. Sullivan-Seigler, and S. L. Gorbach, Microbial synergy in experimental intra-abdominal abscess, Infect. Immun. 13:22 (1976).
9. M. Reznikov, J. J. Finlay-Jones, and P. J. McDonald, Effect of Bacteroides fragilis on the peritoneal clearance of Escherichia coli in mice, Infect. Immun. 32:398 (1981).
10. A. C. Rodloff and H. Hahn, Synergistic lethality in experimental infections with Escherichia coli and Bacteroides fragilis, Zbl. Bakt. Hyg. A258:112 (1984).

11. A. C. Rodloff, P. Gadke, F. Lux, and H. Hahn, Experimentelle
    Pathogenitat von Bacteroides fragilis, FAC

12. R. W. Tofte, P. K. Peterson, D. Schmeling, J. Bracke, Y. Kim, and P. G.
    Quie, Opsonization of four Bacteroides species: role of the
    classicl complement pathway and immunoglobulin, Infect. Immun.
    27:784 (1980).

13. D. W. Watson and Y. B. Kim, Modification of host responses to bacterial
    endotoxins, J. Exp. Med. 118:425 (1963).

THE ROLE OF POST-ENDOTOXIN SERUM COMPONENTS FROM

BCG INFECTED MICE IN THE PROTECTION OF COMPROMISED HOSTS

Renate Urbaschek, Daniela N. Männel[1], Stephan E. Mergenhagen[2], and Bernhard Urbaschek

Department of Immunology and Serology, Institute of Hygiene and Medical Microbiology, Klinikum Mannheim, University of Heidelberg, Mannheim, FRG; German Cancer Research Center Heidelberg, FRG; and National Institutes of Health Bethesda, Maryland

The beneficial effects of bacterial toxins have been demonstrated in various experimental models in which the host was compromised by a great variety of different noxae. The induction of nonspecific tolerance by endotoxins to the toxic effects of endotoxins (5) and of nonspecific resistance to infection (9,18) and to lethal X-irradiation (24,36) was of particular interest in our studies. In recent years it has become more apparent that these effects are mediated by humoral factors which are produced and released from lymphoreticular cells after injection of endotoxin. Freedman (13) reported that endotoxin tolerance was passively transferable with serum from tolerant mice. Using a detoxified endotoxin preparation that induced tolerance to the lethal effects of endotoxin after one single pretreatment (42), we were unable to observe this passive transfer effect. The transfer of enhanced resistance to bacterial infection was achieved with post-endotoxin serum from BCG (Mycobacterium bovis Bacille Calmette Guerin) infected mice (30). Moreover, it was reported that such serum (BCG/ET serum) or postendotoxin serum from zymosan treated mice induced protection against the lethal effects of whole-body X-irradiation (1,48).

In regard to the mechanisms involved in the phenomenon of endotoxin tolerance, recently reviewed (15), accelerated clearance (4,14), increased detoxification of endotoxin by the reticuloendothelial system (41) as well as increased detoxification of endotoxin by plasma factors (27) have been discussed. In recent years a great number of investigations have pointed to the central role of the state of macrophages and their release of mediators in endotoxic effects. A variety of cytotoxic and vasoactive substances has been reported to be released from macrophages in response to endotoxin; among these are lysosomal enzymes, oxygen intermediates, prostaglandins and complement components. These substances may not be released to such an extent or not at all in the status of tolerance towards the lethal effects of endotoxin.

Whereas mechanisms involved in the induction of endotoxin tolerance and nonspecific host defense to bacterial infection by endotoxin may be similar, the radioprotective effect of endotoxin has been related to the early hematopoietic recovery (2,36,38) thus enabling the host to resist the consequences of gut derived bacteremia and endotoxemia following X-ray exposure. We were able to demonstrate that BCG/ET serum is capable of transferring stimulation of granulopoiesis as well as an increase in splenic granulocyte macrophage precursor cells (GM-CFC) and an increase in serum levels of colony-stimulating activity (CSA) (48). CSA is a factor necessary for proliferation and differentiation of CFC in culture (7,32). Serum CSA is known to increase after injection of endotoxin (29) and its polysaccharide fraction (28,34,44). A great variety of mediators are present in BCG/ET serum, also referred to as TNS (tumor necrosis serum) because of its high level of tumor necrosis factor that has been isolated and characterized from such sera (10,16,19,20). Besides TNF or cytotoxic factor, interferon (56), IL_1 (53), CSA (8,45), and many other mediators have been found in much higher concentrations of BCG/ET serum than in post-endotoxin serum. The factor or factors in such serum responsible for the enhancement of nonspecific resistance have not been identified.

An interesting phenomenon is the fact that serum from BCG infected animals (BCG/ET serum) that are proned to die within hours after injection of minute amounts of endotoxin (40) contains factors that transfer beneficial effects into recipients. It is well established that the reactivity of macrophages plays a major role in the underlying mechanism of these events. Endotoxin is extremely toxic for macrophages in BCG infected animals as was shown by Peavy et al. in in vitro experiments (31). Recently, vital microscopic observations of responses of Kupffer cells and the hepatic microvasculature to endotoxin in BCG infected mice revealed the impairment of Kupffer cell function and integrity (21,22). Some mediators that are released from these macrophages that have locally destructive vasoactive and cytotoxic activities may not be in concentrations high enough to also be harmful when serum from these animals is systemically injected into recipients. The released beneficial factors on the other hand cannot be effective in BCG infected mice after endotoxin injection, the responsiveness of their target cells may no longer exist and/or the mechanisms protecting the host need more time to be effective. Moreover, the interacting regulation of several of these mediators (25,26) may be imbalanced. Many questions remain open in this complex and multifactoral sequelae of events of endotoxic effects in normally susceptible, hyperreactive as well as in hyporeactive, tolerant animals, and in regard to the mechanisms and nature of endotoxin-induced mediators involved in the enhancement of nonspecific resistance.

RADIOPROTECTION

The radioprotective effect known to occur after endotoxin pretreatment is mediated by humoral factors present in BCG/ET (48). In our studies BCG/ET serum was obtained from NMRI mice two hours after i.v. injection of 5 µg/-mouse endotoxin 14 days after infection with 5×10^7 viable BCG organisms (Trudeau Institute, Saranac, NY). The endotoxin used throughout the experiments was extracted in our laboratories from E. coli 0111 (Boivin method). When BCG/ET is injected into C3H/HeJ mice--a mouse strain genetically resistant to endotoxin (39)--at 24 hours prior to whole-body X-irradiation an increased survival is observed whereas postendotoxin serum is ineffective (48). If endotoxin-susceptible mice are used as recipients, both postendotoxin serum and BCG/ET serum induce radioprotection. These results, summarized in Table 1, clearly illustrate the necessity to use endotoxin low responder mice when concluding that humoral factors are involved. In the present study, parameters were measured in C3H/HeJ mice that fail to be altered after injection of endotoxin: lethal irradiation (43), increase in GM-CFC and serum CSA (3) and induction of increased nonspecific resistance to bacterial infection (11). By means of a kinetic Limulus-amebocyte-lysate test (12,51) we could demonstrate that considerable amounts of endotoxin are present in serum two hours after endotoxin injection (47), so that it cannot be excluded that endotoxin per se is effective in post-endotoxin or

Table 1. Endotoxin-Independent Serum Factor(s)-Induced Stimulation of
Granulopoiesis and Radioprotection of Post-Endotoxin Serum
from BCG-Infected Mice (BCG/ET serum), and Endotoxin-Dependent
Effect of Post-Endotoxin Serum (ET-serum) in Responder Mice

	Responder Mice (NMRI)			Non-responder Mice (C3H/HeJ)		
Parameters	Control Sera	ET-sera	BCG/ ET sera	Control Sera	ET-sera	BCG/ ET sera
Serum CSA	0	↑	↑	0	0	↑
Splenic GM-CFC	0	↑	↑	0	0	↑
Irradiation lethality	0	↓	↓	0	0	↓

(For further details, see Reference 48)
↑ = increase; ↓ = decrease

BCG/ET serum when transferred to endotoxin susceptible recipients. Therefore, in all of our studies C3H/HeJ mice were used for serum transfer, except
in those experiments of BCG/ET serum transfer after X-irradiation, since it
is known, that the optimal time for the induction of radioprotection by
endotoxin is the pretreatment before irradiation (2,36). Also in some experiments in which the effect of BCG/ET serum fractions with very low endotoxin
content were studied, NMRI mice were used. BCG/ET serum fractions were
prepared in order to study which fraction(s) may be responsible for the
beneficial effects induced by BCG/ET serum transfer. BCG/ET serum fractions were obtained after molecular sieving (column: 2.5 x 100 cm, Sephacryl
S-300; buffer: Tris 0.05 M NaCl 0.1 M; pH 7.4, flow 20 ml/hour). After
injection of the different fractions (0.25 ml i.v. plus 0.25 ml i.p.) into
C3H/HeJ and NMRI mice 24 hours before X-irradiation, lethality rates were
observed for 30 days. Fractions II_1 and III_1 reduced lethality in C3H/HeJ
mice and fraction III_1 in NMRI mice (Tables 2 and 3).

The finding that BCG/ET serum is also radioprotective when given 24
hours or 48 hours after exposure to X-irradiation in C3H/HeJ and NMRI mice
may give further insight into mechanisms involved in radioprotection. We
observed that 0.5 µg of endotoxin injected at 24 hours or 48 hours after
whole-body X-irradiation in NMRI mice had no protective effect (98% resp.,
150% died, as percent of control), whereas BCG/ET serum (0.3 ml) injected
at the same times after irradiation reduced the lethality to 50% (+ 24
hours) and to 83% (+48 hours). In C3H/HeJ mice transfer of 0.5 ml BCG/ET
serum caused a reduction in lethality to 17.5% (as percent of control),
when injected at 48 hours after irradiation (48). These results indicate
that after irradiation-induced damage of the extremely radiosensitive hematopoietic stem cell compartment and small mature lymphocytes has occurred,
humoral factors are capable to enhance host resistance. The ineffectiveness
of endotoxin may be related to the observed hypersensitivity of irradiated
animals to endotoxin (36) which coincides with period during which disruption
of ileal tight junctional barriers was observed (54). This increase in
sensitivity to injected endotoxin and decrease in intestinal barrier integrity is biphasic, reoccurring at days eight through ten post-irradiation
(37,54). The study of the function of phagocytic cells and of the release
of mediators during the different states of reactivity and of the

Table 2. CSA in BCG/ET Serum and Its Fractions (A); CSA in Serum from C3H/HeJ Mice Two Hours After Injection (B); ± sd, and LD/30 (lethality rate at 30 days) After Whole Body X-Irradiation (600R) in Pretreated (−24 h) C3H/HeJ Mice (C)

Substance	Molecular weight kD	Endotoxin ng/ml	CSA*		$\frac{C}{LD/30}$ X-Irradiation LD/30, % of control (42%) n = 12
			A Substance	B Serum** n = 5	
BCG/ET serum		291	51.3 ± 12.6	134.7 ± 7.4	19
FractionI$_1$	Vo − 150	160	42.0 ± 6.0	5.0 ± 6.5	78
Fraction II$_1$	150 − 50	32	0	130.2 ± 8.8	19
Fraction III$_1$	50 − 10	31	0	108.1 ± 7.7	19
Fraction IV$_1$	10 − 5	28	4.7 ± 2.1	0.3 ± 0.3	178
Control Serum			0	0	100

* Expressed as number of colonies (> 50 cells) per 10^5 nucleated femoral bone marrow cells, using 25 µl of the samples.

**Individual serum samples were assayed in triplicates

Table 3. Lethality Rates After Lethal Challenge with Endotoxin
(150 µg/mouse) Respectively After Whole Body X-Irradiation
(600 R) in NMRI mice pretreated (-24 h) with BCG/ET Serum
Fractions I_1 - IV_1 (0.25 ml i.v. plus 0.25 ml i.p.)

Pretreatment	Endotoxin Challenge Lethality, % of Control (50%) n = 8	X-Irradiation LD/30, % of Control (53%) n = 15
Fraction I_1	76	100
Fraction II_1	150	89
Fraction III_1	50	13
Fraction IV_1	76	76
Saline	100	100

responsiveness of hematopoietic stem cells to BCG/ET serum after irradiation will give some insight into the underlying mechanism.

STIMULATION OF GRANULOPOIESIS

BCG/ET serum contains an endotoxin-independent granulopoiesis stimulating factor (48). Three days after injection into C3H/HeJ mice splenic GM-GFC increased as well as CSA at two hours after injection (Table 1). It was of interest to study the effect of fractions of BCG/ET serum in particular in regard to the role of CSA. The results (Table 2) clearly indicate that those fractions that do not contain CSA (fraction II_1 and fraction III_1) induce radioprotection and elevated serum CSA when transferred into C3H/HeJ mice. The CSA inducing capacity of humoral factors present in BCG/ET serum may be involved in the increased host resistance to X-irradiation.

Several substances have been identified that inhibit CSA stimulation of granulocyte macrophage progenitor cells in culture, such as tumor necrosis factor, prostaglandins, and interferon (17,23,35). These factors are present in postendotoxin serum, and in much higher concentrations in postendotoxin sera from BCG-infected mice. This inhibitory activity was detected in the assay used (7,32) by the fact that with increasing dilution of BCG/ET serum in bone marrow cultures the levels of CSA increased (45). It should be emphasized here that it is important to test several dilutions of serum samples when for instance the extent of CSA stimulation by endotoxin is compared with that of detoxified preparations. To elucidate this, results obtained with a nontoxic bacterial polysaccharide preparation should be mentioned. Upon dilution of sera from BCG-infected mice obtained two hours after injection of this preparation CSA levels decrease (45). This also was observed using a nontoxic, LAL-negative, native hapten from E. coli, kindly provided by Dr. Ribi, Immunochem Research, Hamilton, MT. According to our experience the lack of production of these inhibitors is a very sensitive parameter indicating that there is no residual toxicity in the detoxified preparation used. On the other hand, the presence of these CSA inhibitors may represent an indicator of excessive mediator release by activated macrophages after exposure to toxic agents. BCG/native hapten serum (injection of 10 µg native hapten in BCG-infected mice) which did not contain CSA

Table 4. Endotoxin, CSA Inhibition of GM-CFC in Culture, and Lethality Rate in % of Control (100%) After Cecal Ligation and Puncture of C3H/HeJ Mice Pretreated (-24 h) with BCG/ET Serum Fractions I_2 - V_2 (0.25 ml i.p.)

Substances	Molecular weight kD	Endotoxin ng/ml	CSA inhibition %	Total No. of mice	Lethality Rate Post-Surgery % of control	
					Day 2	Day 5
Fraction I_2	Vo – 150	84	100	11	81	100
Fraction II_2	150 – 90	32	72	13	62	69
Fraction III_2	90 – 30	33	60	10	60	90
Fraction IV_2	30 – 15	13	0	12	100	100
Fraction V_2	< 15	10	0	13	0	15

inhibitors did not transfer radioprotection. Injection of native hapten per se induced endotoxin tolerance, and enhancement of nonspecific resistance to septicemia and X-irradiation (unpublished data). This example provides further evidence that CSA transfer--BCG/native hapten serum contained elevated CSA levels--is not responsible for the radioprotective effect, as was described above using BCG/ET serum fractions.

When BCG/ET serum fractions were assayed for CSA inhibitors in culture it became apparent that inhibitory activity was present in the molecular range above 50,000 D. Inhibition of CSA in culture was measured by adding 25 µl of the standard CSA--obtained two hours after injection of 5 µg of endotoxin in NMRI mice--to the bone marrow cultures (1 x 10^5 nucleated femoral bone marrow cells per ml) containing 25 µl of the different fractions. Fractions I_1 and II_1 were inhibitory whereas III_1 and IV_1 did not show CSA inhibition of GM-CFC in culture. The CSA inhibitory capacity of fractions I_2 to V_2 is shown in Table 4. Endotoxin per se added to the bone marrow cultures in the range of concentrations present in the volumes of the fractions tested has no inhibitory effect on the formation of GM colonies. These results show that there was no correlation between the lack of CSA inhibition by the fractions and their nonspecific resistance inducing effect (Tables 2 and 4).

ENDOTOXIN TOLERANCE

The capacity of BCG/ET serum fractions to induce tolerance to the lethal effect of endotoxin was tested in NMRI mice. Twenty-four hours before the challenge injection of endotoxin (150 µg/mouse) NMRI mice received 0.25 ml i.v. plus 0.25 ml i.p. of the fractions I_1 to IV_1. The results of this experiment (Table 4) show that fraction III_1 (10-50 kD), the same fraction that caused radioprotection, induced 50% reduction in lethality. Mechanisms discussed to be involved in endotoxin tolerance mentioned above were mainly concerned with accelerated clearance and detoxification of endotoxin. The responsiveness of macrophages and the extent of mediator release in the status of tolerance or of hyperreactivity to endotoxin seem to play a major role. The complexity and diversity of these events are demonstrated by the fact that 14 days after BCG infection the host is hyperreactive to endotoxin (40), more resistant to bacterial infection (6), normally reactive to X-irradiation (50), and normally susceptible to the endotoxin tolerance (47) and radioprotection inducing capacity of endotoxin (50). In this context experiments are of interest that were performed comparing serum levels of mediators that are present in BCG/ET serum with those from BCG-infected mice that were rendered tolerant to the lethal effects of endotoxin (52). Fourteen days after BCG infection, mice were injected with minute amounts of endotoxin (10 ng) and 24 hours later they received a lethal dose of endotoxin (5 µg). In these tolerant mice CSA and interferon were present two hours after the lethal challenge with endotoxin in concentrations similar to those in animals that were not pretreated. IL 1 levels, however, were significantly decreased and the cytotoxic factor (TNF) was absent in serum from tolerant mice, whereas high levels of these mediators were detected in nontolerant BCG/ET serum. These results point to the significance of macrophages as central target cells and of their status of reactivity to the lethal effects of endotoxin. In the protected animals those mediators that are produced by macrophages are reduced or not detectable and those cytokines, CSA and interferon, originating from several cell sources besides macrophages, are present in the serum of tolerant mice. These results support the concept of Greisman (15) that tolerance to endotoxin lethality is primarily dependent upon induction of resistance of RES macrophages to endotoxin cytotoxicity and mediator-releasing activity.

ENHANCEMENT OF NONSPECIFIC RESISTANCE TO GUT-DERIVED SEPTICEMIA

Septicemia was induced by cecal ligation and puncture, an experimental model (55) in which gut-derived bacteria are involved, simulating clinical situations of peritoneal sepsis. The course of this septicemia until 48 hours after surgery was described recently (49) including lethality rates, and determinations of anaerobic and aerobic bacterial counts and of endotoxin in cardiac blood.

Pretreatment with one single injection of 1 µg endotoxin (0.5 µg i.v. plus 0.5 µg i.p.) 24 hours before surgery resulted in decreased lethality (19% versus 8%) and in a drastic decrease in the number of aerobic bacteria in the initial phase, whereas anaerobic bacterial numbers were unchanged. In separate studies we found that the essential part of the combined application was the i.p. injection, which was effective per se. Pretreatment injected i.v. alone did not result in increased survival. Interesting in these experiments was the result of the quantification of circulating endotoxin. The endotoxin-pretreated group had high levels of endotoxin at the time of surgery (3 ng/ml) which began to decrease after 12 hours. In nonpretreated mice after an initial continuous increase, endotoxin reached a plateau still existing at the end of the experiment.

It was of interest to study whether endotoxin-induced increased nonspecific resistance in this septic shock model is mediated by soluble serum factors. Therefore we pretreated C3H/HeJ mice 24 hours prior to surgery with BCG/ET serum fractions $I_2 - V_2$ (0.25 ml i.p.). It was quite interesting that injection of fraction V_2 (< 15 kD) resulted in 0% lethality at two days and in 15% lethality at five days after surgery (Table 4). After five days no more mice died until weeks later, when abdominal abscesses had formed. In a repeated experiment with five mice per group similar results were obtained. After pretreatment with fraction II_2 60% died and with fraction V_2 20% died, all other groups had a lethality rate of 100%.

CONCLUDING REMARKS

The results of the experiments with BCG/ET serum support the concept that beneficial effects of endotoxins are mediated by humoral serum factors. BCG/ET serum transfer into endotoxin low-responder C3H/HeJ mice is capable of inducing nonspecific resistance to the consequence of lethal irradiation when injected before or after X-ray exposure as was discussed. The successful radioprotection of post-irradiation injection differs from the effects of endotoxin. BCG/ET serum contains a granulopoiesis-stimulating factor that may be closely related to its radioprotective effect. Serum CSA and the number of granulocyte macrophage progenitor cells significantly increase after BCG/ET serum transfer. The in vivo function of CSA, respectively of the different CSF types as GM-poetin has not been clearly demonstrated.

It becomes apparent from recent literature that CSA has a diversity of effects not related to its growth-promoting and GM-CFC stimulating activity. From our experiments with BCG/ET serum fractions it can be concluded that the transferred CSA in the BCG/ET serum is not responsible for the radioprotective effect, because fractions not containing CSA were protective. This does not exclude that CSA does not play a role, as was recently discussed (1). In general, it is doubtful that CSA in the BCG/ET serum has a direct effect in the recipient, because it is unlikely that they reach their target cells in sufficient concentrations. The stimulating effect of those components in BCG/ET serum on the endogenous production of the beneficial factor or factors involved in radioprotective mechanism seems to be important. If CSA production is involved in enhanced nonspecific resistance to irradiation the results reported here would be an example. BCG/ET serum fractions that are radioprotective are CSA inducers.

The endotoxin tolerance inducing BCG/ET serum fraction (fraction III$_1$) has a molecular weight of 10 -50 kD and the fraction V$_2$ (< 15 kD) enhanced nonspecific resistance to gut-derived septicemia. Although the study in order to find the molecular range in which the effective component(s) are present in BCG/ET serum is only a first crude step, it may be helpful to narrow potential candidates. One of these is IL$_1$. In preliminary experiments purified IL$_1$ obtained from Dr. Theresa Krakauer had no effect in regard to stimulation of granulopoiesis or radioprotection after injection into mice. The identification of the factor(s) in BCG/ET serum responsible for the beneficial effects described will be facilitated by the availability of cloned cytokines for in vivo studies.

The results obtained from studies of four different cytokines measured individually in the same serum samples from endotoxin tolerant and nontolerant BCG-infected mice support the central role of macrophages in the mechanisms of endotoxin tolerance. They reflect the importance of the refractory status of the macrophage in response to endotoxin established by the tolerance-inducing injection. This refractoriness is expressed by the absence of TNF in tolerant BCG-infected mice after a lethal challenge with endotoxin. The discrepancy of our results that CSA serum is increased in these tolerant mice and the lack of CSA elevation in tolerant mice described in the literature (33) cannot be explained. The fact that different tolerance induction procedures used (one versus several daily endotoxin injections) may give different results was excluded by earlier studies in which repeated injections resulted in CSA increase in tolerant mice (46).

ACKNOWLEDGMENT

We are grateful to Ruth Breunig and Ute Ell for excellent technical assistance.

REFERENCES

1. P. D. Addison and L. J. Berry, Passive protection against X-irradiation with serum from zymosan-primed and endotoxin injected mice, J. Reticuloendothelial Soc. 30:301 (1981).

2. E. J. Ainsworth and H. B. Chase, Effect of microbial antigens on irradiation mortality in mice, Proc. Soc. Exp. Biol. Med. 102:483 (1959).

3. R. N. Apte and D. H. Pluznik, Genetic control of lipopolysaccharide induced generation of serum colony stimulating factor and proliferation of splenic granulocyte/macrophage precursor cells, J. Cell Physiol. 89:313 (1976).

4. P. B. Beeson, Tolerance to bacterial pyrogens. II. Role of the reticuloendothelial system, J. Exp. Med. 86:39 (1947).

5. B. Benacerraf, M. M. Sebestyen, and S. Schlossman, A quantitative study of the kinetics of blood clearance of P^{32} labelled Escherichia coli and staphylococci by the reticuloendothelial system, J. Exp. Med. 110:27 (1959).

6. G. Biozzi, C. Stiffel, B. N. Halpern, and D. Mouton, Recherches sur le mechanisme de l'immunité non spécifique produite par les mycobactéries, Rev. Franc Etudes Clin. Biol. 5:876 (1960).

7. T. R. Bradley and D. Metcalf, The growth of mouse bone marrow cells in vitro, J. Exp. Biol. Med. Sci. 44:287 (1966).

8. R. C. Butler, A. M. Abdelnoor, and A. Nowotny, Bone marrow colony-stimulating factor and tumor resistance-enhancing activity of postendotoxin mouse sera, Proc. Natl. Acad. Sci. U.S.A. 75:2893 (1978).

9. R. J. Dubos and R. W. Schaedler, Reversible changes in the susceptibility of mice to bacterial infections, J. Exp. Med. 104:53 (1956).

10. E. J. Carswell, L. J. Old, R. L. Kassel, S. Green, N. Fiore, and B. Williamson, An endotoxin-induced serum factor that causes necrosis of tumors, Proc. Natl. Acad. Sci. U.S.A. 72:3666 (1975).

11. L. Chedid, M. Parant, C. Damais, F. Parant, D. Juy, and A. Galleli, Failure of endotoxin to increase nonspecific resistance to infection of lipopolysaccharide low-responder mice, Infect. Immun. 13:722 (1976).

12. B. Ditter, K. P. Becker, R. Urbaschek, and B. Urbaschek, Quantitativer Endotoxin-nachweis. Automatisierter, kinetischer limulus-amobozyten-lysat-mikrotiter-test mit Messung probenabhangiger Interferenzen, Arzneim. Forsch. 33:681 (1983).

13. H. H. Freedman, Passive transfer of protection against lethality of homologous and heterologous endotoxins, Proc. Soc. Exp. Biol. Med. 102:504 (1959).

14. S. E. Greisman, F. A. Carozza, and J. D. Hills, Mechanisms of endotoxin tolerance. I. Relationship between tolerance and reticuloendothelial system phagocytic activity in the rabbit, J. Exp. Med. 117:663 (1963).

15. S. E. Greisman, Induction of endotoxin tolerance, in: "Beneficial Effects of Endotoxins," A. Nowotny, ed., Plenum Press, New York and London (1983).

16. F. C. Kull and P. Cuatrecasas, Preliminary characterization of the tumor cell cytotoxin in tumor necrosis serum, J. Immunol. 126:1279 (1981).

17. J. Kurland and M.A.S. Moore, Modulation of hemopoiesis by prostaglandins, Exp. Hematol. 5:357 (1977).

18. M. Landy and L. Pillemer, Increased resistance to infection and accompanying alteration in properdin levels following administration of bacterial lipopolysaccharides, J. Exp. Med. 104:383 (1956).

19. D. N. Männel, D. L. Rosenstreich, and S. E. Mergenhagen, Mechanism of lipopolysaccharide-induced tumor necrosis: requirement for lipopolysaccharide-sensitive lymphoreticular cells, Infect. Imm. 24:5 (1979).

20. D. N. Mannel, R. N. Moore, and S. E. Mergenhagen, Macrophages as a source of tumoricidal activity (tumor necrosis factor), Infect. Immun. 30:523 (1980).

21. R.S. McCuskey, R. Urbaschek, P. A. McCuskey, and B. Urbaschek, In vivo microscopic studies of the responses of the liver to endotoxin, Klin. Wochenschr. 60:56 (1982).

22. R. S. McCuskey, R. Urbaschek, P. A. McCuskey, and B. Urbaschek, In vivo microscopic observations of the responses of Kupffer cells and the hepatic microcirculation to Mycobacterium bovis BCG alone and in combination with endotoxin, Infect. Immun. 42:362 (1983).

23. T. A. McNeill and I. Gresser, Inhibition of haematopoietic colony growth by interferon preparations from different sources, Nature (London) 244:173 (1973).

24. R. B. Mefferd, D. T. Henkel, and J. B. Loeffer, Effect of piromen on survival of irradiated mice, Proc. Soc. Exp. Biol. Med. 83:54 (1953).

25. R. N. Moore, Regulation of macrophage accessory functions by interaction involving lymphokines and endotoxin, Klin. Wochenschr. 60:754 (1982).

26. R. N. Moore and B. T. Rouse, Enhanced responsiveness of committed macro-phage precursors to macrophage-type colony-stimulating factor (CSF-1) induced in vitro by interferons alpha and beta, J. Immunol. 131:2374 (1983).

27. S. C. Moreau and R. C. Skarnes, Host resistance to bacterial endo-toxemia: mechanisms in endotoxin-tolerant animals, J. Infect. Dis. 128 (Suppl.):122 (1973).

28. A. Nowotny, U. H. Behling, and H. L. Chang, Relation of structure to function in bacterial endotoxins. VIII. Biological activities in a polysaccharide-rich fraction, J. Immunol. 115:199 (1975).

29. D. Metcalf, Acute antigen-induced elevation of serum colony-stimulating factor (CSF) levels, Immunol. 21:427 (1971).

30. M. A. Parant, F. J. Parant, and L. A. Chedid, Enhancement of resistance to infections by endotoxin-induced serum factor from Mycobacterium bovis BCG-infected mice, Infect. Immun. 28:654 (1980).

31. D. L. Peavy, R. E. Baughn, and D. M. Muscher, Effects of BCG infection on the susceptibility of mouse macrophages to endotoxin, Infect. Immun. 24:59 (1979).

32. D. H. Pluznik and L. Sachs, The cloning of normal "mast" cells in tissue culture, J. Cell Physiol. 66:319 (1965).

33. P. Quesenberry, J. Halperin, M. Ryan, and F. Stohlman, Jr., Tolerance to the granulocyte-releasing and colony-stimulating factor elevating effects of endotoxin, Blood 6:789 (1975).

34. J. Rothman, A. G. Johnson, H. Friedman, E. Kovats, P. H. Pham, F. Sanavi, A. M. Nowotny, and A. Nowotny, Biological effects of White-type polysaccharides of gram-negative bacteria, J. Biol. Resp. Modif. 4:169 (1985).

35. R. G. Shah, S. Green, and M.A.S. Moore, Colony-stimulating factor and inhibiting activities in mouse serum after Corynebacterium parvum-endotoxin treatment, J. Reticuloendothelial Soc. 23:29 (1978).

36. W. W. Smith, I. M. Alderman, and R. F. Gillespie, Increased survival in irradiated animals treated with bacterial endotoxin, Am. J. Physiol. 191:124 (1957).

37. W. W. Smith, I. M. Alderman, C. Schneider, and J. Cornfield, Sensitivity of irradiated mice to bacterial endotoxin, Proc. Soc. Exp. Biol. Med. 113:778 (1963).

38. W. W. Smith, G. Brecher, S. Fred, and R. A. Budd, Effect of endotoxin on the kinetics of hemopoietic colony-forming cells in irradiated mice, Radiat. Res. 27:710 (1966).

39. B. M. Sultzer, Genetic control of leucocyte responses to endotoxin, Nature (London) 219:1253 (1968).

40. E. Suter, G. E. Ullman, and R. G. Hoffman, Sensitivity of mice to endotoxin after vaccination with BCG (Bacillus Calmette-Guérin), Proc. Soc. Exp. Biol. Med. 99:167 (1958).

41. R. A. Trejo and N. R. DiLuzio, Influence of endotoxin tolerance on detoxification of Salmonella enteritidis endotoxin by mouse liver and spleen, Proc. Soc. Exp. Biol. Med. 141:501 (1972).

42. B. Urbaschek and A. Nowotny, Endotoxin tolerance induced by detoxified endotoxin (endotoxoid), Proc. Soc. Exp. Biol. Med. 127:650 (1968).

43. R. Urbaschek, S. E. Mergenhagen, and B. Urbaschek, Failure of endotoxin to protect C3H/HeJ mice against lethal X-irradiation, Infect. Immun. 18:860 (1977).

44. R. Urbaschek, Effects of bacterial products on granulopoiesis, in: "Macrophages and Lymphocytes: Nature, Functions and Interaction," M. R. Escobar and H. Friedman, eds., Plenum Publishing Corporation, New York (1979).

45. R. Urbaschek, R. K. Shadduck, C. Bona, and S. E. Mergenhagen, Colony-stimulating factor in nonspecific resistance and in increased suscep-tibililty to endotoxin, in: "Microbiology 1980," D. Schlessinger, ed., American Society for Microbiology, Washington, D.C. (1980).

46. R. Urbaschek and B. Urbaschek, The effects of endotoxic substances on granulopoiesis, in: "Natural Toxins," D. Eaker and T. Wadstrom, eds., Pergamon Press, Oxford, New York (1980).

47. R. Urbaschek and B. Urbaschek, Aspects of beneficial endotoxin-mediated effects, Klin. Wochenschr. 60:746 (1982).

48. R. Urbaschek andk B. Urbaschek, Ability of post-endotoxin serum from BCG-infected mice to induce nonspecific resistance and stimulation of granulopoiesis, Infect. Immun. 39:1488 (1983).

49. B. Urbaschek, B. Ditter, K. P. Becker, and R. Urbaschek, Protective effects and role of endotoxin in experimental septicemia, Circ. Shock 14:209 (1984).

50. R. Urbaschek and B. Urbaschek, Induction of nonspecific resistance and stimulation of granulopoiesis by endotoxins and nontoxic bacterial cell wall components and its passive transfer, Ann. N.Y. Acad. Sci. (in press, 1985).

51. B. Urbaschek, K. P. Becker, B. Ditter, and R. Urbaschek, Quantitation of endotoxin and sample-related interference by using a kinetic limulus amebocyte lysate microtiter test, in: "Microbiology 1985," L. Leive, ed., American Society for Microbiology, Washington, D.C. (1985).

52. R. Urbaschek, D. N. Männel, G. H. Northoff, and H. Kirchner, Release of macrophage mediators in response to endotoxin in hyperreactive and tolerant BCG infected mice, in preparation.

53. S. N. Vogel and D. L. Rosenstreich, LPS-unresponsive mice as a model for analyzing lymphokine-induced macrophage differentiation in vitro, Lymphokines 3:149 (1981).

54. R. I. Walker, Hematologic contributions to increases in resistance or sensitivity to endotoxin, in: "Experimental Hematology Today 1979," S. J. Baum and G. D. Ledney, Springer-Verlag, Berlin (1979).

55. D. A. Wichterman, A. E. Baue, and I. H. Chaudry, Sepsis and septic shock--a review of laboratory models and a proposal, J. Surg. Res. 29:189 (1980).

56. J. S. Younger and W. Stineberg, Interferon appearance stimulated by endotoxin, bacteria or viruses in mice pretreated with Escherichia coli endotoxin or infected with Mycobacterium tuberculosis, Nature (London) 208:456 (1965).

PROTECTION AGAINST LETHAL HAEMOPHILUS PLEUROPNEUMONIAE INFECTION IN

SWINE BY ANTIBODIES TO LPS CORE ANTIGENS

B. W. Fenwick, J. S. Cullor, and H. J. Olander

Department of Veterinary Pathology
School of Veterinary Medicine
University of California
Davis, California

INTRODUCTION

Haemophilus pleuropneumoniae (HP) is the cause of a contagious pneumonia in pigs. The disease is a major problem in the swine industry and is being increasingly diagnosed as a cause of mortality and lowered production. Presently used procedures to control the disease have failed to slow its spread or significantly reduce losses associated with infection. The clinical signs, pathology and immunopathology suggest that bacterial toxins, possibly endotoxin, play a major role in the pathogenesis of the disease (1,12).

Recent studies involving a number of gram-negative infections suggest the involvement of antigenetically similar cell wall components in resistance to infection (9,15). In addition, antibodies against "core" glycolipids of Enterobacteriaceae have been associated with protection against experimental gram-negative sepsis in man (15). These studies and others demonstrate that immunity to core LPS antigens confer protection from a wide variety of gram-negative infections.

To investigate the involvement and potential protective effect of antibodies against shared LPS core antigens during a localized infection, HP was used to induce severe pneumonias in weanling pigs which had been previously immunized with a bacterin of an Rc mutant of E. coli 0111. This strain, termed J5, is genetically stable due to a deficiency in uridine 5'-diphosphate (UDP)-galactose epimerase (7). At peak anti-J5 antibody levels the pigs were intranasally challenged with HP. Sequential clinical, microbiological, serologic and pathologic studies were used to evaluate the severity of the disease in immunized and control animals. The results show that antibodies of the IgG class that are specific for core LPS antigens provide protection from an otherwise lethal pneumonia.

The mechanism by which antibodies to LPS core antigens provide protection from the sequellae of gram-negative infections is unclear. To better define the mechanism of protection, high titer equine anti-J5 antiserum was absorbed to cultured viable HP at hourly intervals. Absorption of J5 antibodies occurred primarily during the log phase of bacterial growth suggesting that the protection provided by anti-J5 antibodies predominantly occurs when bacterial growth is most rapid.

METHODS

Bacteria

J5 is a genetically stable mutant of E. coli 0111 that produces an incomplete lipopolysaccharide due to the lack of UPD–galactose 4–epimerase (7). The organism is equivalent to the Rc mutants of Salmonella in that all sugars distal to the point of galactose incorportion are absent. The result is exposure to the central "core" lipopolysaccharide which are both structurally and antigenetically similar to those of other gram–negative bacteria. J5 bacterins were prepared as described by Ziegler (15).

Haemophilus pleuropneumoniae strain J45 (serotype 5) was isolated from a naturally occurring case of porcine pleuropneumonia. It was cultured in PPLO media containing seven percent neutralized equine serum and supplemented with NAD and fresh yeast extract. The bacteria used in the absorption experiment were grown in a 14 liter bench top fermenter (New Brunswick Science, Edison, New Jersey) under strict conditions.

Serology and Antiserum. J5 antibody titers were determined at four day intervals by direct ELISA using a modification of the procedure described by Ito et al. (8). Equine antiserum was produced by immunization of adult horses at ten day intervals with the J5 bacterin until high specific titers were reached. Swine were immunized with the J5 bacterin and boostered after two weeks.

Bacterial Challenge. Ten week pigs from a HP–free herd were confirmed free of infections by repeated serological and bacteriologic examinations. Twenty-five animals were immunized with 3 ml of E. coli JF bacterin and five animals survived as nonimmunized controls. All were challenged intranasally with 2×10^7 washed viable organisms in 2 ml of saline. Complete serological and pathological examinations were performed on all animals.

Immunoabsorption. During a ten liter fermentation of HP aliquotes of the culture were removed hourly and the number of bacteria per ml determined

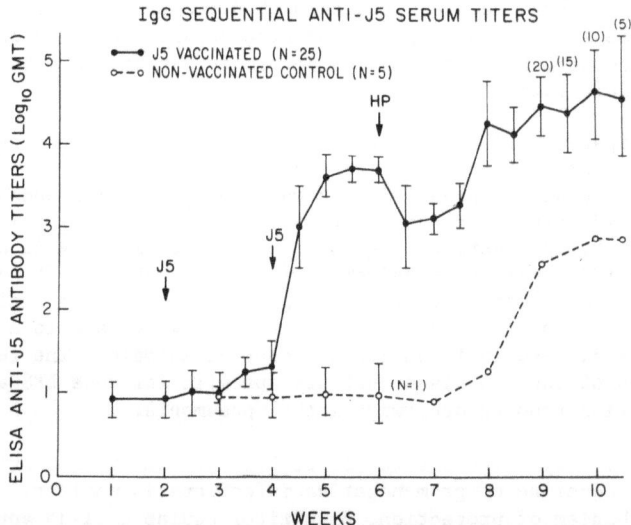

Figure 1. Mean E. coli J5 IgG ELISA titers by treatment group throughout the experiment. (J5 - immunization with J5 bacterin; HP - challenge with Haemophilus pleuropneumoniae.)

by limiting dilution. At the same time enough of the culture was removed so that one gram of washed packed viable cells could be harvested by centrifugation. The cells were held on ice at all times. After resuspending the cells in 9 ml of hyperimmune equine anti-J5 antiserum the mixture was gently rocked for 30 minutes. The cells were removed by centrifugation and the serum antibody titer determined as before.

RESULTS

Immunization with the J5 bacterin produced high antibody titer in both the pigs and horses with no adverse reactions. Low anti-J5 specific titers were present in all animals prior to immunization.

Clinical, bacteriologic, serologic and pathologic findings indicated that all of the pigs were infected with HP. No deaths occurred in J5 immunized pigs (n=25) whereas four out of five of the control animals died within 24 hours of the bacterial challenge. Changes in antibody titer are shown by treatment group in Figure 1. In the J5 immunized group anti-J5 IgG titers dropped during the acute stages of the infection and rebounded to well above prechallenge levels during convalescence. The J5 titer also increased in the single surviving control animal.

The results of the growth phase dependent absorption of equine anti-J5 antisera by HP are shown in Figure 2. The amount of antibody absorbed significantly increased during the period of most rapid bacterial growth.

DISCUSSION

The results show that antibodies against shared LPS core antigens provide considerable protection against an otherwise lethal Haemophilus pneumonia in swine. The findings extend previous studies which demonstrated that antibodies to J5 confer cross-protection against diverse gram-negative bacteremias (13-15) as well as the effects of endotoxin (2,3,5). This is

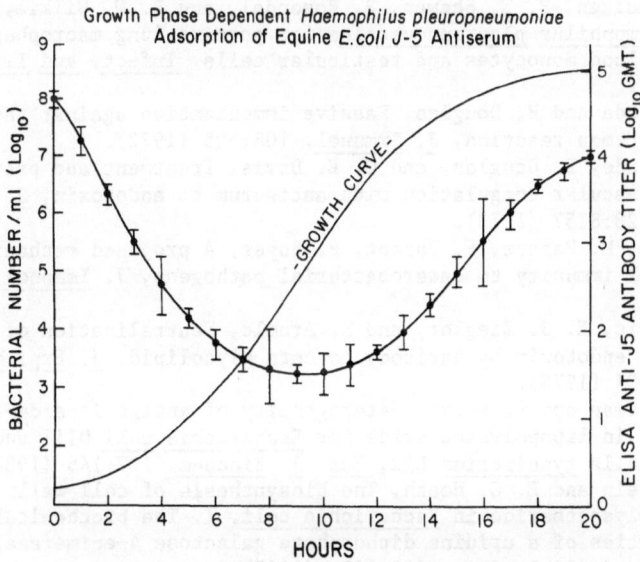

Figure 2. Sequential absorption of hyperimmune equine J5 antiserum with cultured Haemophilus pleuropneumoniae.

the first report of protection from the sequellae of a severe localized gram-negative infection in the natural host. Although the J5 immunization did not prevent infection there was significant improvement in the clinical signs, pathologic lesions and mortality. A marked drop in J5 antibody titer during the acut phases of the infection plus the significantly increased titers during convalescence support the involvement of these antibodies and the presence of cross-reacting antigens during the infection. Additional evidence is the increase in J5 titer in the single surviving non-immunized pig.

The protective antigens provided by the J5 bacterin certainly involve cross-reactive epitopes present in the complex cell wall of gram-negative bacteria. Similar protective antigens are also found in antiserum to other rough mutants (10). The mechanism of the protection is most likely multi-factorial (4). However, there is evidence that J5 antiserum is not a strong nonspecific opsonin and does not increase complement mediated serum bacter-iolysis (13). Prompted by the heterogeneity of LPs side-chain length in gram-negative bacteria (6,11) Ziegler et al. (15) suggested that J5 anti-serum acts to block exposure of biologically active sites within the LPS. They theorized that these components are exposed to a relatively greater degree during rapid bacterial growth when LPS side-chain production is in-complete. Our results support this concept by showing a pronounced growth phase dependent absorption of anti-J5 antibodies. This finding indicates that antibodies to core antigens are likely to be most effective when bac-terial growth is unrestricted, as during the early stages of an infection or in a compromised host. Such immunity, whether actively or passively acquired, may provide the necessary time for activation of host defenses or for medical therapy to be effective.

ACKNOWLEDGMENTS

This research was supported by U.S.D.A. grant 03-CRSR-2-2175.

REFERENCES

1. P. H. Bendixen, P. E. Shewen, S. Rosendal, and B. W. Wilkie, Toxicity of Haemophilus pleuropneumoniae for porcine lung macrophages, periph-eral blood monocytes and testicular cells, Infect. and Immun. 33:673 (1980).
2. A. I. Braude and H. Douglas, Passive immunization against the local Schwartzman reaction, J. Immunol. 108:505 (1972).
3. A. I. Braude, H. Douglas, and G. E. Davis, Treatment and prevention of intravascular coagulation with antiserum to endotoxin, J. Infect. Dis. 128:S157 (1973).
4. L. Chedid, M. Parant, F. Parant, F. Boyer, A proposed mechanism for natural immunity to enterobacterial pathogens, J. Immunol. 100:292 (1968).
5. C. E. Davis, E. J. Ziegler, and K. Arnold, Neutralization of meningo-coccal endotoxin by antibody to core glycolipid, J. Exp. Med. 147:1007 (1978).
6. R. C. Goldman and L. Leive, Heterogeneity of antigenic-side-chain length in lipopolysaccharide for Escherichia coli O111 and Salmonella typhimurium LT2, Eur. J. Biochem. 107:145 (1980).
7. A. D. Elbein and E. C. Heath, The biosynthesis of cell wall lipopolysaccharide in Escherichia coli. I. The biochemical properties of a uridine diphosphate galactose 4-epimeraseless mutant, J. Biol. Chem. 240:1919 (1965).
8. J. I. Ito, A. C. Wunderlich, and J. L. Lyons, The role of magnesium in the enzyme-linked immunosorbent assay for lipopolysaccharide of rough

Escherichia coli strain J5 and Neisseria gonorrhoeae, J. Infect. Dis. 142:532 (1980).

9. M. I. Marks, E. J. Ziegler, H. Douglas, L. B. Corbeil, and A. I. Braude, Induction of immunity against Haemophilus influenzae type b infection by Escherichia coli core lipopolysaccharide, J. Clin. Invest. 69:742 (1982).

10. W. R. McCabe, Immunization with R mutants of S. minnesota. I. Protection against challenge with heterologous gram-negative bacilli, J. Immunol. 108:601 (1972).

11. R. S. Munford, C. L. Hall, and P. D. Rick, Size heterogeneity of Salmonella typhimurium lipopolysaccharides in outer membrane and culture supernatant membrane fragments, J. Bacteriol. 144:630 (1980).

12. S. Rosendal, W. R. Mitchell, and M. Weber, Haemophilus pleuropneumoniae lung lesions induced by sonicated bacteria and sterile culture supernatant, Proc. Internat. Pig Vet. Soc. Cong. 221 (1981).

13. E. J. Ziegler, H. Douglas, J. E. Sherman, C. E. Davis, and A. I. Braude, Treatment of E. coli and Klebsiella bacteremia in agranulocytic animals with antiserum to a UPD-Gal epimerase-deficient mutant, J. Immunol. 111:433 (1973).

14. E. J. Ziegler, J. A. McCutchan, H. Douglas, and A. I. Braude, Prevention of lethal pseudomonas bacteremia with epimerase-deficient E. coli antiserum, Trans. Assoc. Am. Physicians 88:101 (1975).

15. E. J. Ziegler, J. A. McCutchan, J. Fierer, M. P. Glauser, J. C. Sadoff, H. Douglas, and A. I. Braude, Treatment of gram-negative bacteremia and shock with human antiserum to a mutant Escherichia coli, New Eng. J. Med. 307:1225 (1982).

SALMONELLA ANTIGENS AS PROTECTIVE IMMUNOGENS IN SALMONELLA INFECTION

Toby K. Eisenstein[1], Loran M. Killar[2], Barnet M. Sultzer[3], and Marshall Phillips[4]

[1]Temple University School of Medicine, Philadelphia; [2]Yale University of School of Medicine, New Haven, Connecticut; [3]Downstate Medical Center, Brooklyn, New York; [4]National Animal Disease Center, Ames, Iowa

The capacity of nonviable preparations of Salmonella and of immune serum to protect mice against experimental, systemic Salmonella infection has been a subject of considerable controversy in the literature. The controversy is of theoretical importance because it relates to the formulation of hypotheses concerning the relative importance of cellular versus humoral immunity in host defenses to Salmonellae causing enteric fevers, with S. typhi being the organism of greatest medical interest. Protection by nonviable preparations and the capacity to achieve passive protection with immune serum implies that host defense is mainly humoral, whereas lack of protection by these agents, in contrast to protection achieved with viable organisms, implies cellular immunity as the major mechanism of host defense. The correctness of the experimental model chosen has practical implications for the development of an improved vaccine for typhoid fever in man, i.e., whether it should be nonviable or attenuated. The literature on this subject goes back over 60 years and has been reviewed thoroughly by Eisenstein and Sultzer (1).

In the studies to be described in this paper, Salmonella lipopolysaccharides in various states of purity were tested, along with other Salmonella antigens and vaccines, for ability to protect mice against challenge with virulent Salmonellae, and for ability to induce antibody. Several strains of mice which differ in their genetically determined responsiveness to LPS and their innate susceptibilities to Salmonella infection were used, including C3H/HeJ animals which are Salmonella hypersusceptible but endotoxin hyporesponsive; (2-4) C3H/H3NCr1BR mice which are Salmonella resistant and endotoxin responsive (5,6); and C3HeB/FeJ mice which are Salmonella hypersusceptible and endotoxin responsive (7,8).

The results show that the capacity of nonviable preparations to protect mice against Salmonella challenge is markedly influenced by the mouse strain, and correlates with the innate susceptibility of the mouse to Salmonella infection, but not with histocompatibility type, nor with the capacity to respond to endotoxin physiologically or immunologically. If systemic murine Salmonellosis is used as one model of gram-negative infection, then this disease provides an example in which susceptibility to the organism is separable from reactivity to endotoxin.

MATERIALS AND METHODS

Mice

C3H/HeJ and C3HeB/FeJ mice were purchased from Jackson Laboratories, Bar Harbor, Maine. C3H/HeNCr1BR and CD-1 mice were purchased from Charles River Breeding Laboratories, Wilmington, Massachusetts. Salmonella typhimurium, strain SL3235, is an attenuated derivative of a mouse virulent strain that is blocked in aromatic synthesis which renders it avirulent (10). It was derived and characterized by Hoiseth and Stocker (Stanford University), and kindly supplied by them for these studies. Listeria monocytogenes, strain EGD, was used for cross-protection studies.

Bacterial Strains

Salmonella typhimurium strain W118-2 was used in all of these studies for both preparation of vaccines and challenge of vaccinated mice. This organism has been used extensively in previous studies and has 0 antigens, 1, 4, 5, and 12, and H antigens 1 and 2 (12).

Vaccines

All nonviable vaccines were prepared from S. typhimurium W118-2. Trichloroacetic acid extracted lipopolysaccharide (TCA-LPS) was prepared by the procedure of Sultzer and Goodman (9). Endotoxin Protein (EP) and phenol-water purified lipopolysaccharide (PW-LPS) were obtained by hot phenol-water treatment of the TCA-LPS as previously described (11). Acetone-killed cells and ribosomal vaccine used in protection studies were prepared as previously described (12). For experiments carried out in mice of the C3H lineage, single lots of each vaccine were used. Vaccines tested in protection studies in CD-1 mice were usually from different lots than those used in the inbred mice.

Ribosomal vaccine used to study immunomodulation in C3H/HeJ mice was prepared from Salmonella (W118-2) by the method of Fogel and Sypherd (12). Brucella and Aspergillus ribosomes were purified as previously described (13). Phenol-water purified LPS was complexed to the homologous or heterologous ribosomes using 3.8% formaldehyde (pH 7.6), followed by extensive dialysis (13). Uncomplexed LPS was also subjected to 3.8% formaldehyde treatment to control for any effects on immunogenicity.

Immunization and Challenge

Vaccines were administered ip in 0.5 ml of saline without adjuvant. If a booster injection was used, it was given 14 days after the primary inoculation. Challenge was ip 21 days after the last injection. Organisms for challenge were grown in Brain-Heart Infusion Broth according to a standard procedure (12), and numbers of organisms injected determined by duplicate spread-plate counts on blood agar. Survivors were scored at 60 days for inbred mice and for 30 days for CD-1 mice. For groups of animals with greater than 50% mortality, mean time to death (MTD) was also calculated.

Antibody Responses

Agglutinating antibody against whole killed cells was measured using standard techniques (6). IgG was quantitated by ELISA (6), using PW-LPS coated cuvettes and γ-chain specific peroxidase-linked goat anti-mouse antibody (Litton Bionetics, Charleston, South Carolina).

Table 1. LD$_{50}$ Values of $\underline{S.}$ $\underline{typhimurium}$ W118-2 in Various
Mouse Strains

| Mouse Strain | Breeder | No. of cells = 1 LD$_{50}$ | |
		ip	iv
C3H/HeJ	The Jackson Laboratory	<7	<3
C3HeB/FeJ	The Jackson Laboratory	<2	<9
C3H/HeNCrlBR	Charles River Breeding Laboratories	1.2×10^{3}	N.D.
CD-1	Charles River Breeding Laboratories	1.0×10^{4}	8.8×10^{4}

Passive Serum Transfer

Immune serum for transfer into mice of the C3H lineage was raised in
75 C3H/HeNCrlBR mice by giving four weekly ip of 60 µg each of acetone-killed
cells. Sera were collected one week after the last injection and pooled.
Normal serum was collected from 75 unimmunized mice. For transfer, recipients received 0.2 ml of a 1:5 dilution of normal or immune serum iv two
hours before challenge, and an additional 0.2 ml 48 hours post-challenge.
Challenge was iv or ip with organisms suspended in 0.1 ml of saline.

For experiments in CD-1 mice, hyperimmune serum was raised in CD-21
mice by three weekly injections of 250 µg each of acetone-killed cells.
Serum was collected and pooled one week after the last injection, and 0.2
ml was injected into recipients iv at a dilution of 1:10. Normal serum
from unimmunized mice was also pooled as used at a 1:10 dilution.

Statistics

Statistical differences were assessed using Fisher's exact test for 2
x 2 tables (15). LD$_{50}$ and TD$_{50}$ values were determined by the method of
Reed and Muench (16).

RESULTS

Susceptibility of Various Mouse Strains to Salmonella Infection

Three mouse strains with known LD$_{50}$s were selected for use in vaccination and protection studies. CD-1 animals had been shown in our laboratory
to have strong innate resistance to virulent Salmonella with an intraperitoneal LD$_{50}$ of 1×10^{4} cells (12). We had also confirmed the hypersuscepti-
bility of C3H/HeJ mice and the resistance of C3H/HeNCrlBR mice to Salmonella
using this particular organism (6,14). It had been proposed that the hyper-
susceptibility of the C3H/HeJ mice was related to the Lps defect, so we
anticipated that the closely related C3HeB/FeJ mice, which are LPS responders, would be innately resistant. It was found instead, that these
animals were as hypersusceptible as the C3H/HeJ strain (see Table 1) to $\underline{S.}$
$\underline{typhimurium}$ W118-2. This discovery provided an example of a mouse in the
C3H lineage in which LPS responsiveness and Salmonella resistance was not
correlated.

Table 2. Protection and Antibody Responses in C3H/HeJ Mice Immunized with Various Salmonella Antigens

Mice Immunized with:	Dose (μg)[a]		MTD[b]	60-day Survival[c]		Agglutination titer		IgG Log10 ELISA titer[d]	
	1	2		Alive/total	%	Day 14	Day 35	Day 15	Day 35
PW-LPS	100	100	21	0/10	0	<2	2	<2.000	<2.000
TCA-LPS	100	100	31	3/14	21	8	32	3.130	3.463
Ribosomes	250	250	-	7/14	50	16	64	3.061	3.431
Acetone-killed cells	60	60	29	5/14	36	8	64	3.000	3.518
Saline	-	-	10	0/14	0	<2	<2	-	-

[a] Mice immunized ip with dose 1, and 14 days later with dose 2.

[b] Mean time to death (days).

[c] Mice infected ip of 6 cells with W118-2 21 days after dose 2.

[d] Pooled sera of 4 mice.

Table 3. Evaluation of Endotoxin Protein (EP) as a
Protective Vaccine in C3H/HeJ Mice

Mice Immunized with:[a]	30-day Survival[b] (alive/total)	% Survival
50 µg PW-LPS	0/10	0
50 µg EP	0/10	0
50 µg LPS + 25 g EP	0/10	0
50 µg LPS + 50 g EP	1/10	10
50 µg acetone-killed cells	1/9	11
Saline	0/10	0

[a] Mice immunized ip
[b] Mice infected ip with 17 cells of W118-2 21 days post-vaccination. [Copyright by the University of Chicago (6).]

Protection and Antibody Responses of Four Mouse Strains to Various Salmonella Immunogens

Table 1 through 4 show the results obtained when a panel of vaccines was tested in the three mouse strains in the C3H lineage. Note that the challenge doses used in the two hypersusceptible mouse strains were minimal (six cells in the C3H/HeJ and 24 cells in the C3HeB/FeJ). Yet neither of

Table 4. Protection and Antibody Responses in C3HeB/FeJ Mice
Immunized with Various Salmonella Antigens

Mice Immunized with:	Dose (µg)[a]		MTD[b]	60-day Survival[c]		Agglutination Titer[f]	
	1	2		Alive/total	%	Day 14	Day 35
PW-LPS	50	25	19	1/10	10	2	4
	100	100	19	0/7[d]	0	<2	8
Endotoxin protein	50	25	17	0/10	0	<2	2
	100	100	17	2/10	20	<2	2
TCA-LPS	50	25	–	5/10	50	8	64
	100	100	31	0/5[e]	0	8	64
Ribosomes	100	100	–	6/10	60	8	64
Acetone-killed cells	60	60	–	10/10	100	16	64
Saline	–	–	12	0/10	0	<2	<2

[a] Mice immunized ip with dose 1, and 14 days later with dose 2.
[b] Mean time to death (days).
[c] Mice infected ip with 24 cells of W118-2 21 days after dose 2.
[d] 3 mice died from toxicity of immunization.
[e] 5 mice died from toxicity of immunization.
[f] Titers on designated days after primary immunization. Pooled sera of 4 mice. [Copyright by the University of Chicago (6).]

Table 5. Protection and Antibody Responses in C3H/HeNCr1BR Mice Immunized with Various Salmonella Antigens

| Mice Immunized with | Dose (μg)[a] | | 60-day Survival[c] | | Agglutination titer[i] | | $\log_{10}$ ELISA Titer[f] | | | |
| | 1 | 2 | Alive/Total | % | Day 15 | Day 35 | IgM | | IgG | |
							Day 15	Day 45	Day 15	Day 45
PW-LPS	50	25	9/10[d]	90	2	4	2.196	3.022	2.161	2.000
	100	100	5/8	63	2	8	<2.000	3.268	2.029	2.439
Endotoxin protein	50	25	10/10	100	<2	4	<2.000	<2.000	<2.000	2.610
	100	100	10/10	100	<2	8	<2.000	<2.000	2.612	2.953
TCA-LPS	50	25	10/10	100	32	64	2.872	3.321	2.726	3.593
	100	50	4/4[e]	100	32	128	3.240	3.719	2.615	3.616
Ribosomes	100	100	10/10	100	16	128	3.317	4.026	2.765	3.969
	250	250	7/7	100	32	256	3.228	3.863	2.913	3.928
Acetone-killed	60	60	10/10	100	32	128	2.373	3.953	<2.000	3.973
Phosphate-buffered saline (PBS)[b]	-	-	1/9	11	<2	<2				

a Mice immunized ip with dose 1, and 14 days later with dose 2.
b Mean time to death was 7 days for PBS mice; not calculated for other groups.
c Mice infected ip with 2.3 x 10^4 cells of W118-2 (= 19 LD$_{50}$ doses) 21 days after dose 2.
d 2 mice died from toxicity of immunization.
e 6 mice died from toxicity of immunization.
f Pooled sera of 4 mice
[Copyright by the University of Chicago (6).]

Table 6. Titration of Protective Capacity of Salmonella Antigens in CD-1 Mice

Vaccine type[a]	% Survival at various challenge doses				
	1×10^{6}[b] (100 LD$_{50}$)	5×10^{6} (500 LD$_{50}$)	1×10^{7} (1,000 LD$_{50}$)	5×10^{7} (5,000 LD$_{50}$)	1×10^{8} (10,000 LD$_{50}$)
Live cells					
6.5×10^{3}	100[c]		90	90	30
1.3×10^{3}	100		90	50[d]	40[e]
AKC (μg)					
1.0	90		0		
10	100		60		
50	100		90		
100	100		100	10	0
250	100		100	60	0
500	100		100		
RIB (μg)					
0.1	60		0		
1.0	90		0		
25	100		30		
100	100		80	10	0
250	100		80		
1000	100		90	30	0
LPS (μg)					
0.1		30	0		
0.5		40	0		
1.0		70	0		
25		100	0		
100		100	0	0	0
150		100	0		
200		100	38		

[a]Vaccines given intraperitoneally in 0.5 ml of saline, AKC, Acetone-killed cells; RIB, ribosomes.
[b]Number of viable W118-2 given intraperitoneally 3 weeks after immunization.
[c]10 mice per group
[d]1.3×10^{3} live cells versus 250 μg of acetone-killed cells or 1,000 μg of ribosomes, not significant; 1.3×10^{3} live cells or 250 μg of acetone-killed cells versus 100 μg of LPS, P < 0.01.
[e]1.3×10^{3} live cells versus other vaccines, P < 0.05.

these mouse strains were protected by highly purified phenol-water extracted LPS, nor by Endotoxin Protein. TCA-LPS afforded some protection in the C3HeB/FeJ mice, but not in the C3H/HeJ; ribosomes gave partial protection to both strains, and acetone-killed cells protected C3HeB/FeJ but not C3H/HeJ. In light of the very low challenge doses used, it can be concluded that all vaccines, except the acetone-killed cells in the C3HeB/FeJ, were poorly protective in both strains of Salmonella hypersusceptible mice. In contrast, the inherently Salmonella-resistant C3H/HeNCrlBR mice showed solid protection against 19 LD$_{50}$ doses when vaccinated with all of the preparations tested (Table 5). When these results are compared with experiments carried out in inherently resistant CD-1 mice, a similar picture of protection by nonviable vaccines was observed (Tables 5 and 6). The experiments in CD-1 mice were carried out at a different time using different lots of LPS, ribosomes and AKC from those used in mice of the C3H lineage. Further, the CD-1 mice received only a single dose of each vaccine and high challenge doses of 500 or 1000 LD_{50}s. Nevertheless, like the inherently Salmonella resistant C3H/HeNCr1BR mice, they were well protected by all three nonviable vaccines, including the purified PW-LPS and EP (Table 7).

Comparison of antibody responses in the different mouse strains shows that lack of protection cannot be correlated with failure to induce agglutinating and anti-O antibody (Tables 2, 4, and 5). Thus, both C3H/HeNCr1BR and C3HeB/FeJ mice immunized with TCA-LPS gave agglutination titers of 1:64, but C3H/HeNCr1BR were 100% protected, whereas C3HeB/FeJ were only 50% protected. Further, TCA-LPS induced agglutinating titers in C3H/HeNCr1BR equivalent to those induced in C3H/HeJ by acetone-killed cells, and the two strains had ELISA titers which were comparable, yet C3H/HeNCr1BR were protected, but C3H/HeJ were not (100% versus 36%, respectively).

Immune Responses to O Antigens in C3H/HeJ Mice

An interesting aspect of the data collected above, was the observation that C3H/HeJ mice could in fact make apparently normal antibody responses to O antigens as assessed by ELISA, and also confirmed earlier observations that ribosomal vaccine could induce normal hemagglutinating and whole cell agglutinating responses in these mice (14). To explore further the interrelationship of O antigen and ribosomes in immune responsiveness in C3H/HeJ

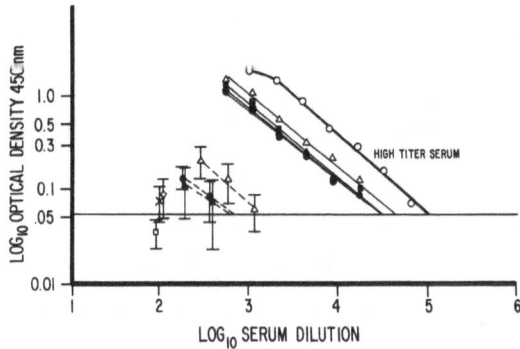

Figure 1. Pooled anti-IgM antibody responses for 6 mice per group receiving various Salmonella antigens. Dotted lines are primary responses and solid lines are secondary responses. O - high titered serum raised to acetone-killed cells which served as a standard; ● - Salmonella ribosomes; ▲- Salmonella LPS formaldehyde complexed to Salmonella LPS; △ - Salmonella LPS formaldehyde complexed to Brucella ribosomes; ■ - Salmonella LPS formaldehyde complexed to Aspergillus ribosomes. There was not detectable primary immune response for this group. Error bars on secondary responses of these groups are not shown due to crowding of the graph. Groups receiving Salmonella LPS alone - x; Brucella ribosomes alone -◇; or Aspergillus ribosomes alone -□; gave no measurable responses after the primary immunization. After secondary immunization, responses at serum dilutions of 1:200 dilution of secondary serum are indicated by the appropriate symbols with error bars.

246

Table 7. Protection in CD-1 Mice Against S. typhimurium Challenge by
Immunization with Endotoxin Protein

Immunizing Agent	Dose (μg)[a]		30-day survival against[b]			
	1	2	50 LD$_{50}$	200 LD$_{50}$	500 LD$_{50}$	1000 LD$_{50}$
S. typhimurium	50	25	10/10	8/10	9/10	4/10
S. typhimurium PW-LPS	50	25	5/10	8/10	-	1/10
PBS	-	-	0/10	0/10	-	-

[a] Mice immunized ip on day 0 with first dose and on day 14 with second dose.
[b] Mice given S. typhimurium W118-2 ip 21 days after the booster injection.
[Copyright by Thieme-Stratton, Inc. (27).]

mice, PW-LPS was complexed to highly purified ribosomes of Aspergillus or
Brucella using 0.38% formaldehyde treatment. Figures 1 and 2 show that
complexed O antigen, but not the purified antigen alone, induce high levels
of anti-O antibody in C3H/HeJ mice. Further, the purified material could
only induce low levels of IgM antibody, but the complexed LPS induced not
only IgM, but also substantial amounts of IgG. As the ribosomes from the
heterologous species had no polyclonal activating effect themselves, the
results show that C3H/HeJ mice can respond to PW-LPS on a carrier.

Passive Serum Transfer

In order to assess more rigorously the contribution of antibody to
protection, hyperimmune mouse serum raised to whole killed cells was passively transferred. As shown in Tables 8 and 9 immune serum protected
C3H/HeNCrlBR and CD-1 mice but not C3H/HeJ. C3HeB/FeJ showed partial protection against very low challenge doses, but even a slight increase in the
challenge overwhelmed the protection. Further, it appears that intravenous
challenge is more stringent than intraperitoneal, particularly when protec-

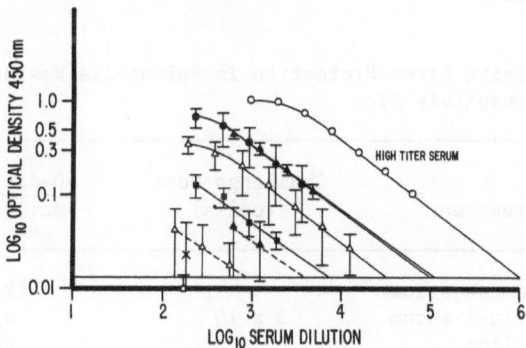

Figure 2. Pooled IgG responses of 6 mice per group receiving the antigens
described in the legend to Figure 1. The median value for the
secondary response of mice given Brucella ribosomes alone ◇ ,
is not shown because it was below the limits plotted on the
graph. There was no measurable primary IgG response induced
by Salmonella ribosomes - ● , nor by Salmonella LPS formaldehyde
complexed to Aspergillus ribosomes - ■ .

Table 8. Protection of Three Mouse Strains in the C3H Lineage by
Passive Serum Transfer

Route of challenge	Dose (LD$_{50}$s)	Treatment[a]	60-day Survival (Survivors/Total) (mean time to death in days)		
			C3H/HeJ	C3HeB/FeJ	C3H/HeNCr1BR
Intra- venous	10	IS	0/14(12.5)	5/14(19.3) p<.025	9/14(ND) p<.005
		NS	0/14 (9.7)	0/14(15.4)	0/14(10.4)
		NaCl	0/14 (8.6)	0/14(18.8)	0/14 (9.2)
	16	IS	–	0/14(19.6)	–
		NS	–	0/14(15.4)	–
		NaCl	–	0/14(10.6)	–
Intra- peritoneal	9	IS	0/6(36.8)NS	2/6(33.2)NS	6/6(ND) p<.01
		NS	0/6 (9.7)	0/6(12.0)	1/6(10.2)
		NaCl	0/6 (9.7)	0.6(10.7)	1/6 (8.0)
	18	IS	1/6(26.0)NS	4/6(ND) p<.05	6/6 (ND) p<.01
		NS	–	–	–
		NaCl	–	–	–

[a]IS, immune serum; NS, normal serum. A 1:5 dilution of serum or saline was given in 0.2 ml iv 2 hr prior to challenge and again 48 hr post-challenge. ND = not done. [Copyright by the University of Chicago (6).]

tion is assessed by mean time to death. Thus, immune serum markedly prolonged the mean time to death of both C3H/HeJ and C3HeB/FeJ mice when challenge was ip. In fact, if protection were to have been measured by number of survivors at 20 or 30 days, instead of 60 days, it would have appeared as if these mice were protected. These studies show that immune serum is differentially able to protect mice which are inherently resistant to Salmonella infection, but unable or less able to protect mice which are inherently hypersusceptible.

Table 9. Passive Serum Protection in Salmonella Resistant and
Susceptible Mice

Mouse Strain	Treatment[a]	Challenge Dose[b] (cells)	60-day survival Survivors/total	%
CD-1	Immune serum		12/14	86
	Normal serum	3 x 10^5	0/14	0
	Saline		0/14	0
C3HeB/FeJ	Immune serum		3/14	21
	Normal serum	17	0/14	0
	Saline		0/14	0

[a]0.2 ml of 1:10 dilution of serum was given iv 2 hr prior to challenge.
[b]Challenge was iv. [Copyright by Plenum Publishing Corporation (1).]

Table 10. Protection of C3H/HeJ and C3HeB/FeJ Mice
by S. typhimurium SL3235

Pretreatment[a]	60-day Survival (survivors/total)					
	C3H/HeJ mice			C3HeB/FeJ		
	Challenge dose (cells)[b]					
	90	900	9000	120	1200	12,000
SL3235	11/12	11/12	8/11	12/12	11/12	12/12
Ribosomes	7/12	1/12	0/12	8/12	7/12	3/11
Saline	0/12	–	–	0/12	–	–

[a]C3H/HeJ mice were immunized ip with 5×10^4 live, avirulent S. typhimurium
SL3235 21 days prior to challenge. C3HeB/FeJ mice received 7×10^4
organisms. Ribosomes were given as 2 doses of 250 μg each 14 days apart.
Challenge was 21 days after the second injection.
[b]Mice were challenged ip with virulent S. typhimurium, strain W118-2.
[Some data copyright Journal of Immunology (19).]

Protection by Live, Avirulent Salmonella

In contrast to the protective capacity of nonviable vaccines in inherently Salmonella susceptible mice, the live attenuated strain SL3235 was
tested. As shown in Tables 10 and 11 this organism was able to confer immunity to about 10,000 LD_{50} challenge doses of virulent Salmonella, even in
the HeJ mice. The ribosomal vaccine, which was the best of the nonliving
preparations, gave measurable protection, but it titrated out at 90 LD_{50}s.
These high levels of protection occurred in the HeJ mice without converting
them to an LPS responsive state as judged by in vivo mortality after endotoxin administration (Table 12). Similarly, SL3235-infected mice did not
exhibit an increase in capacity of their peritoneal cells to ingest C3b-
coated erythrocytes in vitro after LPS injection in vivo (data not shown,
see reference 19), showing in a different way that they remain refractory
to LPS. SL3235 did, however, apparently overcome the macrophage defect of

Table 11. Titration of protection by SL3235 in C3H/HeNCrlBR Mice

Immunization	Challenge dose (cells)[b]	Survivors/total[c]
SL3235[a]	1×10^7	0/6
	1×10^6	6/6[d]
	1×10^5	6/6[d]
Saline	1×10^5	0/6

[a]Mice were injected ip either with 7×10^4 S. typhimurium, strain SL 3235,
suspended in 0.5 ml saline or with saline alone as a control.
[b]Mice were challenged 21 days after immunization with S. typhimurium W118-2.
[c]Survival was scored for 60 days.
[d]Statistically significant differences from saline controls (P < .005).
[Copyright by the American Society for Microbiology (20).]

Table 12. Effect of Avirulent Salmonella Infection on the Sensitivity of Mice to Endotoxin Lethality

Mouse Strain	Pretreatment[a]	Dose LPS[b] (mg)	Dead[c]/total
C3H/HeJ	NaCl	2.0	4/8
		1.0	0/8
		.75	0/8
		.50	0/8 TD_{50} = 2.0[d]
C3H/HeJ	SL3235	2.0	6/8
		1.0	2/8
		.75	3/8
		.50	0/8 TD_{50} = 1.26
C3HeB/FeJ	NaCl	.50	8/8
		.25	8/8
		.10	0/8
		.05	0/8 TD_{50} > .10 < .25
C3HeB/FeJ	SL3235	.50	6/8
		.25	6/8
		.10	3/8
		.05	1/8 TD_{50} = .17

[a] Injected ip with saline or 1.8×10^6 SL3234 21 days prior to challenge.
[b] Difco S. typhimurium PW-LPS injected ip in 0.5 ml saline.
[c] Scored 48 hr post-injection.
[d] Determined by method of Reed and Muench.
[Copyright by the Journal of Immunology (19).]

C3H/HeJ mice, as immunized animals were resistant to Listeria (Table 13), and their splenic macrophages could transfer Salmonella immunity (Table 14).

Table 13. Protection of C3H/HeJ Mice Against Listeria Infection by Immunization with Salmonella SL3235

Immunized with[a]	Challenge Dose (No. of LD_{50}s)[b]	Survivors/Total	% Survival
S. typhimurium SL3235	725	3/10	30
	145	8/10	80
	37	8/10	80
	7	10/10	100
Saline	37	0/10	0
	7	0/10	0

[a] Injected ip with 5×10^5 SL3235 or saline 21 days before Listeria challenge.
[b] L. monocytogenes, strain EGD. [Copyright by Journal of Immunology (19).]

Table 14. Transfer of Protection to C3H/HeJ Mice with Spleen
Cell Populations Enriched for Macrophages

Experiment[a]	Fraction[b]	Total Number of Cells Transferred[c]	Percent Macrophages[d]	Number of Macrophages Transferred	Survivors/Total[e]
1	Nonfractionated	2×10^7		6.0×10^6	$6/6$[f]
		1×10^7	30	3.0×10^6	$4/6$[f]
		5×10^6		1.5×10^6	$2/6$
	Nonadherent	2×10^7		2.8×10^6	$1/6$
		1×10^7	14	1.4×10^6	$1/6$
		5×10^6		7.0×10^5	$1/6$
	Adherent	2×10^7		1.4×10^7	$6/6$[f]
		1×10^7	68	7.0×10^6	$3/6$
		5×10^6		3.5×10^6	$3/6$
	Control	None			$0/7$
2	Nonfractionated	1×10^7	26	2.6×10^6	$2/8$
		5×10^6		1.3×10^6	$1/8$
	Nonadherent	1×10^7	20	2.0×10^6	$0/8$
		5×10^6		1.0×10^6	$1/8$
	Adherent	1×10^7	72	7.2×10^6	$6/8$[f]
		5×10^6		3.6×10^6	$1/8$
	Control	None			$0/8$

[a]In Experiment 1, donors were immunized ip with 1.8×10^6 SL3235 1 week before challenge. In Experiment 2, donors were immunized ip with 5×10^5 SL3235 1 week before challenge.
[b]Cells were fractionated by incubating on gelatin/plasma-coated plates (28).
[c]Cells were suspended in 1 ml RPMI-1640 and injected ip. Controls were injected with RPMI-1640 only.
[d]Percentage of cells that ingested greater than 10 latex particles.
[e]Recipients were challenged ip 24 hr after cell transfer, with virulent S. typhimurium, strain W118-2. In Experiment 1, mice were challenged with 25 LD_{50} doses and in Experiment 2 with 22 LD_{50} doses.
[f]Statistically significant from control ($P < .05$).
[Copyright by the American Society for Microbiology (20).]

DISCUSSION

The experiments presented in this paper permit several important conclusions concerning anti-Salmonella immunity and the relationship between
innate and acquired immunity to this gram-negative infection and endotoxin
sensitivity of the host. First, our observation that C3HeB/FeJ mice are as
Salmonella hypersusceptible as C3H/HeJ mice, yet are endotoxin responsive
(7,8), provided the first evidence that there are other genes in mice of
the C3H lineage besides Ity^r (17) and Lps^d (5) which control Salmonella
susceptibility. The fact that C3H/HeJ mice are endotoxin resistant, but

Salmonella hypersusceptible, would also seem to rule out endotoxemia as a primary mechanism in Salmonella pathogenesis.

In regard to acquired immunity to Salmonella infection, it was found that a live avirulent Salmonella strain used as a vaccine confers very high levels of immunity on C3H/HeJ mice, presumable via macrophage activation, in the face of sustained endotoxin hyporesponsiveness. Therefore, conditions can be found in which the defect of macrophage activation of C3H/HeJ mice can be overcome without simultaneous reversion to endotoxin responsiveness. Further, endotoxin responsiveness is neither a necessary nor a sufficient condition for induction or expression of Salmonella immunity.

It was shown that C3H/HeJ mice are poorly protected by a variety of nonviable Salmonella vaccines which exhibit a spectrum of purity. However, the observed lack of protectability was not primarily related to their LPS defect, as the endotoxin-responsive C3HeB/FeJ mice also exhibited poor protection by many nonviable vaccines. Although the more complex nonviable vaccines did seem to be slightly better in C3HeB/FeJ than C3H/HeJ mice, the levels of protection they conferred were substantially inferior to protection obtained in the inherently Salmonella resistant C3H/HeNCrlBR and CD-1 strains. Comparison of immunization experiments in the three mouse strains in the C3H lineage with results obtained in outbred CD-1 mice, shows that ability to be protected by nonviable vaccines does not correlated with histocompatability type, as all of the C3H mice are H-2^k, yet the C3H/HeNCrlBR give a pattern or protectability like CD-1 animals, rather than like the C3H/HeJ or C3HeB/FeJ. Protectability does not correlate solely with ability to make an antibody response to the vaccines, as levels of anti-O antibody in poorly protected strains were frequently comparable to those observed in protected strains. Further, passive antibody afforded significant protection to inherently Salmonella resistant mouse strains (C3H/HeNCrlBR and CD-1), but was not effective (C3H/HeJ) or marginally effective (C3HeB/FeJ) in protecting inherently Salmonella hypersusceptible mouse strains. Thus, the capacity to be protected by nonviable vaccines seems to correlate best with the inherent susceptibility or resistance of the mouse strain to Salmonella infection.

It is interesting that administration of the avirulent, live strain of Salmonella, SL3235, protects both inherently susceptible and inherently resistant mice against high challenge doses of virulent organism. Yet, these high levels of resistance occur without induction of protective antibody in C3H/HeJ mice as assessed by failure to confer passive protection (20). These observations raise the question as to the mechanism of protection by the attenuated organism, and how it compares with that of the nonviable vaccines. C3H/HeNCrlBR mice immunized with SL3235 develop delayed hypersensitivity to a culture supernatant elicitin, supporting the hypothesis that the organism induces sensitized T cells, and hence cellular immunity (18). Macrophage activation occurs as evidenced by the concomitant, transient cross-protection to another intracellular pathogen, Listeria monocytogenes (19), and ability to transfer immunity with immune macrophages (20). Further, peritoneal macrophages of SL3235-immunized mice are activated to kill tumor cells in vitro (manuscript in preparation). Thus, the interpretation of the mechanism of action of the live organism is that it induces cellular immunity, which seems to work equally well in all the mouse strains.

Why do many nonviable vaccines show a divergence in protective capacity in different mouse strains apparently related to inherent resistance or susceptibility to Salmonella? As it is widely believed that the mechanism of action of most nonviable vaccines is to induce protective antibody, it can be hypothesized that the failure of nonviable vaccines to give high levels of protection in inherently hypersusceptible mice is attributable to a reduced capacity of these animals to utilize the antibody effectively.

252

An attractive hypothesis for why this should be so, and for which there is some supporting evidence, is that their macrophages are in a lower state of activation. Thus, macrophages of C3H/HeJ mice, in contrast to those of C3H/HeNCr1BR mice, have been shown to require two signals for activation to tumoricidal activity with most activating agents (21), and to fail to be activated by BCG to kill Leishmania (22) or Rickettsia (23). Macrophages of C3HeB/Fe mice may also have a defect in that they do not secrete IL-1 in the same manner as responder mice after stimulation with muramyl dipeptide (24). Thus, in the case of Salmonella susceptible mouse strains, antibody induced by nonviable vaccines may place Salmonella inside macrophages which have no less capacity to kill and to be activated to kill. An alternative explanation for why antibody is not protective may be related to the observation that C3H/HeJ mice have fewer Fc receptors and lose them more quickly than C3H/HeN mice (25). Thus, antibody may not be as opsonic in C3H/HeJ as in C3H/HeNCr1BR mice, permitting the organism to multiply extracellularly to greater numbers.

It is interesting that a correlation was observed between the complexity of nonviable vaccines and their protective capacity. In inherently resistant mice, the correlation probably relates to the capacity to form IgG antibodies. Thus, phenol-water purified LPS induces only IgM, which is short-lived (29), whereas protein containing TCA-LPS, acetone-killed cells or ribosomal vaccine induces IgG. However, in the hypersusceptible mice, even high levels of IgG were not protective when passively transferred. Yet acetone-killed cells and ribosomal vaccine did afford measurable levels of protection in C3H/HeJ and C3HeB/FeJ mice. One hypothesis which we have proposed to explain these observations is that the complex vaccines contain immunomodulators in addition to surface antigens. In particular the double-stranded RNA in the ribosomal vaccine may act to release mediators that activate C3H/HeJ macrophages (14). There is some evidence to support this hypothesis (26). It is planned to test the protective capacity of the formaldehyde-LPS complexes as one way to evaluate the validity of this interpretation.

From a theoretical point of view, these studies have explored the relationship of endotoxin responsiveness to innate Salmonella susceptibility and to acquired resistance to Salmonella. They have also illuminated some of the parameters regulating acquired resistance to systemic Salmonella infection in mice, of which the most important seems to be genetically determined innate susceptibility or resistance to the virulent organism. From a practical point of view, it is unequivocally shown that nonviable vaccines can be excellent protective antigens in inherently Salmonella resistant strains of mice. However, identical lots of vaccine were poorly protective in inherently susceptible mice. Thus, if the mouse model is to be used as a standard to evaluate the potential efficacy of typhoid vaccines for humans, a way must be found of assessing which mouse most closely resembles the human. These results also raise the possibility that variation in protective capacity of various nonviable typhoid vaccines observed in past field trials might be due to variations in the host populations tested, analogous to the differences seen in the various strains of mice. Finally, as the live attenuated strain seems to protect all mouse strains tested, use of this type of vaccine may be the most sensible strategy to pursue for human immunization.

REFERENCES

1. T. K. Eisenstein and B. M. Sultzer, Immunity to Salmonella infection, in: "Host Defenses to Intracellular Pathogens," T. K. Eisenstein, H. Friedman, and P. Actor, eds., Plenum Press, New York (1983).
2. B. M. Sultzer, Genetic control of leucocyte responses to endotoxin, Nature (London) 219:1253 (1968).

3. J. Watson and R. Riblet, Genetic control of responses to bacterial lipopolysaccharides in mice, I. Evidence for a single gene that influences mitogenic and immunogenic responses to lipopolysaccharides, J. Exp. Med. 140:1147 (1974).

4. H. G. Robson and S. I. Vas, Resistance of inbred mice to Salmonella typhimurium, J. Inf. Dis. 126:378 (1972).

5. A. D. O'Brien, D. L. Rosenstreich, I. Scher, G. H. Campbell, R. P. MacDermott, and S. B. Formal, Genetic control of susceptibility to Salmonella typhimurium in mice: role of the Lps gene, J. Immunol. 124:20 (1980).

6. T. K. Eisenstein, L. M. Killar, and B. M. Sultzer, Immunity to infection with Salmonella typhimurium: mouse strain differences in vaccine- and serum-mediated protection, J. Inf. Dis. 150:425 (1984).

7. T. K. Eisenstein, L. W. Deakins, and B. M. Sultzer, The C3HeB/FeJ mouse, a strain in the C3H lineage which separates Salmonella susceptibility and immunizability from mitogenic responsiveness to lipopolysaccharide, in: "Genetic Control of Natural Resistance to Infection and Malignancy, E. Skamene, P. A. L. Kongshavn, and M. Landy, eds., Academic Press, New York (1980).

8. T. K. Eisenstein, L. W. Deakins, L. Killar, P. Saluk, and B. M. Sultzer, Dissociation of innate susceptibility to Salmonella infection and endotoxin responsiveness in C3HeB/FeJ mice and other mouse strains in the C3H lineage, Infect. Immun. 36:646 (1982).

9. B. M. Sultzer and G. W. Goodman, Endotoxin protein: a B-cell mitogen and polyclonal activator of C3H/HeJ lymphocytes. J. Exp. Med. 144:821 (1976).

10. S. K. Hoiseth and B. A. D. Stocker, Aromatic-dependent Salmonella typhimurium are non-virulent and effective as live vaccines, Nature 291:238 (1981).

11. G. W. Goodman and B. M. Sultzer, Further studies on the activation of lymphocytes by endotoxin protein, J. Immunol. 122:1329 (1979).

12. C. R. Angerman and T. K. Eisenstein, Comparative efficacy and toxicity of a ribosomal vaccine, acetone-killed cells, lipopolysaccharide, and live cell vaccine prepared from Salmonella typhimurium, Infect. Immun. 19:575 (1978).

13. M. Phillips, T. K. Eisenstein, and J. Meissler, Immunomodulation of the antibody response to LPS in C3H/HeJ mice by complexing with heterologous ribosomes, Infect. Immun. 48:244 (1985).

14. T. K. Eisenstein and C. Angerman, Studies on the protective capacity and immunogenicity of lipopolysaccharide, acetone-killed cells, and a ribosome-rich extract of Salmonella typhimurium in C3H/HeJ and CD-1 mice, J. Immunol. 121:1010 (1978).

15. C. I. Bliss, "Statistics in Biology," Vol. 3, McGraw-Hill Book Company, Inc., New York (1967).

16. L. J. Reed and H. Muench, A simple method of estimating fifty percent endpoints, Am. J. Hyg. 27:493 (1938).

17. J. Plant and A. A. Glynn, Genetics of resistance to infection with Salmonella typhimurium in mice, J. Inf. Dis. 133:72 (1976).

18. L. M. Killar and T. K. Eisenstein, Differences in delayed-type hypersensitivity responses in various mouse strains in the C3H lineage infected with Salmonella typhimurium, strain SL3235, J. Immunol. 133:1190 (1984).

19. T. K. Eisenstein, L. M. Killar, B. A. D. Stocker, and B. M. Sultzer, Cellular immunity induced by avirulent Salmonella in LPS-defective C3H/HeJ mice, J. Immunol. 133:958 (1984).

20. L. M. Killar and T. K. Eisenstein, Immunity to Salmonella infection in C3H/HeJ and C3H/HeNCr1BR mice: Studies using an aromatic-dependent live Salmonella strain as a vaccine, Infect. Immun. 47:605 (1985).

21. L. P. Ruco and M. S. Meltzer, Defective tumoricidal capacity of macrophages from C3H/HeJ mice, J. Immunol. 120:329 (1978).

22. W. T. Hockmeyer, D. Walters, R. W. Gore, J. S. Williams, A. H. Fortier, and C. A. Nacy, Intracellular destruction of _Leishmania donovani_ and _Leishmania tropica_ amastigotes by activated macrophages: Dissociation of these microbicidal effector activities in vitro, _J. Immunol._ 132:3120 (1984).

23. C. A. Nacy and M. S. Meltzer, Macrophages in resistance to Rickettsial infections: Protection against lethal _Rickettsia tsutsugamushi_ infections by treatment of mice with macrophage-activating agents, _J. Leukocyte Biol._ 35:385 (1984).

24. O. Abehsira-Amar, C. Damais, M. Parant, and L. Chedid, Strain dependence of muramyl dipeptide-induced LAF (IL-1) release by murine-adherent peritoneal cells, _J. Immunol._ 134:365 (1985).

25. S. N. Vogel and D. L. Rosenstreich, Defective Fc receptor-mediated phagocytosis in C3H/HeJ macrophages. I. Correction by lymphokine-induced stimulation, _J. Immunol._ 123:2841 (1979).

26. R. C. Butler, H. Friedman, S. C. Specter, and T. K. Eisenstein, Induction of immunoenhancing factors for murine splenocyte cultures by _Salmonella typhimurium_ ribosome and ribonucleic acid extracts, _Infect. Immun._ 32:1123 (1981).

27. T. K. Eisenstein and B. M. Sultzer, Salmonella vaccines: Protection by endotoxin protein and lipopolysaccharide in two different mouse strains, _in_: "Bacterial Vaccines," J. Robbins, J. Hill, and J. Sadoff, eds., Thieme-Stratton, New York (1982).

28. B. Freundlich and N. Avdalovic, Use of gelatin/plasma coated flasks for isolating human peripheral blood monocytes, _J. Immunol. Methods_ 62:31 (1983).

29. C. R. Angerman and T. K. Eisenstein, Immunity to _Salmonella_ infection in mice: Correlation of the duration and magnitude of protection to _Salmonella_ infection afforded by various vaccines with antibody titers, _Infect. Immun._ 27:435 (1980).

PROTECTIVE FUNCTION OF NEUTROPHILS DURING EXPERIMENTAL ENDOTOXIC SHOCK

B. W. Fenwick, J. S. Cullor and A. Kelly

Departments of Veterinary Pathology and Veterinary Medicine
University of California
Davis, California

INTRODUCTION

Dramatic hematologic and hemostatic alterations are common sequelae to generalized gram-negative infections (14). Two of the most consistent aberrations are a rapid decrease in circulating neutrophils (PMN) and consumptive coagulopathies. Purified bacterial endotoxins (lipopolysaccharides, LPS) when given intravenously to experimental animals and human volunteers will induce similar pathophysiological changes (8). It now appears that much of the host response to endotoxin is initiated by damage at the microcirculatory level, and persistant inflammation at this level is associated with the lethal effects of endotoxin (6,14). This is due primarily to disseminated intravascular coagulation (DIC) and vascular collapse. Neutrophils as inflammatory cells have the potential of damaging host tissues and thus may contribute to the initiation of these events.

Neutrophils are a significant component in the inflammatory response to bacterial infections (1). The rapid decrease in circulating neutrophils during septic and endotoxic shock plus their ability to cause tissue damage has led to the suggestion that they are primarily or secondarily involved in host response to endotoxin. In fact, neutrophils appear to be a primary target of endotoxin with more endotoxin binding to neutrophils than other leukocytes (13,15). The relationship between the disappearance of neutrophils and the pathogenesis of endotoxic and septic shock is unclear.

In order to evaluate the involvement of PMNs in the pathogenesis of endotoxic shock, profound neutropenias were induced in calves prior to the slow infusion of purified endotoxin. Clinical signs and coagulation profiles were compared to calves receiving endotoxin but having normal numbers of neutrophils. The results indicate that neutrophils provide a degree of protection from the effects of endotoxin rather than contribute to the severity of the host response.

METHODS

Healthy eight to ten week old calves were utilized in this experiment. Neutropenias were induced in a portion of the animals by the daily intravenous infusion of hydroxyurea (88 mg/kg) dissolved in 250 ml of saline. When circulating neutrophil numbers fell to below ten percent of the

individual's control value endotoxic shock was induced by the intravenous
infusion of purified Salmonella typhimurium LPS (List Biologics). The endotoxin was given at .05 µg/kg total dose diluted in 500 ml of saline over a
five hour period by infusion pump. Clinical evaluations were made hourly
and coagulation profiles performed just prior to the LPS infusion and at 1,
3, 6, and 24 hours after the start of the infusion.

Clinical evaluations were scored on a scale of 0 to 3; 0 = normal; 1 =
depressed; 2 = depressed and unable to stand; 3 = comatose. The other clinical parameters that were measured included heart rate and quality, rectal
temperature and feces quality. The coagulation profile included prothrombin time (PT), partial thromboplastin time (PTT), packed number PTL No.),
protamine sulfate (PSO$_4$), antithrombin III (AT-3), prekallikreins (pKEL)
and fibrin degradation products (FDP).

RESULTS

In the control calves (normal number of neutrophil) the endotoxin treatment caused a rapid decrease in the number of circulating neutrophils followed by a rebound to above control values, Figure 1. Profound neutropenias
occurred in the experimental calves after five daily treatments with hydroxyurea. Figure 2 shows the time-course of the neutrophil depletion and regeneration in the animals receiving hydroxyurea. The effects of hydroxyurea
on blood leukocytes and platelets are shown in Figure 3. Total numbers of
circulating leukocytes and neutrophils fell an average of 80.8 percent and
95.6 precent, respectively. The number of lymphocytes and platelets decreased but the decline was not significantly below baseline values (P<0.05).

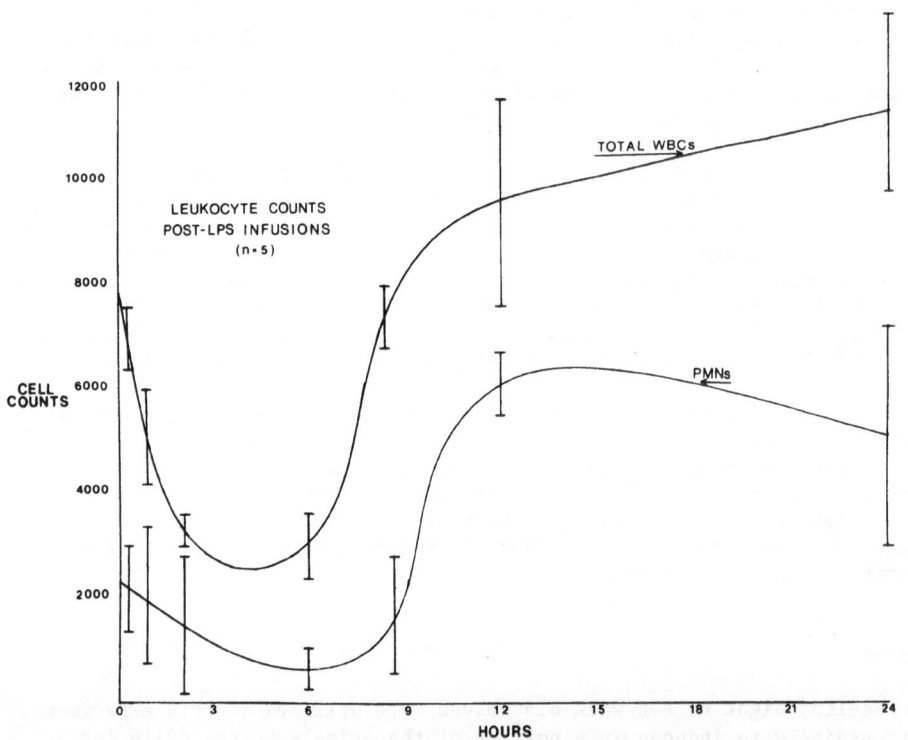

Figure 1. Leukocyte kinetics during and after a five-hour
infusion of endotoxin in normal calves.

Table 1. Effect of Neutrophil Depletion on Coagulation Homeostasis
During Endotoxemia in Calves

	PT	PTT	PCV	PP	FIB	PLT#	PSO$_4$	AT-3 (%)	pKEL (%)	FDP
Normal Animals										
Control Levels	20.4	30.0	27.8	6.30	460	890.0	0.60	96.2	70.6	0
SD	1.8	3.4	1.1	0.14	288	175.9	0.89	8.5	17.6	0
% Change from Control										
1 hr	-5.4	-9.8	-3.5	-1.5	+17.3	-6.8	1.2	0.0	-0.8	0
3 hr	-2.9	-10.8	-1.4	-4.1	+17.3	-22.6	2.8	0.0	-0.8	0
6 hr	-0.6	-0.6	+0.7	-5.3	+17.3	-18.9	2.2	-4.3	-11.0	0
24 hr	+5.9	+0.1	-0.6	-4.7	+30.4	-22.0	0.6	-3.9	-9.6	0
Neutrophil Depleted Animals										
Control Levels	24.6	53.2	31.3	6.03	425	881.6	0.0	93.0	43.0	0
SD	1.6	2.8	1.7	0.12	125	82.3	0.0	3.3	3.7	0
% Change from Control										
1 hr	+4.8	+9.9	+0.6	+0.3	+29.4	-41.6	0.0	0.0	+1.0	0
3 hr	+0.2	+15.6	-2.5	-7.1	-5.8	-43.5	0.0	-2.1	+4.6	0
6 hr	+5.4	+32.2	+0.6	-8.7	+41.1	-54.9	0.0	-10.7	-3.4	0
24 hr	+12.6	+3.6	+0.6	-2.1	+88.2	-58.3	0.5	-16.6	-15.0	0

SD = standard deviation; PT = prothrombin time; PTT = partial thromboplastin time; PCV = packed cell volume; PP = plasma protein; FIB = fibrinogen; PTL # = platelet number; PSO$_4$ = protamine sulfate; AT-3 = antithrombin III; pKEL = prekallikreins; and FDP = fibrin degradation products.

The clinical responses of the neutropenic calves were in all cases more severe than those of controls. Mean clinical score during the endotoxin infusion for the control animals was 1.14 which is significantly lower than the score of the neutrophil depleted calves, 2.47. Changes in the coagulation profiles also indicate that the effect of the endotoxin was more severe in the neutropenic animals, Table 1. Significant differences between the neutropenic calves and the controls were 1) more rapid and severe depletion of coagulation factors; 2) higher plasma fibrinogens; 3) more severe thrombocytopenias; and 4) lower antithrombin III levels.

DISCUSSION

The rapid and dramatic decrease in circulating neutrophil numbers in the early stages of endotoxic and septic shock plus their ability to mediate tissue injury has led to speculation that neutrophils play an important role in the pathogenesis of these conditions (11,17). This notion is supported by the high affinity of neutrophils for endotoxin (13,15) and the findings that endotoxin may stimulate respiratory activity and release of lysosomal constituents from neutrophils (4,16). Other investigators have been unable to demonstrate any direct effect of endotoxin on neutrophils when studied in vitro (12,20,9). These findings suggest that neutrophils, since they are not influenced directly by endotoxin in vitro, are not likely responsible for the biological activity of endotoxin.

In this experiment we examined the effect of neutrophil depletion on the clinical and hematostatic changes that occur during experimental endotoxic

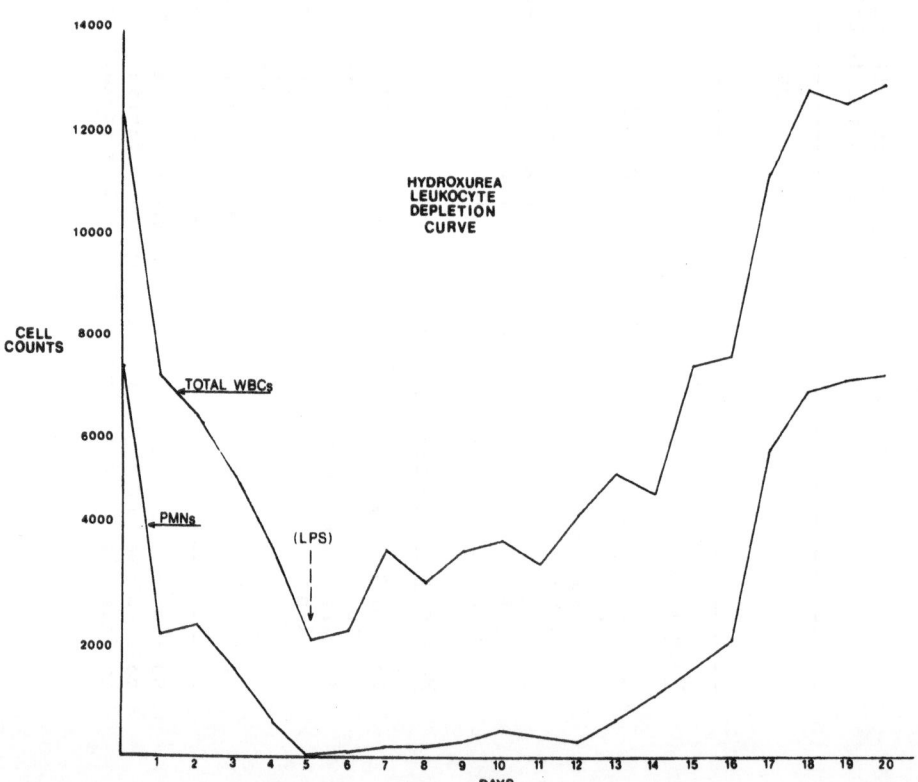

Figure 2. Effect of hydroxyurea on total leukocytes and neutrophil numbers. (LPS - endotoxin challenge.)

shock. Hydrooxyurea was chosen to induce neutropenias because of its relatively specific effect on neutrophils and because it does not interfere with protein synthesis (2,10). Using hydroxyurea we were able to selectively deplete circulating neutrophils without significantly changing other leukocyte or platelet numbers. The hydroxyurea did not alter the parameters measured in the coagulation profiles.

Endotoxic shock was induced by the slow infusion of small amounts of purified LPS because this technique provides the most reproducible effects and more accurately simulates septic shock. The reaction of the control calves to the LPS is typical of experimental endotoxic shock in many animals. The rapid decrease in circulating neutrophil numbers is believed to be due to their magination and subsequent extravasation. The "rebound" effect appears to be the result of increased release of immature neutrophils from bone marrow as well as increased production, possibly due to increased levels of colony stimulating factor.

The primary site involved in the pathogenesis of endotoxic and septic shock is microvascular. Neutrophils, platelets and direct endothelial injury have been suggested as important mediators in endotoxin induced damage to the microvascular circulation (17). The most sensitive measure of damage is alterations in coagulation profiles. Endotoxin induced profound alterations in coagulation including DIC (17). The severity of the endotoxin induced coagulopathies is dose dependent and closely related to lethality (5).

Identical doses of the same endotoxin cause both more severe clinical signs and coagulopathies in the neutropenic calves than in the controls.

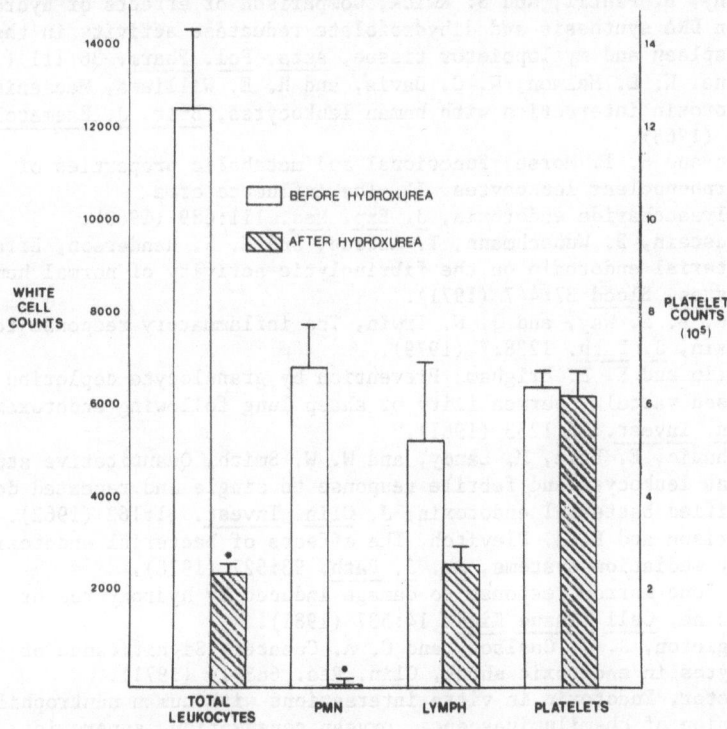

Figure 3. Effects of hydroxyurea on blood leukocytes and platelet counts in calves (* = significant difference from baseline value, p < .05.)

Even though the influence of the hydroxyurea cannot be completely discounted it appears that neutrophils provide protection from endotoxic shock. This conclusion is consistent with the in vitro finding of no direct effect of endotoxin on neutrophil function (9). On the other hand, it has been reported that neutrophils are involved in increasing lung vascular permeability but not pulmonary vasoconstriction during endotoxic shock in sheep (7). Most likely neutrophils depending on the mediator system examined can have protective as well as deleterious functions.

The mechanism of neutrophil mediated protection is unknown. Neutrophils may be acting by way of their superior affinity to endotoxin thereby sequestering it and preventing interactions between endotoxin and more biologically active cellular targets (18,19). Platelets have only a slightly lower affinity for endotoxin than neutrophils (15). This mechanism of endotoxin clearance would help to explain the nonlinear dose response to endotoxin, that is as the amount of endotoxin increases larger numbers of neutrophils become saturated and thus greater amounts of endotoxin is available to bind to more biologically active targets. The apparent ability of neutrophils to sequester endotoxin may explain the dramatic early decrease in circulating neutrophil numbers. The interaction between neutrophils and endotoxin may contribute to the nonspecific defense provided by these cells, and clarify why neutropenic patients are more likely to develop septic shock despite appropriate antibiotic therapy.

REFERENCES

1. B. M. Babior, R. S. Kipnes, and J. T. Curnutte, Biological defense mechanisms. The production by leukocytes of superoxide, a potent bactericidal agent, J. Clin. Invest. 52:741 (1973).
2. S. S. Bitny, B. Pantil, and S. Kwick, Comparison of effects of hydroxyurea on DNA synthesis and dihydrofolate reductase activity in the mouse spleen and myelopoietor tissue, Acta. Pol. Pharm. 36:111 (1979).
3. M. J. Cline, K. L. Melmon, W. C. Davis, and H. E. Williams, Mechanism of endotoxin interaction with human leukocytes, Brit. J. Haematol. 15:539 (1968).
4. Z. A. Cohn and S. I. Morse, Functional and metabolic properties of polymorphonuclear leukocytes. II. The influence of a lipopolysaccharide endotoxin, J. Exp. Med. 111:689 (1960).
5. I. M. Goldstein, B. Wunschmann, T. Astrup, and E. S. Henderson, Effects of bacterial endotoxin on the fibrinolytic activity of normal human leukocytes, Blood 37:447 (1971).
6. M. L. Good, B. A. Way, and J. W. Irwin, The inflammatory response to endotoxin, J. Path. 1238:7 (1979).
7. A. C. Heflin and K. L. Brigham, Prevention by granulocyte depletion of increased vascular permeability of sheep lung following endotoxemia, J. Clin. Invest. 68:1253 (1981).
8. R. C. Mechanic, E. Frie, M. Landy, and W. W. Smith, Quantitative studies of human leukocyte and febrile response to single and repeated doses of purified bacterial endotoxin, J. Clin. Invest. 41:162 (1962).
9. D. C. Morrison and R. J. Ulevitch, The effects of bacterial endotoxins on host mediation systems, Am. J. Path. 93:527 (1978).
10. E. Necas, Bone marrow response to damage induced by hydroxyurea or colchicine, Cell Tissue Kiwet 14:537 (1981).
11. W. W. Pingleton, J. J. Coalson, and C. A. Guenter, Significance of leukocytes in endotoxic shock, Clin. Res. 683:19 (1971).
12. R. A. Proctor, Endotoxin in vitro interactions with human neutrophils: depression of chemiluminescence, oxygen consumption, superoxide production, and killing, Infect. Immun. 25:912 (1979).
13. J. W. Shands, Affinity of endotoxin for membranes, J. Infect. Dis. 128(Supp):9197 (1973).

14. H. Shubin, M. Weil, and H. Nishijima, Clinical features in shock associated with gram-negative bacteremia, in: "Gram-Negative Bacterial Infections," B. Urbaschek, R. Urbaschek, and E. Neter, eds., Springer-Verlag, New York (1975).

15. G. F. Speringer and J. C. Adye, Endotoxin-binding substances from human leukocytes and platelets, Infect. Immun. 12:978 (1975).

16. B. S. Strauss and C. A. Stetson, Studies on the effect of certain macromolecular substances on the respiratory activity of the leukocytes of peripheral blood, J. Exp. Med. 112:653 (1960).

17. B. Urbaschek, R. Urbaschek, and E. Neter, Gram-negative bacterial infections and mode of endotoxin action: pathophysological, immunological and clinical aspects, in: "Immuno-Symposium," Springer-Verlag, New York (1975).

18. R. I. Walker, J. R. Fletcher, and D. A. Walden, Participation of granulocytes and humoral factors in resolution of platelet aggregates during endotoxemia, Experientia 36:255 (1980).

19. R. I. Walker and J. R. Fletcher, Possible contribution of platelets to the pathophysiology of host responses to endotoxin, Prog. Clin. and Biol. Res. 62:173 (1981).

20. M. E. Wilson, P. Munkenback, and D. C. Morrison, Influence of bacterial endotoxins on neutrophilic leukocytes: lack of a correlation between in vivo and in vitro responses, Prog. Clin. and Biol. Res. 62:157 (1981).

ACTIVE IMMUNIZATION WITH E. COLI J5 AND ITS PROTECTIVE EFFECTS

FROM ENDOTOXIC SHOCK IN CALVES

J. S. Cullor, B. Fenwick, B. P. Smith, K. Pelzer,
A. Kelly, and B. I. Osburn

Department of Veterinary Pathology
University of California
Davis, California

INTRODUCTION

Bovine salmonellosis is of worldwide importance due to public health concerns and causes significant economic losses to the agricultural industry. Cattle of all ages can be infected but serious infections and deaths are most often seen in calves up to ten weeks of age. The most common serotypes involved are S. typhimurium and Salmonella dublin (3-6). The major clinical manifestations associated with endotoxic shock are similar in each disease.

A number of studies utilizing active immunization with LPS or whole cells of E. coli 0111 strain J5 and passive administration of antisera to J5 LPS have reported protection in laboratory animals and possibly also in humans against gram negative bacteremia and septicemia (1,2,7). In this study we sought to determine the antibody classes which provide the protective effects of E. coli LPS core antigens during experimental endotoxic shock in calves.

MATERIALS AND METHODS

Healthy five to eight week old holstein bull calves were divided into two groups with one group (five animals) being immunized with an Rc mutant of E. coli (strain J5) in order to develop high antibody titers against common core antigenic components of gram-negative bacterial cell walls. The remaining calves (eight head) acted as nonimmunized controls.

The J5 bacterin was prepared as previously described by Zeigler (8). The immunized calves were given four subcutaneous injections 2 ml J5 and 2 ml Freund's incomplete adjuvant every ten days, then challenged with endotoxin seven to ten days after the last immunization.

Endotoxic shock was induced by slow infusion of Salmonella typhimurium LPS (List Biologics, Campbell, California) at 0.5 µg/kg (total dose) in 250 ml of nonpyrogenic sterile saline. The mixture was administered via infusion pump through a jugular catheter over five hours. Clinical signs were monitored and scored (Figure 1) and class specific anti-J5 antibody activity determined by indirect ELISA at 1, 3, 6 and 24 hours after the start of the infusion.

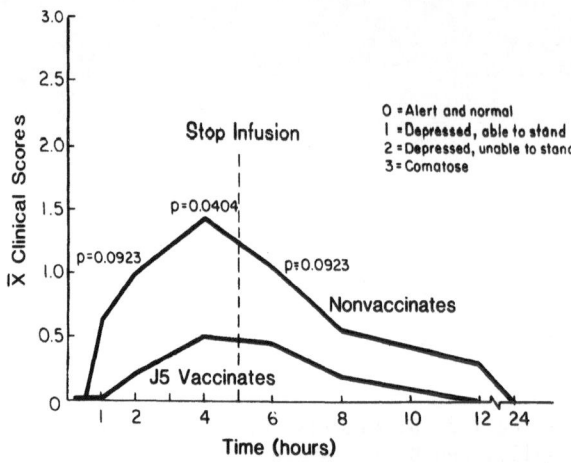

Figure 1. X clinical scores vs time, slow infusion LPS
           challenge. Mean clinical scores of J5 immunized
           and control animals during and after LPS infusions.
           The J5 immunized calves had significantly lower
           clinical scores.

RESULTS/DISCUSSION

Clinical signs of endotoxic shock in the J5 immunized animals were
less severe than in the controls (Figure 1). The anti-J5 IgM and IgG_1 antibody activity did not decrease significantly during or after the infusion
in the nonimmunized control animals (Figure 2). In the J5 immunized group
there was not a significant decrease in anti-J5 IgM activity. However, at
the time periods 3 hrs, 6 hrs and 24 hrs there was a significant decrease
($p < 0.0122$) in antibody activity in the IgG_1 class recognizing the J5 antigens (Figure 3).

Other investigations have reported antibodies of the IgM class to afford
protection against gram-negative endotoxemia in humans and laboratory animals

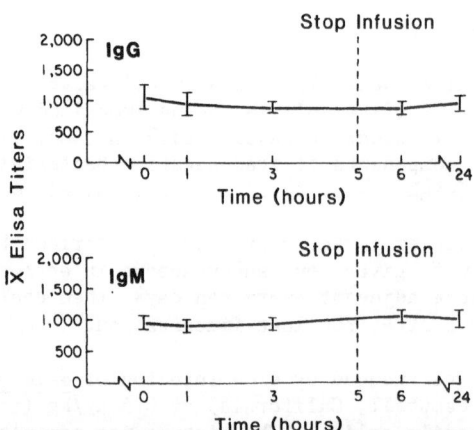

Figure 2. Nonimmunized slow infusion LPS challenge. Mean ELISA
           titers (IgG_1, IgM) of nonimmunized control calves during
           and after the slow LPS infusion challenge. There was no
           significant change in titers at any of sampling times.

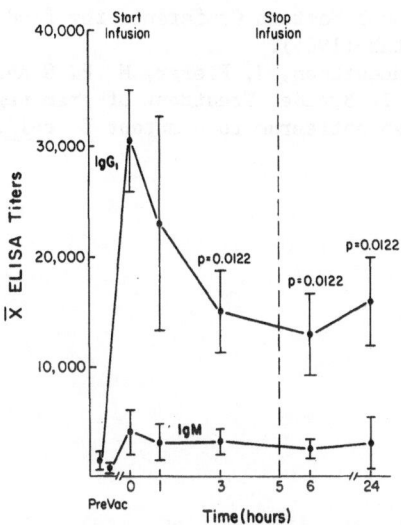

Figure 3. Mean ELISA titers (IgG$_1$, IgM of J5 immunized calves
during and after slow infusion endotoxin challenge.
A significant decrease in anti-J5 IgG$_1$ antibody activity
occurred at 3, 6, and 24 hrs after the start of the LPS
infusion. No significant anti-J5 IgM antibody titer
resulted from immunization and no decrease was detected
during or after the LPS challenge.

(7). In contrast, the results of this bovine experiment indicate a protective effect, during a heterologus endotoxin challenge, associated with antibodies of the IgG$_1$ class that are directed at LPS core antigens. This observation is the first report in food animals: 1) that antibodies raised during
active immunization with J5 can mitigate the effects of experimentally
induced endotoxic shock; and 2) antibodies of the IgG$_1$ class are associated
with this protective effect.

REFERENCES

1. A. I. Braude, E. J. Zeigler, H. Douglas, and J. A. McCutchan, Antibody
    to cell wall glycolipid of gram negative bacteria: induction of
    immunity to bacteremia and endotoxemia, J. Infect. Dis. 136:S167
    (1977).
2. S. C. Bruins, R. Stumacher, M. A. Johns, and W. R. McCabe, Immunization
    with R mutants of Salmonella minnesota. Comparison of the protective
    effect of immunization with lipid A and the Re mutant, Infect.
    Immun. 17:16 (1977).
3. E. H. Johnson, B. P. Smith, and M. Guerra, Diffusion in gel-enzyme
    linked immunosorbent assay (DIG-ELISA) to record the immunoglobulin
    response of calves vaccinated with Salmonella, Vet. Micro. 10:71
    (1984).
4. A. A. Lindberg and J. A. Robertson, Salmonella typhimurium infection in
    calves: cell-mediated and humoral reactions before and after
    challenge with live virulent bacteria in calves given live or
    inactivated vaccines, Infect. Imm. 41:751 (1983).
5. B. P. Smith, Bovine Salmonellosis, Val. Vet. 4:27 (1980).
6. B. P. Smith, F. G. Habasha, M. Guerra, and A. J. Hardy, Immunization of
    calves against Salmonellosis, Am. J. Vet. Res. 40:1947 (1980).

7. M. Wickstrom, J. Hodgson, C. Gay, R. Lee, and L. Corbeil, Protection of calves against colisepticemia with cross-reactive antisera, Abstract, Sixth Annual Western Conference for Food Animal Veterinary Medicine, Logan, Utah (1985).

8. E. J. Zeigler, J. A. McCutchan, J. Fierer, M. P. Glauser, J. C. Sadoff, H. Douglas, and A. I. Braude, Treatment of gram negative bacteremia and shock with human antiserum to a mutant E. coli, N. Eng. J. Med. 307:1225 (1982).

CLINICAL RECOGNITION AND TREATMENT OF ENDOTOXINEMIA

Joseph G. Sinkovics

Community Cancer Center, St. Joseph's Hospital and University of South Florida College of Medicine, Tampa, Florida Department of Virology and Epidemiology, Baylor College of Medicine, Houston, Texas

INTRODUCTION

In health we co-exist with our bowel flora without harm from lipopolysaccharide endotoxins that the gram-negative constituents of this flora produce and release. The lower intestinal tract of patients entering hospitals becomes rapidly colonized by the nosocomial gram-negative flora dominant in that environment. Green salads, faucets, the vase holding cut flowers at the bedside serve as common sources of colonization without tissue invasion. Breaking through natural barriers (indwelling arterial and venous lines, genitourinary and other catheters, mucosal ulcerations of the oral cavity and gastrointestinal tract), leukopenia (including granulocyto- and lymphocytopenia and depletion of monocytes-macrophages) allow for invasion of tissues and blood stream by gut flora containing also the colonizing nosocomial bacteria (Pseudomonas sp., Serratia sp.). Absorption of endotoxin from the compromised gastrointestinal or genitourinary tracts may occur without actual bacterial invasion of the blood stream. Gram-negative bacteria invading soft tissues and blood produce and release endotoxins. These endotoxins exert diverse and most profound biological effects often cascading relentlessly toward irreversible shock and death.

PATHOPHYSIOLOGY

Prostaglandins

The proaggregatory vasoconstrictor thromboxane A2 is formed through three major sequential steps in platelets:

1) Platelet membrane phospholipids release arachidonic acid; 2) platelet cyclooxygenases convert these fatty acids to cyclic endoperoxidases; and 3) thromboxane synthetase converts endoperoxides to thromboxane A2 (1).

Incubation of human platelets with endotoxin results in release of arachidonic acid and in accelerated conversion of arachidonic acid to thromboxane B2 which is the stable end product of the vasoconstrictor thromboxane A2. Endotoxin shock was prevented in animals treated with the thromboxane synthetase inhibitor imidazole or by thromboxane A2 antagonist 13-azaprostanoic acid. Anisodamine, the cholinergic alkaloid with structural relationship to atropine inhibits platelet aggregation and thromboxane synthesis by inhibition of cyclo-oxygenase and thromboxane synthetase (2,3).

Prostaglandins E and F levels rise in endotoxin shock; aspirin and indomethacine decrease the acute hemodynamic response to endotoxin and improve survival (4). Indomethacin could prevent Shwartzman reaction (5). The imidazole derivative UK-37248-01 acted in humans as selective inhibitor of thromboxane synthetase (6).

Kallikrein and Bradykinin

Depletion of prekallikrein occurs in endotoxinemia. Cleavage of the potent hypotensive peptide bradykinin is followed by decrease in peripheral vascular resistance. Bradykinin is generated when kallikrein enzymes act on α-2 globulin kininogens. The decapeptide lysylbradykinin becomes the nonapeptide bradykinin. Bradykinin causes release of prostaglandins (7,8). Bradykinin possibly mediates bronchospasm, carcinoid flush and capillary permeability with vasodilation in endotoxin shock. When catecholamine-resistant endotoxin shock responded to angiotensin drip, a natural antagonism between angiotensin and bradykinin was postulated (9).

Opioids

Endogenous opioid peptides (endorphins) mediate hypotension in endotoxin shock as judged by beneficial therapeutic effect of opiate receptor antagonists in shock (10).

Of morphin-preferring (μ) and enkephalin-preferring (δ) receptors, agents (β-funaltrexamine) antagonizing μ receptors failed, while agents (ICI: M154,129) and antagonizing δ receptors reversed endotoxin shock; however, pretreatment with β-funaltrexamine pre-empted this latter effect (11).

Defective Utilization of Energy Sources

Ineffective oxidization of glucose in skeletal muscle during endotoxinemia results in accumulation of pyruvate not taken up in the Krebs cycle. Glucose metabolism proceeds through the lactic acid (Cori) cycle resulting in lactic acidemia, while lactate:pyruvate ratio remains normal. Glucagon levels rise more than insulin levels leading to decreased insulin:glucagon ratio (12).

Catecholamines and glucagon release fatty acids from adipose tissue and insulin, by its antilipolytic effect, antagonizes this action. It is probably long-chain fatty acid esters that inhibit pyruvate dehydrogenase thus preventing the processing of pyruvate in the Krebs cycle (12).

Ureagenic (proline), sulfur-containing (methionine) and aromatic (phenylalanine) amino acids remain unutilized in septicemia and their high levels in blood predict increased mortality. Branched chain amino acids (valine, leucine, isoleucine) are catabolized best by skeletal muscle as energy source and are recommended for intravenous administration to patients in shock (12).

Tumor Necrosis Factor (TNF)

Monocytes-macrophages primed with BCG or Corynebacterium parvum and exposed to endotoxin release TNF and closely related polypeptides (cachectins) (13). Hemorrhagic tumor necrosis in vivo and tumor cell death in vitro occurred after treatment with TNF. Normal connective tissue cells (fibroblasts) are insensitive to this substance. Recombinant DNA technology revealed 30% sequence homology between TNF and lymphotoxin. Interferon-potentiated the antitumor effect of TNF. TNF/cachectin suppress erythrocyto-poietic precursor cells (14) and inhibit fatty acid biosynthesis in lipo-cytes. mRNAs and lipoprotein lipase in adipocytes are suppressed by these factors (15). The substances may be responsible for anemia and cachexia in patients with certain cancers. Exposure to endotoxin causes release of these substances and antiserum to TNF/cachectin protected mice against the lethal effects of E. coli endotoxin (16).

This author reviewed in the early 1970's 338 cases of gram-negative bacterial sepsis in patients with various malignancies (genitourinary tract, melanoma, sarcomas, etc.) and found only two instances of partial tumor regression (17).

Pseudomonas vaccine added to standard treatment of acute leukemia re-duced death rate in one clinical trial (18). It is not known whether this effect was due to immunization against endotoxin or due to the effects of endotoxin on leukemic cells or on the host. In repeated clinical trials this effect was not readily reproducible. In patients with breast carcinoma, pseudomonas vaccine induced topical and systemic reactions but it did not influence the course of disease significantly (19).

SYMPTOMS AND SIGNS

Hemodynamics and Myocardial Depression

Patients with chills and fever, flushed, warm skin, mental confusion, tachycardia, tachypnea, hypoxia, normal blood pressure and urine output are in the hyperdynamic stage. Cool skin, slight hypotension and oliguria in addition to mental obtundation and hypoxia characterize the normodynamic stage. Cold clammy skin often with cyanosis and severe hypoxemia; hypoten-sion, thready, rapid pulse and anuria signify hypodynamic stage. Metabolic acidosis and respiratory failure contribute significantly to death (20,21,22).

Depressed ejection fraction (< 0.4), despite normal or elevated cardiac index, occurs frequently in septic shock, it is reversible and paradoxically occurs more frequently in survivors than in nonsurvivors (23). Release of a humoral myocardial depressant factor from the ischemic pancreas is sus-pected but not fully documented (24).

Pooling of blood in venous capacitance beds leads to reduction of cir-culatory blood volume and diminished cardiac output; consequentially tissue perfusion and oxygenation are inadequate to meet demand. Anaerobic glycoly-sis and metabolic acidosis set in (20,21,22).

ECG changes are those of ischemic myocardium and arrhythmias appear. Central venous pressure is useful as a guide of volume replacement so that at 12-14 cm of water furosemide may be given to avoid fluid overload. Cardiac output normally consists of a stroke volume of 70 ml ejected about 70 times per minute resulting 5 liters/min. Cold and cyanotic fingers, toes, ears and nose speak for peripheral vasoconstriction, increased periph-eral resistance and decreased cardiac output. The temperature of big toe falls from 28°C to 22°C.

Coagulopathies and Thrombocytopenia

Consumption of activated platelets leading to disseminated intravascular coagulation with or without fibrinolysis or development of platelet-associ-ated immune globulins (PAIgG) or circulating immune complexes contribute to these complications. The case for platelet-bound immune complexes as cause of endotoxin shock-associated thrombocytopenia remains inadequately documented (25).

Complement-depleted rabbits exposed to endotoxin did not develop Shwartzman reaction suggesting a major role for complement in hemorrhagic necrosis. Heparine prevented generalized Shwartzman reaction. Complement (C3) consumption does occur in patients with bacteremic shock (26,27,28). Activation of Hageman factor initiates the cascade of the intrinsic coagulation system and generation of bradykinin from prekallikrein. Depletion of platelets and leukocytes protected animals against endotoxin-induced coagulopathies, re-infusion of platelets did not, but re-infusion of leukocytes restored the animals' susceptibility to endotoxin-induced coagulopathies (29,30).

In an unpublished retrospective review by this author patients with various cancers who were rendered leukopenic and thrombocytopenic by chemotherapy experienced reduced incidence of disseminated intravascular coagulopathy and/or shock lung (2/69 = 2.9%) than patients with various cancers who were not leukopenic and/or thrombocytopenic (7/37 = 18.6%) when both groups of patients suffered gram-negative bacteremia with or without endotoxin shock (Sinkovics, unpublished, 1977). Table 1 shows serial measurements of various parameters in a young man with osteogenic sarcoma during endotoxin shock due to E. coli septicemia (31).

Hypoglycemia

The combination of liver cirrhosis and septicemia often results in hypoglycemia associated with lactic acidosis; rapid onset of circulatory failure follows. When a cirrhotic patient develops hypoglycemia, gram-negative septicemia should be considered as the underlying process (32).

Inappropriate Polyuria

In the hyperdynamic state of gram-negative sepsis, blockade of the distal tubules and collecting ducts preventing salt and water conservation occurs resulting in polyuria, hypovolemia and hypotension. Glomerular filtration rate, renal blood flow, renal blood flow distribution and response to antidiuretic hormone remain normal (33).

Rhabdomyolysis

Myoglobinuria due to rhabdomyolysis and renal failure are exceptional and rare complications of septicemia. Endotoxinemia is the possible cause (34,35).

Shock Lung

Alveolar capillary cells suffer severe damage during endotoxin shock. Adult respiratory distress syndrome develops several days after the onset of septicemia. Gas exchange is impaired by interstitial and alveolar edema. Exsudation of intraalveolar plasma can be so rapid and excessive that large amounts of fluid appear in the bronchi and trachea (36). Cerebral hypoxia can lead to this phenomenon by stimulation of sympathetic centers in the hypothalamus resulting in increased pulmonary venous pressure.

Customarily accepted criteria for endotracheal intubation and assisted ventilation are respiratory rate > 35/min; tidal volume < 5 ml/kg; vital capacity < 15 ml/kg; pO_2 < 60 mmHg at room air; pCO_2 > 55 mmHg; pulmonary shunt > 25% (37).

Table 1. Hypodynamic Endotoxin Shock with Consumption Coagulopathy

Patient: 18 year old MDAH #126533
Diagnosis: Osteogenic sarcoma, metastatic to lungs
Treatment: Adriamycin, ↑MTX-LV rescue, L-PAM

July '78: Cold, clammy, cyanotic, hypotensive
          E. coli grown from blood
Treatment: NaHCO$_3$; Solu-Medrol 1 gm i.v.; tobramycin, carbenicillin,
           cephalothin; albumin; fresh frozen plasma; platelets; O$_2$;
           digoxin; dopamine 800 μ/ml at 45 ml/h
Outcome: Recovered (ischemic hepatocellular damage)

Parameters	Day 1	Day 2	Day 3
Hgb	12.1	9	8
WBC	11000	9000	8000
Plat	6000	108000	103000
pH	7.28	7.5	7.5
pO$_2$	52	62	102
pCO$_2$	16	30	41
CVP	28	25	25
SGOT	1870		1095
Bili tot	5.3		3.3
dir	2.2		1.5
LDH	1182		331
PT	43	19	19
PTT	82	35	29
Fibrinogen	54	160	170
FSP	410	204	102

MTX-LV = methotrexate-leucovorin; PAM = L-phenylalanine mustard; FSP =
fibrin split products

TREATMENT

Volume Replacement

Depletion of intravascular volume requires correction as guided by
flow-directed pulmonary artery catheter for monitoring wedge pressure of
pulmonary artery.

"Fluid challenge" consists of the intravenous infusion of fluid at 5
to 20 ml/min for 10 minutes. This infusion is to be discontinued when wedge
pressure increased more than seven mmHg above the initial preinfusion pressure. The central venous pressure gives information on the ability of the
heart in accepting the fluid load. It is recommended to limit the infusion
rate to 5 ml/min if pulmonary wedge pressure and central venous pressure
exceed 20 mmHg and 15 cm H$_2$O, respectively, but infuse fluids at rate of 20
ml/min when these readings are below 12 mmHg and 10 cm H$_2$O, respectively.

Physiologic or hypertonic saline solution or lactated Ringer's solution
are most commonly used but five percent human serum albumin or plasma protein
fraction or macromolecular dextran (MW 70,000 to 80,000) or hetastarch are
used for correction of low colloid osmotic pressure; however, hypersensitivity reactions and blood clotting defects may occur when colloids are used
instead of crystalloids (20,21,22).

Vasoactive Agents

Endotoxin acts sympathomimetically, therefore α-receptor stimulation
with norepinephrine, metaraminol or levarterenol cannot be justified. Yet
metaraminol has been used in the warm phase of septic shock in gynecologic
patients (38) and levarterenol is used for the reversion of severe, acute
hypotension. When this latter agent is used, phentolamine, atropine and
propranolol should be immediately available to counteract excessive vasoconstriction, bradycardia and arrhythmias, respectively. The use of α-receptor
blockers (phenoxybenzamine or phentolamine) or β-receptor stimulators (isoproterenol) is better justified. Isoproterenol causes vasodilation, lowers
central venous pressure, reduces venous pooling and increases cardiac output
and causes tachycardia; it should be used only after adequate volume replacement. Dilation of certain parts of the vascular bed in endotoxin shock
appears to be consequential to β-receptor stimulation; in this case β-receptor blockade by propranolol may be justified. Since propranolol reduces
cardiac output, rapid digitalization should precede its use.

Dopamine, the natural precursor of norepinephrine, increases myocardial
contractibility and cardiac output, accelerates renal blood flow and glomerular filtration rate and it decreases peripheral resistance. The renal
splanchnic vascular bed is most sensitive to dopamine. Its effect on this
segment of the vascular bed renders dopamine most efficacious in endotoxin
shock. Lower doses act through β-, higher doses through α-receptor stimulation (37,39). Propranolol suppresses ventricular arrhythmias occasionally
induced by rapid, high dose (> 50 μg/kg/min) infusion of dopamine.

Corticosteroids

Dexamethasone 3 mg/kg or methylprednisolone 30 mg/kg intravenously is
frequently used in endotoxin shock in expectation of reduced capillary permeability and lysosomal membrane stabilization. Up to 1976, favorable clinical impression supported the use of corticosteroids based on studies not
satisfying requirements of valid clinical trials (40,41) but a "double blind"
study with low dose betamethasone (1 mg/kg/d) showed no benefit (42). In
1976, both a retrospective and a prospective clinical trial yielded positive
results: mortality of septic shock was reduced from 42% and 38% of controls
to 14% and 10% of patients treated with corticosteroids (43). By the early
1980's it became evident that while shock was reversed, the ultimate mortality rate was not significantly improved in patients treated with corticosteroids even though life was prolonged (44-47). While in animals combinations of corticosteroids and antibiotics clearly protected against mortality
of septicemia, in humans proof of this effect is still not evident. Adverse
effects (increased rate of superinfection in patients treated with dexamethasone alone; increased mortality of pseudomonas sepsis for patients receiving
corticosteroids) are now on the record (47,48). Both at Memorial Sloan-
Kettering Cancer Center and at M. D. Anderson Hospital the death rate of
patients with pseudomonas sepsis increased (45% and 37%) after administration
of corticosteroids in comparison to patients also receiving all other modalities of treatment but not corticosteroids (15% and 8%) (48). Nevertheless,
administration of large doses of corticosteroids early in septic shock is
still recommended (49).

Antibiotics

A semisynthetic penicillin (ticarcillin, piperacillin, azlocillin) and an aminoglycoside (gentamicin, tobramycin, amikacin) or a cephalosporin of second or third generation (cefamandole, cefoxitin, cefotaxime, ceftazidime, cefoperazone, cefonicid, ceftriaxone, cefsulodin, ceftizoxime, cefmenoxime, cefuroxime, etc.) constitute the most commonly used regimens chosen empirically for the treatment of infected and often neutropenic patients (37). In patients allergic to penicillin, intravenous cotrimoxazole may replace semisynthetic penicillins or cephalosporins. Most consultants are still uncomfortable with a third generation cephalosporin antibiotic used alone for the treatment of clinically infected neutropenic patients even though cefoperazone compared favorably with cefoperazone and amikacin (50) and ceftazidime performed as well as cephalothin, gentamicin and carbenicillin (51). These regimens provide 56-84% cure and response rates of septicemias (commonly with gram-negative bacteria) in these compromised patients (52,55).

Since Staphylococcus aureus is emerging as a common pathogen in patients with vascular access devices, nafcillin and tobramycin were compared with cefotaxime alone in non-neutropenic patients with serious bacterial infections. Cefotaxime alone was more effective and less toxic (56).

M. D. Anderson Hospital compared ceftazidime alone with ceftazidime and tobramycin against gram-negative infections of neutropenic patients and observed no advantage of the combination (57). The University of Florida in Gainesville, compared ceftazidime versus cephalothin, gentamicin, carbenicillin versus ceftazidime, vancomycin; bacteriologic cures of bacteremias occurred in 80% versus 80% versus 100% of patients. Gram positive bacteria superinfected patients treated with ceftazidime; gram-negative bacteria superinfected patients treated with cephalothin, gentamicin, carbenicillin; no super-infections occurred in patients treated with ceftazidime, vancomycin (58). The University of Maryland Cancer Center compared double β-lactam plus aminoglycoside (ceftazidime, tobramycin) regimens for febrile granulocytopenic patients and observed 78% response rate for both regimens (59).

Other Measures

Airway and oxygen, digitalization for high central venous pressure and left ventricular failure and chlorpromazine as an α-blocker often are essential or at least beneficial. Angiotensin-2 octapeptide could reverse overwhelming α-1 blockade. Glucagon acts as an inotropic cardiotonic agent, stimulates adenylate cyclase, increases myocardial levels of cyclic adenosine monophosphate, improves perfusion in the splanchnic vascular bed, increases pH and decreases lactic and fatty acid levels. Bolus or infusion of naloxone reverses hypotension and bradycardia induced by endogenous endorphins (opiates) released by endotoxin. Macromolecular dextran, frozen plasma or cryo-precipitate and heparinization are used to treat consumption coagulopathy but elimination of the underlying cause is the most effective measure against this complication.

Fructose diphosphate infusion may serve as energy source to hypoxic tissues. Branched chain amino acids (valine, leucine, isoleucine) are given to be catabolized in skeletal muscle. Ketoacids deriving from leucine are oxidized in the Krebs cycle; alanine is formed from ammonia and pyruvate and serves as substrate for gluconeogenesis in the liver.

Prostaglandin synthetase inhibitors (indomethacin) protected animals (rats) against death due to peritonitis by E. coli when added to standard treatment (gentamicin) but efficacy of prostaglandin inhibitors in human endotoxin shock remains unproven (60).

Antiendotoxin Immunoglobulins

Antibody directed to the lipopolysaccharide core of endotoxin was raised in healthy volunteers immunized with heat-killed E. coli J5 strain. This antibody provided significant protection to patients in endotoxin shock (61,62): for all patients with bacteremia and shock death rate was reduced from 39% of conventionally treated patients to 22% of conventionally treated plus immune serum-receiving patients; in profound endotoxin shock, the immune serum reduced death rate from 77% to 44%.

SUMMARY

A lipopolysaccharide toxin produced by the symbiotic flora in the large intestine elicits profound pathophysiological effects if it enters the blood stream or if it is produced by gram-negative bacteria invading soft tissues or blood. This toxin interacts with the most basic life processes: it activates complement, induces the release of biologically potent molecular mediators and renders the host hypotensive without adequate tissue perfusion resulting from depletion of circulating volume and cardiac output and leading to oliguria and severe metabolic acidosis and death. One molecular mediator released from monocytes-macrophages exposed to endotoxin is tumor necrosis factor, but during uncontrolled endotoxin shock in patients with cancer tumor necrosis occurs very seldom. The standard treatment of endotoxin shock rests on early recognition, volume replacement, administration of vasoactive agents and antibiotics and supportive care (O_2 and airway); digitalization as indicated. Massive doses of corticosteroids, antiendorphins and anti-endotoxin immune globulins remain promising but investigational modes of treatment.

ACKNOWLEDGMENTS

The author is grateful to Dean Andor Szentivanyi, M.D., for his invitation to publish this material in this volume and to Ms. Lana Powell for secretarial assistance.

REFERENCES

1. M. J. Stuart, Effect of endotoxin on arachidonic acid release and thromboxane B2 production of human platelets, Am. J. Hemat. 11:159 (1981).
2. J. A. Cook, W. C. Wise, and P. V. Halushka, Elevated thromboxane levels in the rat during endotoxin shock. Protective effect of imidazole, 13-azaprostanoic acid or essential fatty acid deficiency, J. Clin. Invest. 65:227 (1980).
3. R.-J. Xiu, D. E. Hammerschmidt, P. A. Cappo, and H. S. Jacob, Anisodamine inhibits thromboxane synthesis, granulocyte aggregation and platelet aggregation, JAMA 247:1458 (1982).
4. J. R. Fletcher, P. W. Ramwell, and C. M. Herman, Prostaglandins and hemodynamic course of endotoxin shock, J. Surg. Res. 20:589 (1976).
5. E. L. Howes, M. T. Kwok, and D. G. McKay, The effects of indomethacin on the generalized Shwartzman reaction, Am. J. Pathol. 90:7 (1978).
6. H. M. Tyler, C.A.P.D. Saxton, and M. J. Parry, Administration to man of UK-37248-01, a selective inhibitor of thromboxan synthetase, Lancet 1:629 (1981).
7. D. C. Morrison and R. J. Ulevitch, The effects of bacterial endotoxin on host mediation systems, Am. J. Pathol. 93:527 (1978).
8. J. A. Robinson, M. L. Klodnycky, H. S. Loeb, M. R. Racic, and R. M. Gunnar, Endotoxin, prekallikrein, complement and systemic vascular resistance, Am. J. Med. 59:61 (1975).

9. J. G. Sinkovics, Project M26/gm 9: Life-threatening infections (endotoxic shock included) in patients with cancer, Research Report, The University of Texas M.D. Anderson Hospital and Tumor Institute, Houston, Texas (1967 and 1972).

10. B. Chernow and J. W. Holaday, The pathogenesis of septic shock, JAMA 252:208 (1984).

11. R. D'Amato and J. W. Holaday, Multiple opioid receptors in endotoxin shock: evidence for δ involvement and μ-δ interactions in vivo, Proc. Nat. Acad. Sci. U.S.A. 81:2989 (1984).

12. B. Mizock, Septic shock. A metabolic perspective, Arch. Int. Med. 144:579 (1984).

13. F. M. Torti, B. Dieckmann, B. Beutler, A. Cerami, and G. M. Ringold, A macrophage factor inhibits adipocyte gene expression: an in vitro model of cachexia, Science 229:867 (1985).

14. S. Sassa, M. Kawakami, and A. Cerami, Inhibition of the growth and differentiation of erythroid precursor cells by an endotoxin-induced mediator from peritoneal macrophages, Proc. Nat. Acad. Sci. U.S.A. 80:1717 (1983).

15. P. H. Pekala, M. Kawakami, C. W. Angus, and D. Lane, Selective inhibition of synthesis of enzymes for de novo fatty acid biosynthesis by an endotoxin-induced mediator from exudate cells, Proc. Nat. Acad. Sci. U.S.A. 80:2743 (1983).

16. B. Beutler, I. W. Milsark, and A. G. Cerami, Passive immunization against cachectin/tumor necrosis factor protects mice from lethal effect of endotoxin, Science 229:869 (1985).

17. J. G. Sinkovics, Program M27: Infectious and immunological diseases in patients with cancer, Research Report, The University of Texas, M.D. Anderson Hospital and Tumor Institute, Houston, Texas (1974).

18. S. Passe, V. Mike, R. Mertelsmann, T. Gee, and B. Clarkson, Acute non-lymphoblastic leukemia, Cancer 50:1462 (1982).

19. G. N. Hortobagyi, H.-Y. Yap, C. L. Wiseman, G. R. Blumenschein, A. U. Buzdar, S. S. Legha, J. U. Gutterman, E. M. Hersh, and G. P. Bodey, Chemoimmunotherapy for metastatic breast cancer with 5-fluorouracil, adriamycin, cyclophosphamide, methotrexate, 1-asparaginase, Corynebacterium parvum, and pseudomonas vaccine, Cancer Treatments Reports 64:157 (1980).

20. H. Shubin and W. H. Weil, Bacterial shock, JAMA 236:421 (1976).

21. M. C. Houston, L. Thompson, and D. Robertson, Diagnosis and management, Arch. Int. Med. 144:1433 (1984).

22. M. M. Parker and J. E. Parrillo, Septic shock: hemodynamics and pathogenesis, JAMA 250:3324 (1983).

23. M. M. Parker, J. H. Shelhamer, S. L. Bacharach, M. V. Green, C. Natanson, T. M. Frederick, B. Damske, and J. E. Parrillo, Profound but reversible myocardial depression in patients with septic shock, Ann. Int. Med. 100:483 (1984).

24. R. McConn, J. K. Greineder, F. Wasserman, and G.H.J. Clowes, Is there a humoral factor that depresses ventricular function in sepsis? Circ. Shock S1:22 (1979).

25. T. R. Poskitt, and P. K. F. Poskitt, Thrombocytopenia of sepsis. The role of circulating IgG-containing immune complexes, Arch. Int. Med. 145:891 (1985).

26. L. Thomas and R. A. Good, Studies of the generalized Shwartzman reaction. I. General observations concerning the phenomenon, J. Exp. Med. 96:605 (1952).

27. R. A. Good and L. Thomas, Studies on the generalized Shwartzman reaction. IV. Prevention of the local and generalized Shwartzman reaction with heparin, J. Exp. Med. 97:871 (1953).

28. W. R. McCage, Serum complement levels in bacteremia due to gram-negative organisms, New Eng. J. Med. 288:21 (1973).

29. C. A. Stetson and R. A. Good, Studies on the mechanism of the Shwartzman phenomenon. Evidence for the participation of polymorphonuclear leukocytes in the phenomenon, J. Exp. Med. 93:49 (1951).

30. J.S.C. Fong and R. A. Good, Prevention of the localized and generalized Shwartzman reaction by an anticomplementary agent, cobra venom factor, J. Exp. Med. 134:642 (1971).

31. J. G. Sinkovics, Infectious complications of cancer, J. Inter-American Med. 4:10 (1979).

32. O. Nowel, J. Bernuau, B. Rueff, and J.-P.Benbamou, Hypoglycemia, a common complication of septicemia in cirrhosis, Arch. Int Med. 141:1477 (1981).

33. A. Cortez, J. Zito, C. E. Lucas, and S. J. Gerrick, Mechanism of in-appropriate polyuria in septic patients, Arch. Surg. 112:471 (1977).

34. W. L. Henrick, D. Prophet, and J. P. Knochel, Rhabdomyolysis associated with Escherichia coli septicemia, South Med. J. 73:936 (1980).

35. S. B. Kalish, M. S. Tallman, F. V. Cook, and E. A. Blumen, Polymicro-bial septicemia associated with rhabdomyolysis, myoglobinuria and acute renal failure, Arch. Int. Med. 1142:133 (1982).

36. L. B. Pemberton, Shock lung with massive tracheal loss of plasma, JAMA 237:2511 (1977).

37. J. G. Sinkovics, "Medical Oncology. An Advanced Course," 2nd Edition, Marcel Dekker, New York (1986).

38. D. Cavanaugh, Septic shock in a pregnant or recently pregnant woman, Postgrad. Med. 62:62 (1977).

39. R. C. Tarazi, Sympathomimetic agents in the treatment of shock, Ann. Int. Med. 81:364 (1974).

40. S. Weitzman and S. Berger, Clinical trial design in studies of cortico-steroids for bacterial infections, Ann. Int. Med. 81:36 (1974).

41. J. G. Sinkovics, Severe infectious complications in patients with treated neoplasms, Abstracts XIIIth International Congress of Inter-nal Medicine (1976).

42. J. Klastersky, R. Cappel, and L. Debusscher, Effectiveness of beta-methasone in management of severe infections, New Eng. J. Med. 284:1248 (1971).

43. W. Schumer, Steroids in the treatment of clinical septic shock, Ann. Surg. 184:333 (1976).

44. J. N. Sheagren, Septic shock and corticosteroids, New Eng. J. Med. 305:456 (1981).

45. E. H. Kass, High dose corticosteroids for septic shock, New Eng. J. Med. 311:1178 (1984).

46. L. B. Hinshaw, Corticosteroids for septic shock, New Eng. J. Med. 312:510 (1985).

47. C. L. Sprung, P. V. Caralis, E. H. Marcial, M. A. Gelbard, and R. C. Duncan, Corticosteroids for septic shock, New Eng. J. Med. 312:511 (1985).

48. M. Tapper and D. Armstrong, Bacteremia due to Pseudomonas aeruginosa complicating neoplastic disease: a progress report, J. Inf. Dis. 130 (1974).

49. W. Schumer, Corticosteroid treatment of septic shock, JAMA 253:3165 (1985).

50. M. Piccart, J. Klastersky, F. Meunier, H. Lagast, Y. VanLaethem, and D. Weerts, Single drug versus combination empirical therapy for gram-negative bacillary infections in febrile cancer patients with and without granulocytopenia, Antimicrob. Ag. Chemother. 26:870 (1984).

51. P. A. Pizzo, J. Commers, D. Cotton, J. Gress, J. Hathorn, J. Hiemenz, D. Longo, D. Marshall, and K. J. Robichaud, Approaching the contro-versies in antibacterial management of cancer patients, Am. J. Med. 76:436 (1984).

52. G. P. Bodey, Antibiotics in patients with neutropenia, Arch. Int. Med. 144:1845 (1984).

53. V. Fainstein, G. P. Bodey, R. Bolivar, L. Elting, K. B. McCredie, and J. J. Keating, Moxalactam plus ticarcillin or tobramycin for treatment of febrile episodes in neutropenic cancer patients, Arch. Int. Med. 144:1766 (1984).

54. L. Jadeja, R. Bolivar, V. Fainstein, M. Keating, K. McCredie, M. Hay, and G. P. Bodey, Piperacillin plus vancomycin in the therapy of febrile episodes in cancer patients, Antimicrob. Ag. Chemother. 26:295 (1985).

55. R. Feld, T. J. Louie, L. Mandell, E. J. Bow, H. G. Robson, A. Chow, A. Belch, L. Miedzinski, A. Rochlis, Y. Pater, and A. Willan, A multi-center comparative trial of tobramycin and ticarcillin vs. moxalactam and ticarcillin in febrile neutropenic patients, Arch. Int. Med. 145:1083 (1985).

56. C. R. Smith, R. Ambinder, J. J. Lipsky, B. G. Petty, E. Israel, R. Levitt, E. D. Mellits, L. Rocco, J. Longstreth, and P. S. Lietman, Cefotaxime compared with nafcillin plus tobramycin for serious bacterial infections, Ann. Int. Med. 101:469 (1984).

57. V. Fainstein, Treatment in the U.S. cancer referral center vs. the community-based facility, in: "Infections in the Cancer Patient: Antimicrobial Therapy Today," Health Projects International and Glaxo Pharmaceuticals, San Diego (1985).

58. R. Ramphal, B. S. Kramer, and K. H. Rand, Comparison of ceftazidime versus cephalothin-gentamicin-carbenicillin versus ceftazidime plus vancomycin in febrile granulocytopenic patients, in: "Infections in the Cancer Patient: Antimicrobial Therapy Today," Health Projects International and Glaxo Pharmaceuticals, San Diego (1985).

59. J. Joshi and S. O. Schimpff, Empiric antibiotic therapy for febrile granulocytopenic cancer patients: double β-lactam vs. β-lactam-aminoglycoside regimen that suppresses or preserves alimentary canal anaerobes, in: "Infections in the Cancer Patient: Antimicrobial Therapy Today," Health Projects International and Glaxo Pharmaceuticals, San Diego (1985).

60. Editorial, Host responses in gram-negative septicemia, Lancet 2:693 (1982).

61. E. J. Ziegler, J. A. McCutchan, J. Fierer, M. P. Glauser, J. C. Sadoff, H. Douglas, and A. I. Braude, Treatment of gram-negative bacteremia and shock with human antiserum to a mutant Escherichia coli, New Eng. J. Med. 307:1225 (1982).

62. J. A. McClutchan and E. J. Ziegler, Treatment with anti-gram-negative antibodies, Lancet 2:802 (1983).

TRANSIENT SUPPRESSION OF YEAST PHAGOCYTOSIS INDUCED BY LEGIONELLA

PNEUMOPHILA IN CULTURES OF MURINE RESIDENT PERITONEAL MACROPHAGES

Jeanne Becker, Robert J. Grasso, and Herman Friedman

Department of Medical Microbiology and Immunology
University of South Florida College of Medicine
Tampa, Florida

Macrophages play a central role in host resistance to microbial infections. Ingestion and subsequent killing of pathogenic microorganisms are necessary for prohibiting the spread of infectious agents. However, microorganisms or their products have the potential to severely compromise host immune defenses by altering the normal functioning of macrophages. For example, Legionella pneumophila is a facultative intracellular pathogen that grows within monocytes and macrophages (2,13,18). Proliferation of the organisms within these cells occurs even in the presence of serum antibody and complement (14,15). L. pneumophila appears to evade destruction within phagocytes by inhibiting phagosome-lysosome fusion (12). Thus, the prevention of phagolysosome fusion impairs this normal macrophage function which can lead to decreased host resistance.

There is very little information known about the ability of these bacteria to modulate other macrophage functions. Recently, L. pneumophila has been shown to inhibit macrophage spreading (6), a process that is dependent upon the extension of fluid membranes. In addition to cell spreading in vitro, the process of phagocytosis is also associated with cell membranes. In order for these cells to spread and engulf particles, the membranes must be able to extend and surround the particles to be ingested. Furthermore, numerous studies have indicated that the incidence of acquiring legionellosis is enhanced in individuals receiving glucocorticoid therapy (1,4,9,10,16,17, 20,21). These steroids inhibit both cell spreading and phagocytosis when supplied to macrophage cultures at pharmacological concentrations (7,8). Microbial agents that are able to potentiate the suppressive effects of steroids would further compromise host resistance. Therefore, since L. pneumophila and glucocorticoids both inhibit macrophage cell spreading we examined whether these bacteria would also suppress the process of phagocytosis. In addition, we explored the relationship between L. pneumophila and the steroid-induced suppression of this macrophage function to determine if these organisms are capable of modulating the inhibitory effects produced by the steroid.

In this chapter, we present evidence that L. pneumophila alone produces a transient inhibition of phagocytosis and, when supplied with dexamethasone, potentiates the steroid-induced suppression of phagocytosis, also in a transient manner.

MATERIALS AND METHODS

Animals

Inbred female BDF$_1$ hybrid mice were used for these studies. The animals were obtained from Jackson Laboratory Supply Company, Indianapolis, Indiana. The mice were approximately eight to ten weeks of age and weighed 18-20 grams.

Bacterial Preparations

L. pneumophila serogroup 1, (Philadelphia Strain), originally obtained from Roger McKinney (Centers for Disease Control, Atlanta, Georgia), was maintained on charcoal yeast extract agar (GIBCO, Grand Island, New York) at 37°C. The bacterial colonies were grown for 24-36 hrs, suspended in sterile saline and killed by overnight incubation at room temperature in 0.5% formalin. The killed bacteria were washed several times in sterile saline by centrifugation. The organisms were adjusted to a concentration of 10^8 bacteria/ml by using MacFarland Standards. Escherichia coli (Strain B) was prepared in the same manner expect that the bacteria were grown on trypticase soy agar (GIBCO) for 12-18 hrs prior to harvest. In some experiments, purified E. coli lipopolysaccharide (LPS) serotype 026:B6 (Sigma Chemical Company, St. Louis, Missouri) was used. For the preparation of L. pneumophila sonicates, bacterial suspensions were exposed to six one minute pulses using a Bronwill Biosonik Sonicator (Bronwill Scientific, Inc., Rochester, New York) at an intensity set at 25% of full power. Cellular debris was pelleted by centrifugation at 10,000 x g and the supernatant extracts were collected. The sonic extracts were adjusted with saline to a concentration of 1 mg/ml as determined by absorbance at 280 nm; bovine serum albumin served as the standard.

Resident Peritoneal Macrophage Cultures

Macrophage cultures were established as described previously (7). Briefly, resident peritoneal leukocytes were obtained by lavage and resuspended to a concentration of 2.5 x 10^6 cells/ml in Dulbecco's Modified Eagle Medium (GIBCO) supplemented with antibiotics and heat-inactivated fetal bovine serum. The cells were plated on glass coverslips in 24 well plates (Costar, Cambridge, Massachusetts). After a 2 hr incubation period at 37°C, the nonadherent leukocytes were removed by washing. Day 0 refers to the day that the coverslips were established. The adhered macrophages were cultured for up to five days in the absence (i.e. controls) and the presence of various bacterial preparations or 1 μM dexamethasone (Merck, Sharpe and Dohme Research Laboratories, West Point, Pennsylvania) which was supplied on day 0.

Yeast Phagocytosis Assays

At selected time intervals after the establishment of the macrophage cultures, yeast phagocytosis assays were performed using heat-killed Saccharomyces cerevisiae as described previously (7). Briefly, 0.5 ml of a 4 x 10^7/ml suspension of S. cerevisiae was added directly to control and treated cultures; the preparations were incubated at 37°C for 15 min. The coverslips were then taken from the wells and rinsed vigorously in Hank's Balanced Salt Solution (GIBCO) which removed extracellular yeast particles. The coverslips were fixed and prepared for light microscopy. Upon examination of the coverslips, two types of measurements were recorded: the percentages of phagocytes in the macrophage populations and the phagocytic indices. No less than 300 individual cells were examined for each type of measurement. A phagocyte is defined as a macrophage which had ingested at least one yeast particle. The percentages of phagocytes were calculated by dividing the

282

number of phagocytes by the total number of macrophages counted and multiply-
ing by 100. Phagocytic index measurements refer to one through eight or >8
yeast particles ingested by the phagocytic subpopulation. The data illustrated in the figures represent mean values and only SEM >4% are indicated.

RESULTS

In order to determine if the presence of L. pneumophila in macrophage
cultures modulates the ingestion of yeast particles, the percentage of phagocytes and phagocytic indices were measured in cultures exposed to the bacteria. In addition, the ability of the organisms to modulate the glucocor-
ticoid-induced suppression of this macrophage function was examined in cultures incubated with L. pneumophila together with the steroid. The percentages of phagocytes in control cultures increased continuously from approximately 50% to almost 100% between day 0 and day 5 (Figure 1, curve A).
However, the percentages of phagocytes were <30% after one day of exposure
to L. pneumophila (curve B) as compared to approximately 75% in control
cultures. This effect was transient, since the percentages in these treated
cultures increased by day 2 and were similar to controls by day 5. In cultures exposed to dexamethasone, the percentages of phagocytes remained <25%
between day 1 and day 5, regardless of whether the bacteria were supplied

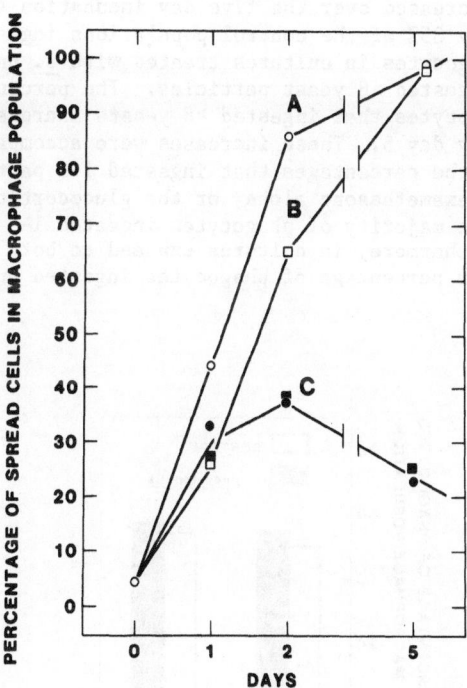

Figure 1. Transient suppression of yeast phagocytosis in macrophage
cultures treated with L. pneumophila. The symbols represent
mean values in control cultures (o), and in cultures treated
with 10^8/ml L. pneumophila (□), 1 μM dexamethasone (●), and
both 10^8/ml L. pneumophila plus 1 μM dexamethasone (■).
N values are as follows: Day 0 (N = 10); Days 1 and 2
(N = 4), and Day 5 (N = 4, 5 or 6).

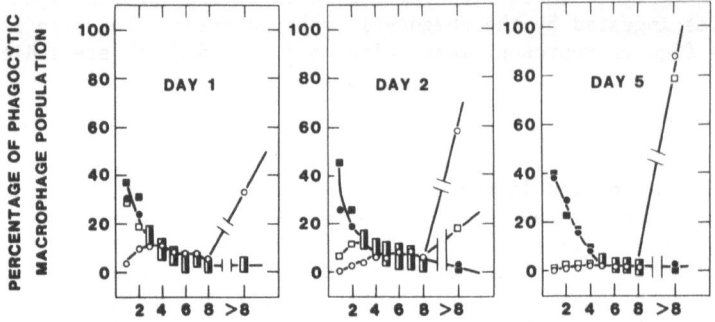

NUMBER OF INGESTED YEAST PARTICLES PER PHAGOCYTIC MACROPHAGE

Figure 2. Phagocytic indices in control and treated cultures.
The symbols and the N values are the same as indicated
in the legend to Figure 1, except for the composite
symbol (■) which represents all treatments and was
designed for clarity.

to the cultures (curve C). Phagocytic index measurements in control
cultures revealed that the percentages of the macrophage populations that
ingested >8 yeasts increased over the five day incubation (Figure 2). After
one day, approximately 35% of the control populations ingested >8 yeasts.
In contrast, most phagocytes in cultures treated with L. pneumophila ingested
1-3 particles; <5% ingested >8 yeast particles. The percentages of these
bacteria-treated phagocytes that ingested >8 yeasts increased to 20% by day
2 and to almost 80% by day 5. These increases were accompanied by corresponding decreases in the percentages that ingested 1-3 particles. In cultures that received dexamethasone alone, or the glucocorticoid together
with the bacteria, the majority of phagocytes ingested 1-3 yeasts over the
five day period. Furthermore, in cultures exposed to both the bacteria and
the steroid, a greater percentage of phagocytes ingested one and two yeast

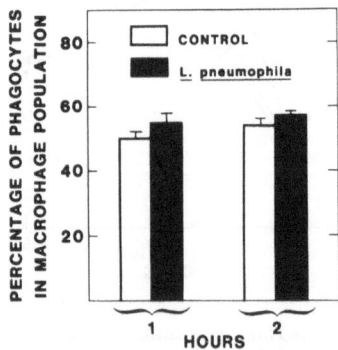

Figure 3. The inability of L. pneumophila to suppress yeast
phagocytosis shortly after its addition to the
macrophage cultures. In this experiment, the bars
represent mean values ± SEM (N = 3 or 4).

particles on days 1 and 2 relative to cultures treated with dexamethasone alone. This potentiation effect was transient and was no longer evident by day 5.

Figures 1 and 2 demonstrate that, relative to controls, the phagocytic capacity of macrophages exposed to L. pneumophila was decreased between day 0 and day 1. We therefore examined whether this decrease occurred in a relatively short period of time after the addition of bacteria to the cultures. Hence, control macrophages cultures and cultures treated with 10^8/ml L. pneumophila were established and the percentages of phagocytes and phagocytic indices were measured after one and two hours. The results illustrated in Figure 3 indicate that approximately 50% of the macrophages in control cultures and in cultures exposed to the bacteria were phagocytic after one and two hours. The majority of these phagocytes ingested 1-3 particles yeasts and 5% ingested >8 yeasts (Figure 4).

We next explored whether the suppressive effects of L. pneumophila on phagocytosis was specific for this bacterial species or if other gram-negative organisms or their products would induce similar responses. Phagocytosis was therefore examined in macrophage cultures treated with either E. coli or purified E. coli LPS. In addition, cultures were exposed to sonicates of L. pneumophila in order to determine if intact organisms were required for the suppression of this macrophage function. Since studies with intact L. pneumophila demonstrated that the inhibition of phagocytosis was transient, these experiments were only conducted over a two day period. In control cultures, the percentages of phagocytes in the macrophage populations increased from >60% to approximately 90% between day 0 and day 2 (Figure 5). However, after one day of exposure to L. pneumophila, E. coli or E. coli LPS, the percentages of phagocytes were <40%. Approximately 50% of the population was phagocytic in cultures exposed to L. pneumophila soni-

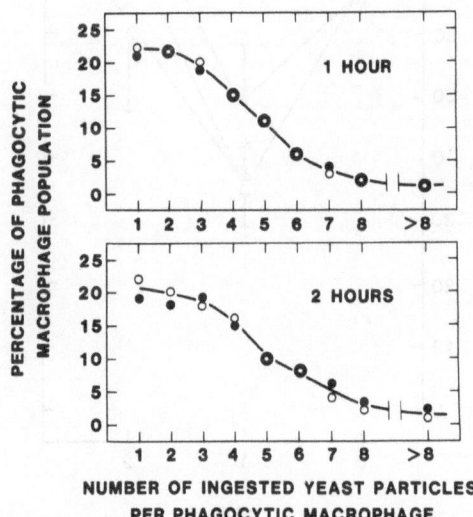

Figure 4. Phagocytic indices shortly after the establishment of cultures in control (o) and L. pneumophila-treated (●) macrophage cultures. The N values are the same as indicated in the legend to Figure 3.

sonicates for one day. By day 2, these inhibitory responses were reversed. Phagocytic index measurements indicated that the percentages of phagocytes that ingested >8 yeasts were inhibited in cultures treated with the microbial preparations for one day (Figure 6). After one day, approximately 20% of the phagocytes in control cultures ingested >8 yeast particles, whereas <5% ingested >8 particles in the treated cultures. Approximately 65% of the phagocytes in the treated cultures ingested 1-3 particles as compared to 35% in control cultures. After two days, the percentages of phagocytes which ingested >8 yeasts increased in both control and treated macrophage cultures, with corresponding decreases in the percentages that ingested 1-3. However, the phagocytic capacity of the treated cultures was decreased relative to controls.

DISCUSSION

These results demonstrate that exposure of macrophages to L. pneumophila decreases the percentages of the cell populations capable of phagocytizing yeast particles, as well as inhibiting the remaining phagocytic subpopulation from ingesting large numbers of yeast. Additional studies revealed that the bacteria also retard the initial rate of macrophage cell spreading which confirms previous findings (6) (data not shown). Both inhibitory effects occur between 2 and 24 hrs after exposing cultures to the organisms and are

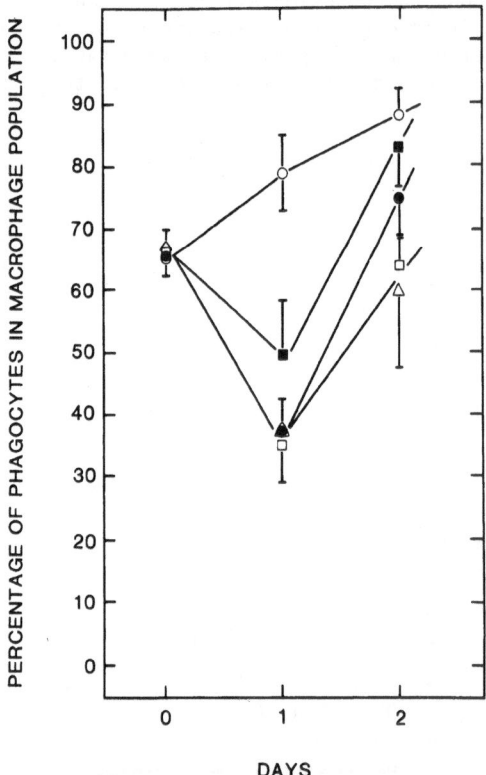

Figure 5. Transient inhibition of yeast phagocytosis in control and treated macrophage cultures. The symbols represent mean values in control cultures (o), and in cultures treated with 10⁸/ml intact L. pneumophila (□), 10⁸/ml intact E. coli (●), 100 g/ml L. pneumophila sonicate (■), and 10 µg/ml LPS (Δ). (N = 4)

286

transient, since suppression of these functions is no longer evident by day 5. L. pneumophila appears to delay the ability of the macrophages to phagocytize for 24 hrs, since the percentages of phagocytes and the phagocytic indices on day 2 in cultures treated with the bacteria are similar to those in control cultures after one day. The inhibition of phagocytosis in macrophage cultures is not specific for only L. pneumophila since similar responses occur when the cultures are exposed to E. coli or purified E. coli LPS. Sonicated preparations of L. pneumophila also produce similar inhibitory responses. These findings thus suggest that LPS may be the component of L. pneumophila that is responsible for mediating the suppression of this macrophage function, since each of the active bacterial preparations contained this microbial product.

Phagocytosis is inhibited to nearly the same degree in cultures exposed to the steroid together with the bacteria relative to cultures treated with only dexamethasone. This indicates that L. pneumophila does not interfere with the steroid-induced suppression of this macrophage function. In fact, on days 1 and 2, higher percentages of phagocytes ingest one and two yeast particles in cultures which receive dexamethasone together with L. pneumophila, as compared to macrophages exposed to either the steroid or the bacteria alone. This suggests that L. pneumophila potentiates the inhibitory effects elicited by the glucocorticoid. The impairment of normal host defenses for one day could result in enhanced susceptibility to infection by L. pneumophila. By day 2, when the suppression of phagocytosis induced by the bacteria alone is beginning to reverse, the increased inhibition of this process in cultures incubated with the bacteria plus the steroid is most pronounced. L. pneumophila may therefore induce a partially additive effect with dexamethasone to suppress further the phagocytic capacity of macrophages.

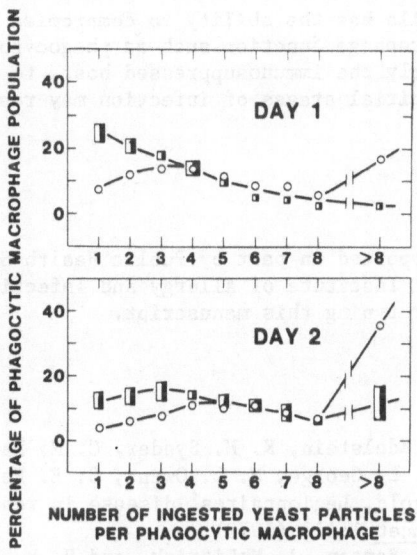

Figure 6. Phagocytic indices in macrophage cultures treated with L. pneumophila, E. coli or LPS. The symbols represent mean values in control cultures (O), and in cultures exposed to all treatments, designated by the composite symbol (◘) which was designed for clarity.

Cell mediated immunity contributes significantly to host defenses against L. pneumophila. During the acute stages of infection in both animal models and humans, polymorphonuclear leukocytes and mononuclear phagocytes predominate in the leukocytic infiltrate (3,19,23). Macrophages, which persist at the site of infection, are essential for the elimination of these organisms. Studies have shown that monocytes are stimulated to inhibit the multiplication of ingested L. pneumophila via cytokines produced by peripheral blood mononuclear cells from patients who have recovered from legionellosis (11). The lymphocytes from these patients can also be stimulated to undergo blastogenesis in response to sonicated preparations of the bacteria (22). These findings suggest that the LPS component of the organisms may be responsible for stimulating these responses, since boiling the bacterial preparations at 100°C does not affect blastogenesis. Furthermore, purified L. pneumophila LPS was shown to induce high levels of blastogenesis by splenocytes from normal and infected animals. Our results suggest that yeast phagocytosis is transiently inhibited by L. pneumophila and that this suppression is enhanced when dexamethasone is also present. This effect may have clinical relevance since immunosuppressed individuals that are treated with steroids are particularly susceptible to infection by this ubiquitous organism (1,4,9,10,16,17,20,21). Studies which investigated the incidence of legionellosis in renal transplant patients reveal that a clear association exists between glucocorticoid suppressive therapy and infection by L. pneumophila (1,10,20). Following initiation of steroid therapy, the onset of disease in these patients corresponds to the incubation period for infection which is approximately two to ten days. Furthermore, patients whose illnesses were treated with steroids during the epidemic in 1976 suffered a 40% higher fatality rate, relative to the overall death rate of infected individuals (4). Although legionellosis in immunosuppressed patients is generally attributed to nosocomial infections, one report suggests that steroid therapy may induce reactivation of a latent infection by L. pneumophila (21). In this study, two asymptomatic individuals were shown to possess high levels of antibodies to L. pneumophila antigens. Several months later, these individuals developed legionellosis following extended administration of high dosages of methylprednisolone.

Thus, L. pneumophila has the ability to compromise host resistance by inhibiting a vital macrophage function such as phagocytosis. The inability of the host, particularly the immunosuppressed host, to eliminate these organisms during the initial stages of infection may result in increased severity of disease.

ACKNOWLEDGMENTS

This study was supported in part by Public Health Service Grant AI 16618 from the National Institute of Allergy and Infectious Diseases. We thank Ms. Sean Bell for typing this manuscript.

REFERENCES

1. B. V. Bock, P. H. Edelstein, K. M. Synder, C. M. Hatayama, R. P. Lewis, B. D. Kirby, W. L. George, M. L. Owens, C. E. Haley, R. D. Meyer, and S. M. Finegold, Legionnaires' disease in renal transplant recipients, Lancet 1:410 (1978).
2. J. A. Daisy, C. E. Benson, J. McKitrick, and H. M. Friedman, Intracellular replication of Legionella pneumophila, J. Infect. Dis. 143:460 (1981).
3. R. B. Fitzgeorge, A. Baskerville, M. Broster, P. Hambleton, and P. J. Dennis, Aerosol infection in animals with strains of Legionella

pneumophila of different virulence: comparison with intraperitoneal and intranasal route of infection, J. Hyg. Camb. 90:81 (1980).

4. D. W. Fraser, T. R. Tsai, O. Orenstein, W. E. Parkin, H. J. Beecham, R. G. Sharrar, J. Harris, G. F. Mallison, S. M. Martin, J. E. McDade, C. C. Shepard, and P. S. Brachaman, Legionnaires' disease. Description of an epidemic of pneumonia, N. Eng. J. Med. 297:1189 (1977).

5. H. Friedman, R. Widen, T. Klein, L. Searls, and K. Cabrian, Legionella pneumophila-induced blastogenesis of murine lymphoid cells in vitro, Infect. Immun. 43:314 (1984).

6. M. Friedman, T. W. Klein, and H. Friedman, Legionella pneumophila-induced suppression of macrophage spreading in vitro, Infect. Immun. 42:421 (1983).

7. R. J. Grasso, T. W. Klein, and W. R. Benjamin, Inhibition of yeast phagocytosis and cell spreading by glucocorticoids in cultures of resident murine peritoneal macrophages, J. Immunopharmacol. 3:171 (1981).

8. R. J. Grasso, L. A. West, R. C. Guay, Jr., and T. W. Klein, Inhibition of yeast phagocytosis by dexamethasone in macrophage cultures: reversibility of the effect and enhanced suppression in cultures of stimulated macrophages, J. Immunopharmacol. 4:265 (1983).

9. D. W. Gump, R. O. Frank, W. C. Winn, Jr., R. S. Foster, Jr., C. B. Broome, and W. B. Cherry, Legionnaires' disease in patients with associated serious disease, Ann. Intern. Med. 90:538 (1979).

10. C. E. Haley, M. L. Cohen, J. Halter, and R. D. Meyer, Nosocomial legionnaires' disease: a continuing common-source epidemic at Wadsworth Medical Center, Ann. Intern. Med. 90:583 (1979).

11. M. A. Horwitz, Cell mediated immunity in legionnaires' disease, J. Clin. Invest. 71:1686 (1983).

12. M. A. Horwitz, The legionnaires' disease bacterium (Legionella pneumophila) inhibits phagosome-lysosome fusion in human monocytes, J. Exp. Med. 158:2108 (1983).

13. M. A. Horwitz and S. C. Silverstein, Legionnaires' disease bacterium (Legionella pneumophila) multiplies intracellularly in human monocytes, J. Clin. Invest. 66:441 (1980).

14. M. A. Horwitz and S. C. Silverstein, Interaction of the legionnaires' disease bacterium (Legionella pneumophila) with human phagocytes. I. L. pneumophila resists killing by polymorphonuclear leukocytes, antibody and complement, J. Exp. Med. 153:386 (1981).

15. M. A. Horwitz and S. C. Silverstein, Interaction of legionnaires' disease bacterium (Legionella pneumophila) with human phagocytes. II. Antibody promotes binding of L. pneumophila to monocytes but does not inhibit intracellular multiplication, J. Exp. Med. 158:398 (1981).

16. E. Jones, P. Checko, A. Dalton, J. Cope, J. Barbaree, G. Klein, W. Martin, and C. Broome, Nosocomial legionnaires' disease associated with exposure to respiratory equipment, in: "Legionella. Proceedings of the Second International Symposium," (1984).

17. B. D. Kirby, K. M. Snyder, R. D. Meyer, and S. M. Finegold, Legionnaires' disease: clinical features of 24 cases, Ann. Intern. Med. 89:297 (1978).

18. R. A. Kishimoto, J. D. White, F. G. Shirey, V. G. McGann, R. F. Berendt, E. W. Larson, and K. W. Hedlund, In vitro response of guinea pig peritoneal macrophages to Legionella pneumophila, Infect. Immun. 31:1209 (1981).

19. G. L. Lattimer, R. A. Rachman, and M. Scarlato, Legionnaires' disease pneumonia: histopathologic features and comparison with microbial and chemical pneumonias, Ann. Clin. Lab. Sci. 9:353 (1979).

20. W. Marshall, R. S. Foster, Jr., and W. Winn, Legionnaires' disease in renal transplant patients, Am. J. Surg. 141:423 (1981).

21. Y. Naot, A. Brown, E. M. Elder, J. Shonnard, B. J. Luft, and J. S. Remington, IgM and IgG antibody response in two immunosuppressed patients with legionnaires' disease. Evidence of reactivation of latent infection, Am. J. Med. 73:791 (1982).

22. J. F. Plouffe and I. M. Baird, Lymphocyte transformation to Legionella pneumophila, J. Clin. Lab. Immunol. 5:149 (1981).

23. K. H. Wong, P. R. B. McMaster, J. C. Feeley, R. J. Arko, W. O. Schalla and F. W. Chandler, Detection of hypersensitivity to Legionella pneumophila in guinea pigs by skin test, Curr. Microbiol. 4:105 (1980).

THE EFFECT OF ENDOTOXIN ON MIGRATION INHIBITORY FACTOR AND INTERFERON

Samuel B. Salvin and Pamela B. Renda

Department of Microbiology
University of Pittsburgh School of Medicine
Pittsburgh, Pennsylvania

Lymphokines are soluble mediators released by lymphocytes. Two lympho-kines frequently associated in vivo in the development of cell-mediated immunity are migration inhibitory factor (MIF) and γ-interferon (γ-IFN) (21). The effect of endotoxin on these two lymphokines is being discussed herein.

ENDOTOXIN (LPS) AND MIGRATION INHIBITORY FACTOR (MIF)

In the inhibition of migration of macrophages in vitro, the lymphokine MIF has received much attention because of its immunologic implications. The factor is produced primarily by T-lymphocytes, and to a lesser extent by B-lymphocytes, in the responses of sensitized lymphocytes to specific antigen or of normal lymphocytes to a mitogen (22,25). Unfortunately, the effect of MIF on migration of macrophages, as indicated by the migration-inhibition assay, can be simulated by other non-specific substances. For example, fetal calf serum has a factor that imitates MIF, even to the point of being inhibited by 1-fucose (7,8). Other substances, such as cyclic AMP (18), sodium periodate (9), and bacterial endotoxin (LPS) (6), may also inhibit the migration of macrophages.

LPS does inhibit in vitro the migration of macrophages, although great variation exists from animal species to animal species, as the source of macrophage (6). The effect of the LPS on the macrophages was direct. When guinea pig peritoneal macrophages were purified by adherence, to the extent that macrophages comprised 95% of the exudate, typical inhibition of migra-tion by LPS followed. In experiments where normal macrophages were exposed to MIF alone and to MIF plus various concentrations of endotoxin, endotoxin potentiated the effect of MIF. If a medium containing endotoxin was used in an assay for MIF, the contaminating endotoxin was probably also present in the controls and therefore, the migration pattern should have been equally affected. However, since endotoxin potentiates the effect of MIF, a falsely high inhibition may be produced.

Relationships between endotoxin and MIF were studied further in the LPS-unresponsive C3H/HeJ mice vs. the closely related, but LPS-responsive C3H/HeN mice (26). When MIF-containing supernatants were added to peri-toneal-exudate cells from C3H/HeJ mice, migration inhibition did not occur. In contrast, peritoneal-exudate cells from C3H/HeN were significantly inhibited.

When, however, the C3H/HeJ mice were inoculated intraperitoneally with 2-3 x 10[6] BCG organisms seven days before the peritoneal-exudate cells were collected, the peritoneal-exudate cells became as susceptible to inhibition by MIF as the peritoneal-exudate cells of normal or BCG-injected C3H/HeN mice. Thus, infection with BCG reversed the refractoriness of C3H/HeJ cells to MIF, with the indication that the failure of normal C3H/HeJ macrophages to respond to MIF was not due to a constitutive defect. This suggestion was further enhanced when the addition of small numbers of peritoneal-exudate cells from C3H/HeJ mice to peritoneal-exudate cells from C3H/HeN mice rendered the mixed population unresponsive to MIF. Even though the macrophages from C3H/HeJ mice did not respond to MIF, their lymphocytes could produce MIF in vitro in response to antigenic stimulation. Thus, the deficiency in responsiveness of macrophages to MIF is correlated with a defective responsiveness to endotoxin, possibly influenced by suppressor cells. When negative results appear in direct assays of migration-inhibitory factor, it should be borne in mind that such results could be due either to a failure of the lymphocytes to produce MIF or to a failure of the macrophages to respond to the lymphokine.

C3H/HeJ mice were also compared with C3H/HeN mice for the capacity to release MIF in vivo into the circulation after sensitization and stimulation with specific antigen. The mice were inoculated intravenously with 300 μg cell walls of Mycobacterium bovis strain BCG in Drakeol-Tween 80; three weeks later, the mice were challenged intravenously with either 50 mg old tuberculin (OT) or 1 μg LPS, and exsanguinated four hours later (20). The C3H/HeJ mice were weakly responsive to both the OT and the LPS in the in vivo release of MIF, whereas the C3H/HeN mice did not release detectable quantities of the lymphokine after challenge with either the OT or the LPS.

Prothymosin α is a thymic peptide which can enhance the capacity of certain inbred mouse strains to release MIF in vivo into the circulation (14,20). When C3H/HeJ and C3H/HeN mice were treated with 160 ng prothymosin daily for three days prior to both sensitization and challenge, the titers of in vivo MIF in the circulations of the C3H/HeJ mice increased. The C3H/HeN mice, however, still remained unresponsive. It has been reported (12) that macrophages from the LPS-resistant C3H/HeJ mice are not killed by LPS in vitro, whereas those macrophages from the closely related LPS-responsive C3H/HeN strain are killed. Accordingly, the suggestion may be made that macrophages may be necessary for the production of MIF in vivo in C3H mice.

Supernatants or extracts from certain bacteria have also affected the migration of macrophages. The incubation fluid from 24-hour cultures of normal guinea pig small intestines was found to contain activity capable of inhibiting the migration of normal guinea pig peritoneal macrophages (10). The belief was held that the inhibitory activity was associated with a low molecular weight (about 25,000 daltons), heat stable (at 56° for 30 minutes) substance, presumably MIF, as well as a high molecular weight (about 55,000 daltons) substance, presumably endotoxin. The presence of MIF, and therefore of T-cell activity, in the intestine suggests that cell-mediated immunity may play a role in local protection against gut infections. The release of MIF may be stimulated by non-specific action of endotoxins from such bacteria as E. coli or Salmonella sp. or by such mitogens as Staphylococcus enterotoxin B. Here, the endotoxins or enterotoxins are mitogenic for both B- and T-lymphocytes (2). The release of MIF might also have been due to a specific immune reaction with endotoxin as the antigen, in which case the donor would have had to have been exposed to and sensitized to the specific antigen.

The in vivo release of MIF was stimulated when Swiss Webster mice infected intravenously with 5 x 10[6] to 5 x 10[7] viable Mycobacterium bovis

292

strain BCG were subsequently challenged with 0.5 µg LPS (29). The injection of the LPS produced a low titer of circulating MIF. This moderate response was completely abrogated by desensitization with LPS, but not with old tuberculin (OT).

ENDOTOXIN AND INTERFERON

Mice inoculated with bacterial endotoxin release the viral-inhibitor interferon (IFN) into the circulation (11,27,31,32,). The endotoxin-induced interferon has a molecular weight of about 90,000 daltons by Sephadex filtration, whereas the virus-induced interferon has a molecular weight of only 26,000 daltons (13,15,17).

Mice inoculated intravenously with 125 µg endotoxin developed peak interferon titers in the plasma two hours later, but had non-detectable interferon levels by six hours (31). When, however, mice were first inoculated intravenously with 125 µg endotoxin and 48 hours later were challenged intravenously with a second dose of 125 µg endotoxin, interferon was not detectable during the first six hours post-injection. Endotoxin reduced, but did not completely eliminate, the development of interferon when the endotoxin was injected intravenously 48 hours before intravenous challenge with 5×10^7 plaque-forming units of Newcastle disease virus (NDV). Prior injection of NDV into mice resulted in hyporeactivity to endotoxin. It appears, therefore, that the decreased appearance of interferon in hyporeactive animals is not limited to the material used to produce this state.

When mice were infected with Mycobacterium bovis BCG, they became hyper-reactive to endotoxin in both interferon response and lethality. In contrast, when BCG-infected mice which were hyperreactive to the lethal and interferon-stimulating effects of endotoxin were inoculated with NDV and assayed for interferon production, significant differences could not be detected in the responses of infected vs. control mice.

This lack of specificity in the induction of interferon by endotoxin vs. NDV appeared when the interferon-inducing properties of endotoxin vs. those of statolon were compared (32). Here, pretreatment with endotoxin failed to reduce significantly the response to statolon, and prior injection of statolon did not affect the titer of interferon induced by endotoxin.

The time and extent of appearance of IFN may vary according to the inducer. Endotoxin-induced IFN in mice reached a peak at two hours and could not be detected in the blood after six hours (31). When NDV was the inducer in mice, IFN could be detected at two hours post-infection, but the titer was still rising at 12 hours. With 1,000 µg statolon as the inducer in mice, the viral inhibitor was present at two hours after inoculation, rose to a peak titer between eight and twelve hours, and by 48 hours was not longer detectable (32). With 10^{10} cells of Brucella abortus as the inducer in chickens, antiviral activity was detected in the sera by three hours after inoculation, reached a peak at about 12 hours, was declining rapidly by 24 hours, and was not detectable by 48 hours (30).

Experiments have been reported wherein reciprocal hyporeactivity was tested in mice inoculated with endotoxin or statolon (32). Groups of mice pretreated intravenously with either 250 µg E. coli endotoxin or 2,500 µg statolon were challenged 48 hours later intravenously with the homologous or heterologous substance. Blood samples were obtained at two hours after endotoxin challenge and at six hours after challenge with statolon. Marked depression of interferon titer occurred only in mice challenged at 48 hours with the homologous substance. Challenge with the heterologous material failed to demonstrate any hyporeactivity.

To determine the hyperreactive effect of BCG on the response of mice to endotoxin vs. the response to statolon, mice were first infected intravenously with a ten-day culture of BCG, and 21 days later were challenged with either endotoxin or statolon. Examination of the sera revealed that BCG-infected mice had an enhanced capacity to produce interferon in response to endotoxin, but did not show any significantly different response from uninfected mice after challenge with statolon.

Early release of interferon was demonstrated in the tissues of intact mice inoculated with bacterial endotoxin (34). Endotoxin from E. coli produced maximum titers of circulating interferon in two hours after injection, whereas injection of Brucella abortus or Newcastle disease virus (NDV) produced maximum levels of the inhibitor in six to twelve hours (1,24,32). Whether this early interferon is preformed or superinduced is not clear (27). Doses of puromycin or cycloheximide which effectively inhibited protein synthesis did not prevent the appearance of circulating interferon in mice inoculated with endotoxin. In contrast, the appearance of interferon after injection of B. abortus or NDV was prevented by the blockade of protein synthesis (34).

Enhancement of interferon production or release has been demonstrated in animals subjected to stimulants of the reticuloendothelial system. For example, mice inoculated with Corynebacterium acnes showed a marked enhancement of serum interferon after injection of endotoxin. In contrast, inoculation with viable or non-viable C. acnes resulted in depressed interferon production after inoculation of mice with NDV, Chikungunya virus, or poly-inosinic-polycytidylic acid (4).

When doses of 1 to 5 mg cortisol were administered to 1 kg rabbits, the appearance of interferon after 10 µg of endotoxin was inhibited (19). The time of appearance of the endotoxin-induced interferon was not affected by the steroid. In contrast, doses of cortisol as high as 25 and 250 mg did not inhibit interferon production after inoculation of the animals with Newcastle disease virus. Since adrenalectomy markedly potentiated the production of interferon by endotoxin, the conclusion was drawn that endogenous steroids suppress the interferon response to endotoxin.

After the demonstration that bacterial endotoxin caused the release of interferon in intact animals, the question arose as to what cells were responsible. Incubation of rabbit peritoneal macrophages at 37°C in the presence of E. coli LPS resulted in high titers of interferon by four to six hours (23). Since macrophages must be present for either T or B lymphocytes to produce interferon after stimulation with mitogens (3), the question may be asked as to whether the interferon induced by the lipopolysaccharide in a macrophage suspension originates from contaminating lymphocytes rather than from macrophages. Macrophages alone apparently can produce endotoxin-induced interferon, since the addition of varying dilutions of anti-lymphocyte serum did not alter the amount of IFN produced in macrophage cultures. Thus, neither T nor B lymphocytes are essential for endotoxin-induced IFN production in macrophage suspensions.

LPS-stimulated interferon production by peritoneal-macrophages vs. spleen-cell cultures differed in that the optimum temperature for the macrophages within the first 24 hours was 23°C, whereas the optimum temperature for the spleen cells was 37°C. Late interferon production by spleen cells between 24 and 48 hours could only be found at 23°C and 27°C (16). Incubation of T-cells with LPS did not induce detectable amounts of interferon, whereas incubation of B-cells with LPS did induce production of interferon. LPS is not inductive in ordinary fibroblast or epithelial cell cultures.

Macrophage interferon was heat labile, whereas B-cell interferon was more heat stable. Also, early serum interferon was more heat labile than later interferon. None of the interferons was completely destroyed at pH 2, 4°C, in 24 hours. With regard to molecular weight, endotoxin-induced interferon has at least two types: an entity of more than 100,000 daltons, which has been reported as heat and acid labile, and low-molecular weight entity of 40,000-50,000 daltons, which is more stable (28).

The murine cells associated with the release of IFN by endotoxin are different from those that release IFN on stimulation with specific antigen (29). Mice with delayed hypersensitivity induced by infection with M. bovis strain BCG were desensitized by a single large dose of specific antigen (old tuberculin, ODT) or LPS. Subsequent challenge of the desensitized animals revealed only a homologous hyporeactivity. Mice desensitized with OT had decreased γ-IFN responses to the specific antigen, which were not affected by desensitization with LPS. Conversely, mice desensitized with LPS had decreased α, βIFN to LPS, which was not affected by desensitization with OT.

Since the LPS molecule is complex, containing O-specific side chains, core polysaccharide, and lipid A, studies have been conducted to determine what part of the molecule is most influential in the production of interferon. The interferon response elicited by Salmonella typhimurium mutants in mice is not dependent on the presence of complete cell-wall lipopolysaccharide (5). A mutant (G30/C21) which does not have the polysaccharide side chains and sugars of the O antigen and contains only 2-keto-3-deoxyoctonate (KDO) and lipid has the same interferon-stimulating capacity as the wild type, which possesses a complete O-antigen with polysaccharide side chains. Interferon production is not induced by O-antigenic hapten lacking the glycolipid moiety. Lipid A which does not contain any KDO is fully active in the release of interferon (14,33). It is thus clear that the lipid portion of the LPS, and not the polysaccharide chains, is the primary site of the interferon-inducing activity.

In conclusion, endotoxin can stimulate release of migration inhibitory factor and of interferon to varying degrees. The conditions for their optimum release vary depending on a variety of factors, such as the strain of animal, the degree of exposure to endotoxin, and the nature and source of the lipopolysaccharide.

ACKNOWLEDGEMENT

The authors thank Dr. J. S. Youngner for his critical review of the manuscript.

REFERENCES

1. S. Baron and C. E. Buckler, Circulating interferon in mice after intravenous injection of virus, Science 141:1061 (1963).
2. P. DeRinaldis, E. Jirillo, R. Pantaleo, and D. Fumarola, Inibizione della migrazione leucocitaria da endotossine su piastre di agarosio secondo clausen, G. Batteriol., Vir. and Immun. 67:238 (1974).
3. L. P. Epstein, M. J. Cline, and T. C. Merigan, PPD-stimulated interferon: in vitro macrophage-lymphocyte interaction in the production of a mediator of cellular immunity, Cell. Immunol. 2:602 (1971).
4. P. A. Farber and L. A. Glasgow, Effect of Corynebacterium acnes on interferon production in mice, Infect. Immun. 6:272 (1972).
5. D. S. Feingold, J. S. Youngner, and J. Chen, Interferon production in

mice by cell wall mutants of <u>Salmonella typhimurium</u>. III. Role of lipid moiety of bacterial lipopolysaccharide in interferon production in animals, <u>Ann</u>. <u>N.Y</u>. <u>Acad</u>. <u>Sci</u>. 173:249 (1970).

6. R. A. Fox and K. Rajaraman, Endotoxin and macrophage-migration inhibition, <u>Cell</u>. <u>Immunol</u>. 53:333 (1980).

7. R. A. Fox, D. S. Gregory, and J. D. Feldman, Macrophage receptors for migration inhibitory factor (MIF), migration stimulatory factor (MSF), and agglutinating factor, <u>J</u>. <u>Immunol</u>. 112:1867 (1974).

8. R. A. Fox, D. S. Gregory, and J. D. Feldman, Migration inhibition factor (MIF) and macrophage stimulation factor (MSF) in fetal calf serum, <u>J</u>. <u>Immunol</u>. 112:1861 (1974).

9. R. A. Fox, L. A. Hernandez, and K. Rajaraman, Migration inhibition produced by sodium periodate oxidation of the macrophage membrane, and reversal by sodium borohydride, <u>Scand</u>. <u>J</u>. <u>Immunol</u>. 6:1151 (1977).

10. N. Gadol, R. H. Waldman, and L. W. Clem, Inhibition of macrophage migration by normal guinea pig intestinal secretions, <u>Proc</u>. <u>Soc</u>. <u>Exp</u>. <u>Biol</u>. <u>Med</u>. 151:654 (1976).

11. A. W. Gledhill, The interference of mouse hepatitis virus with ectromelia in mice and a possible explanation of its mechanism, <u>Brit</u>. <u>J</u>. <u>Exp</u>. <u>Path</u>. 40:291 (1959).

12. L. M. Glode, A. Jacques, S. E. Mergenhagen, and D. L. Rosenstreich, Resistance of macrophages from C3H/HeJ mice to the in vitro cytotoxic effects of endotoxin, <u>J</u>. <u>Immunol</u>. 119:162 (1977).

13. J. V. Hallum, J. S. Youngner, and W. R. Stinebring, Interferon activity associated with high molecular weight proteins in the circulation of mice injected with endotoxin or bacteria, <u>Virology</u> 27:429 (1965).

14. A. A. Haritos, G. J. Goodall, and B. L. Horecker, Properties of the major immunoreactive form of thymosin alpha$_1$ in rat thymus, <u>Proc</u>. <u>Natl</u>. <u>Sci</u>. <u>U.S.A</u>. 81:1008 (1984).

15. Y. J. Ke, M. Ho, and T. C. Merigan, Heterogeneity of rabbit serum interferon, <u>Nature</u> 211:541 (1966).

16. N. Maehara and M. Ho, Cellular origin of interferon induced by bacterial lipopolysaccharide, <u>Infect</u>. <u>Immun</u>. 15:78 (1977).

17. T. C. Merigan, Purified interferons. Physical properties and species specificity, <u>Science</u> 145:811 (1964).

18. E. Pick, Cyclic AMP affects macrophage migration, <u>Nature</u> <u>New</u> <u>Biology</u> 238:176 (1972).

19. B. Postic, C. DeAngelis, M. K. Breinig, and M. Ho, Effects of cortisol and adrenalectomy on the induction of interferon by endotoxin, <u>Proc</u>. <u>Soc</u>. <u>Exp</u>. <u>Biol</u>. <u>and</u> <u>Med</u>. 125:89 (1977).

20. S. B. Salvin, A. A. Haritos, and B. L. Horecker, Immunoenhancing activities of the thymic polypeptide prothymosin alpha (submitted for publication, 1985).

21. S. B. Salvin, E. Ribi, D. L. Granger, and J. S. Youngner, Migration inhibitory factor and type II interferon in the circulation of mice sensitized with mycobacterial components, <u>J</u>. <u>Immunol</u>. 114:354 (1975).

22. S. B. Salvin, G. Sonnenfeld, and J. Nishio, In vivo studies on the cellular source of migration inhibitory factor in mice with delayed hypersensitivity, <u>Infect</u>. <u>Immun</u>. 17:639 (1977).

23. T. J. Smith and R. R. Wagner, Rabbit macrophage interferons. I. Conditions for biosynthesis by virus-infected and uninfected cells, <u>J</u>. <u>Exp</u>. <u>Med</u>. 125:559 (1967).

24. W. R. Stinebring and J. S. Youngner, Patterns of interferon appearance in mice injected with bacteria or bacterial endotoxin, <u>Nature</u> 204:712 (1964).

25. K. Sugane, T. Kasahara, and K. Shioiri-Narano, Release of migration inhibitory factor from mouse T and B Cells activated by soluble phytomitogens, <u>Jpn</u>. <u>J</u>. <u>Exp</u>. <u>Med</u>. 45:19 (1975).

26. A. Tagliabue, J. L. McCoy, and R. B. Herberman, Refractoriness to

migration inhibitory factor of macrophages of LPS nonresponder mouse strains, J. Immunol. 121:1223 (1978).

27. J. Vilcek, Cellular mechanisms of interferon production, J. Gen. Physiol. 56:76 (1970).

28. J. S. Youngner and D. S. Feingold, Interferon production in mice by cell wall mutants of Salmonella typhimurium, J. Virol. 1:1164 (1967).

29. J. S. Youngner and S. B. Salvin, Type I and II interferons and migration inhibitory factor induction in Mycobacterium bovis BCG infected mice desensitized with old tuberculin or lipopolysaccharide, Infect. Immun. 19:912 (1978).

30. J. S. Youngner and W. R. Stinebring, Interferon production in chickens injected with Brucella abortus, Science 144:1022 (1964).

31. J. S. Youngner and W. R. Stinebring, Interferon appearance stimulated by endotoxin, bacteria, or viruses in mice pre-treated with Escherichia coli endotoxin or infected with Mycobacterium tuberculosis, Nature 208:456 (1965).

32. J. S. Youngner and W. R. Stinebring, Comparison of interferon production in mice by bacterial endotoxin and statolon, Virology 29:310 (1966).

33. J. S. Youngner, D. S. Feingold, and J. K. Chen, Involvement of a chemical moiety of bacterial lipopolysaccharide in production of interferon in animals, J. Infect. Dis. 128(Supplement):227 (1973).

34. J. S. Youngner, W. r. Stinebring, and S. E. Taube, Influence of inhibitors of protein synthesis on interferon formation in mice, Virology 27:541 (1965).

SECTION IV. EFFECTS OF ENDOTOXINS ON CELLS OF THE IMMUNE SYSTEM

Regulation of the Cell Cycle of Murine B Lymphocytes by Lipopolysaccharides
Fritz Melchers

Synergistic Effect of Endotoxin with Concanavalin A on DNA Synthesis
in Lymphocytes and the Role of Interleukins 1 and 2
Masayasu Nakano and Toshimasa Nitta

Biochemical, Immunological and Functional Analysis of Lymphocytes
from the LPS Non-Responder C3H/HeJ Mouse
D. C. Morrison, H-W. Wollenweber, S. W. Vukajlovich, and S. A. Goodman

Effects of Endogenous Gut LPS on Cells of the Secretory Immune System
Suzanne M. Michalek, Jerry R. McGhee, Dawn E. Colwell, Shane I. Williamson,
Thomas A. Brown, David M. Spalding, William J. Koopman, and Jiri Mestecky

Molecular Mechanisms in Lymphokine-Induced Macrophage Activation –
Enhanced Production of Oxygen Radicals
Edgar Pick

Induction of Gamma Interferon by Endotoxin in "Aged" Murine
Splenocyte Cultures
D. K. Blanchard, T. W. Klein, H. Friedman, and W. E. Stewart II

Characterization and Localization of Lipopolysaccharides Following
the Ingestion of E. coli by Murine Macrophages In Vitro
Robert L. Duncan, Jr., Vernon Tesh, and David C. Morrison

The Release of Immunopotentiating Mediators from Macrophages
Activated by Endotoxins
R. Christopher Butler, Jeri M. Frier, Mrunal S. Chapekar,
Herman Friedman, and Alois Nowotny

Changes in Macrophage Progenitor Cell Composition in the Bone Marrow
of "Early Phase" Endotoxin-Tolerized Mice
Stefanie N. Vogel and Gary S. Madonna

Production of Colony-Stimulating Factor (CSF) by Bone Marrow Cells
Stimulated with Lipopolysaccharide (LPS) and Phorbol Esters
Dov H. Pluznik and Stephan E. Mergenhagen

REGULATION OF THE CELL CYCLE OF MURINE B LYMPHOCYTES BY LIPOPOLYSACCHARIDES

Fritz Melchers

Basel Institute for Immunology
Basel, Switzerland

INTRODUCTION

The cell cycle of murine B lymphocytes is normally controlled by antigen [acting via surface-bound immunoglobulin (Ig)], by macrophages (producing α factor-type lymphokines) and by helper T lymphocytes (producing β factor-type lymphokines) (5,7). For both types of lymphokines B cells are expected to display so far unknown specific receptors. Most B cells in the immune system are normally resting and must be excited to become susceptible to the action of these lymphokines so that they can enter the cell cycle. In T cell-dependent B cell activation this excitation is affected by the inter-action of helper T cells with B cells that bridges on one side antigen and class II histocompatibility complex (MHC) molecules on B cells with, on the other side, the complementary antigen-specific, MHC-restricted receptors on T cells.

B cell activation into the cell cycle can, therefore, be seen as a three-step process. In the first step, helper T cells recognize antigen presented on macrophages in the context of class II MHC antigens. This leads to an endocrine production of α and β factors. These factors are active on B cells, provided that they have been excited from the resting state. This excitation is the second step that again requires the cooper-ation of a helper T cell. In the third step, α and β factors, together with antigen, regulate the cell cycle of excited, activated B cells.

Lipopolysaccharides (LPS) have multiple ways to act on B cells in helper T cell-independent ways (1,2). First they most likely excite resting B cells to become susceptible to the action of α and β factor lymphokines. They do polyclonally, circumventing the requirements of binding to surface-bound Ig and to surface-bound class II MHC molecules. This is repeated at the beginning of every subsequent cell cycle. Secondly, LPS replace β factor-type lymphokines in their action during the cell cycle to promote mitosis. Finally, they activate macrophages and, thereby, indirectly supply α factor-type lymphokines to the B cells (4). This paper summarizes our current understanding of these processes.

HELPER T CELL-DEPENDENT, ENDOCRINE PRODUCTION OF GROWTH FACTORS FOR B CELLS

When antigen enters the immune system it is thought to be taken up by macrophages (accessory cells, A cells), processed, and then presented on the surface in the context of class II-MHC molecules. These complexes of foreign antigen and class II-MHC molecules can then be recognized by helper T cells that have receptors for these complexes. Recognition leads to the production of lymphokines. Both types of cells recognize their interaction with each other: A cells produce α factors, T cells β factors that are active with B cells. Both types of cells are probably producing other factors as well as show activities with other cell lineages. Thus, helper T cells alone produce IL-2 (T cell growth factor) and CSF (colony stimulating factors for the erythro-myelopoietic cell lineages).

HELPER T CELL-DEPENDENT, ANTIGEN-SPECIFIC, MHC-RESTRICTED

EXCITATION OF B CELLS

Most antigen-sensitive B cells in the immune system are resting, in the Go phase of the cell cycle. In this resting state, they are refractory to the action of α and β factors. Antigen-specific helper T cells must recognize the antigen, presented by surface-bound Ig molecules in context with class II-MHC molecules on B cells. This recognition, again, leads to reactions of both cells. Helper T cells continue to produce β factors (as well as other lymphokines). Resting B cells become excited, susceptible to the action of α and β factors. As in the interaction of T cells with A cells, the interaction of T cells with B cells usually requires specific recognition of both foreign antigen and self MHC molecules.

CELL CYCLE CONTROL OF ACTIVATED B CELLS

Once B cells have been excited, either by helper T cells and antigen, or by T-independent polyclonal activators such as LPS, they can now be stimulated through the cell cycle and through mitosis by α and β factors. This has been investigated recently in greater detail with LPS-activated B cell blasts from C57BL/6J nu/nu mice that were synchronized by size selection using velocity sedimentation under earth gravity (8). Monoclonal μ-heavy chain specific antibodies, coupled to Sepharose, were used as agents acting via surface Ig, as they are known to excite B cells from their resting state in a polyclonal and MHC-unrestricted way (6). It was found that α factors were needed three to five hours into the G1 phase, while β factors were

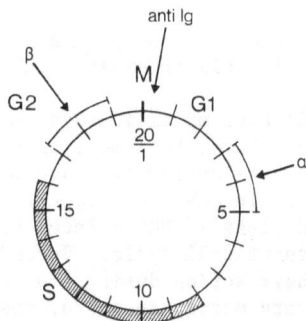

Figure 1. The cell cycle of activated murine B cells is controlled by occupancy of surface Ig (either by antigens, or by Ig-specific antibodies) by α-factors produced by activated macrophages and by β-factors produced by helper T cells (8). M = mitosis; S = S phase, phase of DNA replication G1 and G2 = phases of the 20-hour long cell cycle between S and M.

required two to four hours before mitosis, late in the G2 phase of the cell cycle. When B cells had completed mitosis, occupancy of surface Ig by Ig-specific antibodies was needed to excite the cells again for subsequent action of α- and β-factors. This occupancy of surface Ig, that could be as short as 15 minutes, was needed in each of at least the first six cell cycles early after mitosis. Figure 1 summarizes the growth controlling elements of the B cell cycle.

REQUIREMENT FOR MACROPHAGES OR FOR MACROPHAGE-DERIVED α FACTORS

IN LPS-STIMULATED B CELL RESPONSES

When murine splenic B cells are sufficiently depleted of A cells they no longer respond to LPS by polyclonal proliferation or maturation to Ig secreting cells. Responsiveness can be restored with a very few A cells. Thus, five to 30 CSF-grown bone marrow-derived macrophages from a single colony picked in semi-solid media will restore full polyclonal responsiveness of 10^4 A cell-depleted resting splenic murine B cells to LPS (3). Restoration of responsiveness can also be achieved with factors contained in media conditioned by activated macrophages. Amongst the macrophages that have been used to generate these so called α factors that restore responsiveness are those of the macrophage cell line P388D1. Activation of this cell line to α factor production can be done with polyanions (such as dextransulfate) or with the well known B cell mitogens lipoprotein and LPS. Activation to α factor production is dose-dependent. For LPS, it is optimal between 0.5 and 25 μg/ml. Higher concentrations of LPS induce P388D1 macrophages to condition the supernatant medium with inhibitory factors, as the test with an α factor requiring myeloma cell line indicates (4).

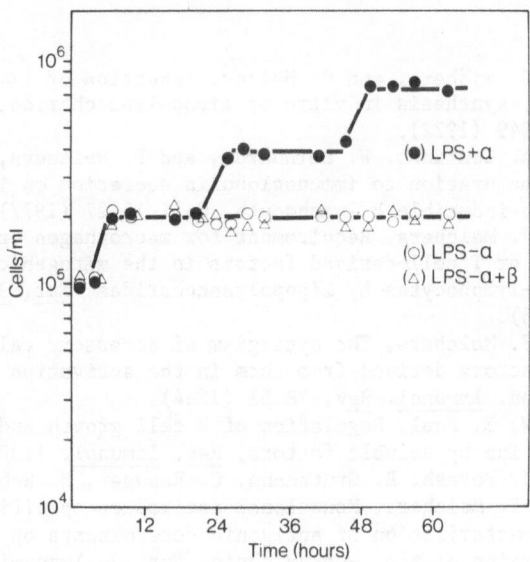

Figure 2. Growth of 48 hrs LPS-activated, synchronized B cells in the presence of only LPS (o), LPS + β-factors (Δ) and LPS + α-factors (●). In the absence of all factors, a growth curve similar to that in the presence of only LPS was obtained. In the presence of LPS α-factors and β-factors a growth curve similar to that obtained in the presence of LPS plus α-factors was obtained. For experimentals see (8).

THE ACTION OF LPS IN THE B CELL CYCLE

In T cell-dependent B cell proliferation the cell cycle is controlled by the occupancy of Ig, by α and by β factors. Since LPS-stimulated B cell growth was found to be α factor dependent, it appeared that LPS replaced Ig occupancy as well as the action of β factors within the cell cycle. This was tested with A cell-depleted B cell blasts, activated either by LPS and α factors, or by Ig-specific antibodies, β factors and α factors. Both types of B cell blasts were synchronized by size selection using velocity sedimentation. The results in Figure 2 show that, in the presence of α factors, LPS stimulates both types of activated B cells through successive rounds of divisions. LPS, therefore, circumvents the occupancy of surface Ig on B cells and, furthermore, replaces the action of β factors. Experiments are now under way to test the dose dependencies of these two actions of LPS on B cells.

In summary, LPS acts in three ways at three restriction points within the B cell cycle. Early after mitosis it acts to excite B cells to susceptibility for α and β factors, replacing the requirement for the occupancy of surface Ig by either antigen and MHC-restricted helper T cells, or by Ig-specific antibodies. It then indirectly activates A cells (macrophages) to α factor production that control the B cell cycle three to five hours after mitosis within the G1 phase. It finally replaces the action of helper T cell-derived β factors late in the cell cycle, two to four hours before mitosis in the G2 phase of the cell cycle.

ACKNOWLEDGMENTS

The Basel Institute for Immunology was founded and is supported by F. Hoffman-LaRoche Ltd. Company, Basel, Switzerland.

REFERENCES

1. J. Andersson, O. Sjöberg, and G. Möller, Induction of immunoglobulin and antibody synthesis in vitro by lipopolysaccharide, Eur. J. Immunol. 2:349 (1972).
2. J. Andersson, A. Coutinho, W. Lernhardt, and F. Melchers, Clonal growth and maturation to immunoglobulin secretion on in vitro of every growth inducible B lymphocyte, Cell 10:27 (1977).
3. C. Corbel and F. Melchers, Requirement for macrophages or for macrophage- or T cell-derived factors in the mitogenic stimulation of murine B-lymphocytes by lipopolysaccharides, Eur. J. Immunol. 13:528 (1983).
4. C. Corbel and F. Melchers, The synergism of accessory cells and of soluble α factors derived from them in the activation of B cells to proliferation, Immunol. Rev. 78:51 (1984).
5. M. Howard and W. E. Paul, Regulation of B cell growth and differentiation by soluble factors, Rev. Immunol. 1:307 (1983).
6. M. Leptin, M. J. Potash, R. Grutzmann, C. Heuwser, M. Schulmann, G. Kohler, and F. Melchers, Monoclonal antibodies specific for murine IgM. I. Characterization of antigenic determinants on the four constant domains of the -heavy chain, Eur. J. Immunol. 14:534 (1984).
7. F. Melchers, and J. Andersson, B cell activation: three steps and their variations, Cell 37:715 (1984).
8. F. Melchers and W. Lernhardt, Three restriction points in the cell cycle of activated B lymphocytes. Cell submitted (1985).

SYNERGISTIC EFFECT OF ENDOTOXIN WITH CONCANAVALIN A ON DNA

SYNTHESIS IN LYMPHOCYTES AND THE ROLE OF INTERLEUKINS 1 and 2

Masayasu Nakano and Toshimasa Nitta

Department of Microbiology, Jichi Medical School
Tochigiken, Japan and Department of Bacteriology
Tohoku Dental University, Koriyama, Japan

INTRODUCTION

Bacterial lipopolysaccharide (LPS) extracted from the cell walls of gram-negative bacteria has a number of effects on the cells of the immune system (see 21 for review). The effects of LPS on macrophages (Mϕ) and B lymphocytes are very prominent. Mϕ activated by LPS increase their phagocytic ability (20,21), pinocytosis (5,30), oxidative metabolism (15,25), synthesis of cellular proteins including lysosomal enzymes (19,20), secretion of collagenase (36) and arginase (4), tumor cytotoxicity (33), microbicidal activity (27), interleukin (IL) 1 (18), colony-stimulating factor, and interferon (21). When murine splenic B lymphocytes are stimulated by LPS, about one-third of the B lymphocytes initiates DNA synthesis (13), and subsequently, these activated B lymphocytes produce antibodies polyclonally (21). However, its effect on human peripheral B lymphocytes is not obvious (21,26,29), unless these cells are cultured for long periods of time (seven to nine days) with prescreened lots of fresh human serum (17). On the contrary, LPS has no obvious mitogenic effect on T lymphocytes, except on a very small percentage of cells (35), and it is incapable of initiating T lymphocytes to produce lymphokines. However, if T lymphocytes are activated by T cell stimulants, the T lymphocytes seem to be able to accept the stimulus of LPS. The combination of some kinds of phytomitogens with LPS can synergistically enhance the blastogenic responses of thymocytes in mice (7,24,32), spleen lymphocytes of rats (8) or peripheral blood T lymphocytes of humans (12,29) as measured by increased [^{3}H]thymidine ([^{3}H]TdR) uptakes. Analysis of the synergy of LPS and Concanavalin A (Con A), one of the T lymphocyte mitogens, may provide us with some valuable information on the complexity of T cell activation. In this chapter, we deal with the synergistic effect of LPS and Con A on the [^{3}]TdR uptake of lymphocytes in cultures in relation to productions of IL 1 and 2, and demonstrate that the effect is mainly due to an elevation of IL 2 production from T cells costimulated with LPS and Con A.

SYNERGY OF LPS AND CON A ON PROLIFERATIVE RESPONSES OF MURINE LYMPHOCYTES

LPS is a powerful B cell mitogen and Con A is also a well-known T cell mitogen. Either of these mitogens can increase the incorporation of [^{3}H]TdR into B or T lymphocytes in murine splenic cell cultures. When a very small amount of LPS (0.6 μg/well) was added together with small amounts of Con A (0.05 to 3.2 μg/well) to the cultures, a marked synergy of Con A and LPS was observed in the proliferative responses of these cultured cells (Figure 1). The synergic response was the most obvious when the optimal dose of Con A (0.2 μg/ml) was added to the cultures. The synergy was quite obvious when 0.1 to 5 μg of LPS was added to cultures with 2.5 μg of Con A.

T CELLS ARE THE RESPONSIVE CELLS TO CO-STIMULATION OF LPS AND CON A

Which cells are proliferative cells by co-stimulation of LPS and Con A? In order to clarify this, the spleen cells were previously treated with rabbit anti-mouse IgM and IgG sera or mouse anti-Thy1 serum plus complement

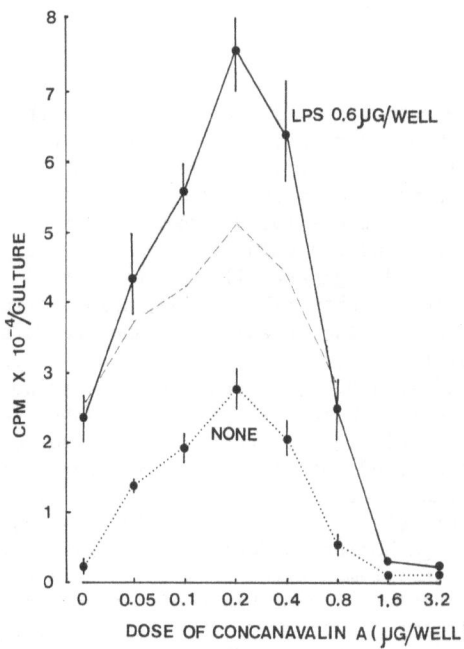

Figure 1.   Synergistic effect of LPS and Con A on DNA synthesis in mouse spleen cells in culture. Spleen cells were prepared from the spleens of C3H/HeN mice. The cells were cultured in RPMI 1640 medium supplemented with 5% heat-inactivated fetal calf serum, penicillin (100 M/ml), and streptomycin (100 μg/ml) (1 x 10⁵ cells/0.1 ml/well, Microtest plate of A/S Nunc) in the presence of various concentrations of Con A (E. Y. Laboratories) with or without LPS (0.6 μg/well) for 48 hrs in a humidified 5% CO_2-95% air chamber. The LPS used was extracted from the LT2 strain of <u>Salmonella typhimurium</u> with hot phenol according to method of Westphal and Luderitz (38). [³H]TdR was added to cultures 20 hr before the end of culture. When incubation was completed the culture cells were harvested on a filter, and radioactivity was assessed by a liquid scintillation counter. Each symbol represents the mean value of triplicate cultures and standard deviation. The dashed line represents the expected values if the effects of LPS and Con A are additive.

(C), and then these cells were cultured with or without the mitogens. The cells, which had been treated with anti-IgM and IgG sera with C, were still able to respond to the mitogens and showed the synergistic effect, while no synergy could be seen in the cultures of the cells which had been previously treated with anti-Thyl serum and C. The results clearly indicated that the synergy was due to the over-proliferation of T cells, but not B cells.

Distinct subsets of murine T lymphocytes have been characterized with respect to their membrane surface antigens by in vitro tests using antibodies having specificities towards the antigens such as Lyt-1, Lyt-2 and Lyt-3, and peripheral T lymphocytes can be divided into three subpopulations by their surface membrane markers of the Lyt system (2). In order to clarify which T cell subpopulation is responsive to the synergistic action of LPS and Con A, the spleen cells were previously treated with either anti-Lyt 1 serum or anti-Lyt 2 serum with C, and then these cells were cultured in the presence of the mitogens. The previous treatment of the cells with anti-Lyt 1 serum and C blocked the response, while the treatment of the cells with anti-Lyt 2 serum and C did not affect the response to the concomitant stimuli of LPS and Con A. The treatment of the cells with anti-Lyt2 serum and C would be expected to destroy the subpopulations with cell surface antigens of $Lyt-123^+$ and $Lyt-1^-23^+$. Therefore, the responding cells to the co-stimulation should belong to the subpopulation with $Lyt-1^+23^-$ marker on their cell surface membranes, which are equivalent to an inducer/helper cell population.

LOW SYNERGY OF LPS AND CON A ON PROLIFERATION OF PREMATURE T LYMPHOCYTES

The thymus is responsible for the development of T-lymphocytes. This gland contains a large number of premature T lymphocytes and a small number of mature ones. When mice are injected with an adequate dose of hydrocortisone, it is known that the premature T lymphocytes in the thymus disappear (3,6,37). In order to study the synergy of LPS with Con A on the proliferative response of T lymphocytes in the different stages of the maturation, we examined thymus cells which had been obtained from either normal or hydro-

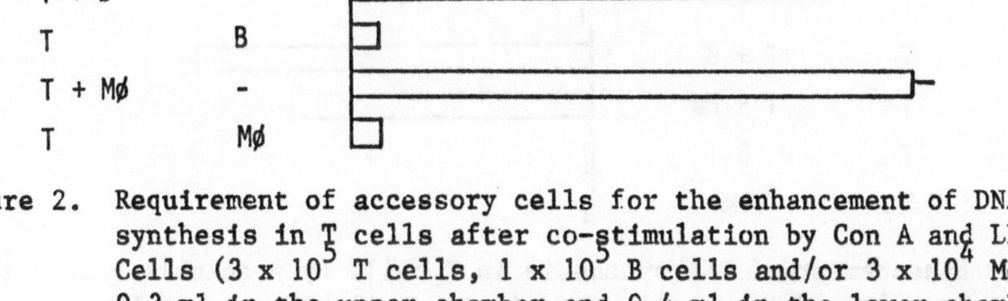

Figure 2.   Requirement of accessory cells for the enhancement of DNA synthesis in T cells after co-stimulation by Con A and LPS. Cells (3 x 10^5 T cells, 1 x 10^5 B cells and/or 3 x 10^4 Mø 0.3 ml in the upper chamber and 0.4 ml in the lower chamber/ culture) were cultured in Marbrook-type vessels for 48 hrs. Con A (0.4 µg/0.1 ml) and LPS (0.1 µg/0.1 ml) were added in both chambers. [³h]TdR was added into the upper chamber 20 hrs. before the end of culture, and the incorporation into the cells in the upper chamber was examined at the end of culture. Each bar represents the mean cpm of triplicate cultures ± SD.

cortisone-treated mice (2.5 mg of hydrocortisone acetate/mouse, intraperitoneally, three days before sacrifice). The thymus cells were cultured in
the presence of the mitogens, and the [³H]TdR uptakes of the cells were
examined. The results obtained indicate that the cells from normal mice
responded quite poorly to simultaneous stimulation by LPS and Con A, while
the cells from hydrocortisone-treated mice responded quite well to the co-
stimulation. These results indicate that the premature T lymphocytes as
well as the mature T lymphocytes are capable of responding to Con A, but
the premature T lymphocytes seem to be incapable of accepting the stimulatory
signal of LPS in the presence of Con A.

THE ROLE OF ACCESSORY CELLS FOR THE SYNERGISTIC EFFECT OF LPS AND CON A ON

DNA SYNTHESIS IN T CELLS

For further examination of the synergistic effect, single cell populations of T lymphocytes, B lymphocytes and M∅ are desirable for cell culture.
T cells were isolated by passing spleen cells through a nylon wool column
(14), and then Ia-positive cells in the column-passed cells were eliminated
by treatment with anti-Ia^k serum and C. B cells were prepared from the
spleen cells by passing them through a Sephadex G-10 column after the treatment with anti-Thy1.2 serum and C. The M∅ population was obtained from the
peritoneal exudate cells of mice by collecting the adherent cells on fetal
calf serum-coated plastic dish. These cell populations were used for cultures in Marbrook type culture vessels. In each experiment, some cell popu-
lation(s) were put in the upper chamber and others were placed in the lower
chamber. Both chambers were separated by a membrane filter (pore size
0.4 μm) which did not allow passing of the cells, but was porous to the
medium. The mitogens were added into both of the chambers, and the cells
were cultured for 48 hours. [³H]TdR was added into the upper chamber 20
hours before the end of the culture, and the [³H]TdR uptakes of the cultured
cells in the upper chamber were examined at the end of the culture.

When T lymphocytes were cultured together with B lymphocytes or M∅ in
the same chamber, [³H]TdR uptakes of the T lymphocytes were obviously higher
(Figure 2). However, if T lymphocytes were separated from B lymphocytes or

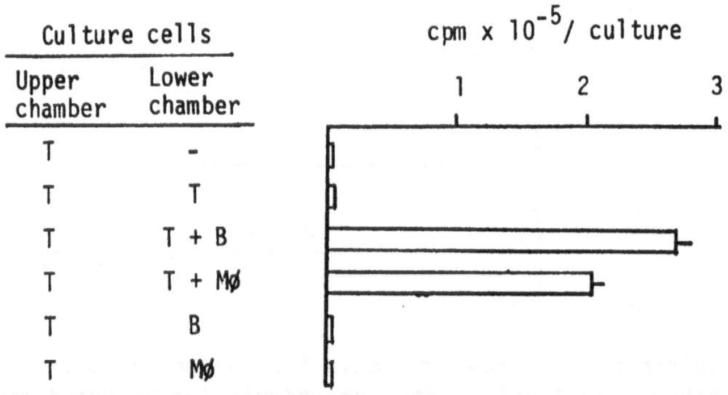

Figure 3. Enhancement of DNA synthesis in T cells by a soluble factor that
is released from cells co-stimulated with Con A and LPS. Purified cell populations (3 x 10^5 T cells, 1 x 10^5 B cells and/or
3 x 10^4 M∅) were cultured separately in each chamber (0.3 ml in
upper chamber and 0.4 ml in lower chamber) of a Marbrook-type
vessel in the presence of LPS (0.1 μg/chamber) and Con A (0.4
μg/chamber for 48 hrs and [³H] TdR uptake by the T cell population in the upper chamber was assessed.

Mφ by the membrane filter during the culture, the proliferative response of
T lymphocytes to the mitogens was very weak. For the proliferative response
of mature T lymphocytes to Con A, Ia-positive accessory cells are necessary
(9,11). The mitogens can act on T lymphocytes only under the condition
where T lymphocytes are in direct contact with Ia-positive accessory cells
such as B lymphocytes or Mφ. The helper function of the accessory cells
disappeared when these cells were treated with anti-Ia serum and C before
the culture.

INVOLVEMENT OF SOLUBLE FACTOR

        Some soluble factor(s) is involved in the proliferative response of T
cells by co-stimulation of LPS and Con A. When T lymphocytes were cultured
in the upper chamber of the Marbrook-vessel separately from T cells and
accessory cells in the lower chamber in the presence of LPS and Con A,
[3H]TdR uptakes of T lymphocytes in the upper chamber increased quite well
(Figure 3). T cells or accessory cells alone in the lower chamber did not
enhance the response of T cells in the upper chamber. These results indicate that T lymphocytes in the upper chamber can proliferate by receiving
some soluble factor which has been produced from the mitogenically stimulated cells in the lower chamber, and the cells stimulated with Con A and
LPS together produce the factor efficiently.

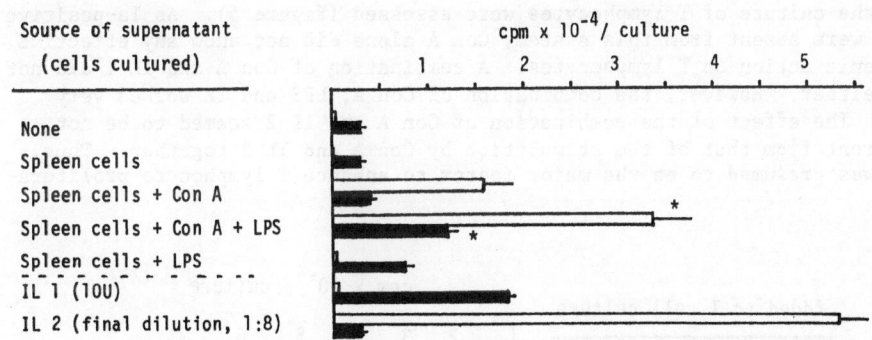

Figure 4. Production of IL 1 and IL 2 from spleen cells by co-stimulation
           of LPS and Con A. Supernatants collected from 48 hr cultures
           (1 x 10^6 spleen cells/ml/dish; Model 3047, Falcon Plastics) with
           or without LPS (0.75 μg/ml) and/or Con A (1 μg/ml) were added to
           cultures (50% V/V) of IL 2-dependent CTLL cells (1 x 10^4 cells/
           0.2 ml/well) for the IL 1 assay (28), and were cultured for 48
           or 72 hrs, respectively. As the control, IL (human IL 1, Genzymo
           Co., or human IL 2, Biotest Serum Institute) instead of the
           supernatants was added to CTLL or CRL-1445 cells. [3H]TdR was
           added 20 hrs before the end of culture, and incorporation into
           the cells were assayed after the end of culture. The activities
           of ILs in the supernatants were expressed as the incorporation
           of [3H]TdR into the cells. Open column: [3H]Tdr uptake of CTLL
           cells (activity of IL 2). Shaded column: [3H]TdR uptake of
           CRL-1445 cells (activity of IL 1). *Values that are signifi-
           cantly different (p < 0.01, Student's t test) from those of
           cultures with Con A or LPS alone.

ENHANCEMENT OF IL PRODUCTION BY CO-STIMULATION OF LPS AND CON A, AND THE

ROLE OF IL ON THE T CELL PROLIFERATION

The results described above indicate some participation of a soluble factor(s) in the synergistic effect of LPS and Con A. IL 1 and IL2 are widely accepted growth factors for T lymphocyte proliferation. IL 2 is known to be one of the lymphokines produced by T cells. When T cells are stimulated by Con A, the cells can produce some amount of IL 2, and the Con A-activated T cells become capable of responding to IL 2, consequently resulting in the proliferation of T cells (31). The IL 2 production of Con A-stimulated T cells depends on the presence of Ia-positive accessory cells (1,16). After Con A stimulation, IL 2 is predominantly produced from the Lyt-1$^+$23$^-$ cell population while there is little production by the Lyt-1$^-$23$^+$ cells (10). IL 1 is also a growth factor for lymphocytes, which is produced by Mϕ (34).

We examined the production of the ILs in the spleen cell cultures in the presence of the mitogens. At the end of culture, supernatants of the cultures were collected, and the amounts of IL 1 and IL2 in the supernatants were assessed by using IL 1-dependent and IL 2-dependent cell lines, respectively. As shown in Figure 4, co-stimulation by LPS and Con A produced significantly higher amounts of ILs, when compared with those of the cultures stimulated by Con A or LPS alone.

If the higher amounts of ILs produced by the co-stimulation result in the overproduction of T lymphocytes, additions of ILs instead of the culture supernatants to T cell cultures must enhance the proliferative response of T cells. To ascertain this, a T cell population was cultured in the presence of the mitogen and commercially available ILs, and [^{3}H]TdR incorporation into the culture of T lymphocytes were assessed (Figure 5). As Ia-positive cells were absent from this system, Con A alone did not show any effective mitogenic action on T lymphocytes. A combination of Con A and IL 1 did not work either. However, the combination of Con A, LPS and IL worked very well. The effect of the combination of Con A and IL 2 seemed to be not different from that of the stimulation by Con A and IL 2 together. Thus, IL 2 was presumed to be the major factor to enhance T lymphocyte prolifera-

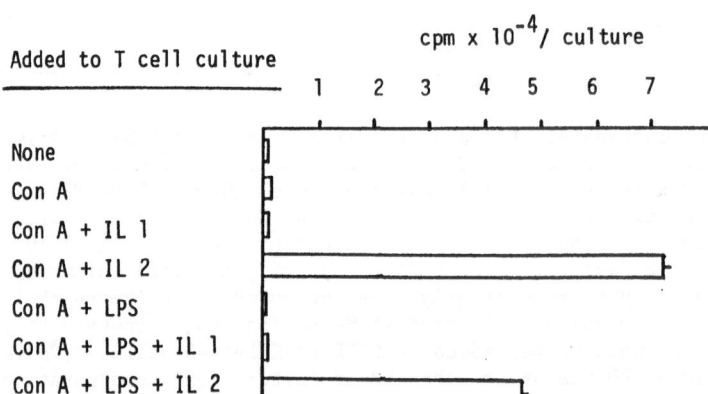

Figure 5. Enhancement of Con A-induced T cell proliferation by IL 2. T cells (3 x 10^5 cells/0.3 ml/well) were cultured with or without Con A (0.4 μg/ml), LPS 0.15 μg/well), IL 1 (10 U/well) or IL 2 (final dilution, 1:8) for 48 hrs, and [^{3}H]TdR uptakes were assessed at the end of culture. IL 2 alone does not enhance the [^{3}H]TdR uptake.

tion in the presence of Con A, but LPS seemed not to be required for the
proliferation of T lymphocytes in the presence of Con A and IL 2.

Taking these results together, we may speculate that the IL 2 produced
from T lymphocytes in support of Ia-positive accessory cells by the co-stimulation of LPS and Con A must play some important role in the synergy. Con
A itself is capable of initiating both IL 2 production from T cells and
proliferation of T cells with the help of Ia-positive accessory cells. An
additional stimulation by LPS on the Con A-activated T cells enhances IL 2
production. Ia-positive accessory cells are also necessary in this first
stage. Then, increase of IL 2 under these circumstances strengthens the
proliferative response of T lymphocytes. In the second stage, LPS and the
accessory cells are not required, but Con A is still necessary for the T
lymphocyte proliferation. Thus, there is a synergistic effect of LPS and
Con A on the proliferative response of lymphocytes.

THE SYNERGY OF LPS AND CON A ON DNA SYNTHESIS OF HUMAN PERIPHERAL

BLOOD LYMPHOCYTES

IL 2-dependent synergy of LPS and Con A can also be seen on the proliferative responses of human peripheral blood T lymphocytes. Human T and B
lymphocyte populations were separated from peripheral blood which has been
obtained from healthy persons according to the methods described in a previous paper (22). The T cells were precultured in culture dishes in the
presence or absence of Con A for 20 hours. After washing, these T cells
were again cultured with B cells in the presence or absence of LPS, and the

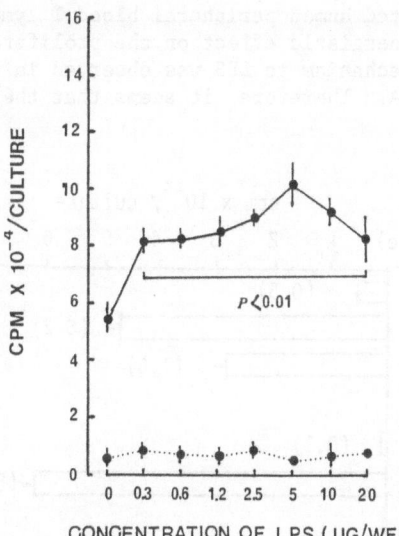

Figure 6. Synergistic effect of LPS on Con A-induced human lymphocyte
proliferation. Mixed cells of a human purified peripheral
blood B lymphocyte population (5 x 10^4) and a Con A-treated
(——) (1 x 10^6 T cells/25 μg Con A/ml at 37°C for 20 hrs) or
untreated T (---) lymphocyte population (1 x 10^5) were cultured (1.5 x 10^5 cells/0.2 ml/well) for 4 days in the presence of various doses of LPS. [^{3}H]TdR was added to the cultures 20 hrs before the end of culture, and the incorporation
into the cells was assessed at the end of culture.

[³H]TdR incorporation into the cells were assessed at the end of culture. As shown in Figure 6, the culture cells that were Con A-treated T cells responded to LPS, while control cells did not. In our culture system, LPS alone does not show any obvious mitogenic effect on both human T cell- and B cell-populations as well as human peripheral blood lymphocytes. The synergistic effect of LPS can be seen in the cultures of Con A-treated T cells and mitomycin C-treated B cells or x-irradiated B cells instead of normal B cells. Therefore, the responsive cells should be T cells. Furthermore, the supernatants obtained from the cultures which had consisted of Con A-treated T cells and B cells with LPS contained a high amount of IL 2 activity (Figure 7), and an addition of exogenous IL 2 into the Con A-treated T lymphocyte cultures resulted in the proliferation of T cells similar to those seen in the cultures having Con A-stimulated T cells, B cells and LPS. These results are quite consistent with those obtained from murine cell cultures.

CONCLUSION

Our results clearly demonstrate that LPS has the ability to enhance the production of T cell growth factor (IL 2) from Con A-stimulated T cells of mice and humans, and the synergy of Con A and LPS on the proliferation of T lymphocytes is presumably related to the production of IL 2. The spleen cells cultured with LPS and Con A produce greater amounts of IL 1 than the control, but the IL 1 seems not to take part directly in the synergy. In mice, this synergy enhances the proliferative response of Lyt 1^+23^- lymphocytes.

We also found that other murine B cell mitogens such as BuWSA (22), dextran sulfate and muramyl dipeptides (23) were capable of enhancing IL 2 production of Con A-stimulated human peripheral blood T lymphocytes, consequently resulting in the synergistic effect on the proliferation of T lymphocytes, and a very similar mechanism to LPS was observed in the combinations of these mitogens with Con A. Therefore, it seems that the enhancement of

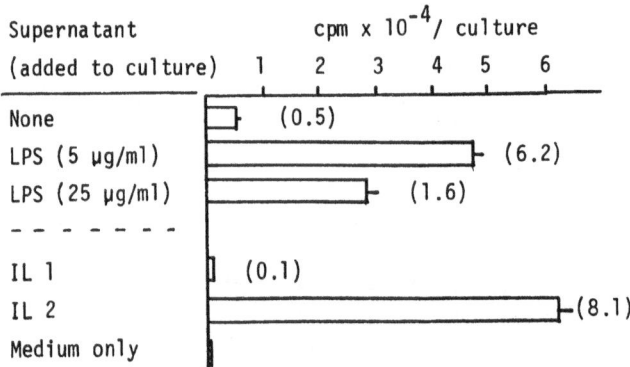

Figure 7. IL 2 activity in the supernatants of human mixed cell cultures containing B cell and Con A-treated cell populations which have been stimulated by LPS. Supernatants were obtained from the mixed cell cultures with B cells and Con A-treated T cells which had been cultured in the presence of LPS for 4 days. The IL 2 activity in the supernatants were estimated by their ability to enhance DNA synthesis in IL 2 dependent CTLL cells. The numbers in parenthesis shoe the calculated units of IL 2 activity in the supernatants, which were made by comparing measured activity to a standard human IL 2 preparation.

IL 2 production by co-stimulation of B cell mitogen and T cell mitogen and the increased DNA synthesis in mitogen-activated T lymphocytes via IL 2 are very common phenomena.

REFERENCES

1. J. Andersson, K. O. Gronvik, E. L. Larsson, and A. Coutinho, Studies on T lymphocyte activation. I. Requirements for the mitogen-dependent production of T cell growth factors, Eur. J. Immunol. 9:581 (1979).

2. H. Cantor and E. A. Boyse, Functional subclasses of T lymphocytes bearing different Ly antigens. I. The generation of functionally distinct T cell subclasses is a differentiative process independent of antigen, J. Exp. Med. 141:1376 (1975).

3. H. N. Claman, Corticosteroids and lymphoid cells, N. Eng. J. Med. 287 (1972).

4. G. A. Currie, Activated macrophages kill tumor cells by releasing arginase, Nature 273:758 (1978).

5. P. J. Edelson, R. Zwiebel, and Z. A. Cohn, The pinocytic rate of activated macrophages, J. Exp. Med. 142:1150 (1975).

6. C. G. Fathman, M. Small, L. A. Herzenberg, and I. L. Weissman, Thymus cell maturation. II. Differentiation of three "mature" subclasses in vivo, Cell. Immunol. 15:109 (1975).

7. J. T. Forbes, Y. Nakao, and R. Smith, T mitogens trigger LPS responsiveness in mouse thymus cells, J. Immunol. 114:1004 (1975).

8. Y. Fradet, R. Roy, and F. Daguillard, Regulation of in vitro lymphocyte responses. I. Adjuvant effect of lipopolyssacharide (LPS) on low-zone unresponsiveness to concanavalin A, Cell. Immunol. 27:94 (1976).

9. J. A. Frelinger, Ia-bearing cells promote the concanavalin A mitogenic response of Ia-negative T cells, Eur. J. Immunol. 7:447 (1977).

10. M. Gullberg and E. Larsson, Con A-induced TCGF-reactivity is selectively acquired by Lyt-2-positive T cell precursors, J. Immunol. 131:19 (1983).

11. S. Habu and M. C. Raff, Accessory cell dependence of lectin-induced proliferation of mouse T lymphocytes, Eur. J. Immunol. 7:451 (1977).

12. S. M. Hatfield and J. K. Schmidtke, Synergistic effects of lipopolysaccharide on phytohemagglutinin- and concanavalin A-induced deoxyribonucleic acid synthesis in human peripheral blood lymphocytes: participation of T lymphocytes, Infect. Immun. 31:1007 (1981).

13. G. Janossy, M. F. Greaves, M. J. Doenhoff, and J. Snajdr, Lymphocyte activation. V. Quantitation of the proliferative response to mitogen using defined T and B cell population, Clin. Exp. Immunol. 14:581 (1973).

14. M. Julius, E. Simpson and L. A. Hertzenberg, A rapid method for the isolation of functional thymus-derived murine lymphocytes, Eur. J. Immunol. 3:645 (1973).

15. M. Kaku, K. Yagawa, S. Nagao, and A. Tanaka, Enhanced superoxide anion release from phagocytes by muramyl dipeptide or lipopolysaccharide, Infect. Immun. 39:559 (1983).

16. E. L. Larsson and A. Coutinho, The role of mitogenic lectins in T-cell triggering, Nature 280:239 (1979).

17. R. A. Miller, S. Gartner, and H. S. Kaplan, Stimulation of mitogenic responses in human peripheral blood lymphocytes by lipopolysaccharide: serum and T helper cell requirements, J. Immunol. 121:2160 (1978).

18. S. B. Mizel, Interleukin I and T cell activation, Immunol. Rev. 63:51 (1982).

19. B. Mørland, Studies on selective induction of lysosomal enzyme activities of mouse peritoneal macrophages, J. Reticuloendothel. Soc. 26:749 (1979).

20. B. Mørland and G. Kaplan, Macrophage activation in vivo and in vitro, Exp. Cell Res. 108:279 (1977).

21. D. C. Morrison and J. L. Ryan, Bacterial endotoxins and host immune responses, Adv. Immunol. 28:293 (1979).

22. T. Nitta, S. Okumura, and M. Nakano, Synergistic effect of concanavalin A and Bu-WSA on DNA synthesis in human peripheral blood lymphocytes, J. Immunol. 134:808 (1985).

23. T. Nitta, H. Konno-Egiri, S. Okumura, A. Ozawa, and M. Nakano, Role of Interleukin 2 on enhancement of concanavalin A-induced human peripheral blood lymphocyte proliferation by murine B cell mitogen, Microbiol. Immunol. 30 (in press, 1985).

24. K. Ozato, W. H. Alder, and J. D. Ebert, Synergism of bacterial lipopoly-saccharides and concanavalin A in the activation of thymic lympho-cytes, Cell. Immunol. 17:532 (1975).

25. M. J. Pabst and R. B. Johnston, Jr., Increased production of superoxide anion by macrophages exposed in vitro to muramyl dipeptide or lipo-polysaccharide, J. Exp. Med. 151:101 (1980).

26. D. L. Peavy, W. H. Alder, and R. T. Smith, The mitogenic effects of enterotoxin and staphylococcal enterotoxin B on mouse spleen cells and human peripheral lymphocytes, J. Immunol. 105:1453 (1970).

27. M. Sasada and R. B. Johnston, Jr., Macrophage microbicidal activity. Correlation between phagocytosis-associated oxidative metabolism and the killing of Candida by macrophages, J. Exp. Med. 152:85 (1980).

28. J. A. Schmidt, S. B. Mizel, D. Cohen, and G. Ira, Interleukin 1, a potential regulator of fibroblast proliferation, J. Immunol. 128:2177 (1982).

29. J. R. Schmidtke and J. S. Najarian, Synergistic effects on DNA synthesis of phytohemagglutinin or concanavalin A and lipopolysaccharide in human peripheral blood lymphocytes, J. Immunol. 114:742 (1975).

30. R. D. Schubert and J. R. David, Stimulation of guinea pig macrophage pinocytosis by lipopolysaccharides (LPS): evidence that LPS acts directly on the macrophage, Cell. Immunol. 55:166 (1980).

31. K. A. Smith, T cell growth factor, Immunol. Rev. 51:337 (1980).

32. M. Tanabe and M. Nakano, Lipopolysaccharide-induced mediators assisting the proliferative response of C3H/HeJ thymocytes to concanavalin A, Microbiol. Immunol. 23:1097 (1979).

33. D. Taramelli and L. Varesio, Activation of murine macrophages. I. Dif-ferent pattern of activation by Poly I:C than by lymphokine or LPS, J. Immunol. 127:58 (1981).

34. E. R. Unanue, The regulatory role of macrophages in antigenic stimula-tion. Part Two: symbiotic relationship between lymphocytes and macrophages, Adv. Immunol. 31:1 (1981).

35. S. N. Vogel, M. L. Hilfiker, and M. J. Caulfield, Endotoxin-induced T lymphocyte proliferation, J. Immunol. 130:1774 (1983).

36. L. M. Wahl, S. M. Wahl, S. E. Mergenhagen, and G. R. Martin, Collagenase production by endotoxin-activated macrophages, Proc. Natl. Acad. Sci. U.S.A. 71:3598 (1974).

37. I. L. Weissman, Thymus cell maturation. Studies on the origin of cortisone-resistant thymic lymphocytes, J. Exp. Med. 139:504 (1973).

38. O. Westphal and O. Luderitz, Chemische Erforschung von Lipopolysacchari-den gram-negativer Bakterien, Angew. Chem. 66:407 (1954).

BIOCHEMICAL, IMMUNOLOGICAL AND FUNCTIONAL ANALYSIS OF LYMPHOCYTES

FROM THE LPS-NON-RESPONDER C3H/HeJ MOUSE

D. C. Morrison, H-W. Wollenweber, S. W. Vukajlovich, and
S. A. Goodman

University of Kansas Medical Center
Microbiology Department
39th and Rainbow Boulevard
Kansas City, Kansas

INTRODUCTION

Since its discovery in 1968 by Sultzer (1) as a mutant mouse strain
which displayed aberrant peritoneal inflammatory cell responses to endotoxin,
the C3H/HeJ mouse has served as one of the dominant experimental models by
which to define both in vivo and in vitro the mode of action of endotoxin.
Extensive studies have documented that the phenotypic characteristic of
endotoxin unresponsiveness is specific for the LPS fraction of endotoxin,
and more specifically, the lipid A component (reviewed in 2). Further, the
available evidence would suggest that the defect responsible for LPS/lipid
A unresponsiveness may be a characteristic of all cells derived from this
mouse in that lymphocytes, macrophages and fibroblasts are all refractory
to LPS stimulation (3). Genetic evidence has clearly linked this mutation
to a locus on chromosome 4. This locus appears to be codominantly expressed
and LPS unresponsiveness is inherited as a single gene trait in the appropriate F2 backcrosses (5).

In spite of rather intensive investigation, however, the gene product
responsible for LPS unresponsiveness in B-lymphocytes and other cells from
this mouse strain has remained elusive. Recent evidence has pointed to a
functional defect in C3H/HeJ B-lymphocytes which is demonstrable at the
cytoplasmic membrane (6). Several investigators have reported altered LPS
binding characteristics of lymphocytes from C3H/HeJ mice in comparison to
C3H congenic responder strains (7,8); however, comprehensive studies in our
own laboratory (9) as well as other laboratories (10,11) have not confirmed
these results. Evidence has also been presented to suggest the presence of
an immunologically defined surface antigen which was present on lymphocytes
from LPS-responder strains but absent from lymphocytes of LPS-non-responder
mice (12,13). These studies suggested the presence of specific LPS receptors
on murine B-lymphocytes; however, they have yet to be confirmed by other
laboratories and the available published data (14) are not in agreement
with these earlier studies.

As a consequence, the biochemical basis for the LPS unresponsiveness
of cells from the C3H/HeJ mouse remains undefined. In the studies to be
reported here, we have employed a variety of experimental approaches to

address this important question. First we carried out experiments to define biochemically those molecules on the lymphocyte cytoplasmic membrane which interact with LPS, in order to explore potential differences between C3H/HeJ lymphocytes and those from congenic C3H responder lymphocytes. For these studies we have synthesized an LPS derivative possessing a radiolabelled photoactivatable and cleavable cross-linking group. A second approach has been to investigate immunologic differences between C3H congenic responder and non-responder lymphocytes using adoptive transfer of viable lymphocytes. Immunologic rejection of transplanted cells has been employed as an assay system to detect potential recognition of "LPS receptors" and results have been compared with equivalent rejection of cells differing at a minor (H-Y) histocompatibility locus. Finally, we have assessed the relative capacity of LPS non-responder lymphocytes to proliferate in vitro in response to a variety of LPS preparations with restricted LPS subunit heterogeneity. These experiments have been predicated upon our published data (15) showing that C3H/HeJ lymphocytes manifest relatively normal responses to a subfraction of LPS containing subunits with limited O-antigen polysaccharide.

The results of these combined experimental approaches do not lend support to the concept that lymphocytes from the LPS-non-responder C3H/HeJ mouse differ from their congenic LPS-responder counterparts by virtue of a dominant high affinity membrane localized binding molecule (LPS receptor). Our evidence would suggest that the major binding proteins/glycoproteins on the surface of responder and non-responder lymphocytes are virtually indistinguishable. Further, if an antigenic difference does exist in these various congenic mouse strains, it is less immunogenic than a minor histocompatibility antigen. Finally our demonstrated ability to elicit high levels of proliferation in lymphocyte cultures from C3H/HeJ mice with a variety of protein free R-LPS chemotypes would suggest that these cells can, in fact, respond to LPS. These combined data, therefore, would not be consistent with the concept of the C3H/HeJ mutational defect as manifest in the absence of expression of specific LPS receptors on B-lymphocytes and other LPS responsive cells. Rather these data point to a defective "triggering" signal which occurs subsequent to the binding of LPS to appropriate target molecules on the B-lymphocyte surface.

MATERIALS AND METHODS

Lipopolysaccharides

LPS was prepared from E. coli K 235 by the phenol water extraction procedure of McIntyre et al. (16) as modified by Skidmore et al. (17). LPS was also extracted from E. coli 0111:B4 and E. coli 055:B5 by the phenol water procedure described by Westphal et al. (18) and further purified by digestion with RNase and pronase followed by gel filtration chromatography according to Morrison and Leive (19). For the photoaffinity experiments reported here, the lower molecular weight LPS II fraction separated from E. coli 0111:B4 by chromatography was used. The LPS from S. minnesota R595 was purified by the phenol-chloroform-petroleum ether procedure of Galanos et al. (20). Various S. minnesota R-chemotype LPS preparations were purchased from List Biological Labs, Inc. (Campbell, California) and were stated by the manufacturer to contain less than two percent of protein by weight.

Photoactivatable Iodinated LPS

The LPS from E. coli 0111:B4 was employed to prepare a cross-linking LPS probe using sulfosuccinimidyl-2-(p-azidosalicidylamido)-1,3'dithiopropionate (SASD, Pierce Chemical Co., Rockford, IL). Briefly, 1.0 mg LPS was incubated at room temperature in borate buffer with 400 μg SASD and then dialyzed extensively against phosphate buffered saline. The derivatized

LPS was subsequently radiolabelled with ^{125}I using a modification of procedures developed by Ulevitch (21). We have shown (Wollenweber and Morrison, submitted for publication) that the LPS is derivatized with SASD primarily at the phosphorylethanolamine residues of the LPS and that the resulting LPS-ASD-^{125}I product has a specific activity of approximately 3 µCi/µg LPS.

Analytical Polyacrylamide Gel Electrophoresis (PAGE)

Electrophoresis was carried out essentially as described by Laemmli (22) using 10-20% gradient gels with a five percent stacking gel. Electrophoresis buffer was 62.5 mM tris, pH 6.8 containing three percent SDS, 4 M urea and ten percent glycerol. In some instances the reducing agent dithioerythritol (0.1 M) was added. Proteins were stained using Coomassie Blue, dried on a gel slab dryer (Biorad Labs, Richmond, California) and autographed using Kodak X-Omat XAR-5 or XKI film (Eastman Kodak, Rochester, New York) and Dupont Cronex Lighting Plus Intensifier screens.

Preparative Polyacrylamide Gel Electrophoresis

Techniques have been developed for the purification of R-chemotype LPS preparations by electrophoresis. Because of difficulties in quantitatively removing SDS from LPS samples, all electrophoresis was carried out in the presence of 0.30% taurodeoxycholate since it has been shown by us (15) and others (23) that intrinsic LPS activity can be fully reconstituted by removal of deoxycholate following electrophoresis. Approximately 2.0 ml of LPS solution (2.5 mg/ml) in sample buffer was layered onto 4.5 mm 22% polyacrylamide slab gels and electrophoresis carried out at 100 volts (approximately 10 mA) for three hours at 4°C. Gels were then sliced horizontally into 5 mm sections and LPS eluted into tris buffer, pH 7.8 overnight at 4°C. Individually fractions were then assayed for the presence of LPS by determination of 2-keto-3-deoxyoctulosonate (24) and LPS containing fractions dialyzed extensively against distilled water and lyophylized.

Mice

Both male and female mice of the C3HeB/FeJ (LPS-responder) and C3H/HeJ (LPS-non-responder) strains were purchased from Jackson Labs (Bar Harbor, Maine) at four to six weeks of age and used for experiments between five to 16 weeks of age. In some experiments, a second LPS non-responder strain C57Bl/10ScN (Nu/nu) and (Nu/+) and the congenic responder C57Bl/10SN were used. Breeding pairs of the former were generously provided by Dr. Carl Hanson (NIH, Bethesda, Maryland) and were bred in our laboratory facilities. Mice of the latter strain were purchased from Jackson Labs.

Lymphocytes

Single cell suspensions of splenocytes were prepared as described earlier (25). For proliferation assays, cells were suspended in RPMI 1640 supplemented with glutamine and penicillin/streptomycin (see below). For adoptive transfer experiments, cells were suspended in Hanks basal salts solution. Splenocytes free of T-lymphocytes were prepared by treatment with anti-thy 1.2 and rabbit low-tox complement two times as described (26). Purified T-lymphocytes were prepared by adherence to nylon wool columns (27). Purification was assessed by proliferative responses to LPS and Con A respectively.

Proliferative Assays

To assess lymphocyte proliferation to LPS, 200 µl aliquots of splenocyte suspensions (2.5 x 10[6]/ml) were pipetted into individual wells of 96 well Limbro flat-bottom culture plates and dilutions of LPS added to triplicate wells. After 24 hours 0.5 µCi of ^{3}H thymidine (40-60 µCi/mM, Amersham Searle, Arlington Heights, Illinois) was added to each well. After an additional 16-20 hours, cells were harvested on a Mash II (Microbiological Associates, Walkersville, Maryland) and counted in a Rack beta 1216 liquid scintillation counter (I.KB Instruments, Rockville, Maryland).

RESULTS

Immunologic Determination of Differences Between C3HeB/FeJ and C3H/HeJ Lymphocytes

It has been postulated, and there exists published evidence to support the concept that responder and non-responder mice may be distinguished immunologically on the basis of a surface antigen tentatively identified as an LPS receptor (12,13) present in B lymphocytes of the former but not the latter. We have adopted a strictly congenic immunization regimen to attempt to detect the presence of this receptor using an experimental protocol which capitalizes on the only generally accepted difference between C3H/HeJ and other C3H substrains; namely the phenotypic capacity of lymphocytes from the latter strains to respond to LPS in vitro. Based upon this difference, we have designed experiments to adoptively transfer viable lymphocytes from C3H/FeJ (LPS-responder) mice into immunologically intact C3H/HeJ mice. Using a hyperimmunization regimen in which cells were transferred at weekly intervals, we subsequently assessed whether such cells were accepted or rejected by the C3H/HeJ host, by determinations of in vitro LPS proliferative capacity of splenocytes from control or immunized C3H/HeJ mice. As controls for these experiments, we took advantage of the known antigenic difference

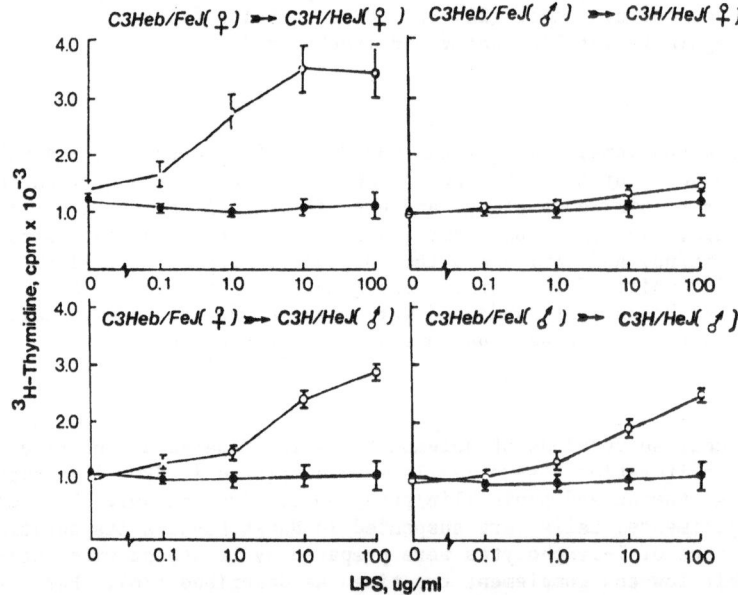

Figure 1. Adoptive transfer of responder C3HeB/FeJ splenocytes into C3H/HeJ recipients. Approximately 2 x 10⁷ viable splenocytes from either male or female C3HeB/FeJ mice were injected intraperitoneally at weekly intervals into male and female recipients. Five days after the fourth injection, recipient splenocytes were assayed in vitro for proliferative responses to protein free E. coli K235 LPS. Results are expressed as the average of quadruplicate determinations + SEM.

defined by the sex-linked minor histocompatibility locus (H-Y), which has been shown to evoke immune responses, even under otherwise strictly syngeneic situations (28).

For these experiments, donor C3HeB/FeJ lymphocytes of either male or female origin were prepared and injected intraperitoneally (2×10^7 viable cells) into either male or female C3H/HeJ recipients. Immunizations were carried out at weekly intervals using fresh splenocytes for a period of five weeks. Five days after the last immunization, splenocytes were harvested and assayed in vitro for their capacity to proliferate in response to LPS. The results of one such experiment are shown in Figure 1 and illustrate several significant points. First, the adoptive transfer of female C3HeB/FeJ lymphocytes into female C3H/HeJ recipients results in a highly reproducible and significant detection of LPS-responder lymphocytes. Although the stimulation index is only about three to four fold, we have shown by the appropriate in vitro mixing experiments using C3HeB/FeJ and C3H/HeJ splenocytes that this response is equivalent to that which one obtains when between five and ten percent of the lymphocytes are of responder origin. A second major point is that qualitatively similar mitogenic responses can be detected when female or male C3HeB/FeJ cells are adoptively transferred into C3H/HeJ male recipients. It is noteworthy that the proliferative response of splenocytes from male recipients is routinely somewhat lower, although always significantly different than background, than female recipients. The reason for this is unknown. Of importance, however, is the LPS response observed when male C3HeB/FeJ cells are adoptively transferred into female recipients. Under these conditions, the LPS responses are indistinguishable from background, suggesting that immunologic differences at the H-Y minor histocompatibility locus are sufficient to mediate graft rejection.

We have considered the possibility that the hyperimmunization regimen employed does not favor the induction of immunity against the putative LPS receptor. However, we have determined that the observed presence of LPS responder cells in C3H/HeJ recipients is remarkably stable, and no loss of activity is observed as late as two months (29) following the last adoptive transfer. To address the question of possible induction of T-suppressor cells, we have repeated the experiments described above for female C3HeB/FeJ cells into female C3H/HeJ recipients in the presence of cyclophosphamide. In these latter experiments, mice were injected intraperitoneally with 500 mg/kg of cyclophosphamide (30) three days prior to each adoptive transfer of viable splenocytes, and again assayed five days after the last immunization. The results of this experiment are shown in Figure 2 and suggest

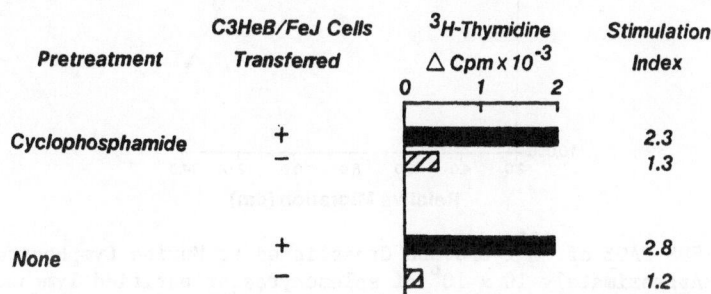

Figure 2. Effect of cyclophosphamide on adoptive transfer of female C3HeB/FeJ splenocytes into C3H/HeJ female recipients. Adoptive transfers were carried out exactly as described in the legend to Figure 1 except that recipients were treated intraperitoneally with cyclophosphamide (500 mg/kg) three days prior to each injection with splenocytes.

that the observed lack of rejection of LPS-responder lymphocytes by LPS-non-responder mice may not be due to the presence of T-suppressor cell induction. These combined data, therefore, would suggest that, if a biochemical difference does exist between lymphocytes responsive to LPS and those refractory to LPS, and which serves the function of an LPS receptor, that difference is less immunogenic than a minor histocompatibility antigen.

Characterization of LPS Binding Sites on Lymphocyte Membranes from C3HeB/FeJ and C3H/HeJ Mice

As a second experimental approach to explore differences between LPS-responder and non-responder mice we have employed a photoactivatable radio-labelled LPS probe to determine the biochemical characteristics of lymphocyte antigens which bind with relative selectivity to LPS in in vitro culture. For these experiments, 10×10^6 spleen cells from a variety of mouse strains were incubated in 100 μl of medium with the photoactivatable radiolabeled LPS (LPS-ASD-^{125}I) for 30 minutes at 37°C. Cells incubated in this experiment were splenocytes from C3HeB/FeJ responder and C3H/HeJ non-responder mice, C3HeB/FeJ responder LPS-hyperimmune mice, C57Bl/10SN responder and C57Bl/10ScN non-responder mice, C57Bl/10ScN athymic (Nu/nu) mice as well as purified B and T lymphocytes from C3HeB/FeJ mice. Following incubation, cells and LPS were irradiated with short wave length UV light (4 watts max emission at 254 nm) to effect covalent cross-linking of LPS to targets on the lymphocyte surface. Cells were then washed three times by centrifugation, and resuspended in electrophoresis running buffer containing dithio-

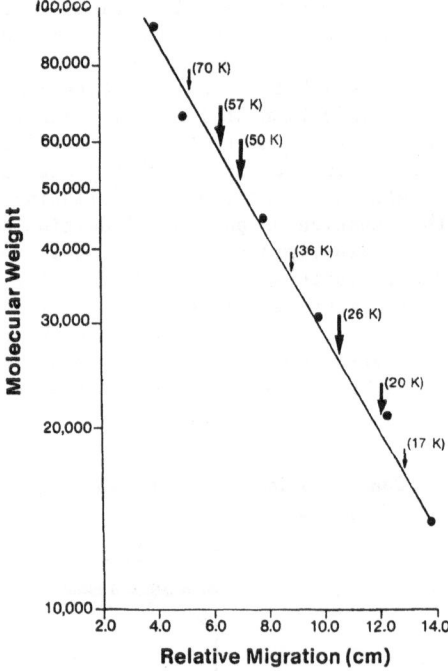

Figure 3a. SDS PAGE of ^{125}I-LPS-ASD Crosslinked to Murine Lymphocytes. Approximately 10×10^6 of splenocytes or purified lymphocyte subpopulations were incubated with ^{125}I-LPS-ASD and then UV irradiated as described in the text. Aliquots of cell lysates were electrophoresed on 10-20% gels, dried and autoradiographed. Molecular weight standards were obtained from Biorad Corp., Richmond, CA and included phosphorylase B (92K) bovine serum albumin (66K) ovalbumin (45K) carbonic anhydrase (31K) soybean trypsin inhibitor (21K) and lysozyme (14K).

erythritol (DTE). Aliquots were then electrophoresed as described and autoradiographed.

The results of one such experiment are shown in Figure 3a and 3b and indicate several relevant findings. First there appear to be three major LPS binding proteins in splenocyte preparations which are resolved by electrophoresis with approximate molecular weights of 57K, 50K, and 26K. In addition there appear to be four minor proteins with molecular weights of approximately 70K, 36K, 20K and 17K. While it has not been possible to resolve whether the major band appearing at 57K is distinct from immunoglobulin heavy chains, we can state with reasonable certainty on the basis of immunoblot analysis (data not shown) that the band appearing at 26K is clearly resolved from light chains which, in this gel system, migrate with an apparent molecular weight of 28K.

Of significance, however, is our observation that there is at least a partial selectivity in the lymphocyte surface antigens which appear to bind to LPS. We have shown that these same banding patterns are observed if lymphocytes are pretreated with NP-40 to solubilize selectively membrane localized antigens. Further, the absence of significant differences in the LPS binding profiles of LPS-hyperimmune vs naive C3HeB/FeJ splenocytes would suggest strongly that the majority of the binding is not to immunoglobulin receptors on antigen specific B-cells. Finally with the exception of a lack of detection of the 26K protein in purified T cells, there are no qualitative differences between binding of LPS to T and B lymphocyte.

Perhaps one of the most significant results of these experiments, however, is the fact that no obvious differences exist between the binding of LPS to lymphocytes from mice which can respond to LPS and those that cannot (compare lanes a vs. b and f vs. g). We have, in addition, shown that

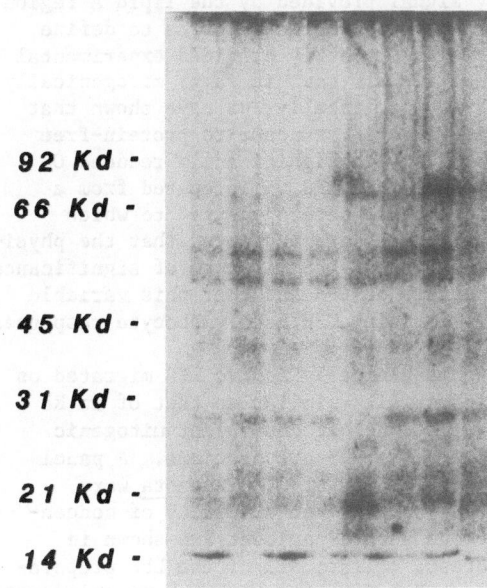

Lymphocyte Source:

1. C3Heb/FeJ
2. C3H/HeJ
3. C3Heb/FeJ LPS Immune
4. C3Heb/FeJ B-cells
5. C3Heb/FeJ T-cells
6. C57Bl/10Sn
7. C57Bl/10ScN
8. C57Bl/10ScN (Nu/nu)

92 Kd -
66 Kd -
45 Kd -
31 Kd -
21 Kd -
14 Kd -

1 2 3 4 5 6 7 8

Figure 3b. Determination of apparent molecular weights of lymphocyte LPS binding proteins. Data were obtained from the relative mobilities of the proteins shown in Figure 3a.

Table 1. Splenocyte Proliferative Response to LPS Preparations

LPS Preparation		C3HeB/FeJ	C3H/HeJ	C57Bl/10ScN
S. minnesota	w.t.	18.8	1.1	1.4
"	Ra	40.1	19.1	21.6
"	Rb_2	38.2	17.4	13.9
"	RcP	33.5	8.1	9.2
"	Rd_1P^-	39.1	18.4	24.3
"	Re	24.2	4.0	6.3
"	lipid A	12.0	1.5	3.1
none	(bkgd)	1.1	0.8	0.7

[1]5×10^5 splenocytes were cultured in 200 µl of medium containing 5 µg/ml of the indicated LPS preparatiave for 48 hrs. ^{3}H-thymidine was added (0.5 Ci/culture) for the last 6 hrs.

neither a kinetic difference in binding profiles nor a change in the concentration of LPS relative to cells will result in detectable differences between responder and non-responder lymphocytes although both of these variables contribute to some extent to the precise distribution of LPS binding targets as assessed by SDS-PAGE analysis (Wollenweber and Morrison, manuscript in preparation).

Demonstration of LPS-Dependent Mitogenic Responses of C3H/HeJ Lymphocytes

As discussed above, the refractory state of B lymphocytes from the C3H/HeJ mouse to the immunostimulatory signal provided by the lipid A region of LPS serves as one of the prototypic in vitro characteristics to define this mutant mouse strain. We have, however, recently provided experimental evidence to suggest that C3H/HeJ lymphocytes are not, in fact, mitogenically unresponsive to all preparations of LPS. Specifically, we have shown that C3H/HeJ lymphocytes will proliferate in vitro in response to protein-free macromolecular LPS composed of LPS subunits with significantly reduced O-antigen polysaccharide content (15). Since this LPS was prepared from a wild type E. coli LPS with considerable subunit heterogeneity, to which C3H/HeJ lymphocytes were not responsive, these data suggested that the physical chemical state and/or subunit composition of LPS might be of significance in defining the immunostimulatory properties of LPS and that this variable might be of more critical importance in defining C3H/HeJ lymphocyte responses.

It is noteworthy that the C3H/HeJ mitogenically active LPS migrated on SDS-PAGE with a subunit electrophoretic mobility similar to that of an Ra chemotype LPS (15). We, therefore, queried whether equivalent mitogenic activity might be present in other R-chemotype LPS preparations. A panel of LPS preparations from various R-mutant strains of S. minnesota were, therefore, examined for immunostimulatory activity at a variety of concentrations. A portion of the results of one such experiment are shown in Table 1 and several points are of interest. First, all of the LPS preparations tested including lipid A are potent B-lymphocyte mitogens in splenocyte cultures from the LPS responder strain C3HeB/FeJ. Similarly, as has been shown earlier by this and numerous other laboratories, C3H/HeJ lymphocytes are refractory to the mitogenic activity of wild type LPS and lipid A. Of interest, however, is the fact that induction of lymphocyte proliferation in spleen cell cultures of C3H/HeJ mice can be effected by most of the R-chemotype LPS preparations tested, including some activity with the Re-LPS.

No response above background, however, is detectable with isolated lipid A. Although the results are shown only for a single concentration of LPS, qualitatively similar results were obtained over the entire dose-response range of LPS. That this response pattern to R-chemotype LPS preparations is neither unique to the C3H/HeJ mouse nor the result of non-B lymphocyte mitogenic responses was suggested by the fact that virtually identical results are obtained with splenocytes from a second LPS-non-responder C57Bl/10ScN strain even on an athymic (Nu/nu) background. Further evidence that this observed response is not unique to S. minnesota LPS is suggested by the fact that high levels of mitogenic activity may also be detected with R-chemotype LPS from E. coli J-5 LPS, S. thyphimurium Rb and Re LPS and Yersinia pseudotuberculosis LPS (data not shown). It would thus appear that C3H/HeJ lymphocytes are, in general, not refractory to stimulation with LPS derived from the appropriate bacterial source.

It is, of course, well recognized that any interpretation of the data presented above must take into consideration potential contributions of components other than LPS to the observed mitogenic responses. In this respect, earlier experiments from this laboratory (31) clearly documented that the mitogenic activity in C3H/HeJ lymphocytes of many LPS preparations was due to the presence of contaminating protein, which we have termed LAP, lipid A associated protein (25). Firm conclusions on the intrinsic mitogenic activity of R-chemotype LPS preparations summarized in Table 1, therefore, require that potential contributions of LAP be unequivocally excluded. We have determined (15) by direct amino acid analysis that the C3H/HeJ mitogenic subcomponent derived from E. coli 055 LPS contains less than 0.1 percent by weight of protein and that the amino acids detected do not correlate with the amino acid distributions for either LAP or the outer membrane lipoprotein.

While the evidence presented above would support the concept that the C3H/HeJ mitogenic response elicited by at least one LPS preparation is independent of the presence of protein, such a conclusion is more tentative with the R-LPS chemotypes. We have, therefore, developed techniques to isolate R-LPS by preparative gel electrophoresis in deoxycholate. These procedures have been premised upon the fact that detergent treatment of LPS dissociates these macromolecules into their constituent monomeric subunits which have a molecular weight (3-4,000) considerably less than any of the contaminating LAP (18,000 - 70,000) (32). Thus, both Ra-LPS and Re-LPS preparations were dissociated in deoxycholate and electrophoresed as described in Materials and Methods. Following elution from the gels, individual fractions were assayed for KDO and LPS containing fractions dialyzed extensively to remove detergent and reform high molecular weight LPS aggregates. Suitable dilutions were then assayed for mitogenic activity in both C3HeB/FeJ and C3H/HeJ splenocytes. The results of these studies are summarized in Figure 4 and indicate clearly that mitogenic activity coelectrophoreses with the KDO-containing LPS fractions in both LPS responder and non-responder mice. These data provide strong evidence, therefore, that the observed mitogenic activity of R-LPS chemotype is actually due to LPS and not to some non-LPS contaminant.

DISCUSSION

Extensive experiments have been carried out to examine in detail the antigenic and biochemical basis for the genetic defect in the C3H/HeJ mouse which results in a phenotypic functional unresponsiveness of B-lymphocytes to mitogenic stimulation by LPS. On the basis of intensive immunizations of C3H/HeJ mice with congenic LPS responder lymphocytes, we conclude that, if this mutation results in an altered gene product expressed on the B-cell surface, it is significantly less immunogenic than a minor histocompatibility antigen. These results are in full agreement with biochemical studies using

a photoactivatable radiolabelled and cleavable LPS probe to detect differences in LPS binding sites on responder and non-responder lymphocytes. As
assessed by polyacrylamide gel electrophoresis of LPS binding targets, indistinguishable patterns were obtained for C3HeB/FeJ and C3H/HeJ lymhocytes.
These results are not consistent with the requirement for LPS induced mitogenic responses as being due to the presence or absence of a high affinity
LPS binding glycoprotein (LPS receptor). Finally we have presented data
which confirms that C3H/HeJ lymphocytes are not refractory to a wide variety
of LPS preparations with R-chemotype structures. We have documented that
such activity is not due to the presence of contaminating lipid A-associated
protein and this represents a true LPS-induced proliferative signal. Thse
combined data would provide strong support for the concept that the genetic
defect in the C3H/HeJ mouse which is manifest in an inability to respond
mitogenically to some preparations of LPS may not be due to the absence of
a specific cytoplasmic membrane localized high affinity binding site or LPS
receptor.

It is clear from our experiments that LPS binding proteins and/or glycoproteins do exist on the surface of lymphocytes and that, at least one of
these proteins is different in B and T lymphocytes. These results confirm
and extend earlier studies by Yokoyama et al. (34) who defined LPS binding

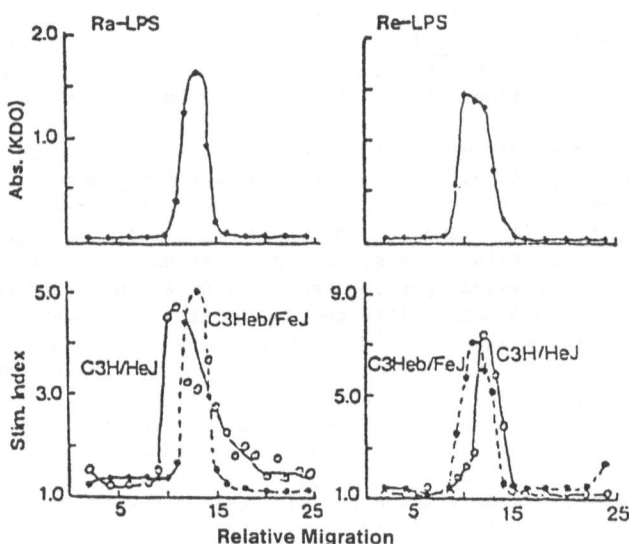

Figure 4.   Preparative polyacrylamide gel electrophoresis of R-chemotype
            LPS preparations. LPS from S. minnesota R60 (Ra) and R595 (Re)
            were solubilized in 0.3% taurodeoxycholate and electrophoresed
            on 22% polyacrylamide gels as described in Materials and
            Methods. Following electrophoresis, the slab gels were sec-
            tioned into 0.5 cm slices, the LPS eluted into tris buffer and
            dialyzed extensively first against tris buffer and then against
            distilled water. Aliquots of 100 μl were assayed for the pres-
            ence of KDO and serial ten-fold dilutions assayed in triplicate
            for mitogenic activity in C3HeB/FeJ and C3H/HeJ splenocytes.
            The results are presented as the average of triplicate deter-
            minations which, in general varied by less than 10%. The data
            are presented for a 1:100 dilution of LPS. Stimulation indices
            were calculated by dividing the experimental values by the con-
            trol values which were approximately 1,500 and 900 cpm for
            C3HeB/FeJ and C3H/HeJ splenocytes, respectively.

proteins on lymphocytes on the basis of affinity chromatography. The identity of these molecules remains to be defined as does the contribution of these molecules to the proliferation signal delivered to the B-lymphocyte by the LPS macromolecule. In the former respect, we cannot exclude the possibility that the LPS binding 70K and 57K molecules may be heavy chains. However, our inability to detect crosslinking to light chains would indicate that, if this does represent LPS binding to surface Ig, it is most likely heavy chain specific and not to antigen receptors. This latter conclusion is supported by studies with splenocytes from LPS hyperimmune mice. There is, in fact, good evidence for Fc dependent non-specific interactions between the lipid A region of LPS and aggregated immunoglobulin (33).

In the absence of specific LPS receptors, it is, therefore, necessary to postulate alternative mutational defects which might manifest phenotypically as an inability to translate the presence of LPS into a cellular triggering signal. Our demonstrated ability to elicit relatively normal mitogenic responses with a variety of R-chemotype LPS preparations, as well as an Ra-like subfraction isolated from wild type LPS, may suggest that some physical-chemical constraints (i.e., molecular rearrangements of LPS within the lymphocyte cytoplasmic membrane phospholipid bilayer) may be essential for a triggering signal and that such events are abnormally regulated in the LPS-non-responder mouse. In this respect, earlier experiments from this laboratory have provided good evidence for selective binding of R-LPS subunits to B-lymphocytes as an initial event (9). It is noteworthy that this selective binding event was characteristic of both C3HeB/FeJ and C3H/HeJ lymphocytes, however, a differential regulatory role for polysaccharide-rich LPS subunits has not yet been excluded.

It must be acknowledged that the evidence summarized here, while providing negative evidence from a variety of experimental systems for specific LPS receptors on lymphocytes, does not totally exclude the existence of such receptors. Thus while our photoactivation studies clearly suggest that the electrophoretic mobilities of LPS binding macromolecules are similar for both responder and non-responder lymphocytes, such studies would not distinguish structural changes in a bifunctional receptor molecule which retained its capacity to bind LPS but lacked the triggering function which putatively might occur subsequent to binding. It is noteworthy that Ulevitch et al. (35) have recently reported that C3H/HeJ lymphocytes will respond mitogenically to LPS when treated concomitantly with trypsin. In addition, our lack of immunologic recognition of an LPS receptor on responder lymphocytes by non-responder mice is subject to multiple interpretations, only one of which would require the lack of such an antigenic structure. It has been documented in other systems in the mouse, that immune responses against minor tumor antigens often require multiple immunizations over a period of months to years in order to elicit detectable immunity (30) and it may be that the putative LPS receptor behaves as an antigen in an analogous manner. Finally, although unlikely, it is possible that the C3H/HeJ cell proliferative response to R-chemotype LPS may be the result of recognition of a structural region other than the traditional lipid A determinant. Additional studies will be required to fully address this point. Thus while we can conclude that, while the mutational defect responsible for the refractory state of C3H/HeJ and C57Bl/10ScN lymphocytes to LPS is not due to any easily detectable structural alteration with the properties of an LPS receptor, the precise biochemical alteration in these LPS-non-responder mouse strains remains to be defined.

REFERENCES

1. B. M. Sultzer, Genetic control of leukocyte responses to endotoxin, *Nature* 219:1253 (1968).

2. D. C. Morrison and J. L. Ryan, A review - bacterial endotoxins and host immune responses, in: "Advances in Immunology," Volume 28, F. J. Dixon and H. G. Kunkel, eds., Academic Press, New York (1979).

3. D. C. Morrison, Bacterial endotoxins and pathogenesis, in: "Reviews of Infectious Diseases," 5:S733 (1983).

4. J. Watson, K. Kelly, M. Largen, and B. A. Taylor, The genetic mapping of a defective LPS response gene in C3H/HeJ mice, J. Immunol. 129:422 (1978).

5. J. Watson and R. Riblet, Genetic control of responses to bacterial lipopolysaccharides in mice. I. Evidence for a single gene that influences mitogenic and immunogenic responses to lipopolysaccharides, J. Exp. Med. 140:1147 (1974).

6. A. Jakobovits, N. Sharon, and I. Zan-Bar, Acquisition of mitogenic responsiveness by nonresponding lymphocytes upon insertion of appropriate membrane components, J. Exp. Med. 156:1274 (1982).

7. H. Nygren, G. Dahlen, and G. Moller, Bacterial lipopolysaccharides bind selectively to lymphocytes from lipopolysaccharide high responder mouse strains, Scand. J. Immunol. 10:555 (1979).

8. R. Chaby, R. Metezeau, and R. Girard, Binding of a rhodamine-labelled lipopolysaccharide to lipopolysaccharide-responder and non-responder lymphocytes, Cell Immunol. 85:531 (1984).

9. S. A. Goodman and D. C. Morrison, Selective association of lipid-rich R-like lipopolysaccharide subunits with murine spleen cells, Molecular Immunol. 21:689 (1984).

10. B. M. Sultzer, Genetic analysis of lymphocyte activation by lipopolysaccharide endotoxin, Infect. Immun. 13:1579 (1976).

11. D. M. Jacobs, D. B. Roberts, J. H. Eldridge, and A. J. Rosenspire, Binding of bacterial lipopolysaccharide to murine lymphocytes, Ann. NY Acad. Sci. 409:72 (1983).

12. L. Forni and A. Coutinho, An antiserum which recognizes lipopolysaccharide-reactive B cells in the mouse, Eur. J. Immun. 8:56 (1978).

13. A. Coutinho, L. Forni, and T. Watanabe, Genetic and functional characterization of an antiserum to the lipid A-specific triggering receptor on murine B lymphocytes, Eur. J. Immun. 8:63 (1978).

14. J. Watson, K. Kelly, and C. Whitlock, Genetic control of endotoxin sensitivity, in: "Microbiology," D. Schlessinger, ed., American Society for Microbiology, Washington, D. C. (1980).

15. S. W. Vukajlovich and D. C. Morrison, Conversion of lipopolysaccharides to molecular aggregates with uniform subunit composition: demonstration of LPS-responsiveness in "endotoxin-unresponsive" C3H/HeJ B-lymphocytes, J. Immunol. 130:2804 (1983).

16. F. C. McIntire, H. W. Sievert, G. H. Barlow, R. A. Finley, and A. Y. Lee, Chemical, physical and biological properties of a lipopolysaccharide from Escherichia coli K-235, Biochem. 6:2363 (1967).

17. B. J. Skidmore, D. C. Morrison, J. M. Chiller and W. O. Weigle, Immunologic properties of bacterial lipopolysaccharide (LPS). II. The unresponsiveness of C3H/HeJ mouse spleen cells to LPS-induced mitogenesis is dependent on the method used to extract LPS, J. Exp. Med. 142:1488 (1975).

18. O. Westphal, O. Lüderitz, and F. Bister, Uber der Extraktion von Bakterien mit phenol Wasser, Z. Naturforsch. 7b:148 (1952).

19. D. C. Morrison and L. Leive, Fractions of lipopolysaccharide from Escherichia coli 0111:B4 prepared by two extraction procedures, J. Biol. Chem. 250:2911 (1975).

20. C. Galanos, O. Lüderitz, and O. Westphal, A new method for the extraction of R lipopolysaccharides, Eur. J. Biochem. 9:245 (1969).

21. R. J. Ulevitch, The preparation and characterization of a radioiodinated bacterial lipopolysaccharide, Immunochemistry 15:157 (1978).

22. U. K. Laemmli, Cleavage of structural proteins during the assembly of the head of bacteriophage T4, Nature (London) 227:680 (1970).

23. J. A. Rudbach, R. L. Anacher, W. T. Haskins, A. G. Johnson, K. C. Milner, and E. Ribi, Physical aspects of reversible inactivation of endotoxin, <u>Ann. NY Acad. Sci.</u> 133:629 (1966).

24. V. S. Waravdekar and L. D. Saslaw, A sensitive colorimetric method for the estimation of 2-deoxy sugars with the use of malonaldehyde-thio-barbituric acid reaction, <u>J. Biol. Chem.</u> 234:1945 (1959).

25. D. C. Morrison, S. J. Betz, and D. M. Jacobs, Isolation of a lipid A bound polypeptide responsible for "LPS-initiated" mitogenesis of C3H/HeJ spleen cells, <u>J. Exp. Med.</u> 144:840 (1976).

26. M. H. Julius, E. Simpson, and L. A. Herzenberg, A rapid method for the isolation of functional thymus-derived murine lymphocytes, <u>Eur. J. Immunol.</u> 3:645 (1973).

27. L. L. Perry and M. I. Greene, Conversion of immunity to suppression by in vivo administration of I-A subregion specific antibodies, <u>J. Exp. Med.</u> 156:480 (1982).

28. W. Fierz, M. Brenan, A. Mullbader, and E. Simpson, Non-H-2 and H-2-linked immune response genes control the cytotoxic T-cell response to H-Y, <u>Immunogenetics</u> 15:261 (1982).

29. S. A. Goodman and D. C. Morrison, Lipopolysaccharides receptors on lymphocytes. I. Lack of immunologic recognition of a putative LPS receptor on LPS responder lymphocytes by non-responder mice, <u>J. Immunology</u>, in press (1985).

30. P. O. Livingston, A. B. DeLeo, M. Jones, and H. F. Oettgen, Comparison of approaches for augmenting the serologic response to the individually specific methylcholanthrene-induced sarcome-neth A: pretreatment with cyclophosphamide is most effective, <u>J. Immunol.</u> 131:2601 (1983).

31. S. J. Betz and D. C. Morrison, Chemical and biological properties of a protein-rich fraction of bacterial lipopolysaccharides. I. The in vitro murine lymphocyte response, <u>J. Immunol.</u> 119:1790 (1977).

32. D. C. Morrison, M. E. Wilson, S. Raziuddin, S. J. Betz, B. J. Curry, Z. G. Oades, and P. Munkenbeck, The influence of lipid A-associated protein on endotoxin stimulation of non-lymphoid cells, <u>in</u>: "Microbiology," D. Schlessinger, ed., American Society for Microbiology, Washington, D. C. (1980).

33. M. H. Ginsberg and D. C. Morrison, The selective binding of aggregated IgG to lipid A-rich bacterial lipopolysaccharides, <u>J. Immunol.</u> 120:317 (1978).

34. K. Yokoyama, J. Mashimo, N. Kasai, T. Terao, and T. Osawa, Binding of bacterial lipopolysaccharide to histocompatibility-2-complex proteins of mouse lymphocytes, <u>Hoppe Seyler's Z. Physiol. Chem.</u> 360:587 (1979).

35. K. Kuus, A. R. Johnston, and R. J. Ulevitch, Enhancement of C3H/HeJ B-cell responsiveness to lipopolysaccharide (LPS) with Trypsin, <u>Fed. Proc.</u> (Abst.) 44:974 (1985).

EFFECTS OF ENDOGENOUS GUT LPS ON CELLS OF THE SECRETORY IMMUNE SYSTEM

Suzanne M. Michalek, Jerry R. McGhee, Dawn E. Colwell,
Shane I. Williamson, Thomas A. Brown, David M. Spalding,
William J. Koopman, and Jiri Mestecky

Departments of Microbiology and Medicine, Division of
Clinical Immunology and Rheumatology, University of
Alabama at Birmingham, Birmingham, Alabama

INTRODUCTION

Lipopolysaccharide (LPS), the major constituent of the outer membrane·
of gram-negative bacteria, is a potent biostimulant for lymphoreticular
cells of the host's immune system (14). The major source of LPS in mammalian
species is the indigenous gram-negative microflora of the gastrointestinal
tract, of which Bacteroides species becomes the predominant genera during
the first few weeks of life. The first major lymphoid tissue affected by
gut LPS is the gut-associated lymphoreticular tissue (GALT), which is mainly
represented by distinct lymphoid nodules, termed Peyer's patches (PP), along
the small intestine. The GALT or PP are major inductive sites for IgA responses to ingested environmental antigens and intestinal pathogens.

The dome region of the PP consists of an epithelium which contains a
unique cell type termed a microfold cell [M cell (16)]. M cells, which are
actively pinocytotic and phagocytic, sample gut lumenal antigens for presentation to lymphoid cells in the dome region and in follicles (B cell zones)
and parafollicular sites (T cell zones) of the PP (10). The gut microflora
strongly influences the development of PP since it has been shown that PP
of germfree mice have poorly developed B cell zones with no germinal centers
(15).

Each follicle of the PP normally possesses one to two germinal centers
containing precursor IgA B cells [surface IgA positive (sIgA$^+$)]; however,
significant differentiation of these cells into mature IgA-producing plasma
cells does not occur in this tissue. Instead, antigen-sensitized T and B
cells leave this tissue via the efferent lymphatics —→ mesenteric lymph
nodes —→ thoracic duct —→ blood stream and migrate to distant mucosal
sites where final differentiation into IgA-secreting plasma cells occurs,
with polymeric IgA being transported into the external secretions. This
network, collectively termed the common mucosal immune system (2,11,12),
provides the mucosa with a continual supply of sensitized B cells for local
IgA responses to environmental antigens and pathogens.

In this brief review, we have focused on the effects of gut LPS on
cells in GALT and on subsequent IgA responses. In addition, we briefly
describe ongoing studies which have assessed the ability of monoclonal

antibodies directed against the O-polysaccharide, R core, or lipid A of Salmonella LPS to inhibit LPS-induced effects and to protect mice against lethal infection with Salmonella.

LPS AND IgA RESPONSES

Since LPS is ubiquitous in the environment, it has been difficult to determine the precise effect of gut endotoxin on the host's secretory immune system. We have minimized this problem by comparing immune responses to orally administered T-dependent (TD) antigens in LPS-nonresponsive C3H/HeJ mice and syngeneic, LPS-responsive C3H/HeN animals. Our previous studies have shown that higher IgA responses occur in C3H/HeJ mice given various TD antigen by the oral route than in identically treated C3H/HeN mice (1,10). Oral immunization induced greater T helper (Th) cell activity in GALT and spleen of C3H/HeJ mice when compared with various LPS-responsive strains, and this accounted for the elevated IgA response pattern seen (8). Genetic studies with F1, F2 and backcross mice derived from parental C3H/HeJ and C3H/HeN animals showed that elevated IgA responses are directly linked to expression of the Lps^d gene (13).

LPS-Induced Suppression of Polyclonal IgA Responses

In order to understand how LPS affects GALT cells and their subsequent responses to TD antigens, it was necessary to develop a uniform method for generating PP cells in high yield. For these studies, isolated PP are digested with the enzyme Dispase[TM], a neutral protease, which releases the entire lymphoreticular cell population (5,9). Approximately 40% of the cell population are B cells, of which a high frequency are $sIgA^+$. Approximately equal numbers of T cells are also obtained (35-38%), including large numbers of $Lyt-1^+$ (inducer and helper phenotypes) and $Lyt-2^+$ (cytotoxic and suppressor phenotypes) cells. The PP also contain significant numbers of accessory cells, including macrophages (5-9%) and functional dendritic cells (17).

Recent studies (18) have shown that murine PP dendritic cells (DC) together with $Lyt-1^+$ T cells can induce polyclonal IgA responses in B cell cultures. When PP DC were mixed with T cells and stimulated with the oxidative mitogen sodium periodate, large DC-T cell clusters formed. The clusters were separated from single cells, incubated for three days, and then fractionated on a density gradient. The low density (p< 1.078) fraction collected consisted of DC and T cells of the Lyt-1 phenotype. The incubation of splenic or PP DC-T cell clusters with splenic or PP B cells resulted in a two to three-fold increase in polyclonal IgM synthesis. However, only PP DC-T cell clusters preferentially induced B cells to polyclonal IgA synthesis (18). These results show that PP DC-T cell interactions are required for polyclonal IgA responses and provide additional evidence that cells in GALT are unique and render this tissue a major IgA inductive site.

The influence of LPS on polyclonal IgA responses was assessed during PP DC-T cell cluster formation and in B cell cultures containing DC-T cell clusters. In these studies, PP DC-T cell clusters were derived from LPS-responsive C3H/HeN mice and B cells were obtained from PP of C3H/HeJ mice to insure that the observed LPS effects were not due to direct interactions of LPS with B cells. LPS had no effect on polyclonal IgA synthesis in B cell cultures containing DC-T cell clusters (Table 1). However, if DC-T cell clusters were generated in the presence of LPS and were added to normal B cell cultures, a 37% reduction in IgA synthesis was observed when compared with responses induced with untreated DC-T cell clusters. When LPS was present during DC-T cell cluster formation and during co-culture with B

Table 1. Influence of LPS on the Induction of Polyclonal IgA Responses by PP Dendritic Cell-T Cell Clusters

PP DC-T Cell Cluster Formation[a]	LPS Added to DC-T, B Cell Clusters[b]	Immunoglobulin Synthesis (ng/ml)[c]	
		IgM	IgA
In absence of LPS	−	1,339	4,709
	+	2,584	4,900
In presence of LPS	−	3,765	2,984
	+	3,094	1,461
None	−	375	98
	+	450	82

[a]Clusters of C3H/HeN PP DC-T cells were formed by sodium periodate stimulation, either in the presence or absence of 20 μg of E. coli K235 LPS.
[b]DC-T cell clusters were added to C3H/HeJ splenic B cells and were cultured for 7 days either in the presence or absence of 20 μg of E. coli K235 LPS.
[c]Culture supernatants were assayed for total IgM and IgA (ng/ml) by radioimmunoassay.

cells, a 50% reduction in IgA synthesis was noted when compared with LPS-pretreated DC-T cell clusters and a 70% reduction when compared with cells never exposed to LPS (Table 1). These results suggest that LPS influences the inductive phase of DC-T cell cluster formation. However, LPS had no apparent effect on mature DC-T cell clusters which contain predominantly Lyt-1[+] T cells. Thus, we have shown that endogenous LPS can influence both TD antigen-specific and polyclonal IgA responses in a manner which could occur following gut LPS stimulation of GALT cells.

Human IgA1 and IgA2 Responses to LPS in Serum and Saliva

Since LPS in the gut could influence the IgA response, we have initiated studies to assess human IgA subclass responses to this macromolecule. In human serum, the IgA1 subclass represents approximately 90% of the IgA, whereas in external secretions IgA1 and IgA2 are found in approximately equal amounts (4). We have developed a radioimmunoassay (RIA) using monoclonal anti-alpha 1 and anti-alpha 2 reagents to assess IgA1 and IgA2 antibody levels to LPS in human serum and saliva. The saliva of most subjects contained predominantly IgA2 antibodies to either intact smooth LPS from Escherichia coli or purified lipid A from Salmonella minnesota (Table 2). However, an opposite pattern was seen in serum from most subjects, with the response to LPS and lipid A being mainly restricted to the IgA1 subclass. Interestingly, serum IgA responses to smooth LPS indicated that approximately half the subjects had mainly IgA1 and the other half largely IgA2 responses. However, all individuals tested exhibited only IgA1 responses to purified lipid A. This finding suggested that IgA2 responses noted in serum were directed to the intact polysaccharide moiety of LPS, whereas the major serum anti-lipid A response was restricted to the IgA1 subclass. These results further suggest that IgA subclass responses to LPS differ in serum and in external secretions and that IgA1 in serum and IgA2 in secretions are the predominant anti-LPS antibodies found. This would imply that serum and secretory IgA responses are under separate regulatory control.

Table 2. Ratios of IgA1/IgA2 Anti-LPS Antibodies in Sera and
Saliva of Normal Human Subjects

| | LPS Antigen Used[a] | | | |
| | Smooth E. coli 0127 LPS | | Salmonella minnesota R595 (Re) Lipid A | |
Subject	Serum	Saliva	Serum	Saliva
#1	1.07	0.93	0.99	0.74
#2	0.14	0.40	1.97	0.10
#3	2.63	2.57	A1 only	1.04
#4	3.35	0.52	A1 only	0.40
#5	A1 only	0.88	A1 only	1.18
#6	0.33	0.75	1.14	0.75
#7	1.54	0.60	A1 only	0.40
#8	0.13	0.10	A1 only	0.14

Major IgA Subclass

IgA1	5/8	1/8	7/8	2/8
IgA2	3/8	7/8	1/8	6/8

[a]Values expressed as the ratio of the amount of IgA1 to IgA2 antibodies to
the appropriate LPS as determined by a radioimmunoassay.

Cellular Responses to Lipid A from Bacteroides in the Gut are Regulated by the Lps Gene

Our past studies of the effects of LPS on GALT cells and IgA responses
were performed using LPS derived from E. coli or Salmonella, which are members of the Enterobacteriaceae and which comprise less than one percent of
the gut microflora. Bacteroides species predominate in the gut, and studies
have indicated that LPS of Bacteroides fragilis induces B cell mitogenic
responses in classical LPS-nonresponsive mice (7). This finding would make
it difficult to explain our past studies which imply that gut LPS influences
GALT cells and subsequent IgA responses, including elevated IgA responses
in the C3H/HeJ mouse strain. Therefore, we have reexamined this issue and
have tested phenol-water extracted LPS of B. fragilis for various in vitro
and in vivo responses in LPS-sensitive C3H/HeN (Lps^n/Lps^n) and LPS-resistant
C3H/HeJ (Lps^d/Lps^d) mouse strains. When B. fragilis LPS (B-LPS) was added
to C3H/HeN or C3H/HeJ spleen cell cultures at either high or low cell density, C3H/HeJ spleen cells responded only at high cell density, and the
response was less than that seen in C3H/HeN cultures. This finding suggested
that C3H/HeJ mice may be low responders to B-LPS in a manner analogous to
that seen with LPS from Enterobacteriaceae (19).

The difference in responsiveness to Bacteroides LPS between the two
mouse strains was most apparent when purified splenic B cells were stimulated
with B-LPS (Table 3). Purified B cells from C3H/HeN mice were fully responsive to B-LPS, while C3H/HeJ splenic B cells were unresponsive (19). However, whole spleen cell cultures from C3H/HeJ mice did respond to B-LPS,
albeit at levels significantly lower than those seen in C3H/HeN spleen cell
cultures (Table 3). These results suggest that direct triggering of B cells
by B-LPS is regulated by the Lps gene and indicate that the low responses

Table 3. Mitogenic Responses to <u>Bacteroides</u> LPS and Its Lipid A
and Carbohydrate Moieties[a]

| Bacteroides | Mitogenic Response (E/C) | |
LPS Preparation	C3H/HeN	C3H/HeJ
<u>B Cell Cultures</u>		
B-LPS (Ph)	22.0	1.8
Lipid A	12.3	1.1
Carbohydrate	1.4	1.3
<u>Spleen Cell Cultures</u>		
B-LPS (Ph)	24.0	10.1
Lipid A	8.2	1.4
Carbohydrate	21.2	17.9

[a]10 µg of the appropriate LPS preparation was added to microculture wells
containing 5×10^5 purified splenic B cells or unfractionated spleen cell/
well. Cultures were incubated for 48 hours and were pulsed with tritiated
thymidine (^{3}H-TdR) for the last six hours of incubation. The values are
expressed as the stimulation ratio, where E/C = the mean ^{3}H-TdR uptake by
cultures containing LPS/mean ^{3}H-TdR uptake by cultures containing no LPS.

which occur in the C3H/HeJ spleen cell cultures result from the presence of
another cell type, perhaps macrophages. In separate studies, purified lipid
A and carbohydrate were prepared by mild acid hydrolysis of B-LPS, and their
mitogenic properties were tested (18). <u>Bacteroides</u> lipid A induced mitogenic
responses in both purified B cell and whole spleen cell cultures only from
C3H/HeN mice (Table 3). On other other hand, the carbohydrate moiety failed
to induce responses in B cell cultures derived from either C3H/HeN or C3H/HeJ
mice, but induced full mitogenic responses in whole spleen cell cultures
derived from both mouse strains. These results suggest that <u>Bacteroides</u>
LPS contains two biologically active components, i.e., lipid A and carbohydrate and that responses to the former are regulated by the <u>Lps</u> gene (19,20).

To more precisely determine the accessory cell requirement for <u>Bacter-
oides</u> carbohydrate responses, splenic B cells from C3H/HeJ mice were incubated with carbohydrate in the presence or absence of adherent cells (macrophages). Good B cell mitogenic responses were seen when splenic macrophages
(MØ) from C3H/HeN or C3H/HeJ mice were added to C3H/HeJ B cell cultures
prior to stimulation with B-LPS (Table 4). The B cell responses observed
were clearly due to a mitogenic component in the carbohydrate of B-LPS,
since the polysaccharide moiety obtained from <u>E. coli</u> K235 LPS by identical
procedures failed to stimulate C3H/HeJ B cells even in the presence of MØ.

We conclude from these experiments that the lipid A of <u>B. fragilis</u> LPS
is similar in biologic function to the lipid A from <u>Enterobacteriaceae</u> and
that responses to the lipid A are regulated by the <u>Lps</u> gene. On the other
hand, B cell mitogenic responses induced in C3H/HeJ spleen cell cultures
are due to a biologically active carbohydrate moiety. These responses
require MØ and do not appear to be regulated by the <u>Lps</u> gene. Our past
studies which showed that IgA responses to oral TD antigens are influenced
by <u>Lps</u> gene are consistent with the observation that <u>Bacteroides</u> lipid A,
which would be the predominant type present in the gut, fails to stimulate
lymphoreticular cells from C3H/HeJ mice.

Table 4. Mitogenic Responses of C3H/HeJ B Cells to the Carbohydrate
Component of Bacteroides LPS Require Macrophages[a]

| LPS Preparation | Mitogenic Response (E/C) | |
	HeJ B Cells	HeJ B Cells + MØ
B. fragilis		
LPS (Ph)	1.6	29.7
Lipid A	1.2	1.1
Carbohydrate	1.0	27.9
E. coli K235		
LPS (Ph)	1.7	2.1
Lipid A	1.4	1.6
Carbohydrate	1.2	1.0

[a]See Table 3 for footnote.

FUNCTIONAL STUDIES OF MONOCLONAL ANTIBODIES TO THE MAJOR REGIONS OF THE

SALMONELLA LPS MOLECULE

A current experimental approach for treatment of patients with gram-negative septicemia involves immunotherapy, e.g., administration of antisera
to common LPS determinants which are present on a number of gram-negative
bacteria (6). Past studies have shown that serum containing antibodies to
the R core region of LPS is protective against gram-negative bacterial infection and septicemia in experimental animals and humans; however, the exact
specificity of the protective antibody has not been determined. Our laboratory has approached this problem by generating monoclonal antibodies to the
three major regions of Salmonella LPS (O-polysaccharide, R core, and lipid
A) for use in a more precise analysis of the isotype and specificity of
anti-LPS antibody involved in protection against Salmonella infection. A
total of eight monoclonal antibodies to the O-polysaccharide, two to R core,
and five to lipid A regions of LPS have been produced. All of the monoclonal
O-polysaccharide antibodies reacted strongly with LPS from smooth Salmonella
but not with LPS from rough S. minnesota R345 (Rb) or R595 (Re) mutants
(Table 5). The two monoclonal antibodies to the R core region exhibited
highest antibody titers to LPS from S. minnesota Rb and showed less reactivity to smooth Salmonella LPS. Both of these monoclonal antibodies failed
to react with Re LPS, clearly indicating their specificity for the R core
region of Salmonella LPS. All five monoclonal anti-lipid A antibodies exhibited low reactivity to smooth LPS, higher reactivity to Rb LPS, and
greatest reactivity to Re LPS (Table 5).

To test the effectiveness of these monoclonal antibodies in protection
against Salmonella infection, groups of C3H/HeN mice were given monoclonal
antibody by intraperitoneal (i.p.) injection and then were challenged (i/p.)
with virulent S. typhimurium SR-11. Untreated control mice challenged with
S. typhimurium SR-11 (2×10^4 bacteria) had a median length of survival of
five days. Of the eight monoclonal anti-O-polysaccharide antibodies, three
exhibited a significant degree of protection of mice, while one showed some
protection (Table 6) (3). The monoclonal antibody 10-5-6 extended the median
length of survival to nine days, with 20% of the treated mice surviving
over 20 days, while the monoclonal antibody 10-5-47 extended the median
length of survival to 13 days and allowed over 50% of the mice to survive

Table 5. Specificity of Murine Monoclonal Antibodies to <u>Salmonella</u> LPS[a]

| Clone Number | Source of LPS | | | |
| | S. typhimurium | | S. minnesota | |
	SR-11	LT-2	R345(Rb)	R595(Re)
Anti-O-Polysaccharide				
10-5-3	++++	++++	–	–
10-5-6	++++	++++	–	–
10-5-7	++++	++++	–	–
10-5-9	+++	+++	–	–
10-5-26	+++	+++	–	–
10-5-35	+++	+++	–	–
10-5-47	++++	++++	–	–
ST-1	+++	++++	–	–
Anti-R-Core				
RC-9	++	++	++++	–
RC-16	++	++	++++	–
Anti-Lipid A				
LA-1	+	+	++	++
LA-2	+	+	++	++
LA-3	+	+	++	++++
LA-4	+	+	++	+++
LA-5	+	+	++	+++

[a]As determined by ELISA using plates coated with the appropriate LPS preparation. All monoclonal antibodies, except ST-1 (IgM), were of the IgG isotype.

challenge. The IgM monoclonal antibody ST-1 gave the greatest degree of protection, resulting in a median length of survival of greater than 20 days, with 76% of the mice surviving the infection. On the other hand, the other monoclonal anti-O-polysaccharide antibodies exhibited high titers to smooth LPS, but afforded no protection (Table 6).

In contrast to these findings, none of the monoclonal antibodies to the R core or lipid A protected mice against <u>S. typhimurium</u> SR-11 infection. In these latter studies, mice were challenged with one log fewer (2×10^5) bacteria than the number used for challenge of mice given anti-O-polysaccharide antibody, but the antibodies failed to control the infection. Even a pool of anti-R core and anti-lipid A monoclonal antibodies was not protective. These results suggest that certain monoclonal antibodies to the O-polysaccharide of LPS are protective; however, antibodies to the R core or lipid A do not affect the outcome of infection with highly virulent smooth <u>Salmonella</u> bacteria.

In additional studies, we have assessed the ability of monoclonal antibodies to lipid A to alter in vitro mitogenic responses induced by LPS. Each anti-lipid A antibody and a pool of all five monoclonal antibodies inhibited in vitro mitogenic responses induced by smooth <u>Salmonella</u> LPS or Re LPS (Table 7). Monoclonal anti-O-polysaccharide antibodies and monoclonal antibodies of extraneous specificities were ineffective in abrogating mitogenic responses to LPS (data not shown). These results suggest that the monoclonal anti-lipid A antibodies are directed against biologically active

Table 6. Protection of C3H/HeN Mice from Lethal S. typhimurium Infection by Monoclonal Antibodies to LPS[a]

Clone Number	Challenge Dose	Median Days Survival	Percent Surviving	Protection
Anti-O-Polysaccharide				
10-5-3	2×10^4	3	0	None
10-5-6		9	20	Yes
10-5-7		5	0	None
10-5-9		4	0	None
10-5-26		7	10	Some
10-5-35		5	0	None
10-5-47		13	52	Yes
ST-1		> 20	76	Yes
None		5	0	None
Anti-R-Core				
RC-8	2×10^3	7	0	None
RC-16		8	0	None
Anti-Lipid A				
LA-1	2×10^3	8	0	None
LA-2		7	0	None
LA-3		7	0	None
LA-4		7	0	None
LA-5		7	0	None
None		7	0	None

[a]Mice were given the appropriate monoclonal antibody by the intraperitoneal (i.p.) route followed one hour later by i.p. challenge with the indicated dose of virulent S. typhimurium SR-11. Mice were monitored and deaths recorded for 20 days.

Table 7. Monoclonal Antibodies to Lipid A Inhibit LPS-Induced Mitogenic Responses

Antibody Tested[a]	Mitogenic Responses (E/C)	
	Re LPS	S LPS
None	26.8	21.1
LA-1	4.6	5.0
LA-2	4.8	3.4
LA-3	5.0	4.6
LA-4	4.2	3.5
LA-5	5.4	3.8
Pool of all five monoclonal antibodies	5.8	8.5

[a]C3H/HeN spleen cells (8×10^5/culture) were incubated in the presence or absence of the indicated monoclonal antibody (1 µg/culture) and LPS (1 µg/ml culture). See Table 3 footnote.

sites on the lipid A molecule. In this regard, we have recently observed that mice given monoclonal anti-lipid A antibody and challenged with a lethal dose of LPS survived significantly longer than mice given LPS alone.

The results of our studies demonstrate that monoclonal anti-O-poly-saccharide antibodies can contribute to host defense against infection with homologous Salmonella. In addition, although others have reported that antisera containing antibodies directed against common R core and lipid A determinants of the LPS molecule were cross-protective against different species and genera of gram-negative bacteria, we found that seven monoclonal antibodies to the R core or lipid A of Salmonella LPS were unable to protect mice against lethal bacteremia resulting from challenge with a highly virulent strain of S. typhimurium. Monoclonal anti-lipid A antibodies were, however, shown to have biological activity in vitro and in vivo, inhibiting LPS-induced mitogenic responses and lethal toxicity, respectively. These results suggest that the relevance of anti-lipid A antibody in host defense may reside more in its anti-endotoxic activity than in its anti-bacterial activity.

ACKNOWLEDGMENTS

This work was supported in part by U. S. Public Health Service Grants AI 19674, AI 18958, DE 02670, DE 00092, and AI 18745. We wish to thank Ms. Yvonne Noll for typing this manuscript.

REFERENCES

1. J. L. Babb and J. R. McGhee, Mice refractory to lipopolysaccharide manifest high immunoglobulin A responses to orally administered antigen, Infect. Immun. 29:322 (1980).
2. J. Bienenstock and A. D. Befus, Review. Mucosal immunology, Immunology 41:249 (1980).
3. D. E. Colwell, S. M. Michalek, D. E. Briles, E. Jirillo, and J. R. McGhee, Monoclonal antibodies to Salmonella lipopolysaccharide: anti-O-polysaccharide antibodies protect C3H mice against challenge with virulent Salmonella typhimurium, J. Immunol. 133:950 (1984).
4. S. S. Crago, W. J. Kutteh, I. Moro, M. R. Allansmith, J. Radl, J. J. Haaijman, and J. Mestecky, Distribution of IgA1-, IgA2-, and J chain-containing cells in human tissues, J. Immunol. 132:16 (1984).
5. M. V. Frangakis, W. J. Koopman, H. Kiyono, S. M. Michalek, and J. R. McGhee, An enzymatic method for preparation of dissociated murine Peyer's patch cells enriched for macrophages, J. Immunol. Methods 48:33 (1982).
6. W. L. Hand and J. W. Smith, Immunology of enterobacterial infections, in: "Comprehensive Immunology. Volume 8. Immunology of Human Infection. Part 1," A. J. Nahamias and R. J. O'Reilly, eds., Plenum Publishing Corporation, New York (1981).
7. K. A. Joiner, K.P.W.J. McAdam, and D. L. Kasper, Lipopolysaccharide from Bacteroides fragilis are mitogenic for spleen cell from endotoxin responder and nonresponder mice, Infect. Immun. 36:1139 (1982).
8. H. Kiyono, J. L. Babb, S. M. Michalek, and J. R. McGhee, Cellular basis for elevated IgA responses in C3H/HeJ mice, J. Immunol. 125:732 (1980).
9. H. Kiyono, J. R. McGhee, M. J. Wannemuehler, M. V. Frangakis, D. M. Spalding, S. M. Michalek, and W. J. Koopman, In vitro immune responses to a T cell-dependent antigen by cultures of disassociated murine Peyer's patch, Proc. Natl. Acad. Sci. U.S.A. 79:596 (1982).
10. J. R. McGhee, H. Kiyono, and C. D. Alley, Gut bacterial endotoxin: influence on gut-associated lymphoreticular tissue and host immune function, Sur. Immunol. Res. 3:241 (1984).

11. J. R. McGhee and JS. M. Michalek, Immunobiology of dental caries: microbial aspects and local immunity, Ann. Rev. Microbiol. 35:595 (1981).

12. J. Mestecky, J. R. McGhee, S. S. Crago, S. Jackson, M. Kilian, H. Kiyono, J. L. Babb, and S. M. Michalek, Molecular-cellular interactions in the secretory IgA response, J. Reticuloendothel. Soc. 28:45S (1980).

13. S. M. Michalek, H. Kiyono, J. L. Babb, and J. R. McGhee, Inheritance of LPS nonresponsiveness and elevated splenic IgA immune responses in mice orally immunized with heterologous erythrocytes, J. Immunol. 125:2220 (1980).

14. D. C. Morrison and J. L. Ryan, Bacterial endotoxins and host immune responses, Adv. Immunol. 28:293 (1979).

15. M. Pollard and N. Sharon, Responses of Peyer's patches in germfree mice to antigenic stimulation, Infect. Immun. 2:96 (1970).

16. R. L. Owen and A. L. Jones, Epithelial cell specialization within human Peyer's patches: an ultrastructural study of intestinal lymphoid follicles, Gastroenterology 66:189 (1974).

17. D. M. Spalding, W. J. Koopman, J. H. Eldridge, J. R. McGhee, and R. M. Steinman, Accessory cells in murine Peyer's patch. I. Identification and enrichment of a functional dendritic cell, J. Exp. Med. 157:1646 (1983).

18. D. M. Spalding, S. I. Williamson, W. J. Koopman, and J. R. McGhee, Preferential induction of polyclonal IgA secretion by murine Peyer's patch dendritic cell-T cell mixtures, J. Exp. Med. 160:941 (1984).

19. M. J. Wannemuehler, S. M. Michalek, E. Jirillo, S. I. Williamson, M. Hirasawa, and J. R. McGhee, LPS regulation of the immune response: Bacteroides endotoxin induces mitogenic, polyclonal, and antibody responses in classical LPS responsive but not in C3H/HeJ mice, J. Immunol. 133:299 (1984).

20. S. I. Williamson, M. J. Wannemuehler, E. Jirillo, D. G. Pritchard, S. M. Michalek, and J. R. McGhee, LPS regulation of the immune responses: separate mechanisms for murine B cell activation by lipid A (direct) and polysaccharide (macrophage-dependent) derived from Bacteroides LPS, J. Immunol. 133:2294 (1984).

MOLECULAR MECHANISMS IN LYMPHOKINE-INDUCED MACROPHAGE ACTIVATION-

ENHANCED PRODUCTION OF OXYGEN RADICALS

Edgar Pick

Laboratory of Immunopharmacology, Department of Human Microbiology, Sackler School of Medicine, Tel-Aviv University
Tel-Aviv, Israel

Macrophages (MPs) are key effector cells in the eradication of a variety
of unicellular or multicellular pathogens and, most likely, in the limitation
of malignant growth. The principal cytotoxic mechanism of MPs is exerted
on microorganisms internalized by phagocytosis such as bacteria, fungi and
protozoa. A second mechanism is represented by damage inflicted to extracellular targets ranging from tumor cells to large metazoan parasites. In
addition, MPs exhibit suppressor activity on cells of the immune system,
such as T lymphocytes.

The constitutive capacity of MPs to cause cellular damage is limited.
The complex differentiation process resulting in a markedly enhanced ability
to destroy intracellular or extracellular targets was termed "MP activation".
No fully satisfactory definition of this term is available and serious doubts
were expressed concerning its vague and, frequently, improper use (8,25).
The best way of avoiding terminological inaccuracy is to make a habit of
providing additional information related to the nature of the effector function that is enhanced. As an example, it is perfectly legitimate to talk
about "MPs activated for enhanced killing of Toxoplasma gondii" or "MPs
activated for tumor cell cytotoxicity." This supplementary information has
to be provided because of the following reasons: (a) MPs exhibiting an
augmented capacity to destroy phagocytosed unicellular organisms are not
necessarily more active in killing tumor cells; (b) the cytotoxic mechanisms
responsible for intra- and extracellular killing are widely different and
more heterogeneity certainly exists within each category; (c) activated MPs
may have a detrimental influence on immune function due to their suppressor
effects (that are, probably, heterogeneous, too); and (d) MP activation,
even though enhancing one or another effector function, may be associated
with no change or even a decrease in another key MP task--that of processing
and presenting antigen to T cells.

Activation of MPs can be induced by multiple mechanisms, all these
belonging to one or the other of two large categories: T cell-dependent or
T cell-independent activation. The classical example of T cell dependent
MP activation is that seen in animals immunized with an antigen eliciting a
cell-mediated immune response and exposed repeatedly to the same antigen
(13). It was soon found that the process can be mimicked in vitro by co-
culturing antigenically stimulated lymphocytes with MPs (34) and, finally,
that activation results from the action on the MP of a T cell-derived soluble
product (lymphokine, LK) (9) named operationally "MP activating factor"

(MAF). T cell-independent activation is effected by agents acting directly on MPs and does not involve the preliminary production of an MP activating LK. Thus, bacterial lipopolysaccharide (LPS) or bacterial cell wall peptidoglycan (or synthetic muramyl dipeptide, MDP) are capable of activating MPs in the absence of T cells. The relative importance of these two main pathways of activation is difficult to estimate and it can only be said that it is likely that the T cell-dependent, LK-mediated pathway represents an important element in the physiologic response of animals to infection. Although MP activation was initially described in the intact animal, research into the molecular basis of the phenomenon gained impetus only when it became possible to induce MP activation in vitro under controlled experimental conditions and with the aid of increasingly well defined agents.

MEDIATORS OF CYTOTOXICITY

Activated MPs damage or destroy phagocytosed or extracellular targets by making use of a multiplicity of cytotoxic mechanisms. It is customary to classify these into two categories: oxygen-dependent and oxygen-independent. The emphasis in this chapter is on the production of toxic oxygen metabolites by activated MPs. Oxidative damage has been definitely established to be of paramount importance in the destruction of bacteria, fungi and protozoa within the phagocytic vacuoles of activated MP but there is considerable uncertainty about the role of oxygen metabolites in the contact-dependent killing of tumor cells and multicellular parasites. A definite case for oxidative destruction of extracellular targets by MPs could only be made when pharmacologic agents known to stimulate oxygen radical (OR) production by MPs, such as phorbol myristate acetate (MPA), were added to the system. On the other hand, oxygen-independent mechanisms, while also participating in post-phagocytic damage, appear to be dominant in extracellular killing. Examples of oxygen-independent cytotoxic mediators are: lysozyme, lysosomal acid hydrolases, microbicidal cationic peptides, arginase, cytolytic serine protease, tumor necrosis factor and other cytotoxic factors, thymidine and arachidonate metabolites. Cooperation among oxidative nonoxidative cytotoxic mechanisms in causing damage to the same target was also described. Finally, activated MPs secrete factors that recruit and activate other cell types such as neutrophils and platelets and may also exert more complex systemic or local effects by the production of factors affecting a variety of target structures ranging from the central nervous system to blood vessels.

THE OXIDATIVE BURST

MPs subject to stimulation by certain agents demonstrate a marked increase in oxygen uptake that is accompanied by the production of superoxide (O_2^-) anions and hydrogen peroxide (H_2O_2). In close association with these events, the cells sharply increase glucose utilization via the hexose monophosphate shunt (HMPS). This coordinated complex of reactions is known as the "oxidative" or "respiratory" burst although oxygen is not utilized for energy production and is not processed in the mitochondrial system. The oxidative burst (OB) is initiated by the one-electron reduction of oxygen to O_2^- performed by a membrane localized enzyme utilizing NADPH as the electron donor (NADPH oxidase) (reviewed in 1 and 2):

$$2O_2 + NADPH \longrightarrow 2O_2^- + NADPH^+ + H^+$$

Most of this O_2^- is dismutated spontaneously to H_2O_2:

$$2O_2^- + 2H^+ \longrightarrow H_2O_2 + O_2$$

An alternative source of H_2O_2 is by the action of either cytoplasmic or particulate superoxide dismutase (SOD), enzymes that convert O_2^- to H_2O_2 at a rate 10^4-fold higher than the spontaneous one. There is also evidence for the production by MPs of hydroxyl radicals (OH·), probably by way of a metal-catalyzed reaction between O_2^- and H_2O_2. MPs possess efficient detoxifying systems meant to prevent damage to cellular components by self-generated ORs. Thus, SOD removes O_2^- that has seeped into the cell interior. H_2O_2 is catabolized by two enzyme systems: catalase and the glutathione peroxidase-glutathione reductase tandem. Glutathione peroxidase converts H_2O_2 to H_2O in a reaction consuming reduced glutathione; glutathione reductase restores the level of reduced glutathione by utilizing NADPH and generating $NADP^+$. Activation of the HMPS is due to the accumulation of $NADP^+$; this is derived from two sources, one is the generation of O_2^- itself by NADPH oxidase, the second is by the action of glutathione reductase.

The purpose of the OB is, therefore, to provide toxic oxygen metabolites capable of killing microorganisms. The relative importance of specific metabolites is subject to argument but it is likely that H_2O_2 is the major toxic oxygen-derived reagent produced by MPs. Because mature MPs do not contain myeloperoxidase, it is unlikely that H_2O_2-derived oxidized halogens (such as hypochlorite or chloramines) are produced in significant amounts. H_2O_2 might, therefore, exert a direct toxic effect or serve as substrate for the formation, in conjunction with O_2^-, of the oxidizing radical, OH·.

The OB can be elicited in MPs by an unusually wide variety of stimuli. The list includes the phorbol ester PMA; opsonized bacteria; lectins, such as concanavalin A (Con A) and wheat germ agglutinin (WGA); the Ca^{2+} ionophore A23187; phospholipase C (but not A_2); formylated peptides, such as formyl-Met-Leu-Phe; NaF; microtubule disrupting drugs, such as colchicine; Na nitroprusside and certain anionic detergents (31). In looking for a common denominator in the mechanism by which these stimulants elicit O_2^- production, we were impressed by the fact that most OB elicitors also caused arachidonic acid (C20:4) liberation from cellular phospholipids. Also, inhibiting endogenous phospholipase activity prevented the elicitation of an OB by the majority of stimuli with the notable exception of PMA (6). Finally, exogenously applied long chain unsaturated fatty acids (UFAs), such as C20:4, were found to be potent initiators of O_2^- production by MPs (6). It was also found that oxygenation of C20:4 produces metabolites with a markedly variable capacity to elicit an OB. Thus, the cyclo-oxygenase derived prostaglandin E_2 (PGE_2) was totally inactive while certain lipoxygenase-derived metabolites such as 15-hydroxyeicosatetraenoic acid (15-HETE) or leukotrienes C_4 and B_4 were active. These findings lead us to suggest that the liberation of UFA from membrane phospholipids by endogenous phospholipases is an obligatory step in the activation of the O_2^- forming enzyme by most stimulants. In other words, UFAs act as second messengers in the elicitation of an OB in MPs.

The O_2^- forming NADPH oxidase has not been isolated in pure form and most information related to it originates from studies performed on neutrophils. It is likely that the enzyme found in MPs is very similar to that of neutrophils. The neutrophil enzyme contains FAD and cytochrome b and the present view of it is that of a multicomponent system transferring electrons from NADPH to oxygen by sequential reduction and reoxidation of FAD and cytochrome b (5). In both MPs and neutrophils, the activated enzyme has been clearly localized to the plasma membrane (or to its invaginated form, the phagosome membrane) but uncertainty exists concerning the localization of the enzyme, or its individual components, in resting cells. We have recently succeeded in demonstrating the activation of the O_2^- forming enzyme in homogenates derived from resting MPs by UFAs (7). Using this system we found that enzyme activation requires the cooperation between membrane-associated and cytosolic components (7). Recent data in neutrophils indicate that elicitation of an OB in neutrophils involves translocation of a cytosolic or intracellular granule-associated component to the plasma membrane. The Km for NADPH of the C20:4- activated cell-free O_2^- forming enzyme derived from resting guinea pig MPs (0.05 mM) was found to be very similar to the Km for NADPH of the enzyme extracted from stimulated MPs (0.03-0.09 mM) (7). The similarity of the C20:4-activated cell free enzyme to the enzyme recovered from intact stimulated cells further supports the hypothesis that endogenously produced UFAs are the physiologic second messengers of O_2^- production. The mechanism by which UFAs activate the NADPH oxidase enzyme complex is a key unresolved question and is at the core of contemporary research in this field. Of significance for the understanding of the modulation of OB activity in MPs is the concept that its elicitation is a two stage process. The first stage involves the interaction of the stimulus with a membrane receptor, the activation of a phospholipase and the liberation of UFA; the second stage consists of the activation of the NADPH oxidase by the UFA. Modulatory effects can, therefore, be exerted at multiple points along this sequence of steps.

LYMPHOKINE MODULATES OXYGEN RADICAL PRODUCTION BY MACROPHAGES

MPs obtained from immunized and antigen-boosted mice display a parallel increase in the ability to kill intracellular microorganisms (Trypanosoma cruzi, Toxoplasma gondii or Candida) and in the production of H_2O_2 in response to an antigenically unrelated stimulus, such as PMA. Nathan et al. (23) were the first to find that in vitro culture of MPs with LK-containing supernatant results in enhanced production of H_2O_2 in response to PMA that is closely allied to the augmented cytotoxicity against trypanosoma. The enhanced capacity of LK-activated MPs to liberate H_2O_2, or both H_2O_2 and O_2^-, in response to a nonspecific stimulus and its correlation with optimal killing of certain intracellular pathogens was repeatedly confirmed in the mouse (18) and man (17,21). The true picture is, however, considerably more complex. This complexity is the result of: (a) the extreme variety of cells simplistically categorized as "macrophages;" (b) the type of OR that is produced; (c) the cellular site of its production; (d) the existence of intracellular scavenging systems; (e) the multiplicity of triggers used to elicit an OB; and (f) the variable susceptibility of targets to oxygen-derived metabolites. Thus, LK effects on oxidative metabolism are commonly studied on MPs with a low intrinsic capacity to produce O_2^- and H_2O_2, such as resident peritoneal MPs, MPs elicited by weak stimulants (such as proteose-peptone) (18,23) or blood monocytes kept in culture for several days, during which period the H_2O_2 producing capacity declines markedly (17,21). Some murine and human MP cell lines are also excellent candidates for such studies because of their poor oxidative response in the absence of conditioning by LK. The frequently reported LK-enhanced oxidative response to PMA could, however, not be demonstrated in guinea pig MP elicited by mineral oil that have a high constitutive capacity to produce O_2^- and H_2O_2 in response to PMA (10). The suitability of resident MPs and of cell lines mimicking their characteristics for the study of LK effects on oxidative metabolism is questionable because resident cells, as opposed to blood-derived inflammatory MPs, are unlikely to play a major role in anti-infectious defense and are also known to exhibit a lower responsiveness to LKs. Little is known on the effect of LK on the oxidative metabolism of tissue MPs (alveolar, Kupffer cells, glial cells). It is also commonly found that the LK effect is more readily demonstrable when H_2O_2 production, rather than O_2^- liberation, is measured (11,21). The most likely explanation for this finding is a preferential effect of LK on "intracellular" O_2^- generation that is followed by rapid dismutation to H_2O_2 (see later section of this chapter). The assay of nitroblue tetrazolium (NBT) reduction (always an intracellular event) also, occasionally permitted demonstration of a LK effect, when simultaneous measurements of O_2^- and even H_2O_2 proved fruitless (18). LK-activated

MPs exhibiting enhanced extracellular release of H_2O_2 were, sometimes, incapable of destroying intracellular pathogens but became active in the presence of catalase inhibitors or when derived from acatalasemic animals (19). Putting in evidence the modulatory effect of LK is also highly dependent on the type of OB eliciting trigger that is employed. An enhanced response of LK-treated MPs to strong elicitors, such as PMA or opsonized zymosan, is mostly associated with low intrinsic responsiveness of the cells before LK exposure (23) or loss of responsiveness by in vitro maintenance in the absence of LK (21). This phenomenon is also clearly illustrated by results obtained with oil-elicited guinea pig MPs; these cells exhibited a markedly LK-augmented O_2^- production in response to Na nitroprusside (a weak stimulant for untreated cells) but unchanged or even reduced O_2^- production in response to PMA and zymosan (10). A similar example is that of mouse resident MPs that demonstrate good zymosan-elicited NBT reduction but failed to reduce NBT after ingestion of toxoplasmas: incubation with LK led to strong toxoplasma-elicited NBT reduction (18). This issue is further complicated by the well known property of LK to promote the expression of a number of membrane receptors on MPs, an effect that is, in all likelihood, distinct from the modulatory influence on oxidative metabolism. While, as it will be discussed later, exposure to LK appears to affect the enzymatic apparatus responsible for OR production independently of receptor modulation, a successful bactericidal effect requires both effective receptor triggering by the pathogen and, independently, an enhanced capacity of the MPs to generate toxic ORs. This double requirement is well illustrated in the killing of leishmania amastigotes by LK-activated MPs but not by normal or in vivo activated cells (16). Finally, the modulatory effect of LK on MP oxidative metabolism, is sometimes, obscured by excessive reliance on biological assay systems such as cytotoxicity or cytotaxis. This is the result of a widely different susceptibility of various targets to ORs produced by LK-activated cells. Examples for this are the resistance of toxoplasma (19) and leishmania amastigotes (16) to oxygen-dependent killing by MPs, mediated by high concentrations of parasitic catalase and glutathione peroxidase. In addition, endogenous MP scavenging enzymes or artefactual scavengers (such as thioglycollate medium) affect the final outcome of the microbicidal process. The finding that LK treatment itself can affect the level of endogenous antioxidant enzymes (11,19) provides an additional element of complexity.

THE EFFECT OF LYMPHOKINE ON O_2^- AND H_2O_2 PRODUCTION BY GUINEA PIG

INFLAMMATORY MACROPHAGES

We focused our studies on mineral oil-elicited peritoneal MPs of guinea pigs. These cells are characterized by a high intrinsic capacity to respond by O_2^- and H_2O_2 production to a wide variety of stimuli (31). We first examined the effect of exposing such MPs to a crude LK-containing preparation for one to three days on O_2^- production in response to nine OB elicitors of various potencies. It was found that LK treatment for three days enhanced O_2^- production elicited by WGA, formyl-Met-Leu-Phe, phospholipase C and Na nitroprusside but not that provoked by PMA, zymosan, Con A, A23187 and NaF (10). The response to this latter group of stimuli was, in fact, inhibited by LK treatment; this inhibitory effect was more evident at higher concentrations of LK. The enhancing effect of LK on the response to the first group of stimuli was best expressed on day three, a time at which the response of MPs maintained in medium was considerably reduced, having peaked after one day in culture. On the contrary, the inhibitory effect of LK on O_2^- production, evoked by the second category of stimuli coincided with the time when the response of untreated cells to these was at its peak. At present, it is not known whether the opposite effects are caused by distinct molecular entities present in crude LK preparations. These findings demonstrate anew that: (a) the enhancing effect of LK is best expressed on cells that have a low intrinsic capacity of generating ORs; and (b) modulation by LK is a slow process, normally requiring two to three days of contact with MPs. The direction of the modulatory effect of LK on OR production in response to specific stimuli is not always predictable. Thus, in vivo activated mouse peritoneal MPs were reported to show an augmented response to PMA and no response to WGA (37), a situation diametrically opposed to that found by us with LK-activated guinea pig MPs.

Because of the common finding that liberation of H_2O_2 by activated MPs largely exceeds that of O_2^-, we also examined the effect of LK treatment on stimulus-elicited H_2O_2 production. We found that elicited guinea pig MPs cultured in LK-containing medium produce large amounts of H_2O_2 in the absence of additional stimulation (11). H_2O_2 generation was not accompanied by O_2^- liberation. Spontaneous LK-dependent H_2O_2 production reached a peak after three days in culture (Figure 1) and could only be demonstrated by employing conditions permitting the uptake of the assay reagents (phenol red and horse-radish peroxidase, see reference 30) into intracellular vesicles, probably via pinocytosis (11,29). This and other arguments supported the hypothesis that the LK-induced H_2O_2 production is limited to the intracellular sector

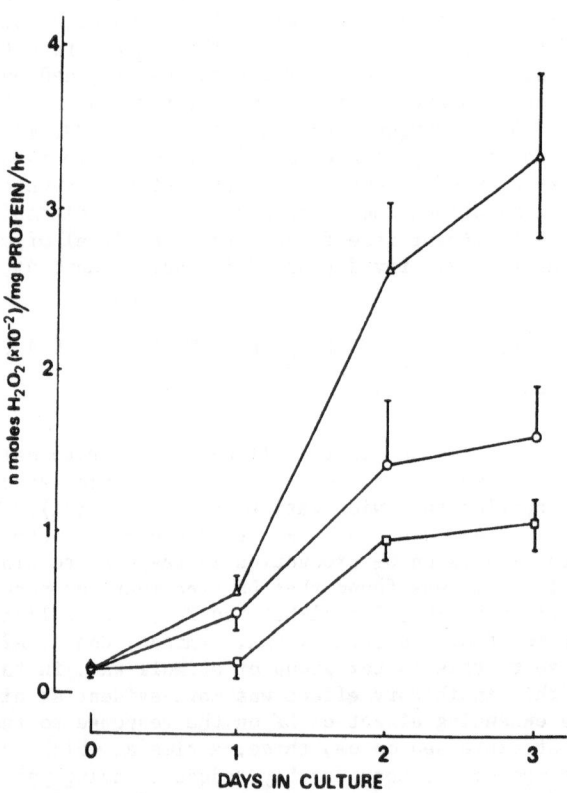

Figure 1. Time kinetics of H_2O_2 production by lymphokine-treated guinea pig peritoneal macrophages. Results represent means of 7 experiments ± SEM. Macrophages were treated with medium (□), control supernatant (o) or lymphokine-containing supernatant (Δ) for the indicated time intervals.

and probably takes place within the endocytic vesicles. The fact that no O_2^- could be detected under these conditions is, probably, the consequence of the inaccessibility of intravesicular O_2^- to the customary assay reagent, ferricytochrome c. The mechanism by which intracellular O_2^- is converted to H_2O_2 is unsettled. In the guinea pig model, we offered evidence for spontaneous (nonenzymatic) dismutation being the dominant mechanism (11,29). On the other hand, Tsunawaki and Nathan (38) found that mouse MPs activated in vivo and stimulated by PMA produce augmented amounts of H_2O_2 that is predominantly derived via enzymatic dismutation effected by Cu,Zn-dependent SOD.

The possible reasons for the preferential activation of intracellular H_2O_2 production by LK and the apparent noninvolvement of the plasma membrane were recently discussed (29). It is noteworthy that such a mechanism makes good biological sense; the principal candidates for cytotoxic attack by H_2O_2 are phagocytosed pathogens and a silent plasma membrane reduces the opportunities for concomitant nonspecific tissue damage. The finding that LK treatment by itself can elicit H_2O_2 liberation stands out among numerous reports of LK augmenting H_2O_2 liberation elicited by common membrane stimulants. It is, however, significant that the relatively few descriptions of a direct LK effect are based on techniques that preferentially detect intracellular OR production (NBT reduction) or are sufficiently nonspecific so as not to be limited to detection of extracellular oxygen metabolites (stimulation of HMPS and augmented oxygen consumption). The fact that our experiments (as well as the majority of work performed by other authors) utilizes crude LK preparations, raises the possibility that the induction of intracellular H_2O_2 production is the result of cooperation among two molecules present in the crude culture fluid: one that acts as a priming agent; the second as a membrane trigger (similar to PMA, zymosan, etc). Such a requirement for two signals in order to activate MPs to the full expression of their effector function is a common finding, especially in the induction of MP cytotoxicity towards tumors. In fact, the existence of molecularly distinct priming and trigger LKs present in the same culture supernatants was proposed by Meltzer (15).

Work on the molecular characterization of the LK responsible for the modulation of MP oxidative metabolism is in its infancy and has been overshadowed by the recent explosion of information indicating that IFNγ is, at least, one of the LKs responsible for the effect (see later section of this chapter). Although there is overwhelming evidence indicating that human and murine IFNγ are capable of enhancing stimulus-elicited H_2O_2 production by MPs, we know little about the existence of LKs other than IFNγ that express such activity. That this might well be so is indicated by recent findings from our laboratory showing that the guinea pig material inducing spontaneous H_2O_2 production in elicited MPs is stable at pH 2 for over 48 hours and resistant to heating at 56°C for three hours (D. Shaag and E. Pick, in preparation). These properties are markedly different from the reported characteristics of murine and human IFNγ (albeit no information is available on the properties of guinea pig IFNγ) and suggest that the active material is not IFNγ. It is, of course, possible that the LK inducing H_2O_2 production in the absence of a second stimulus and that enhancing the response to such stimuli are distinct entities.

THE ENZYMATIC BASIS OF MACROPHAGE ACTIVATION

The mechanism by which exposure of MPs to LK leads to the production of elevated amounts of ORs is still a mystery and its elucidation represents one of the major challenges in this area of cell biology. The following alterations in cell function could explain the augmented oxidative response: (a) a change in the number, properties or intramembranal mobility of receptors for ligands eliciting an OB; (b) a more efficient receptor – NADPH oxidase coupling mechanism; (c) an increase in the content of NADPH oxidase or a change in the kinetic properties of the enzyme; (d) an increase in the amount of NADPH or better accessibility of NADPH to the enzyme.

Although modulation of membrane receptors might contribute to enhanced OR production, it is not the central event. Thus, the markedly increased PMA-elicited H_2O_2 generation of BCG-activated mouse MPs occurs in the absence of any significant change in the number or affinity of membrane receptors for PMA (39). Similarly, LPS and MDP-activated guinea pig MPs exhibiting increased O_2^- release in response to WGA did not bind more WGA than control cells (12). The existence of a more efficient receptor-enzyme coupling mechanism was suggested by us (26,27) in the light of findings that UFAs function as second messengers in the activation of NADPH oxidase in MPs (6,7). The essence of this hypothesis is that LK treatment causes changes in MP phospholipid composition and/or topography or in the activity of enzymes involved in phospholipid degradation or regeneration (phospholipases A_2 and C and acyltransferases). The net result of these changes is that LK-treated MPs respond to OB elicitors by a more pronounced increase in the level of unesterified UFAs in the membrane, most likely in circumscribed microdomains adjacent to NADPH oxidase. The elevated concentration of free UFA leads to activation of the O_2^- forming NADPH oxidase. Work from our laboratory has indeed shown that LK-treated guinea pig MPs prelabelled with ^3H-C20:4 liberate increased amounts of ^3H-C20:4 in response to certain OB elicitors (WGA, formyl-Met-Leu-Phe and phospholipase C) (E. Pick, in preparation). LK-dependent increases in ^3H-C20:4 liberation were limited to those stimuli that also elicited more O_2^- production in LK-treated cells (10). Another level at which LK might modulate the activity of NADPH oxidase is by influencing the pathways of C20:4 oxygenation. We have shown that products of lipoxygenase can activate NADPH oxidase of MPs while cyclooxygenase metabolites are inactive (6,7). Another explanation for the LK effect is that LK influences the O_2^- forming NADPH oxidase itself. It is a general finding that when cell homogenates were prepared from MPs activated in vivo (32,38) or in vitro (12) and stimulated by OB elicitors, the velocity of NADPH-dependent O_2^- production (NADPH oxidase activity) was found to be significantly higher than that of nonactivated cells. The oxidase from mouse MPs activated with LPS in vivo stimulated in vitro by PMA had a higher Vmax and a lower Km for NADPH than the oxidase of normal cells (32). When maintained in vitro overnight, the difference in Km and velocity between activated and control MPs was even more pronounced but Vmax values were similar. It was also found that LPS-activated MPs have a higher content of NADPH (32). In similar studies performed on mouse MPs activated in vivo by casein or periodate and stimulated by PMA, it was found that activation is associated with a marked decrease in the Km for NADPH but little change in Vmax (38). No correlation was found between the lowered Km value and cytochrome b content of activated MPs (38). The data available so far suggest that the predominant change resulting from activation is a lowering of Km permitting a more efficient utilization of intracellular NADPH. An increase in the cellular content of NADPH oxidase cannot be the sole explanation for such a change and it seems likely that we are dealing with a new type of enzyme or a modification of the existing enzyme. The possibilities, therefore, include a change in gene expression or an enzyme modified by allosteric change, phosphorylation, removal of an inhibitor or an altered membrane (phospholipid) environment. In the light of the recent evidence in favor of the multicomponent nature of NADPH oxidase in neutrophils (5,14) and MPs (7), more complex events can be envisaged. One possibility is suggested by the LK-like enhancing effect formyl-Met-Leu-Phe on O_2^- production by neutrophils that is accompanied by the translocation of a cytosolic factor to the cell membrane (14). The interaction of the cytosolic factor with a membrane-localized component was shown to be required for the activation of NADPH oxidase in MPs (7) and neutrophils (14). It is conceivable that LK treatment promotes the assembly of the components of NADPH oxidase in the membrane generating a complex that is more readily activated by OB elicitors. A likely candidate for the yet unidentified cytosolic factor, transferred to the membrane, is protein kinase C. We have recently suggested that phosphorylation of a component of NADPH oxidase by protein kinase C may be the mechanism by which the enzyme is activated (7).

No information is yet available on changes in the properties of NADPH oxidase in MPs immunologically activated in vivo (by a process involving T cells); paradoxically, the published reports (12,32,38) are limited to MPs activated by LPS, MDP, periodate or casein. Also, there are no data on changes in the O_2^- forming enzyme in MPs treated with LK in vitro. Such information is urgently required.

DOES LYMPHOKINE-MEDIATED ACTIVATION INVOLVE CHANGES IN ANTIOXIDATIVE ENZYMES

In the search for the enzymatic basis of LK action, it was also suggested that LK might affect the level of endogenous OR scavenging enzymes. A decrease in intracellular scavenging capacity might result in a net increase in ORs, some of which could be liberated from the cell. The principal experimental findings are the following: SOD activity increased moderately during in vitro maintenance of mouse (19), and guinea pig (M. Freund and E. Pick, unpublished) MPs and human monocytes (22). LK treatment had a moderate enhancing effect in mouse MPs (19) but did not significantly affect SOD levels in man (21) and guinea pig (M. Freund and E. Pick, unpublished). The increase in SOD activity found in mouse MPs might be of significance in the light of the recent claim that H_2O_2 produced by activated mouse MPs is principally derived by intracellular enzymatic dismutation mediated by Cu,Zn-containing SOD (38). Further work is required in order to elucidate in more detail the effect of LK on the two forms of mammalian SOD (Cu,Zn-containing and Mn-containing enzymes).

Glutathione peroxidase activity was unchanged in cultured mouse MPs (19) but tended to decrease in guinea pig (28) MPs and human monocytes differentiating in vitro (22). LK treatment slightly elevated glutathione peroxidase in mouse (19) and guinea pig MPs (28) but did not affect its level in human MPs (21). The possibility that LK might influence levels of the selenium-independent form of the enzyme was little explored. The activity of glutathione reductase was unaffected by LK treatment in both human (21) and guinea pig (28) MPs. Also, the cellular level of glutathione was unaffected by incubation of human (21) and guinea pig (28) MPs with LK.

Considerable interest has focused on changes in catalase levels in LK-treated MPs following the finding that inhibition of catalase by drugs enables LK-activated mouse MPs to express antitoxoplasma activity (19). Diametrically opposite results were reported in different animal species. Thus, cultured mouse MPs display a spontaneous increase in catalase levels that is further enhanced by LK exposure (19). Elicited guinea pig MPs (11) and human monocytes differentiating in culture (22) exhibit a fall in catalase. This fall is further enhanced by LK treatment of guinea pig (11) but not of human (21) MPs.

The principal conclusion to be derived from these results is that the enhanced OR production characteristic of LK-activated MPs is not the consequence of a lowered activity of scavenging enzymes but rather of an increase in the activity of the O_2^- generating enzyme. It is, however, possible that some changes in the activity of antioxidant systems are under LK control and might exert a modulatory influence on the final balance of intracellular and excreted ORs. It remains to be found out whether the effects on NADPH oxidase and on the scavenging enzymes are exerted by the same LK or by distinct molecular entities.

347

ROLE OF INTERFERONS IN THE MODULATION OF THE OXIDATIVE

METABOLISM OF MACROPHAGES

There is an ever increasing amount of experimental evidence demonstrating that the LK responsible for priming MPs for tumor cell killing is identical to IFNγ. This discovery was soon followed by the report that the LK that enhances the oxidative metabolism of human MPs (derived by in vitro differentiation of blood monocytes) is also identical to IFNγ (24). The IFNγ-enhanced production of H_2O_2 correlated well with an increased capacity to kill toxoplasmas (24). The identity between the LK enhancing H_2O_2 production and toxoplasma killing and IFNγ was proven by two approaches: first, both activities were removed from LK-containing culture supernatants with monoclonal antibody to IFNγ but not with antibodies to IFNα and IFNβ; secondly recombinant DNA-derived IFNγ displayed all the properties of the lymphocyte derived LK. In similar studies, it was reported that IFNγ is the active agent in antigen- or mitogen-induced human LK preparations capable of activating human monocyte-derived MPs to kill Leishmania donovani promastigotes and amastigotes (20). In this latter case it was found, however, that killing of leishmania cannot be attributed exclusively to the IFN-activated OR production; MPs of a patient with chronic granulomatous disease (CGD) also responded to IFNγ by augmented killing of leishmania promastigotes (20). The finding that IFNγ also enhances oxygen-independent microbicidal mechanism in MPs appears less unexpected when one considers the following findings: (a) LK-containing culture supernatants enhance the microbistatic activity of MPs from CGD patients and of mouse fibroblasts against toxoplasma and chlamydia; (b) IFNγ blocks the growth of chlamydia and rickettsia in mouse fibroblasts (that do not produce ORs); and (c) the IFNγ induced tumoricidal activity of MPs is also oxygen-independent.

In spite of the wealth of data supporting a key role for IFNγ in all aspects of MP activation and in the modulation of oxidative metabolism, in particular, the existence of MP activating LKs other than IFNγ cannot be excluded. Demonstrating their presence is not always simple, especially because lack of antiviral activity is not a valid argument; MP activation can be demonstrated at dilutions of IFNγ that cannot be measured in the antiviral assay. The use of specific monoclonal antibodies against IFNγ represents a more fruitful approach although the existence of identical or closely related antigenic epitopes on two molecules with identical biological activity cannot be excluded. We have discussed earlier the preliminary characterization of an acid- and heat-resistant guinea pig LK with potent H_2O_2 inducing capacity.

INHIBITION OF THE OXIDATIVE METABOLISM OF MACROPHAGES

LK-induced inhibition of O_2^- production by MPs in response to some stimulants was described by us in the guinea pig (10) and discussed earlier in this chapter. Evidence for a suppressive factor in human LK preparations counteracting the positive effect of IFNγ was also presented by Nathan et al. (24). Medium originating from tumor cells and certain nontumoral cells was also found to contain a factor that suppressed the capacity of MPs to produce O_2^- and H_2O_2 and also prevented the enhancement of OR production by LK (36). The factor also suppressed the destruction of toxoplasma and leishmania by MPs activated in vivo and in vitro (35). The biochemical mechanism by which this deactivation is effected is unknown.

A possible clue to the mechanism of suppression is offered by work showing that IFNβ (3) and IFNα (4) suppress stimulated O_2^- and H_2O_2 production by mouse MPs. This effect may be related to the inhibition of C20:4 liberation described in MPs treated by IFNβ and IFNα (4). In light of the suggestion that C20:4 acts as the second messenger of NADPH oxidase activation (6,7), it is tempting to causally relate that inhibition of C20:4 liberation to the reduced OR production. The true relationship between C20:4 metabolism and OR production of activated MPs is likely to be considerably more complex, as indicated by the fact that in vivo activated MPs have reduced stimulus-dependent C20:4 liberation but exhibit increased OR production (33). It is possible that the relative amounts of the various products of C20:4 oxygenation pathways are at least as important as the absolute quantity of C20:4 liberated, in determining the intensity of the OB. Thus, preferential channelling of C20:4 through the cyclooxygenase pathway would result in the inhibition of OR production while a dominant lipoxygenase pathway would have an enhancing effect.

CONCLUSION

The principal message of this chapter is that immunological activation of MPs, the essential elements of which are evident in MPs exposed to LK in vitro, involves major changes in OR production. The essential function of these is to mediate cytostasis or killing of phagocytosed pathogens while causing minimal damage to adjacent cells. It should be clear that this metabolic change is not synonymous to MP activation and that quite a number of other cytotoxic mechanisms directed against both intra- and extracellular targets exist. The two most common explanations for this multiplicity of effector pathways are: MP heterogeneity and the involvement of more than one LK. Defining MP heterogeneity is being rapidly advanced by the availability of a wide array of monoclonal antibodies defining MP surface determinants and a clearer picture is certain to emerge soon. At the other pole of the issue, the "omnipotency" of IFNγ came as a surprise to many of the oldtimers in the field. Its involvement in both intra- and extracellular killing by both oxygen-dependent and independent mechanisms, not to mention its antiviral and general immunoregulatory functions, makes IFNγ the most "versatile" LK discovered so far. At the breath-taking rate at which research in this area is progressing, it is not unlikely that quite a number of statements made in this chapter might become obsolete by the time that it comes off the printing press.

ACKNOWLEDGMENTS

The author's work described in this chapter was supported by grants from the Arpad Plesch Research Foundation and the United States-Israel Binational Science Foundation, grant no. 2730. We thank Mrs. Patricia Bar-On for excellent secretarial assistance.

REFERENCES

1. B. M. Babior, Oxygen-dependent microbial killing by phagocytes, New Eng. J. Med. 298:659 (1978).
2. J. A. Badwey and M. L. Karnovsky, Active oxygen species and the functions of phagocytic leukocytes, Ann. Rev. Biochem. 49:695 (1980).
3. D. Boraschi et al., IFN beta-induced reduction of superoxide anion generation by macrophages, Immunology 45:621 (1982).
4. D. Boraschi, S. Censini, and A. Tagliabue, Interferon gamma reduces macrophage suppressive activity by inhibiting prostaglandin E_2 release and inducing interleukin 1 production, J. Immunol., in press (1985).
5. N. Borregaard and A. I. Tauber, Subcellular localization of the human neutrophil NADPH oxidase, J. Biol. Chem. 259:47 (1984).

6. Y. Bromberg and E. Pick, Unsaturated fatty acids as second messengers of superoxide generation by macrophages, Cell. Immunol. 79:240 (1983).

7. Y. Bromberg and E. Pick, Unsaturated fatty acids stimulate NADPH-dependent superoxide production by cell-free system derived from macrophages, Cell. Immunol., in press (1985).

8. Z. A. Cohn, The activation of mononuclear phagocytes: fact, fancy and future, J. Immunol. 121:813 (1978).

9. R. E. Fowles, et al., The enhancement of macrophage bacteriostasis by products of activated lymphocytes, J. Exp. Med. 138:952 (1973).

10. M. Freund and E. Pick, Biochemistry of lymphokine action on macrophages-modulation of macrophage superoxide production by lymphokine, in: "Thymic Hormones and Lymphokines," A. L. Goldstein, ed., Plenum Press, New York, in press (1985a).

11. M. Freund and E. Pick, The mechanism of action of lymphokines. VIII. Lymphokine-enhanced spontaneous hydrogen peroxide production by macrophages, Immunology in press (1985b).

12. M. Kaku et al., Enhanced superoxide anion release from phagocytes by muramyl dipeptide or lipopolysaccharide, Infect. Immun. 39:559 (1983).

13. G. B. Macaness, The influence of immunologically committed lymphoid cells on macrophage activation in vivo, J. Exp. Med. 129:973 (1969).

14. L. McPhail, C. C. Clayton, and R. Snyderman, Evidence that activation of human neutrophil NADPH oxidase involves association of a cytosolic factor with membrane components, Clin. Res. 32:315A (1984).

15. M. S. Meltzer, Tumor cytotoxicity by lymphokine-activated macrophages: development of macrophage tumoricidal activity requires a sequence of reactions, Lymphokines 3:319 (1981).

16. H. W. Murray, Cell-mediated immune response in experimental visceral leishmaniasis. II. Oxygen-dependent killing of intracellular leishmania donovani amastigotes, J. Immunol. 129:351 (1982).

17. H. W. Murray and D. M. Cartelli, Killing of intracellular Leishmania donovani by human mononuclear phagocytes. Evidence for oxygen-dependent and independent leishmanicidal activity, J. Clin. Invest. 72:32 (1983).

18. H. W. Murray and Z. A. Cohn, Macrophage oxygen-dependent antimicrobial activity. III. Enhanced oxidative metabolism as expression of macrophage activation, J. Exp. Med. 152:1596 (1980).

19. H. W. Murray, C. F. Nathan, and Z. A. Cohn, Macrophage oxygen-dependent antimicrobial activity. IV. Role of endogenous scavengers and oxygen intermediates, J. Exp Med. 152:1610 (1980).

20. H. W. Murray, B. Y. Rubin, and C. D. Rothermel, Killing of intracellular Leishmania donovani by lymphokine-stimulated human mononuclear phagocytes. Evidence that interferon gamma is the activating lymphokine, J. Clin. Invest. 72:1506 (1983).

21. A. Nakagawara et al., Lymphokines enhance the capacity of human monocytes to secrete reactive oxygen intermediates. J. Clin. Invest. 70:1042 (1982).

22. A. Nakagawara, C. F. Nathan, and A. Z. Cohn, Hydrogen peroxide metabolism in human monocytes during differentiation in vitro, J. Clin. Invest. 68:1243 (1981).

23. C. F. Nathan et al., Activation of macrophages in vivo and in vitro. Correlation between hydrogen peroxide release and killing of trypanosoma cruzi, J. Exp. Med. 149:1056 (1979).

24. C. F. Nathan et al., Identification of interferon gamma as the lymphokine that activates human macrophage oxidative metabolism and antimicrobial activity, J. Exp. Med. 158:670 (1983).

25. R. J. North, The concept of the activated macrophage, J. Immunol. 121:806 (1978).

26. E. Pick and Y. Bromberg, Quo vadis macrophage activation - role of phospholipids in the elicitation of the oxidative burst in macrophages, Transpl. Proc. 14:570 (1982).

27. E. Pick and Y. Bromberg, Regulation of macrophage function by lympho-kines - role of membrane phospholipids, in: "Advances in Immuno-pharmacology 2," J. W. Hadden et al., eds., Pergamon Press, Oxford (1983).

28. E. Pick, Y. Bromberg, and M. Freund, Extrinsic regulation of macrophage function by lymphokines. Effect of lymphokines on the stimulated oxidative metabolism of macrophages, Adv. Exp. Med. Biol. 155:471 (1982).

29. E. Pick and M. Freund, Biochemical mechanisms in macrophage activation by lymphokines: Intracellular peroxide production by lymphokine-treated macrophages, in: "Progress in Immunology," Volume 5, Y. Yamamura and T. Tada, eds., Academic Press, New York (1983).

30. E. Pick and Y. Keisari, A simple colorimetric method for the measure-ment of hydrogen peroxide produced by cells in culture, J. Immunol. Meth. 38:161 (1980).

31. E. Pick and Y. Keisari, Superoxide and hydrogen peroxide production by chemically elicited peritoneal macrophages. Induction by multiple nonphagocytic stimuli, Cell. Immunol. 59:301 (1981).

32. M. Sasada, M. J. Pabst, and R. B. Johnston, Activation of mouse peri-toneal macrophages by lipopolysaccharide alters the kinetic para-meters of the superoxide-producing NADPH oxidase, J. Biol. Chem. 258:9631 (1983).

33. W. A. Scott et al., Regulation of arachidonic acid metabolism by macrophage activation, J. Exp. Med. 155:1148 (1982).

34. H. B. Simon and J. N. Sheagren, Enhancement of macrophage bactericidal capacity by antigenically stimulated immune lymphocytes, Cell. Immunol. 4:163 (1972).

35. A. Szuro-Sudol, H. W. Murray, and C. F. Nathan, Suppression of macro-phage antimicrobial activity by a tumor cell product, J. Immunol. 131:384 (1983).

36. A. Szuro-Sudol and C. F. Nathan, Suppression of macrophage oxidative metabolism by products of malignant and nonmalignant cells, J. Exp. Med. 156:945 (1982).

37. H. Tomioka and H. Saito, Hydrogen peroxide-releasing function of chemically elicited and immunologically activated macrophages: dif-ferential response to wheat germ lectin and concanavalin A, Infect. Immun. 29:469 (1980).

38. S. Tsunawaki and C. F. Nathan, Enzymatic basis of macrophage activation. Kinetic analysis of superoxide production in lysates of resident and activated mouse peritoneal macrophages and granulocytes, J. Biol. Chem. 259 (1984).

39. J. B. Weinberg and M. A. Misukonis, Phorbol diester-induced H_2O_2 produc-tion by peritoneal macrophages. Different H_2O_2 production by macro-phages from normal and BCG-infected mice despite comparable phorbol diester receptors, Cell. Immunol. 80:405 (1983).

INDUCTION OF GAMMA INTERFERON BY ENDOTOXIN IN "AGED" MURINE

SPLENOCYTE CULTURES

D. K. Blanchard, T. W. Klein, H. Friedman, and
W. E. Stewart II

Department of Medical Microbiology and Immunology
University of South Florida College of Medicine
Tampa, Florida

Escherichia coli lipopolysaccharide (LPS) is a lymphocyte mitogen and stimulator of the immune response. It is known to induce alpha/beta interferon (IFN) in murine splenocyte cultures when added at culture initiation. However, in the present study, "aged" populations of splenocytes were seen to produce gamma IFN when stimulated by LPS. Whole spleen cell cultures, which were allowed to incubate 24, 48, or 72 hours before the addition of 10 µg/ml endotoxin, elaborated approximately 150 units of IFN per ml of culture fluid. All but about 20% of this antiviral activity was neutralized by monoclonal antibody to gamma IFN. Furthermore, the addition of 10 µg/ml polymyxin B to 48-hour cultures resulted in the abrogation of gamma IFN and alpha/beta IFN production, demonstrating that LPS is responsible for the induction of both types of IFNs. Aged adherent spleen cell cultures produced a mixture of IFNs, the majority of which was not neutralizable by anti-IFN-gamma, suggesting that macrophages produced predominantly alpha/beta IFN in response to LPS regardless of time in culture. In other experiments, whole spleen cell cultures were depleted of T cells by treatment with anti-Thy 1.2 plus complement and then stimulated with endotoxin. The addition of LPS to T-depleted populations at the start of incubation resulted in a predominantly alpha/beta response, as expected. However, the production of gamma IFN in response to LPS in aged spleen cell cultures was abrogated by T cell depletion. These results indicate that the production of gamma IFN in response to LPS is dependent upon an aged population of T cells. Since interleukin 2 (IL-2) was noted to spontaneously increase in the supernatants of unstimulated cultures, it is possible that in vitro "activation" of T cells by IL-2 is responsible for gamma IFN production by LPS-stimulated lymphocytes. The interaction between LPS, gamma IFN, and other lymphokines may provide a mechanism for the adjuvancy and immunomodulatory properties reported for endotoxin.

INTRODUCTION

While IFNs have traditionally been considered to be anti-viral and anti-proliferative proteins, recent studies have demonstrated their potential for altering numerous immune responses. Gamma IFN appears to be particularly effective in activating macrophages (3), enhancing natural killer cell activity (2), and increasing cell surface Ia antigen expression of many cell types (15). There have also been many reports of the induction of IFNs by bacteria and bacterial products, e.g., Listeria monocytogenes (8), Coryne-bacterium parvum (12), Streptococcal OK-432 antigen (10), Staphylococcal enterotoxins (4), and endotoxin (5). Because of these various immunomodu-latory functions, it is tempting to speculate that IFNs play a role(s) in the resistance to bacterial infections. The induction of gamma IFN may be particularly relevant in the host's response to invasion by microorganisms, initially by activating macrophages as a non-specific line of defense, and secondly by modulating surface antigen expression of specifically responding lymphocytes.

E. coli endotoxin has also demonstrated immunomodulatory properties, activation of macrophages (9), suppression of antibody formation (14), and induction of tumor necrosis factor in immunized mice (1). Under some condi-tions, depending on time and dose effects, LPS displays adjuvancy effects (6). It has also been considered to be primarily a B cell mitogen and macro-phage stimulator, and can induce alpha/beta IFN from these cell populations (5). Recently, however, several T cell clones and splenic populations have been reported to respond to LPS with proliferation (7,13), indicating that some T cells are also stimulated by LPS. In the present study, the induc-tion of gamma IFN, primarily considered to be a T cell product by LPS was demonstrated to require a 24- to 48-hour "aged" or "activated" population of splenocytes.

EXPERIMENTAL METHODS

Animals

Inbred BDF1 and C3H/HeJ mice (Jackson Laboratory, Bar Harbor, Maine) and C3H/HeN (heterozygous or homozygous for the nude trait; colony main-tained at the University of South Florida College of Medicine) were used for these studies. The animals were eight to ten weeks old at the time of each experiment and were fed and cared for according to the guidelines set forth in the Guide of the Care and Use of Laboratory Animals.

Lymphoid Cell Preparations

Mice were killed by cervical dislocation and their spleens were removed. Cell suspensions were prepared by macerating individual spleens in RPMI 1640 (Gibco, Grand Island, New York) containing 10% fetal calf serum (Hyclone, Logan, Utah) and antibiotics (Penicillin/Streptomycin) with 5 x 10^{-5} M 2-mercaptoethanol (Sigma Chemical Company, St. Louis, Missouri). The cells were washed several times in Hanks balanced salt solution (HBSS; Gibco), centrifuged at 200 g at 4°C, and used either unfractionated or after separations were performed by allowing whole spleen cell preparations to adhere to wells of plastic plates for 2, 24, 48, or 72 hours at 37°C, then washing the wells to remove the nonadherent cells with warm medium and adding fresh medium to a final volume of one ml per well. The nonadherent cells were centrifuged and resuspended with fresh medium to a final volume of one ml per well. Cells were used at a concentration of 10^7 per ml in one ml volumes per well of a 24-well culture plate (Costar).

T-cell Depletion

T cell depleted cultures were obtained by incubating 10^7 splenocytes with anti-Thy 1.2 (Cappell Laboratories, Westchester, Pennsylvania) for 30 minutes. Cells were then lysed by adding low-tox (rabbit) complement (Cedar-lane Laboratories, Ontario, Canada) and incubating an additional 30 minutes. Remaining cells were washed three times with HBSS and resuspended in complete RPMI. T-cell depletion was assessed by lymphoblastogenic response to 5 µg/ml purified phytohemagglutinin (Wellcome Laboratories, Research Triangle,

Table 1. Characterization of IFNs Induced by E. Coli LPS in
Murine Splenocyte Cultures

Cell Population	Age of Culture (Units IFN/ml)[a]			
	2 hours	24 hours	48 hours	72 hours
Whole spleen[b]	100 (100)[c]	214 (66)	214 (30)	214 (30)
Adherent cells	53 (53)	65	65 (53)	66 (33)
Nonadherent cells	100 (66)	100	167 (33)	214 (33)

[a]Cells were incubated at 37°C for indicated periods before fractionation
and addition of LPS.
[b]Cultures were stimulated by 10 µg/ml of LPS for 24 hours.
[c]Number in parentheses is IFN activity after neutralization by monoclonal
anti-IFNγ.

North Carolina) or 10 µg/ml E. coli LPS. Cultures were considered to be
depleted of T cells if response to PHA was not above media controls while
the LPS response was not depressed.

IFN Assay

Serial half-log$_{10}$ dilutions of cell-free culture supernatant fluids
were made in 96-well flat-bottomed microtiter plates (Costar). To each
well was then added 2×10^4 fresh L929 cells, as described previously (11).
The plates were incubated at 37°C in 5% CO_2 for 24 hours. Vesicular stoma-
titis virus (VSV), at a concentration of 4000 FPU per well, was then added
and the plates were incubated for 18-24 hours. One unit of IFN was calculated as the reciprocal of the dilution of the spleen cell supernatant fluid
in a well which protected 50% of the monolayer from the virus-induced cyto-
pathogenic effects. All units are expressed in international reference
units, as calibrated against reference mouse IFN alpha/beta reagent #6002-
902-26, obtained from NIAID. Neutralization of gamma IFN was performed
using monoclonal anti-IFN-gamma, a gift from Dr. Edward A. Havel, Trudeau
Institute, Saranac Lake, New York.

EXPERIMENTAL RESULTS

The induction of IFNs by LPS differed according to cell population and
age of the culture, as shown in Table 1. For both the whole splenocytes
and the nonadherent population, LPS induces mostly alpha/beta IFN in two
hour cultures, with a switch to gamma IFN production in 24 to 72 hour cultures. LPS-stimulation of adherent cells results in predominantly alpha/beta
IFN, presumably due to macrophages. A similar experiment using C3H/HeJ
splenocytes, either unfractioned or fractionated, resulted in no detectable
IFN at any incubation time (data not presented).

To further characterize the production of IFNs by LPS stimulation,
two-hour or 48-hour whole spleen cell cultures were incubated with LPS in
the presence of 10 µg/ml polymyxin B, an LPS-inactivator (Table 2). Polymyxin B was seen to abrogate the induction of IFNs in both fresh and 48-hour
cultures, with the loss of both alpha/beta and gamma IFNs from the respective
cultures. Heating of the endotoxin did not affect the induction of IFNs,
nor was the susceptibility to polymyxin B altered. These data further support the induction of gamma IFN by LPS in aged, but not freshly prepared,
cultures.

Table 2. Effect of Polymyxin B and Heating on the Induction
of IFNs by LPS in Murine Splenocyte Cultures

Age of Cell Population[b]	Units IFN/ml	
	+ LPS	+ ΔLPS[a]
2 hours		
whole spleen	76 (65)[c]	100 (65)
+ Polymyxin B (10 μg/ml)	<10	<10
48 hours		
whole spleen	117 (10)	100 (15)
+ Polymyxin B	<10	15 (<10)

[a] Heated at 100°C for 30 minutes.
[b] Splenocytes incubated for 2 to 48 hours before stimulation by LPS
[c] Number in parentheses is IFN activity after neutralization by
monoclonal anti-IFNγ.

Since gamma IFN is considered to be primarily a product of T cells,
the role of T lymphocytes in the production of gamma IFN by LPS-stimulation
was explored. Spenocytes from C3H/HeN mice which were heterozygous or homozygous for the nude trait were used. The nude homozygous mice were demonstrated to be depleted of mature T cells. Both heterozygous and homozygous
splenocytes produced alpha/beta IFN when fresh cultures were stimulated by
LPS (Table 3), but only the 48 hour heterozygous cultures elaborated gamma
IFN. Further evidence of T cell dependency is presented in Table 4. Whole
spleen cell cultures were depleted of T cells by incubation with anti-Thy
1.2 and low-tox complement. Neither two-hour nor 48-hour T-depleted cultures
produced demonstrable gamma IFN in the presence of LPS. Stimulation of
T-depleted cultures with PHA, a T-cell mitogen, confirmed the loss of T
cells in the cultures.

DISCUSSION

From these experiments, E. coli endotoxin was seen to induce a mixture
of IFNs, depending upon the age and cell populations in the cultures. Fresh
cultures and adherent cells responded to LPS stimulation by producing alpha/

Table 3. IFN Induction in C3H/HeN (Nude or Heterozygous Nude)
Splenocytes and Effect of Aging

Splenocytes	Units IFN/ml[a]	
	2 hours	48 hours
C3H/nu[+] + LPS	300 (300)[b]	650 (100)
nu/nu + LPS	300 (200)	300 (300)

[a] Splenocytes incubated 2 or 48 hours before stimulation with LPS.
[b] Number in parentheses is IFN activity after neutralization by
monoclonal anti-IFNγ.

Table 4. Effect of T Cell Depletion on Induction of IFNs by LPS

Cell Population[a]	Units IFN/ml	
	+ LPS	+ PHA[b]
2 hours		
whole spleen	100 (100)[c]	200 (<10)
T–depleted	42 (42)	<10
48 hours		
whole spleen	167 (30)	200 (<10)
T–depleted	30 (30)	<10

[a] Splenocytes incubated 2 or 48 hours before stimulation
[b] PHA added at 2 μg/ml.
[c] Number in parentheses is IFN activity after neutralization by monoclonal anti-IFNγ.

beta IFN, while aged, nonadherent cells elaborated predominantly gamma IFN. The induction of gamma IFN by LPS was sensitive to polymyxin B and was unaffected by heating the endotoxin preparation. Studies involving the nude or heterozygous nude mice demonstrated that T cells (which are absent in the nude trait) were required for gamma IFN production, and are likely the cells which produce the IFN. Furthermore, in vitro depletion of T cells from whole splenocyte cultures resulted in the loss of gamma IFN production in 48 hour cultures, while levels of alpha/beta IFN were still present.

Since interleukin 2 is found to increase in culture supernatant fluids upon aging of unstimulated splenocytes (data not shown), the involvement of IL-2 in the induction of gamma IFN by LPS is suggested. The interaction between gamma IFN, IL-2, and other lymphokines may be responsible for the adjuvancy effect noted for endotoxin. Studies to delineate these possible interactions are currently underway.

REFERENCES

1. E. A. Carswell, L. J. Old, R. L. Kassel, S. Green, and B. Williamson, An endotoxin-induced serum factor that causes necrosis of tumors, Proc. Natl. Acad. Sci. USA 72:3666 (1975).
2. H. Claeys, J. van Danone, M. deLey, C. Vermylen, and A. Billiau, Activation of natural cytotoxicity of human peripheral blood mononuclear cells by interferon: a kinetic study and comparison of different interferon types, Brit. J. Haematol. 50:85 (1982).
3. Y. Fukazawa, K. Kagaya, H. Miura, T. Shinoda, K. Natori, and S. Yamazaki, Biological and biochemical characterization of macrophage activating factor (MAF) in murine lymphocytes: physiochemical similarity of MAF to gamma interferon, Microbiol. Immunol. 28:691 (1984).
4. H. M. Johnson, G. J. Stanton, and S. Baron, Relative ability of mitogens to stimulate production of interferon by lymphoid cells and to induce suppression of the in vitro immune response, Proc. Soc. Exp. Biol. Med. 154: 138 (1977).
5. N. Maehara and M. Ho, Cellular origin of interferons induced by bacterial lipopolysaccharide, Infect. Immun. 15:78 (1977).
6. J. R. McGhee, J. J. Farrar, S. M. Mihalek, S. E. Mergenhagen, and D. L. Rosenstreich, Cellular requirements for lipopolysaccharide

adjuvancy: a role for both T lymphocytes and macrophages for in vitro responses to particular antigens, J. Exp. Med. 149:793 (1979).

7. A. Mukaida, T. Kasahara, K. Shioirinakano, and T. Kawai, Interleukin 2-dependent T cell line acquires responsiveness to phorbal myristate acetate and lipopolysaccharide in the course of long term culture, Immunol. Comm. 13:475 (1984).

8. A. Nakane and T. Minagawa, The significance of alpha/beta interferons and gamma interferon produced in mice infected with Listeria monocytogenes, Cell. Immunol. 88:29 (1984).

9. G. Peri, N. Polentarutti, C. Sessa, C. Mangioni, and A. Mantovani, Tumoricidal activity of macrophages isolated from human ascitic and solid ovarian carcinomas: augmentation by interferon, lymphokines and endotoxin, Int. J. Cancer 28:143 (1981).

10. M. Saito, T. Ebina, M. Koi, T. Yamaguchi, Y. Kawade, and N. Ishida, Induction of interferonγ in mouse spleen cells by OK-432, a preparation of Streptococcus pyogenes, Cell. Immunol. 68:187 (1982).

11. W. E. Stewart II, "The Interferon System," 2nd Edition, Springer-Verlag, New York (1981).

12. M. Sugiyama and L. B. Epstein, Effect of Corynebacterium parvum on human T-lymphocyte interferon production and T-lymphocyte proliferation in vitro, Cancer Res. 38:4467 (1978).

13. S. N. Vogel, M. L. Hilfiker, and M. J. Caulfield, Endotoxin-induced T lymphocyte proliferation, J. Immunol. 130:1774 (1983).

14. S. M. Walker and W. O. Weigle, Effect of bacterial lipopolysaccharide on the in vitro secondary antibody response in mice. II. Abrogation of the suppressive capacity of endotoxin, Cell. Immunol. 67:168 (1982).

15. G. H. W. Wong, I. Clark-Lewis, A. W. Harris, and J. W. Schrader, Effect of cloned interferon on expression of H-2 and Ia antigens on cell lines of hemopoietic, lymphoid, epithelial, fibroblastic and neuronal origin, Eur. J. Immunol. 14:52 (1984).

CHARACTERIZATION AND LOCALIZATION OF LIPOPOLYSACCHARIDES FOLLOWING THE INGESTION OF E. COLI BY MURINE MACROPHAGES IN VITRO

Robert L. Duncan, Jr.[1], Vernon Tesh[2] and David C. Morrison[2]

Department of Oral Biology[1] and Department of Microbiology and Immunology[2], Emory University School of Medicine Atlanta, Georgia

INTRODUCTION

Lipopolysaccharides (LPS) isolated from the outer membranes of gram negative bacteria have been demonstrated to modulate a spectrum of pathological and immunological events both in vitro and in vivo (14,15,23). Several lines of evidence have suggested that macrophages are not only stimulated by bacterial LPS (3,4,7,16) but may also be involved in the detoxification of this bacterial product (9,17,18). In recent studies designed to follow the fate of E. coli LPS after the uptake and catabolism of whole E. coli by macrophages, we demonstrated that LPS are exocytosed slowly from the macrophage. Furthermore, the LPS remaining within the macrophage at 72 hours, and the LPS released from the macrophage over that period, retain endotoxic activity (5,6). In this chapter, we provide evidence that the LPS remaining within the macrophage at 48-72 hours is associated with macrophage phagocytic vacuoles, and that both the retained LPS and the exocytosed LPS have a significantly enhanced capacity to stimulate splenocytes. Parenthetically, our results also indicate that, as assessed by SDS-PAGE, macrophage processed LPS is enriched for higher molecular weight subunits relative to control LPS. Although the precise macrophage-induced biochemical alterations in LPS which contribute to enhanced immunostimulatory activity remain to be elucidated, these data nevertheless provide evidence that macrophage processing of LPS may contribute to the local amplification of immunologically relevant events.

MATERIALS AND METHODS

Labeling of E. coli

The LPS of a galactose epimerase deficient E. coli J5 (gift of Dr. Loretta Leive, NIH, Bethesda, Maryland) was specifically labeled with ^{3}H-galactose by allowing the organisms to grow in a basic salts medium containing 0.005% unlabeled galactose and 4.0 µCi/ml ^{3}H-galactose with a specific activity of 3.7 Ci/mmol (New England Nuclear, Boston, Massachusetts) as described previously (5). Greater than 80% of the radiolabeled galactose incorporated by the E. coli could be recovered in the LPS isolated from the E. coli by hot phenol-water extraction.

Mice

Female C3Heb/FeJ (The Jackson Laboratory, Bar Harbor, Maine) were maintained in our animal facility on Purina mouse chow and water ad libitum. Mice were eight to 12 weeks old when used in this investigation.

Macrophage Cultures

Macrophage monolayers were derived from peritoneal exudate cells (PEC) harvested from mice three days after an i.p. injection of 3 mls of a 4% solution of Brewer's thioglycollate medium or 1 ml of a 10% solution of proteose peptone (Difco Laboratories, Detroit, Michigan). PEC (1.5×10^6 cell/ml in RPMI 1640 + 2.6% FCS) were added to 24 well tissue culture dishes (1 ml/well), incubated (2 hrs at 37°C) then washed to remove nonadherent cells. Fresh medium was added to the resulting macrophage monolayer. In some experiments, PEC were added to the wells of a 24 well tissue culture dish in which each well contained a round glass cover slip.

Preparation of Macrophage "Processed" and Control LPS

^{3}H-galactose labeled E. coli (2×10^7 bacteria/well) were added to macrophage monolayers, centrifuged (800 x G) for ten minutes, and subsequently incubated (37°C and 5% CO_2, 95% air) for 90 minutes. E. coli not taken up by macrophages were removed by several washings with warm 37°C RPMI 1640, and 1 ml of RPMI 1640 + 2% FCS was added back to each well. We have previously demonstrated that the E. coli which remain associated with macrophage after this procedure are almost exclusively internalized (5). After an additional incubation period of 48 hours at 37°C, the culture medium was collected, pooled and concentrated five fold by ultrafiltration through a YM 5 filter (Amicon Corporation, Lexington, Massachusetts). Macrophages were harvested by disruption by pyrogen free distilled water followed by brief sonication. The resulting macrophage lysates were pooled and particulate material in the macrophage lysates or culture medium were sedimented by low speed centrifugation (300 x G for ten minutes). The LPS in the culture medium or macrophage lysates was concentrated and partially purified by isopycnic density gradient ultracentrifugation in CsCl either before or after a phenol-water extraction step exactly as described previously (5). Control LPS was isolated directly from E. coli by phenol-water extraction and CsCl gradients as described previously (5). Fractions containing peak radioactivity as determined by liquid scintillation counting were pooled and dialyzed against pyrogen-free distilled water.

Preparation of Lipid A

Lipid A was prepared from macrophage processed and control LPS following mild acid hydrolysis (0.05 N HC1; up to 60 min 100°C) and removal of the polysaccharide by dialysis. In an identical experiment, the turbidity (O. D. 650 nm) of the LPS samples and the phosphate content (2) of the soluble supernatant from centrifuged (1000 x g, 5 min) samples was determined at ten minute intervals during the acid hydrolysis. Under these conditions of hydrolysis, 30 minutes was shown to result in maximum turbidity and minimal soluble phosphate. Acid hydrolysis for longer periods of time resulted in an increased soluble phosphate in the supernatant suggesting lipid A was being degraded. For this reason, a 30 minute period of acid hydrolysis was used in the preparation of the lipid A for this investigation. After 30 minutes, 1.0 M TRIS buffer pH 8 (1/10 the sample volume) was added to each sample and the pH was adjusted to 7.5 using NaOH. The lipid A was solubilized by adding 0.1% triethanolamine and sonicating the sample. Polysaccharide was removed by dialysis of the sample against PBS.

Sodium Dodecyl Sulfate Polyacrylamide Gel Electrophoresis (SDS-PAGE) of Macrophage Processed and Control LPS

Samples of macrophage processed or control LPS were boiled for two minutes in SDS. Glycerol and bromophenol blue were added to the samples which were loaded into the stacking gel (4%) of a cylindrical 12% polyacrylamide gel. Following electrophoresis at 200V for six hours, gels were sectioned into 1.0 mm slices using an Autogeldivider (Savant Instruments, Hicksville, New York) and the radioactivity in each fraction was determined by liquid scintillation counting.

Fluorescein Conjugated LPS (Fl-LPS) Labeled E. coli

Fluorescence microscopy was used to localize the LPS within macrophages which had phagocytosed Fl-LPS labeled E. coli. Fluorescein conjugated LPS was prepared by incubating purified E. coli LPS with fluorescein isothiocyanate (0.05 mg/ml) in a 0.25 M carbonate-bicarbonate buffer (pH 9.0) overnight at 4°C. Unconjugated fluorescein was removed from the LPS preparation by extensive dialysis against multiple changes of PBS over a three day period. E. coli were labeled with the Fl-LPS by incubating E. coli with Fl-LPS in PBS at 4°C for 24 hours. After 24 hours E. coli were washed several times by centrifugation to remove LPS which did not become reassociated with the E. coli. Fl-LPS labeled E. coli were added to macrophage monolayers which had been prepared in 24 well tissue culture dishes, the wells of which contained glass coverslips. At various intervals (30 min, 1 hr, 2 hr, and every 24 hrs up to 5 days), coverslips were removed from the wells of the tissue culture dish, washed in medium, and inverted on a microscope slide (on which had been placed a drop of glycerol glycine buffer). The coverslip was sealed to the slide using fingernail polish, and then examined with the aid of a fluorescence microscope.

Splenocyte Proliferation Assay

Splenocyte proliferation was carried out as described previously (13). Briefly, 200 μl of a single cell suspension of C3HeB/FeJ splenocytes (2.5 x 10^6 cells/ml in RPMI 1640) were added to the wells of a 96 well tissue culture dish. Macrophage "processed" LPS or control LPS preparations at various dilutions were added to the cultures in a volume of 20 μl. The amount of LPS added to each culture was estimated by the radiolabel associated with the ^{3}H-galactose labeled LPS preparations. We have previously demonstrated by two distinct assay systems (lethality in mice and Limulus amebocyte lysate clotting) that a precise relationship exists between the amount of radiolabel in macrophage processed and control LPS and their biological activities (5,6). Cultures were incubated for 48 hours (37°C, 5% CO_2, 95% air). ^{3}H-thymidine (0.5 μCi/well) was added to cultures for the last 16 hours of incubation, cells were harvested on glass fiber filters and ^{3}H-thymidine uptake was determined by liquid scintillation counting. Results are expressed as the mean of triplicate determinations. Standard deviations were routinely less than ten percent of the mean.

RESULTS

Localization of LPS Retained by Macrophages

We have recently reported that, up to 48 hours after ingestion of E. coli by macrophages in vitro, at least 50% of the LPS from these organisms remain associated with the macrophage. The cellular location of the macrophage retained LPS was therefore of interest. E. coli labeled with fluorescein-labeled LPS (Fl-LPS) were utilized to enable us to follow the postphagocytic fate of the LPS by fluorescent microscopy. Immediately after

the uptake of F1-LPS labeled E. coli by macrophages, the macrophages appeared
intensely uniformly fluorescent. By 48 hours after the ingestion of F1-LPS
labeled E. coli by macrophages, the fluorescence appeared to be associated
with phagocytic vacuoles rather than distributed throughout the cytoplasm
or in association with the membrane. A representative photomicrograph is
shown in Figure 1.

Mitogenicity of Macrophage Processed and Control LPS

The capacity of LPS to stimulate B-lymphocyte proliferation in in vitro
cultures of murine splenocytes, is perhaps, the prototype immunostimulatory
response attributed to LPS (14). We have, therefore, employed this assay
as a more direct determination of the relative immunostimulatory activity
of the macrophage "processed" LPS. As shown in Figure 2, macrophage processed LPS isolated from both macrophage lysates and macrophage culture
supernatants resulted in the proliferation of murine splenocytes at concentrations of LPS which were several orders of magnitude lower than the concentrations of control LPS required to stimulate similar ^{3}H-thymidine uptake.
No differences in mitogenicity were detected between phenol-water extracted
and CsCl purified macrophage processed LPS and macrophage processed LPS
isolated in CsCl gradients without prior phenol-water purification. Also,

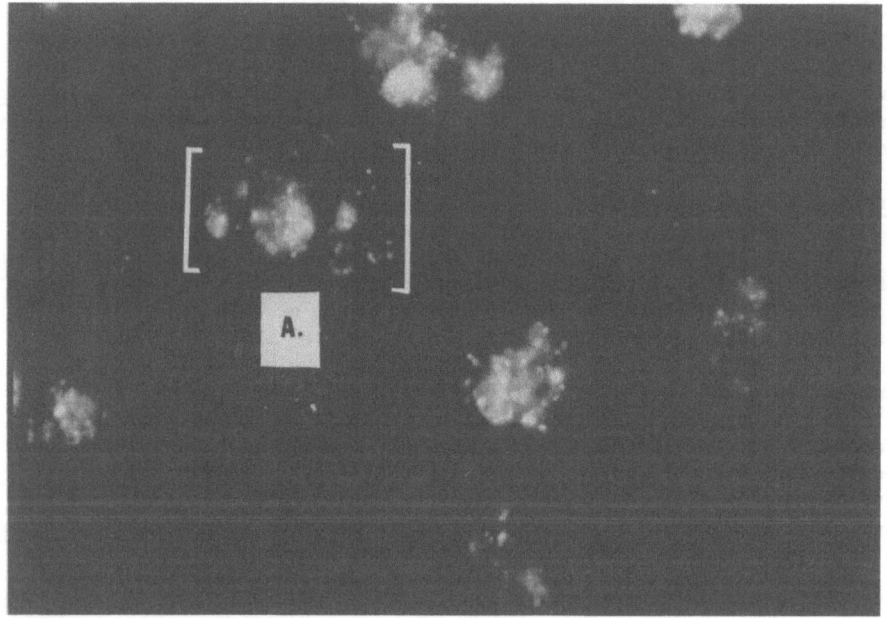

Figure 1. Macrophage phagosome associated F1-LPS. Purified LPS was fluo-
rescenated, dialyzed, then incubated with E. coli. The E. coli
were washed to remove F1-LPS which did not reassociate with E.
coli. The f1-LPS labeled E. coli were added to macrophage monolayers attached to glass coverslips. Cultures containing the
glass coverslips were centrifuged, then incubated at 37°C.
After 90 minutes, E. coli not taken up by macrophages, were removed by washing with warm medium. After 48 hrs to 72 hrs, the
coverslips were removed, washed with PBS, and inverted on a
microscope slide. Photomicrographs were taken of macrophages
under oil emersion illuminated by incident UV light. The braces
(A.) demark a single macrophage.

Table 1. Mitogenicity of Macrophage Processed and Control LPS Before and After Mild Acid Hydrolysis

Dilution[1] of Starting Material	^{3}H-thymidine Uptake (CPM x 10^{-3}) Phenol-water Extracted LPS		Macrophage Lysate LPS		Culture Supernatant LPS	
	LPS	Lipid A	LPS	Lipid A	LPS	Lipid A
1: 10	2.2	1.9	41.3	20.5	ND	ND
1: 50	1.8	1.6	40.2	17.5	30.5	17.4
1:100	1.6	1.3	30.6	13.5	22.2	15.0

[1]LPS preparations were purified on CsCl gradients. All samples were adjusted to an equal concentration of ^{3}H-LPS and divided into two equal aliquots. One aliquot was acid hydrolyzed and neutralized with NaOH as described in the Materials and Methods section. All samples were then dialysed against water and the alliquots assayed in triplicate for mitogenicity. Standard deviations were less than or equal to 10% of the mean CPM.

[2]ND = not done

the lipid A prepared from macrophage processed LPS by mild acid hydrolysis was significantly more mitogenic than lipid A prepared from control LPS in an identical fashion (Table 1).

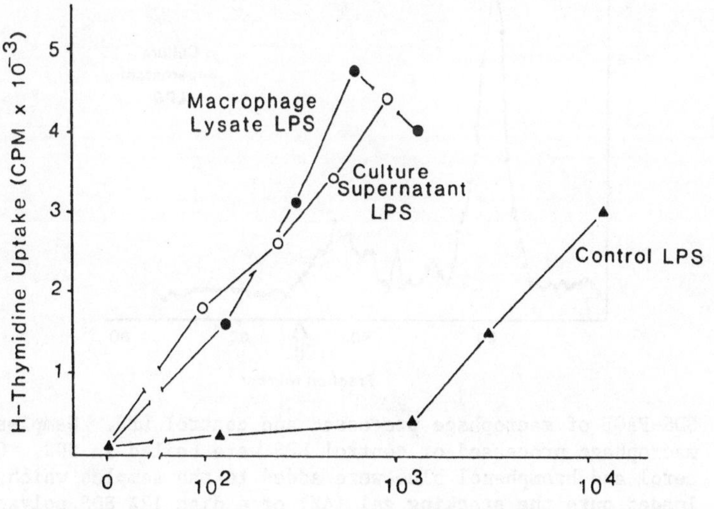

Figure 2. Induction of murine splenocyte proliferation by macrophage processed and control LPS. LPS exocytosed from macrophages which had ingested E. coli (Supernatant LPS O-O) or the LPS retained by macrophages (Macrophage Lysate LPS ●-●), or Control LPS (▲-▲) was added to C3Heb/FeJ murine splenocyte cultures at the LPS concentrations indicated. ^{3}H-thymidine was added to cultures during the last 16 hrs of a 48-hr incubation period. Cells were harvested on glass fiber filters and ^{3}H-thymidine incorporation was determined by liquid scintillation counting.

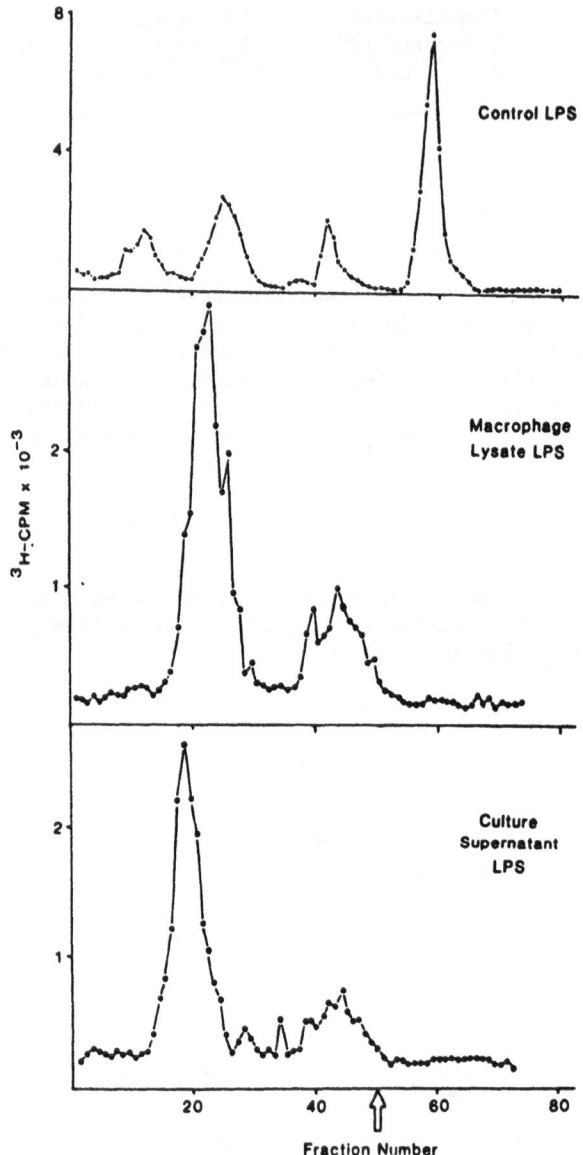

Figure 3. SDS-PAGE of macrophage processed and control LPS. Samples of
           macrophage processed or control LPS were boiled in SDS. Gly-
           cerol and bromphenol blue were added to the samples which were
           loaded onto the stacking gel (4%) of a disc 12% SDS polyacryl-
           amide gel. Gels were electrophoresed at 200 volts for 6 hrs.
           Fractions were collected from the top of the gels and the radio-
           activity in each fraction was determined by liquid scintillation
           counting. The bromphenol blue dye peak is indicated by an
           arrow at approximately fraction number 50.

A second factor which can contribute to the mitogenic activity of LPS is the LPS subunit composition of the LPS macromolecule (21,22). We have therefore considered the possibility that subsequent to E. coli degradation by macrophages the assembly of LPS subunits results in selective association of lipid A rich (R-LPS) subunits. To investigate this possibility, LPS preparations were dissociated in SDS and analyzed for the presence of ^{3}H-galactose by PAGE. Consistent with earlier observations from this laboratory (10,11), control phenol-water extracted, CsCl gradient isolated LPS produced four distinct peaks in SDS-Page (Figure 3). The lowest molecular weight peak was the most prominent, containing roughly twice the amount of radiolabeled material as the other three separate peaks. The LPS isolated from macrophage lysates or from culture supernatants yielded two prominent peaks which roughly corresponded with the two intermediate molecular weight peaks observed with control LPS. No peak corresponding with the prominent low molecular weight peak or the minor high molecular weight peak was observed in the macrophage processed LPS. These data are consistent with the hypothesis that the lower molecular weight R-LPS subunits have been selectively removed, altered and/or partially degraded by the macrophage.

DISCUSSION

We have recently demonstrated that up to 48 hours after the ingestion of E. coli by macrophages in vitro, greater than 50% of the bacterial LPS remains associated with the macrophage (5). The LPS exocytosed from the macrophage and the LPS remaining within the macrophage retain certain classical endotoxic activities (5,6). This investigation was carried out to determine the location of the LPS within the macrophage and to further characterize certain physical and immunopharmacological properties of the macrophage processed LPS.

It is generally accepted that, following the ingestion of microorganisms by macrophages, phagosome lysosome fusion takes place and the lysosomal constituents cause the enzymatic degradation of the microorganism. It has earlier been demonstrated, using radiolabeled Listeria monocytogenes (1,24) and more recently by using radiolabled E. coli (5), that most of the bacterial constituents are rapidly released from the macrophage. In the case of L. monocytogenes, some of the bacterial constituents retained by the macrophage are expressed on the macrophage surface. These surface bound antigens are then presented to lymphocytes in an immunologically relevant fashion (19). Because of the immunopharmacological properties of bacterial endotoxins, it was of interest to determine the cellular location of LPS retained by macrophages after the catabolism of E. coli. By examining macrophages which had ingested Fl-LPS labeled E. coli, it was possible to determine that initially, the LPS appeared to be distributed throughout the cytoplasm and/or the membrane of the macrophage. After 24 hours, the LPS within the macrophage appeared to be associated almost exclusively with discrete granules. While the intensity of the fluorescense associated with these granules diminished in a time dependent manner, it was still detectable up to five days after the macrophage had ingested the E. coli. This is consistent with our previous observation that radiolabled LPS remains associated with macrophages for long periods after other bacterial constitutents have been degraded and released (5).

Because of our previous observations that macrophage processed LPS was indistinguishable from control LPS with respect to its ability to clot Limulus amebocyte lysates or induce lethality in mice (5,6) we queried whether the macrophage processed LPS was likewise immunopharmacologically active. We used LPS mediated splenocyte proliferation to determine the relative immunomodulatory activity of control versus macrophage processed LPS. The concentration of macrophage processed LPS (based on radiolabel associated with these LPS preparations and on Limulus amebocyte clotting) required to induce splenocyte proliferation was significantly less than the concentration of control LPS required. Phenol-water extraction of macrophage processed LPS prior to CsCl gradient isolation did not affect the biological activity of the macrophage processed LPS suggesting that the observed immunoproliferative response was not due to the mitogenicity of lipid associated proteins (LAP) which often contaminate LPS isolated by less harsh means. In addition, we have also shown that macrophage processed LPS remains incapable of stimulating lymphocytes from LPS-low responder C3H/HeJ mice (data not shown). Thus providing additional evidence that mitogenic activity is limited to the LPS preparation.

In order to ascertain if the enhanced activity of the macrophage processed LPS was caused by changes in the carbohydrate or the lipid A region of the LPS molecule, lipid A prepared from macrophage processed and control LPS by subjecting these LPS to mild acid hydrolysis, was tested for its ability to stimulate splenocyte proliferation. While the mitogenicity of LPS were somewhat greater than that of lipid A prepared from an equivalent amount of LPS, the mitogenicity of lipid A prepared from macrophage processed LPS was greater than the mitogenicity of lipid A prepared from control LPS. These data provide evidence that lipid A remains a major contributing component of the immunostimulatory activity of the macrophage processed LPS.

It is noteworthy that R-LPS subunits are not detectable in the macrophage processed LPS. In view of earlier studies from this laboratory demonstrating the relative importance of these subunits to the immunostimulatory activity of LPS (21,22), it is difficult, at the present time, to reconcile the absence of these LPS subunits with increased mitogenic activity of the macrophage processed LPS. However, it has been demonstrated that, under some conditions, LPS can migrate in PAGE as multimer aggregates, which are completely resolved only under rigorous conditions of dissociation (21). It is possible, therefore, that the absence of detectable R-LPS subunits is artifactual, and is, in fact, reflective of macrophage induced changes in lipid A and/or core LPS structures which enhance mitogenic activity and promote subunit aggregation. Additional experiments are currently in progress to distinguish these possibilities.

REFERENCES

1. P. M. Allen, D. I. Beller, J. Braun, and E. R. Unanue, The handling of Listeria monocytogenes by macrophages: the search for an immunologic molecule in antigen presentation, J. Immunol. 132:323 (1984).
2. B. N. Ames and D. T. Dubin, The role of polyamines in the neutralization of pacteriophage deoxyribonucleic acid, J. Biol. Chem. 135:769 (1960).
3. W. F. Doe and P. M. Henson, Macrophage stimulation by bacterial lipopolysaccharides. I. Cytolytic effect on tumor target cells, J. Exp. Med. 148:544 (1978).
4. W. F. Doe, S. T. Yang, D. C. Morrison, S. J. Betz, and P. M. Henson, Macrophage stimulation by bacterial lipopolysaccharides. II. Evidence for differential signals delivered by lipid A and by protein rich fractions of LPS, J. Exp. Med. 148:557 (1978).
5. R. L. Duncan, Jr., and D. C. Morrison, Fate of Escherichia coli lipopolysaccharide following the uptake of E. coli by murine macrophages in vitro, J. Immunol. 132:1416 (1984).
6. R. L. Duncan, Jr., and D. C. Morrison, Activity of macrophage processed endotoxin, in: "Pathogenesis of Bacterial Infections," G. G. Jackson

and H. Thomas, eds., Proc. Grosse Ledder Symposium VIII, Springer-Verlag, Berlin (1985).

7. A. C. Eaves and W. R. Bruce, In vitro production of colony stimulating activity. I. Exposure of mouse peritoneal cells to endotoxins, Cell Tissue Kinet. 7:19 (1974).

8. A. G. Farr, M. E. Dorf, and E. R. Unanue, Secretion of mediators following T lymphocyte-macrophage interaction is regulated by the major histocompatibility complex, Proc. Natl. Acad. Sci. USA 74:3542 (1977).

9. J. P. Filkins, Comparison of endotoxin detoxification by leukocytes and macrophages, Proc. Soc. Exp. Biol. Med. 137:1396 (1971).

10. D. C. Morrison, Z. G. Oades, S. Vukajlovich, S. A. Goodman, and R. L. Duncan, Jr., Mechanism of action of endotoxin at the cellular level, in: "Mechanisms of Hepatocyte Injury and Death," D. Keppler, H. Popper, L. Bianchi, and W. Reutter, eds., MTP Press, Lancaster (1983).

11. S. A. Goodman and D. C. Morrison, Selective association of lipid-rich R-like lipopolysaccharide (LPS) subunits with murine spleen cells, Mol. Imunol. 21:689 (1984).

12. D. C. Morrison, R. L. Duncan, Jr., and S. A. Goodman, In vivo biological activities of endotoxin, in: "Bacterial Endotoxins: Structure, Biomedical Significance and Detection with Limulus Amebocyte Lysate Test," J. W. ten Cate, H. R. Buller, A. Struk, and J. Levin, eds., in press (1985).

13. D. C. Morrison, S. J. Betz, and D. M. Jacobs, Isolation of a lipid A bound polypeptide responsible for "LPS-initiated" mitogenesis of C3H/HeJ spleen cells, J. Exp. Med. 144:840 (1976).

14. D. C. Morrison and J. L. Ryan, Bacterial endotoxins and host immune responses, Adv. Immunol. 28:293 (1979).

15. D. C. Morrison and R. J. Ulevitch, The interaction of bacterial endotoxins with cellular and humoral mediation systems, Am. J. Pathol. 93:527 (1978).

16. L. P. Ruco and M. S. Meltzer, Macrophage activation for tumor cytotoxic activity requires completion of a sequence of shortlived intermediary reactions, J. Immunol. 121:2035 (1978).

17. S. H. Rutenburg, F. B. Schweinburg, and J. Fine, In vitro detoxification of bacterial endotoxin by macrophages, J. Exp. Med. 112:801 (1960).

18. R. A. Trejo and N. R. DiLuzio, Comparative evaluation of macrophage inactivation of endotoxin, Proc. Soc. Exp. Biol. Med. 144:901 (1973).

19. E. R. Unanue, Antigen presenting function of the macrophage, Ann. Rev. Immunol. 2:395 (1985).

20. E. R. Unanue, J. M. Kiely, and J. Calderon, The modulation of lymphocyte functions by molecules secreted by macrophages. II. Conditions leading to increased secretion, J. Exp. Med. 144:155 (1976).

21. S. Vukajlovich and D. C. Morrison, Conversion of lipopolysaccharides to molecular aggregates with reduced subunit heterogeneity: demonstration of LPS responsiveness in "endotoxin-unresponsive" C3H/HeJ splenocytes, J. Immunol. 130:2804 (1983).

22. S. W. Vukajlovich and D. C. Morrison, Lipid-A dependent lymphocyte proliferation in "endotoxin nonresponder" mice, Rev. Infect. Dis. 6:528 (1984).

23. J. J. Wannemuehler, H. Kiyomo, J. L. Babb, M.S. Michalek, and J. R. McGhee, Lipopolysaccharide (LPS) regulation of the immune response: LPS converts germfree mice to sensitivity to oral tolerance induction, J. Immunol. 129:959 (1982).

24. H. K. Ziegler and E. R. Unanue, Decrease in macrophage antigen catabolism by ammonia and chloraquin is associated with inhibition of antigen presentation to T cells, Proc. Natl. Acad. Sci. USA 79:175 (1982).

THE RELEASE OF IMMUNOPOTENTIATING MEDIATORS FROM MACROPHAGES

ACTIVATED BY ENDOTOXINS

R. Christopher Butler[1], Jeri M. Frier[1], Mrunal S. Chapekar[1],
Herman Friedman[2], and Alois Nowotny[3]

Departments of Medical Microbiology and Immunology at
[1]Arlington Hospital, Arlington, Virginia; [2]University of
South Florida College of Medicine, Tampa, Florida;
[3]University of Pennsylvania Medical Center for Oral Health
Research, Philadelphia, Pennsylvania

The infection of mice with murine leukemia viruses often results in an acquired immunodeficiency state in which active immune responses are impaired. The Friend leukemia virus (FLV), which consists of a lymphatic leukemia virus complexed with a spleen focus-forming virus, is a murine oncornavirus which suppresses both humoral and cell mediated responses (2,10,11). In the past it was generally accepted that the suppression of the ability to develop specific antibody responses was due to the interaction of the virus with antibody producing plasma cells or their precursors.

Recent studies, however, have provided evidence that B cells from FLV-infected hosts retain the ability to produce specific antibodies under appropriate conditions. For example, even though the development of an antibody response to sheep erythrocyte antigens (SRBC) is suppressed, the "background" antibody response of unsensitized FLV-infected splenocytes remains at normal levels (1). Therefore, the B cells do not lose the capacity to produce specific antibody, but for some reason they do not respond to antigenic stimulation. Second, the treatment of FLV-infected splenocytes with immuno-stimulants such as lipopolysaccharides (LPS) or muramyl dipeptide (MDP) can enhance the development of an antibody response to SRBC (6). This indicates that with proper stimulation the FLV-infected splenocytes can mount an immunologic response. A third observation was that the addition of normal macrophages to FLV-infected cultures could partially restore the response capacity, thus indicating that the mechanism for this unresponsiveness might be related to these accessory cells (15).

Since the mechanism for the suppression of antibody responses by FLV is now apparently not due solely to infection of B cells by the virus but may involve macrophages, we sought to determine the mechanism by which FLV might exert its suppressive effects via macrophages. In previous studies we have demonstrated that the development of antibody responses by normal splenocytes is greatly enhanced by the production of antibody response helper factor(s) by macrophage (7). These factors are released "spontaneously" by cultured macrophages but their production is greatly enhanced by treatment with LPS. This study was designed to further determine the nature of these endotoxin-induced mediators and to determine what role they might play in the mechanism of acquired immunodeficiency caused by FLV infection.

MATERIALS AND METHODS

Experimental Animals

Inbred male BALB/c mice, six to eight weeks of age, were obtained from Cumberland View Farms, Clinton, TN. Mice were infected by intraperitoneal injection of a 100 LD_{50} dose of Friend Leukemia Virus (FLV) contained in 0.1 ml of a one percent clarified homogenate of infected splenocytes. The virus has been maintained by passage through adult BALB/c mice and contains both the spleen focus-forming and lymphatic leukemia virus components of the Friend complex.

Lipopolysaccharide

Serratia marcescens lipopolysaccharide (LPS) was prepared by the tri-chloroacetic acid extraction procedure as previously described (14).

Antigen

Sheep red blood cells (SRBC) in Elsevier's solution were obtained from Baltimore Biological Laboratories, Baltimore, MD. The erythrocytes were washed several times in medium and resuspended to a 0.1% concentration.

In Vitro Immunization

Covered plastic Linbro plates were used as culture chambers. Spleen cells from normal or FLV preinfected mice were washed in media and the numbers of viable nucleated cells determined by the trypan blue dye exclusion technique with a hemacytometer. A suspension of 8×10^6 viable splenocytes suspended in 2.0 ml of complete tissue culture medium enriched with a standard nutrient cocktail and 20% fetal bovine serum were cultured in the Linbro plate wells as described elsewhere (12). For in vitro immunization 0.1 ml of the 0.1% suspension of SRBC was added to each culture (approximately 2×10^6 erythrocytes). All cultures were incubated for five days at 37°C in a humidified atmosphere containing 10% CO_2.

Assay for Antibody-forming Cells

The numbers of direct hemolytic plaque forming cells (PFC) to SRBC were determined by the micro method of Cunningham and Szenburg (9). The numbers of PFC were enumerated for at least eight to 24 cultures prepared from two to four spleen cell preparations, and the average number of PFC per million viable cells calculated. In all cases only direct nonfacilitated plaques were enumerated, and these were considered to be due to IgM antibody producing cells.

Post-LPS Serum

Normal or FLV-infected mice were injected i.p. with 20 µg LPS and exsanguinated two hours later by aseptic cardiac puncture. Serum was separated from the erythrocytes and kept on ice until use.

In Vitro Factor Production

Suspensions of 10^7 splenocytes per ml from normal or FLV-infected mice were incubated in RPMI 1640 plus 10% fetal bovine serum and antibiotics at 37°C under CO_2. Experimental cultures received 10 µg of LPS per ml at the time of culture initiation. Supernatants were collected after five days and either stored on ice or frozen at -70°C until tested.

Table 1. Stimulation of the Antibody Response by LPS and PS

Culture Treatment[a]		% Control PFC[b] (P)
None		100
LPS	0.1 µg	104
	1 µg	139 (0.01)
	10 µg	267 (0.005)
	20 µg	181 (0.005)
PS	0.1 µg	102
	1 µg	120 (0.05)
	10 µg	231 (0.005)
	20 µg	169 (0.005)

[a] Indicated dose of LPS or PS added to cultures of 8×10^6 normal spleen cells sensitized in vitro with 2×10^6 SRBC.
[b] Average plaque forming cell (PFC) response for 8 cultures each group 5 days after sensitization.

Antibody Response Helper Factor Assay

The presence of antibody response helper factor activity was determined by adding 0.01 ml of post-LPS serum or 0.1 ml of stimulated culture supernatants to the in vitro antibody cultures at the time of sensitization with SRBC. The helper factor activity of each preparation was considered to be proportional to the degree of enhancement of the antibody response over that of the normal untreated control cultures.

RESULTS

It has long been recognized that LPS can enhance T-cell dependent antibody responses. The results in Table 1 demonstrate the stimulation of a

Table 2. Enhancement of LPS and PS Activity by Muramyl Dipeptide (MDP)

Culture Treatment[a]	% Control[b] (P)
None	100
20 µg LPS	295 (0.001)
20 µg PS	188 (0.005)
10 µg MDP	193 (0.005)
20 µg LPS + 10 µg MDP	601 (0.001)
20 µg PS + 20 µg MDP	278 (0.001)

[a] Cultures of 8×10^6 normal mouse splenocytes were treated in vitro with indicated doses of LPS, PS and/or muramyl dipeptide.
[b] Average results from eight cultures per group assayed 5 days after culture initiation.

Table 3. In Vivo Production of Helper Factors in Response to LPS and PS

Culture Treatment[a]	Volume (ML)	% Control PFC[b] (P)
None	--	100
Normal Serum	0.001	110
	0.01	109
	0.1	106
Post-LPS Serum	0.001	121
	0.01	251 (0.001)
	0.1	180 (0.05)
Post-PS Serum	0.001	128 (0.05)
	0.01	226 (0.001)
	0.1	160 (0.01)

[a] Splenocyte cultures were treated with graded quantities of normal mouse serum or 2 hour post-LPS or post-PS serum at time of culture initiation.
[b] Average PFC response after 5 days.

primary antibody response to sheep erythrocyte antigens in vitro by LPS. The nontoxic PS derivative of LPS enhanced this antibody response to an extent similar to the parent LPS. The combined treatment of respondent cells with either LPS or PS together with another immunostimulant, muramyl dipeptide (MDP), produced additive or synergistic effects as demonstrated in Table 2.

Although the adjuvant effect of LPS has been studied for decades the precise mechanism of action is still not agreed upon. In the case of the other beneficial effects of LPS, mediators such as tumor necrotizing factor (TNF), colony stimulating factor (CSF), and tumor resistance enhancing factors were all released into the serum following LPS or PS treatment. We therefore examined post-LPS serum for the presence of antibody response helper factors. Table 3 shows that both post-LPS sera and post-PS sera contain a factor or factors which mediate the adjuvant effect. To rule out the possibility that the observed effects of post-LPS serum were due to the transfer of residual LPS we assayed the serum for endotoxicity. The results in Table 4 demonstrate the absence of significant quantities of LPS in post-LPS serum.

Having demonstrated the presence of an immunostimulatory mediator we attempted to determine its source and target. To assess the role of T cells

Table 4. Residual Endotoxin Levels in Two-Hour Post-LPS Serum

Endotoxin Assay	Residual LPS
1. Limulus Lysate	< 0.001 µg/ml
2. Chick Embryo Lethality	< 0.001 µg/ml
3. Rabbit Pyrogenicity	< 0.001 µg/ml

Table 5. In Vivo Production of Helper Factors by Athymic BALB/c nu/nu Mice

Serum Source[a]	Responder Cell[b] Source	% Control[c] (P)
None	BALB/c	100
BALB/c Normal	"	113
Nude Normal	"	116
BALB/c Post-LPS	"	213 (0.01)
Nude Post-LPS	"	182 (0.02)
None	Nude	100
BALB/c Normal	"	150
Nude Normal	"	84
BALB/c Post-LPS	"	419 (0.01)
Nude Post-LPS	"	447 (0.01)

[a] Two hour post-LPS serum was collected from normal or BALB/c nu/nu nude mice.
[b] Cultures of normal BALB/c or BALB/c nu/nu nude mouse spleen cells were treated with 0.01 ml of normal or post-LPS serum.
[c] Average PFC response 5 days after sensitization.

in this effect we prepared post-LPS sera from athymic BALB/c nu/nu nude mice which lack mature T cells. If the LPS induced mediator were a T cell product, then it should not be produced by these mice. As demonstrated in Table 5, the nude mouse post-LPS serum activity was comparable to that from normal mice. Therefore, the factor is probably not a T cell product.

From the opposite approach, if the target of this LPS mediator were the helper T cell, then nude mouse splenocytes should not show a stimulatory effect when treated with the factor. Table 5 also shows that nude mouse splenocyte cultures respond strongly to both post-LPS and post-PS sera.

Table 6. Production of Helper Factors by Athymic Nude Mouse Macrophages

Cell Type[a]	Treatment	% Control PFC[b] (P)
None (control)		100
Unfractionated	--	107
"	LPS	301 (0.01)
Nonadherent	--	87
"	LPS	119
Macrophages	--	129
"	LPS	291 (0.02)

[a] Splenocytes (10^7/ml) were fractionated into adherent and nonadherent cell populations. These cell populations were treated with 10 μg/ml LPS and incubated for 5 days. The supernatants from these cultures were added at a 5% concentration to normal splenocyte cultures sensitized with SRBC.
[b] Average PFC response after 5 days of culture.

Table 7. Helper Factor Production by the P388D1 Macrophage Cell Line

Culture Treatment[a]		% Control PFC[b] (P)
None		100
10 μg LPS		298 (0.005)
P388D1 Supernatant -	3 day	125
	4 day	172 (0.025)
	5 day	210 (0.01)
P388D1 + LPS SUPn -	3 day	113
	4 day	229 (0.001)
	5 day	443 (0.001)

[a]Splenocyte cultures were treated with 0.1 ml of P388D1 culture supernatants collected after 3, 4 or 5 days of culture from untreated cells or cultures incubated with 10 μg/ml LPS.
[b]Average PFC response after 5 days.

In addition to the in vivo production of the factor in post-LPS serum, stimulatory activity can be generated through the treatment of splenocytes in vitro. When 10^7 splenocytes were cultured in the presence of LPS for five days the resultant cell-free supernatant contained the endotoxin mediator. When these splenocytes were fractionated into adherent and nonadherent cell populations, as shown in Table 6, it was found that only the adherent cell population had the capacity to produce the factor. This adherent cell population was comprised of approximately 90 percent macrophages based on cell morphology.

To confirm that the factor was a macrophage product and not due to contaminating cells, we attempted to stimulate production by the P388D1 macrophage cell line. As shown in Table 7, the P388D1 cell line spontaneously releases antibody stimulating factors. Treatment of the cells with LPS enhanced this activity.

Concurrent with this study of LPS-induced helper factors has been a study of the role of this factor in the mechanism of immunosuppression by FLV. Table 8 demonstrates that splenocytes from FLV-infected mice gradually

Table 8. Suppression of the Antibody Response by Friend Leukemia Virus

Source of Responder Cells[a]	% Control PFC[b] (P)	
Normal spleen	100	--
7 Day FLV	79	--
14 Day FLV	51	0.01
21 Day FLV	10	0.001

[a]Mice were pretreated with 100 ID_{50} of FLV from 7 to 21 days before collecting the spleen cells. Splenocytes were sensitized with SRBC in vitro.
[b]Average PFC after 5 days of culture.

374

Table 9. FLV Infection Depresses Helper Factor Production In Vivo

Serum Source[a]	Treatment of Serum Donor (-2 hr)[b]	% Control PFC[c] (P)
None	--	100
Normal	--	98
	LPS	298 (0.005)
7 Day FLV	--	103
	LPS	207 (0.01)
14 Day FLV	--	89
	LPS	137
21 Day FLV	--	105
	LPS	116

[a]Mice were preinfected with 100 ID_{50} of FLV from 7 to 21 days before the collection of serum.
[b]Normal and FLV-infected mice received an intraperitonal injection of 20 µg LPS 2 hours before the collection of serum.
[c]Splenocyte cultures were treated with 0.5% serum. Average PFC after 5 days of culture.

Table 10. Friend Leukemia Suppresses Antibody Helper Factor Production In Vitro

Source of Cells for Factor Production[a]	Treatment of Cells for Factor Production[b]	% Control PFC[c] (P)
--	--	100
Normal	--	139
"	LPS	302 (0.005)
1 Day FLV	--	149
"	LPS	285 (0.001)
7 Day FLV	--	118
"	LPS	188 (.05)
21 Day FLV	--	98
"	LPS	118

[a]Mice were preinfected with 100 ID_{50} of FLV 1 to 21 days before sacrifice.
[b]Cultures of 10^7 splenocytes/ml were incubated for 5 days with or without 10 µg/ml LPS.
[c]Splenocyte cultures were treated with 5% supernatant from the FLV splenocyte cultures at the time of in vitro stimulation with SRBC.

Table 11. Depression of Helper Factor Production is Not Due to
          Transfer of FLV

Source of Cells for Factor Production[a]	Treatment of Cells[b]	300K MW Filtration[c]	% Control PFC[d] (P)
None	--		100
Normal	LPS	--	302 (0.001)
	LPS	++	261 (0.001)
10 Day FLV	LPS	--	176 (0.005)
	LPS	++	195 (0.005)
30 Day FLV	LPS	--	131
	LPS	++	124

[a] Mice were preinfected with 100 ID_{50} of FLV from 10 to 30 days before
sacrifice. All cultures were treated with 10 µg/ml of LPS for 5 days.
[b] Supernatants were filtered through a 300,000 MW Nucleopore filter to
remove infectious virus.
[c] Cultures of 8×10^6 normal BALB/c splenocytes were treated with 5% supernatant from the FLV splenocyte cultures at the time of in vitro primary
stimulation with SRBC. After 5 days the cultures were collected and
assayed for PFC.

lose their antibody response capability. Antibody responses were depressed
prior to significant spleen enlargement. Parallelling this decreasing
response capacity was a decrease in the production of helper activity in
the post-LPS serum. Table 9 demonstrates that by 14 days post-infection
the FLV leukemic mice could no longer produce significant helper activity.
Similarly, FLV-infected splenocytes lost the ability to produce the factor
in response to LPS when cultured in vitro as shown in Table 10.

    To rule out the possibility that the decrease in activity of the culture
supernatants might be an artifact due to the inadvertent transfer of immunosuppressive FLV, the factor preparations were prefiltered through a 100,000
MW Nucleopore filter prior to addition to antibody-forming cell cultures.
This procedure has previously been demonstrated to remove infections (13).
Filtering the LPS-induced supernatants to remove FLV did not affect the
activity as shown in Table 11.

    Since FLV suppressed factor production, and the macrophage was the
direct source of the LPS-induced factor, cells from the P388D1 macrophage
cell line were treated with FLV to determine if the FLV could exert its
suppressive effect directly on macrophages. Table 12 shows that the addition
of viable FLV to macrophage cell cultures inhibited both spontaneous and
LPS-induced production of the factor. These supernates were also filtered
to remove residual virus. Using indirect immunofluorescent antibody label-
ling with an FLV-specific antiserum it was observed that after five days of
incubation with the virus, FLV antigens were present on the surface of most
P388D1 cells. While the FLV had apparently infected the macrophage cell
line, it did not affect P388D1 cell viability (trypan blue exclusion) or
obvious cell morphology during this time.

Table 12. FLV Suppresses Factor Production by P388D1 Macrophage Cell Line

P388D1 Culture Treatment[a] (5 day Supernate)	% Control PFC[b] (P)
None	100
Untreated	170 (0.05)
FLV	110
LPS	466 (0.001)
FLV + LPS	210 (0.01) (0.05)

[a]P388D1 cultures were treated by the addition of FLV and/or LPS 5 days prior to supernatant collection. All culture supernatants were filtered through a 300,000 MW Nucleopore filter and added 5% to normal splenocytes stimulated with SRBC.
[b]Average PFC results after 5 days.

DISCUSSION

For many years our laboratories have collaborated in studying a variety of the immunostimulatory effects of LPS and post-LPS serum, including the production of myeloid colony-stimulating factor (5), tumor necrotizing factor (4), and tumor resistance enhancing factors (4,5). In view of the stimulatory effects of LPS on antibody responses, we questioned whether LPS also induced the production of factors that could enhance the development of antigen-specific antibody responses. To test this hypothesis we selected the in vitro antibody response as a model for the development of specific primary immune responses.

When small doses of post-LPS sera were added to in vitro splenocyte cultures immunized with SRBC, the peak primary IgM response was elevated. Normal mouse serum had no effect in this system. As described in the results, extensive measurements of endotoxin levels in post-LPS sera demonstrated that this stimulatory effect was not due to the presence of residual LPS in the serum. Therefore, this stimulatory activity appeared to be due to the release of one or more subcellular factors into the serum in the response to LPS.

To investigate the mechanism of action of this factor we searched for the cellular source of the activity. When athymic BALB/c nu/nu nude mice were tested for the ability to produce stimulatory post-LPS serum factors it was observed that the post-LPS serum from nude mice was just as effective as that from normal mice. Therefore, mature T cells did not appear necessary for factor production. Even if LPS treatment does induce the appearance of theta antigen on cells in the nude mouse, it would not be expected that this phenomenon could be responsible for full serum activity in two hour post-LPS serum. It was also demonstrated that splenocytes from nude mice were highly responsive to treatment with post-LPS serum factors, thus indicating that the T cell does not play a role in functional expression of this factor.

The separation of splenocytes into adherent and nonadherent populations revealed that the lymphocyte rich nonadherent cell populations did not have the capacity to produce antibody response enhancing factors in response to LPS. In contrast, the macrophage-rich adherent cell population retained all of the factor-producing capacity of the whole spleen cell suspensions.

These studies implicated the macrophage as the cell that responds directly to LPS in this system. To determine whether pure cultures of macrophages could produce the factors, the P388D1 transformed macrophage cell line was examined. In the presence of LPS the P388D1 cell cultures were highly active in factor production, thereby confirming that the macrophage can be stimulated directly by LPS to produce the factor.

The immunosuppressive effects of Friend Leukemia Virus (FLV) on antibody responses to sheep erythrocyte antigens have been demonstrated in vivo (1,11) and in vitro (2,3,8,15). The results shown here demonstrated that the ability of murine splenocytes to mount an antibody response to SRBC in vitro decreases steadily with time following infection with FLV. The mechanism for this loss of response capacity has not been clearly defined. However, these studies have established a strong relationship between the loss of ability to produce antibody response helper factors either spontaneously or in response to stimulation with LPS. This decrease in helper factor activity was demonstrated to occur both in vivo (post-LPS serum) and in vitro (splenocyte culture supernatants).

In view of the role of these helper factors in antibody responses by normal splenocytes, it is feasible that the decrease in helper factor production by FLV-infected cells could be causally related to the depressed antibody response. If this were the case, then the addition of exogenous helper factor should restore the antibody response of FLV cells to normal levels. Previous studies have shown that this reversal does occur and that B lymphocytes from FLV spleens retain the capacity to produce a normal antibody response under the condition of adequate levels of helper factors (2,3,6,8).

Other studies have suggested that the suppression of antibody responses in FLV mice is due to a suppression of macrophages (15). They demonstrated that the addition of normal macrophages to FLV splenocytes could restore the antibody response capacity. Our results clarify the mechanism for a macrophage-related suppression by indicating that it is the inability of cultures to produce sufficient quantities of a specific macrophage product-- an antibody response helper factor--which inhibits the development of antibody responses by FLV-infected splenocytes.

This suppression of macrophage function appears due to the direct infection of macrophages by FLV as demonstrated by the appearance of FLV antigens on infected P388D1 cells. While this infection did not affect macrophage viability or obvious cell morphology, it did inhibit both the spontaneous and the LPS-induced production of helper factors. Therefore, the infection of macrophages by FLV appears to contribute strongly to the immunosuppressive effects of the virus on primary immune responses.

REFERENCES

1. M. Bendinelli and H. Friedman, B and T lymphocyte activation by murine leukemia virus infection, Adv. Exp. Biol. Med. 121B:91 (1980).
2. R. C. Butler and H. Friedman, Leukemia virus-induced immunosuppression: reversal by subcellular factors, Ann. N.Y. Acad. Sci. 332:446 (1979).
3. R. C. Butler and H. Friedman, Restoration of leukemia cell immune responses by bacterial products, in: "Current Chemotherapy and Infectious Disease, J. D. Nelson and C. Grassi, eds., American Society for Microbiology, Washington, D. C. (1980).
4. R. C. Butler and A. Nowotny, Colony-stimulating factor (CSF)-containing serum has antitumor effects, Med. Sci. 4:206 (1976).
5. R. C. Butler, A. Abdelnoor and A. Nowotny, Bone marrow colony stimulating factor and tumor resistance-enhancing activity of post-endotoxin mouse sera, Proc. Nat. Acad. Sci. U.S.A. 75:2893 (1978).

6. R. C. Butler, H. Friedman and A. Nowotny, Restoration of depressed antibody responses of leukemic splenocytes treated with LPS-induced factors, Adv. Exp. Biol. Med. 121A:315 (1980).

7. R. C. Butler, A. Nowotny, and H. Friedman, Macrophage factors that enhance the antibody response, Ann. N.Y.. Acad. Sci. 332:564 (1979).

8. R. C. Butler, J. M. Frier, M. S. Chapekar, M. O. Graham, and H. Friedman, Role of antibody response helper factors in immunosuppressive effects of Friend Leukemia Virus, Inf. and Immun. 39:1260 (1983).

9. A. J. Cunningham and A. Szenburg, Further improvements in the plaque technique for detecting single antibody forming cells, Immunology 14:599 (1968).

10. H. F. Friedman and W. S. Ceglowski, Leukemia virus-induced immunosuppression. VIII. Rapid depression of in vitro leukocyte migration after infection of mice with Friend leukemia virus, J. Immunol. 107:1673 (1971).

11. S. Hirano, H. Friedman, and W. S. Ceglowski, Immunosuppression by leukemia viruses. VII. Stimulatory effects of Friend leukemia virus on pre-existing antibody-forming cells to sheep erythrocytes and Escherichia coli in nonimmunized mice, J. Immunol. 107:1400 (1971).

12. I. Kamo, S.-H. Pan, and H. Friedman, A simplified procedure for in vitro immunization of dispersed spleen cell cultures, J. Immunol. Methods 11:55 (1976).

13. J. R. Kately, I. Kamo, G. Kaplan, and H. Friedman, Suppressive effect of leukemia virus-infected lymphoid cells on in vitro immunization of normal splenocytes, J. Natl. Cancer Inst. 53:1371 (1974).

14. A. Nowotny, K. R. Cundy, N. L. Neale, A. M. Nowotny, P. Radvany, S. P. Thomas, and D. J. Tripodi, Relation of structure to function in bacterial O-antigens. IV. Fractionation of the components. Ann. N.Y. Acad. Sci. 133:586 (1966).

15. S. Specter, N. Patel, and H. Friedman, Restoration of leukemia virus-suppressed immunocytes in vitro by peritoneal exudate cells, Proc. Soc. Exp. Biol. Med. 151:163 (1976).

CHANGES IN MACROPHAGE PROGENITOR CELL COMPOSITION IN THE BONE MARROW

OF "EARLY PHASE" ENDOTOXIN-TOLERIZED MICE

Stefanie N. Vogel and Gary S. Madonna

Department of Microbiology
Uniformed Services University of the Health Sciences
Bethesda, Maryland

INTRODUCTION

For many years, it has been recognized that prior exposure to sublethal doses of LPS can result in decreased toxicity in response to a subsequent LPS challenge. This has been referred to as a state of "endotoxin tolerance" (reviewed in 3). Classically, two phases of endotoxin tolerance have been described. "Early endotoxin tolerance" refers to the transient period of LPS hyporesponsiveness which occurs within the first few days after initial exposure to LPS and wanes in approximately seven to eight days. This appears to be independent of the presence of anti-O-specific antibodies (3), and has been shown to be transferable with spleen cells (11). In contrast, "late phase" tolerance is delayed, persistent, and antibody-mediated (3).

In the experiments described in this chapter, we have examined early endotoxin tolerance with respect to alterations in macrophage precursor frequency in the bone marrow. The results of these studies indicate that the tolerant state is associated with increased numbers of macrophage progenitors in the bone marrow.

MATERIALS AND METHODS

Mice

Male and female outbred ICR mice (Sprague-Dawley, Indianapolis, IN), six to eight weeks old, were used throughout these studies. C3H/HeJ mice (Jackson Laboratories, Bar Harbor, ME) were used only as a source of bone marrow cells for the CSF assay. Mice were housed conventionally and were fed standard lab chow and water ad libitum.

Reagents

Protein-free, phenol-water extracted E. coli K235 LPS was prepared by the method of McIntire et al. (4). Colony stimulating factor (CSF) was partially purified as described elsewhere (10).

Measurement of CSF Activity in Serum

CSF activity was measured in a bone marrow colony assay in semi-solid agar as described (11). Colonies were counted seven days after culture.

Measurement of the Numbers of Macrophage Progenitor Cells in the Bone Marrow

Femurs from injected mice were very carefully removed and were homogenized using a mortar and pestle. Bone fragments were allowed to settle and bone marrow cells were sized and counted using a Coulter Model ZM Channelyzer (Coulter Electronics Ltd., England). To determine the number of macrophage colony forming units (M-CFU), 10^5 nucleated bone marrow cells were incubated for seven days in the presence of an excess of CSF (approximately 7000 U/ml) in semi-solid agar as described elsewhere (6). After seven days, the number of M-CFU colonies were counted under an inverted microscope. In one study, the number of progenitors with high proliferative potential (HPP) was similarly measured. To detect HPP's, the bone marrow cells were incubated in a combination of CSF and human spleen cell conditioned medium (HSCM) as previously described (1). HPP colonies fail to grow in the presence of CSF alone and require the HSCM as a source of "synergistic activity" (1). HPP colonies are counted after 12 days in culture and are considerably larger and morphologically distinct from M-CFU (1).

Induction of "Early Phase" Endotoxin Tolerance

The basic protocol which was used to induce a state of "early phase" endotoxin tolerance was adapted from recent work by Williams et al. (11; Table 1). Three experimental groups of outbred ICR mice were injected on Day 0, either with saline or LPS. On Day 3, these mice were again injected such that one group, the saline controls received saline both times, the non-tolerized group received LPS for the first time on Day 3, and the tolerized group received LPS on both Days 0 and 3. At various time intervals after the second injection, various parameters were measured as indicators of tolerance.

RESULTS

Establishment of "Early Phase" Tolerance

Preliminary studies indicated that an initial exposure to 25 µg of the E. coli K235 LPS (either i.v. or i.p.) resulted in maximal tolerance as assessed by the production of colony stimulating factor following 25 µg LPS challenge on Day 3 or 4. These findings were completely confirmatory of those previously reported by Williams et al. (11). The results of a typical experiment are shown in Table 1. As can be seen from this experiment, mice that received LPS only three hours prior to being bled produced high levels of CSF activity. In contrast, mice that had been tolerized by an initial exposure to LPS on Day 0 had only approximately 20 percent of the control CSF activity in response to a second exposure. This state of tolerance gradually wanes through Day 8.

To further ensure the validity of this tolerizing regimen, we also examined the LD_{50} and symptoms in control and tolerized mice (Table 1). We found that for ICR mice, this tolerance regimen more than doubled the LD_{50}, and very significantly reduced the overt symptoms of LPS-mediated toxicity in response to a sublethal dose of LPS. In other experiments, not shown here, we further confirmed these findings using heterologous combinations of E. coli, S. typhimurium, and Ps. aeruginosa LPS. Lastly, the non-toxic, monophosphorylated lipid A (Ribi Immunochem) was found to be capable of inducing tolerance with respect to CSF production.

Effect of Tolerance Regimen on Thioglycollate-Elicited Macrophages

Our earlier work with LPS hyporesponsive C3H/HeJ mice had suggested that a relationship between the state of macrophage differentiation and LPS

Table 1. Induction of "Early Phase Endotoxin Tolerance"[a]

Experimental Group	Day of Injection Day 0	Day 3	CSF (CFU/ml)[b]	LD_{50} (µg)[c]	Symptoms[d] Diarrhea	Ruffled Fur	Conjunctival Discharge
Saline Control	Saline	Saline	0	—	—	—	—
Non-Tolerized	Saline	LPS	6370 ± 476	453	9/16	13/16	11/16
Tolerized	LPS	LPS	1510 ± 175	1059	0/16	0/16	0/16

[a] Mice were injected with either 0.5 ml saline or phenol-water extracted E. coli K235 LPS (25 µg) on Day 0 and Day 3.
[b] Mice were injected and serum was collected 3 hr after the second injection.
[c] LD_{50} was calculated by the method of Reed and Muench (7). Results represent the mean of two separate experiments.
[d] The proportion of mice clearly exhibiting the indicated symptom 6 hr after injection of 25 µg LPS on Day 3.

Table 2. Effect of Tolerance Induction on Recovery of Thioglycollate-
Induced Peritoneal Exudate Cells

| Experimental Group[a] | Macrophages | % of WBC Recovered[b] | | |
		Neutrophils	Lymphocytes	Other
Saline Control (Sal/Sal)	92	2	5	1
Non-Tolerized (Sal/LPS)	86	11	0	2
Tolerized (LPS/LPS)	31	65	1	2

[a]Mice were treated on Days 0 and 3 as indicated. Thioglycollate was
administered on Day 1 and peritoneal exudate cells collected by lavage on
Day 3, 3 hr after the second injection.
[b]Cytospin preparations were stained using a modified Wright's stain, and
differential counts performed on $\geq$ 200 cells per slide.

sensitivity might exist (9). Since several reports had suggested that macrophages from tolerized mice failed to produce factors when stimulated in
vitro with LPS (2,8), we had originally planned to examine the differentiation state of thioglycollate-induced peritoneal exudate macrophages from
these mice. When fluid thioglycollate was administered to mice during the
tolerizing period (as had been done by others (11), two very surprising
findings were observed. First, the number of exudate cells recoverable
from the peritoneal cavity was markedly reduced in tolerized mice as compared
to controls (data not shown). Even more striking was the preferential decrease in the proportion of macrophages which were elicited (Table 2).
This finding, in conjunction with the marked reduction in circulating CSF
levels in tolerized mice, suggested to us the possibility that the turnover
of macrophage progenitor cells in the bone marrow might differ in control
vs. tolerized mice.

Table 3. Effect of Tolerance Induction on the Number of Macrophage
Colony Forming Units (M-CFU) in the Bone Marrow[a]

Experimental Group	M-CFU/10^5 WBC	M-CFU/Femur
Saline Control (Sal/Sal)		
0 hr	155	1.78×10^4
3 hr	175	1.90×10^4
24 hr	125	1.44×10^4
Non-Tolerized (Sal/Sal)		
0 hr	115	1.25×10^4
3 hr	105	$.88 \times 10^4$
24 hr	75	$.25 \times 10^4$
Tolerized		
0 hr	360	3.52×10^4
3 hr	515	3.76×10^4
24 hr	485	3.64×10^4

[a]Macrophage colonies were grown in the presence of partially-purified
fibroblast-derived CSF. Colonies were counted at 7 days.

Table 4. Effect of Tolerance Induction on the Number of Macrophage CFU
and Progenitors with High Proliferating Potential (HPP)

Experimental Group	$CFU/10^5$ WBC[a] (CFU/Femur)	$HPP/10^5$ WBC[b] (HPP/Femur)
Saline Control (Sal/Sal)	240 (5.7×10^4)	96 (2.3×10^4)
Non-Tolerized (Sal/LPS)	154 (2.9×10^4)	72 (1.3×10^4)
Tolerized (LPS/LPS)	426 (7.0×10^4)	242 (4.0×10^4)

[a]Colonies grown in CSF only.
[b]Colonies grown in CSF supplemented with human spleen conditioned medium as
a source of synergistic activity.

Changes in the Numbers of Macrophage Progenitors in the Bone Marrow of Control vs. Tolerized Mice

Bone marrow cells from control and tolerized mice were prepared and
the number of progenitors which would form colonies in the presence of fibroblast-derived CSF assessed. To do this, bone marrow cells were derived
from the femurs of mice immediately (0 hr), three hours, or 24 hours after
the second injection. The total number of white blood cells (WBC) per femur,
as well as the number of M-CFU/10^5 WBC, were determined and allowed us to
calculate the number of M-CFU per femur (Table 3). The number of CSF-sensi-
tive progenitors is markedly reduced within the first 24 hours of an initial
LPS exposure. However, by 72 hours, there has been a highly significant
increase in the number of M-CFU, which is not significantly altered by an
additional exposure to LPS.

A more primitive population of macrophage progenitor cells with high
proliferative potential (HPP) has been identified (1). These cells failed
to develop colonies in the presence of CSF alone, but will form colonies in
the presence of CSF plus "synergistic activity." Table 4 illustrates that
not only is there an increase in the number of M-CFU/femur in tolerized
animals, but also a significant increase in the number of HPP's/femur.

Changes in Cell Sizing Profiles in Bone Marrow of Control vs. Tolerized Mice

We next examined the cell sizing profiles of bone marrow cells from
control vs. tolerized mice using a Coulter Channelyzer (Figure 1). Two
major populations of cells exist for both experimental groups. Peak I refers
to the smaller population and Peak II to the larger. In control mice, the
number of cells in the two peaks is approximately equal. In contrast, in
tolerized mice, Peak I is much smaller than Peak II, which is displaced in
size relative to control Peak II. By density gradient sedimentation using
a 15-30 percent fetal calf serum gradient, Peaks I and II are separable,
and in both sets of mice, all of the macrophage progenitor activity is contained within Peak II (data not shown). When cytospin preparations of density gradient-purified Peak II were prepared and stained, an enrichment of
very large, blast-like mononuclear cells were seen in the cells derived
from tolerized mice.

Lastly, the development of the denser, larger Peak II correlated well
with the acquisition and maintenance of the tolerant state (Figure 2).
Within three hours of initial LPS exposure, Peak II is significantly reduced
in size. This trend persists through 48 hours, at which time no well-defined
second peak can be observed. By 72 hours, the larger, denser Peak II is
clearly present. The arrow indicates the original position of Peak II. As
tolerance is progressively lost over Days 5-8, there is a progressive return
of Peak II to its original position (data not shown).

DISCUSSION

Several years ago, work by Michalek et al. (5) indicated that the cell
type responsible for endotoxin-induced lethality was bone-marrow derived,
radiation-sensitive, and lymphoreticular in origin. Further work by Rosenstreich and Vogel (9) provided evidence to suggest that the macrophage played
a central role in mediating the toxic effects of LPS. These findings were
based on work which demonstrated a correlation between the state of macrophage activation and LPS sensitivity in vivo and in vitro. Both studies
utilized the C3H/HeJ mouse as a model for genetically-conferred hyporesponsiveness to LPS. Another model which provides a state of endotoxin hyporesponsiveness in a genetically susceptible animal is an "endotoxin tolerance"
model. Injection of sublethal doses of LPS renders normally sensitive
animals refractory to a subsequent challenge (reviewed in 3). However, the
mechanisms which underlie the induction of tolerance to gram negative endotoxin have not been well-defined. Early studies have shown that the transient form of induced tolerance which occurs soon after initial exposure to
LPS and wanes rapidly (i.e., "early phase" tolerance) is not immunoglobulin-
mediated (3). These findings were recently extended by Williams et al.
(11) who provided the first convincing evidence that "early phase" tolerance
to LPS could be transferred into irradiated recipients with spleen cells
from tolerized animals. The findings that macrophages from endotoxin-tolerized mice failed to respond to LPS in vitro to produce products associated
with LPS sensitivity (2,8), again focuses attention to the macrophage as a
principal cellular participant.

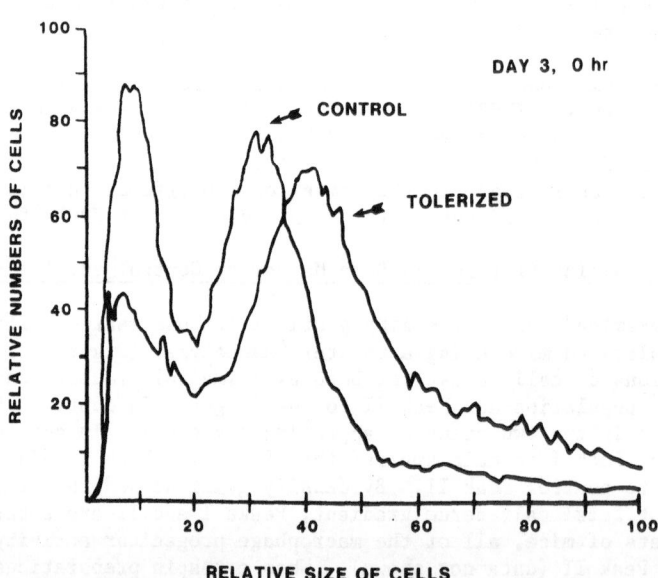

Figure 1. Cell sizing profiles of bone marrow cells derived from control
vs. tolerized mice.

The studies presented in this chapter were undertaken to define further the cellular mechanisms which underlie "early phase" tolerance. The major findings can be summarized as follows: (1) the tolerizing regimen employed was found to be efficacious as assessed by reduced circulating CSF levels, increased LD_{50} and mitigated symptoms in response to an endotoxin challenge. In addition, this state of induced hyporesponsiveness was found to be transient and heterologously inducible. By these criteria we feel confident that "early phase endotoxin tolerance," as defined previously (3), has been established; (2) the state of "early phase" endotoxin tolerance is associated

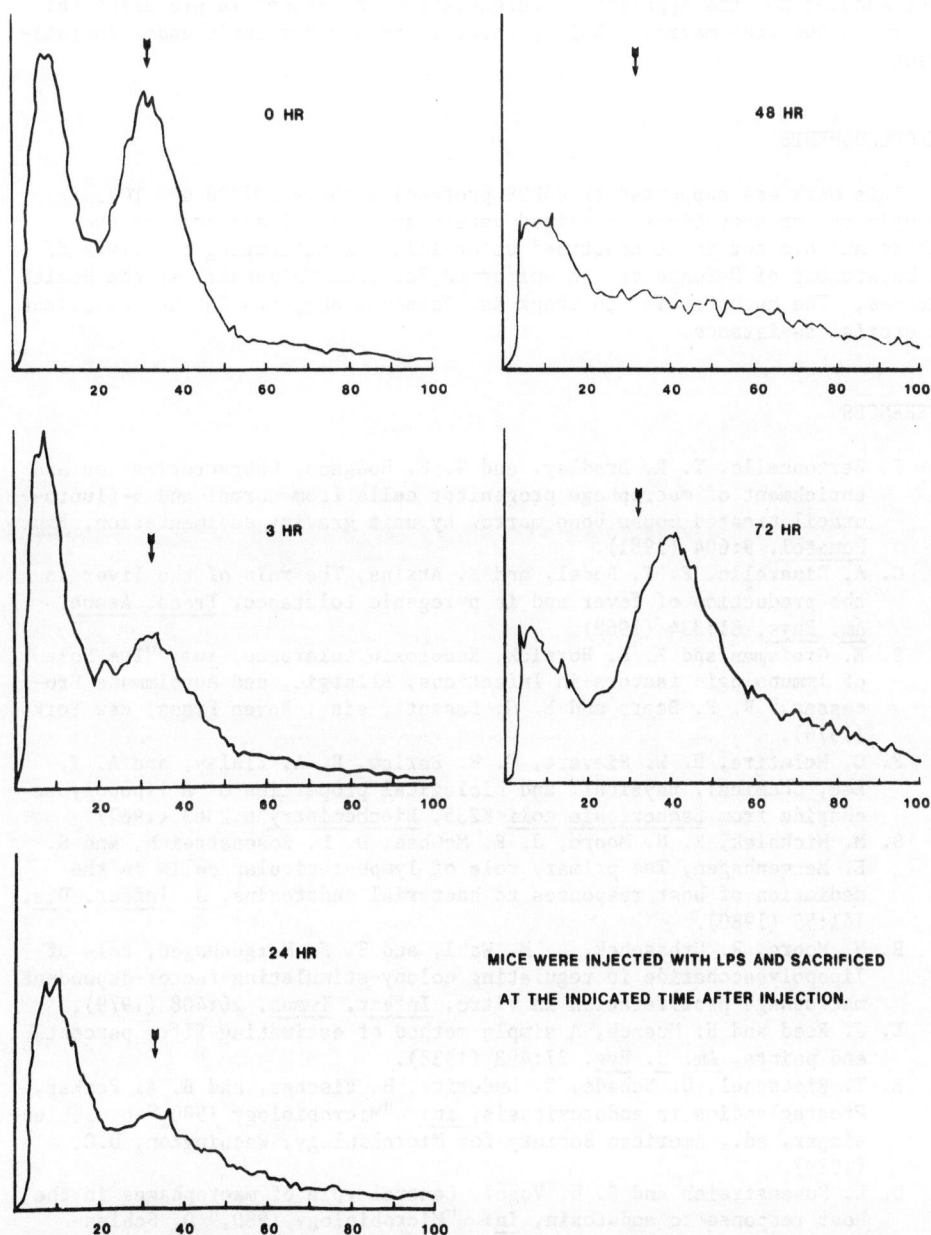

Figure 2. Cell sizing profiles of bone marrow cells derived from mice at various times after the injection of LPS. The arrow indicates the original position of Peak II.

with increased numbers of macrophage progenitor cells (both M-CFU and HPP) in the bone marrow; and (3) by Coulter Channelyzer sizing analysis, in conjunction with density gradient separation, there is an enrichment in the bone marrow for a larger, denser, mononuclear blast-like cell type.

These findings suggest several possibilities which could possibly provide an underlying mechanism for LPS-induced "early phase tolerance." If the macrophages which are released into the circulation are relatively immature, they might not be able to respond to LPS to produce those soluble factors associated with the endotoxemia. Alternatively, if there is a failure in tolerized mice to release macrophages from the bone marrow, this might account for the apparent differentiation "blockade" in precursor cell numbers in the bone marrow. Both possibilities are currently under investigation.

ACKNOWLEDGMENTS

This work was supported by USUHS protocol numbers RO7338 and TO7355. The opinions or assertions contained herein are the private ones of the authors and are not to be construed as official or reflecting the views of the Department of Defense or the Uniformed Services University of the Health Sciences. The authors wish to thank Ms. Jeanette Shepherd for her excellent secretarial assistance.

REFERENCES

1. I. Bertoncello, T. R. Bradley, and G. S. Hodgson, Characterization and enrichment of macrophage progenitor cells from normal and 5-fluorouracil treated mouse bone marrow by unit gravity sedimentation, Exp. Hematol. 9:604 (1981).

2. C. A. Dinarello, P. T. Bodel, and E. Atkins, The role of the liver in the production of fever and in pyrogenic tolerance, Trans. Assoc. Am. Phys. 81:334 (1968).

3. S. E. Greisman and R. B. Hornick, Endotoxin tolerance, in: "The Role of Immunologic factors in Infectious, Allergic, and Autoimmune Processes," R. F. Beers and E. G. Bassett, eds., Raven Press, New York (1976).

4. F. C. McIntire, H. W. Sievert, G. H. Barlow, R. A. Finley, and A. Y. Lee, Chemical, physical, and biological properties of a lipopolysaccharide from Escherichia coli K235, Biochemistry 6:2363 (1967).

5. S. M. Michalek, R. N. Moore, J. R. McGhee, D. L. Rosenstreich, and S. E. Mergenhagen, The primary role of lymphoreticular cells in the mediation of host responses to bacterial endotoxins, J. Infect. Dis. 141:55 (1980).

6. R. N. Moore, R. Urbaschek, L. M. Wahl, and S. E. Mergenhagen, Role of lipopolysaccharide in regulating colony-stimulating factor-dependent macrophage proliferation in vitro, Infect. Immun. 26:408 (1979).

7. L. J. Reed and H. Muench, A simple method of estimating fifty percent end points, Am. J. Hyg. 27:493 (1938).

8. E. T. Rietschel, U. Schade, O. Luderitz, H. Fischer, and B. A. Peskar, Prostaglandins in endotoxicosis, in: "Microbiology 1980," D. Schlessinger, ed., American Society for Microbiology, Washington, D.C. (1980).

9. D. L. Rosenstreich and S. N. Vogel, Central role of macrophages in the host response to endotoxin, in: "Microbiology 1980," D. Schlessinger, ed., American Society for Microbiology, Washington, D.C. (1980).

10. M. K. Warren and S. N. Vogel, Bone marrow-derived macrophages: development and regulation of differentiation markers by colony-stimulating factor and interferons, J. Immunol. 134:982 (1985).

11. Z. Williams, C. F. Hertzogs, and D. H. Pluznik, Use of mice tolerant to lipopolysaccharide to demonstrate requirement of cooperation between macrophages and lymphocytes to generate lipopolysaccharide-induced colony-stimulating factor in vivo 41:1 (1983).

PRODUCTION OF COLONY-STIMULATING FACTOR (CSF) BY BONE MARROW CELLS

STIMULATED WITH LIPOPOLYSACCHARIDE (LPS) AND PHORBOL ESTERS

Dov H. Pluznik and Stephan E. Mergenhagen

Cellular Immunology Section, Laboratory of Microbiology
and Immunology, National Institute of Dental Research
National Institutes of Health
Bethesda, Maryland

INTRODUCTION

Colony stimulating factors (CSF) are glycoproteins which regulate the proliferation and differentiation of committed hemopoietic stem cells into mature granulocytes and macrophages. Murine spleen cells are one of the best sources for CSF and bacterial lipopolysaccharides (LPS) are effective stimulants of these cells to generate CSF. The injection of LPS into mice is followed by a rapid increase of CSF in the serum (15,23,3), and addition of LPS to spleen cultures induces release of CSF into the supernatants (4). In a previous study (2), we reported that lipid A is the moiety of the LPS molecule which is responsible for the generation of CSF. In a recent study we have shown that some preparations of polysaccharides, obtained from LPS after carefully controlled hydrolytic conditions, can also induce the generation of CSF (19). By using the LPS-low responder strain of mice, C3H/HeJ, we analyzed the genetic control involved in LPS-induced generation of CSF (3) and interferon (5,8). CSF is generated by macrophages and lymphocytes stimulated with LPS (6). Employing adherent and non-adherent spleen cells from LPS-responsive mice and C3H/HeJ mice we demonstrated that both subpopulations of cells are required for the generation of CSF by LPS. Moreover, LPS interacts with the adherent cells which release a soluble factor which in turn stimulates the non-adherent cells to release CSF (7). Interaction between adherent and non-adherent cells is also required in vivo; mice tolerant to LPS which do not generate CSF after a challenging injection of LPS, produce CSF when injected with a mixture of adherent and non-adherent spleen cells from non-tolerant mice prior to the challenging injection of LPS (27).

Most of the data reported in our previous studies describe the response of spleen cells to LPS. The question arises whether bone marrow (BM) cells, which contain the target cells for CSF activity, are also capable of generating CSF. Indeed, murine BM cells do not produce CSF after exposure to LPS. Can other substances stimulate CSF production by BM cells or is this lack of production limited only to LPS? A partial answer to these questions was obtained in studies reported a few years ago in which addition of tumor promoting phorbol esters (TPA) to BM cells in soft agar cultures was followed by growth of small colonies of granulocytes and macrophages (10). The appearance of such colonies was interpreted as follows: a) TPA replaces CSF in inducing clonal growth of BM cells; b) TPA renders the membrane of the

committed hemopoietic stem cells more sensitive to subthreshold amounts to
CSF present in the serum which is supplemented in the agar culture; or c)
TPA induces release of CSF from BM cells (25). Therefore, we examined
whether TPA and/or a combination of TPA with T or B cell mitogens, which
independently induce generation of CSF by spleen cells, will stimulate BM
cells to produce CSF. The mitogens PHA, Con A, PWM and LPS were tested
together with TPA for their capacity to stimulate BM cells to generate CSF.
Interestingly, only LPS was active with TPA in inducing clonal growth of
these cells and in stimulating the generation of CSF by BM cells. The present study describes the results of our experiments with LPS and TPA as synergistic inducers of CSF generation by murine BM cells. These studies led to
the hypothesis that LPS may exert an influence on cellular activity by mobilizing calcium which is required for metabolic processes and mediator
production.

MATERIALS AND METHODS

Mice

CBA/J male mice 8-16 weeks old were used in all experiments (Jackson
Laboratory, Bar Harbor, Maine).

Phorbol Esters

12-0-Tetradecanoly-phorbol-13-acetate (TPA) and 4-0-Methyl-TPA were
purchased from Consolidated Midland Company, Brewster, New York and dissolved
in dimethyl sulfoxide (DMSO) to 2×10^{-3}M and stored at -70°C. Before addition to cells the TPA and 4-0-Me-TPA were diluted in growth medium. The
final concentration of DMSO was always less than one percent and usually
0.05%.

LPS

LPS from Salmonella abortus equi was purchased from Difco, Detroit,
Michigan. LPS breakdown products were a gift from Dr. C. Galanos, Max
Planck Institute fur Immunobiologie, Freiburg, Federal Republic of Germany.

Growth Media

Dulbecco's modified Eagle's medium (DMEM) heat inactivated horse serum
were purchased from GIBCO, Grand Island, New York.

Preparation of CSF from BM Cells

BM cultures were prepared by seeding 5×10^6 cells/ml in DMEM supplemented with 25% horse serum. TPA (10^{-6}M) and LPS (12.5 µg/ml) were added
to the cells for four hours at 37°C, after which the cells were washed with
DMEM and resuspended in fresh DMEM containing 25% horse serum for an additional 20 hours at 37°C. After incubation, the cells were centrifuged and
the supernatants assayed for CSF activity.

Assay of CSF Activity

Two assays were used to quantitate CSF activity: a) cloning of BM
cells in soft agar cultures, as previously described (20,21). Briefly,
various concentrations of TPA and LPS or supernatants containing CSF (final
concentration 25%) were incorporated in the hard layer agar base (0.5% agar)
on top of which 10^5 BM cells were seeded in soft agar medium (0.3% agar).
Colonies developing in the soft agar were counted seven days after incubation
at 37°C. The number of colonies is proportional to the amount of CSF in
the plate; b) proliferation of cells of a CSF-dependent cell line, PT-18,

as previously described (22,26). Briefly, supernatants containing CSF were
added to PT-18 cells for 40 hours after which the cells were pulse labelled
with ^{3}H-TdR incorporation into cellular DNA was measured. CSF activity was
expressed as stimulation index which is the ratio between cpm obtained in
cells stimulated with CSF to cpm obtained in cells incubated in medium alone.

RESULTS AND DISCUSSION

Effect of LPS and TPA on Clonal Growth of BM Cells in Agar

In order to determine whether incubation of BM cells with LPS and TPA
could enhance the clonal growth of committed hemopoietic stem cells, these
agents were incorporated directly in the hard agar base on top of which BM
cells were seeded in soft agar. After seven days of incubation the colonies
developing in the soft agar were elevated. Figure 1 shows the morphology
of single colonies developing in the presence of a) LPS and TPA with CBA/J
BM cells; b) TPA alone with CBA/J BM cells; c) LPS and TPA with C3H/HeJ BM
cells and; d) LPS and 4-0-Me-TPA, with CBA/J BM cells. The simultaneous
addition of LPS and TPA to the agar cultures induced large colonies containing granulocytes and macrophages. The addition of TPA alone to the agar
cultures induced small colonies (clusters) containing 50 or less cells.
The addition of LPS alone to the agar cultures induced aggregates of cells.
When TPA was replaced with the non-tumor promoting phorbol ester, 4-0-Me-TPA,
in the presence of LPS, no colonies appeared in the agar, similar to the
effect of LPS alone. When BM cells from C3H/HeJ mice were substituted for
BM cells from BCA/J mice and LPS and TPA were added to the cultures, only
clusters appeared in the soft agar. This was similar to the clusters appearing in response to TPA alone added to CBA/J BM cells. Using varying concen-

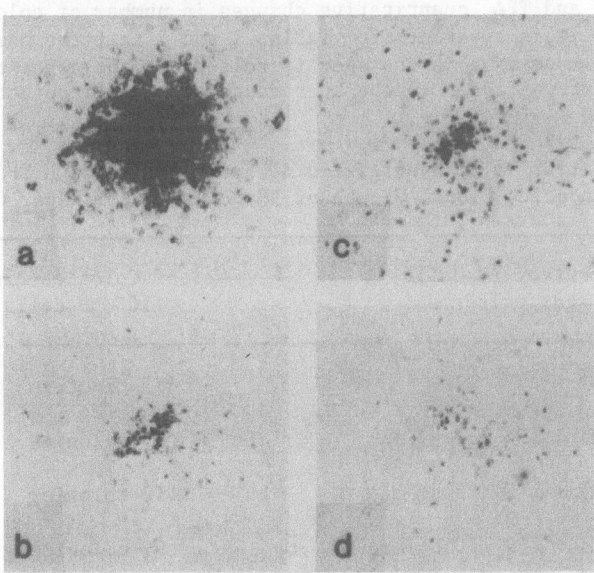

Figure 1. Morphology of single colonies/clusters in soft agar cultures,
after 7 days of incubation at 37°C. 10^5 BM cells were cloned
per plate. The hard agar layer was supplemented with:
a) LPS 25 μg/ml and TPA 10^{-6} M; CBA/J BM cells, (colony) x50;
b) TPA 10^{-6} M; CBA/J BM cells, (cluster) X50; c) LPS 25 μg/ml
and TPA 10^{-6} M; C3H/HeJ BM cells, (cluster) X50; d) LPS 25 μg/
ml and 4-0-Me-TPA 10^{-6} M; CBA/J BM cells, (aggregate) X50.

Table 1. Synergistic Activity of TPA and LPS in Inducing
Clonal Growth of BM Cells in Soft Agar Cultures

Treatment of BM Cells in agar cultures	No. of colonies-clusters/10^5 BM cells[1]	
	CBA/J mice	C3H/HeJ mice
TPA (10^{-6}M) + LPS (25 µg/ml)	126 colonies	41 clusters
TPA (10^{-7}M) + LPS (25 µg/ml)	131 colonies	25 clusters
TPA (10^{-8}M) + LPS (25 µg/ml)	80 colonies	4 clusters
4-0-Me-TPA (10^{-6}M) + LPS (25 µg/ml)	10 aggregates	3 aggregates
4-0-Me-TPA (10^{-7}M) + LPS (25 µg/ml)	8 aggregates	3 aggregates
4-0-Me-TPA (10^{-8}M) + LPS (25 µg/ml)	12 aggregates	2 aggregates
TPA (10^{-6}M)	135 clusters	32 clusters
TPA (10^{-7}M)	130 clusters	19 clusters
TPA (10^{-8}M)	69 clusters	5 clusters
LPS (25 µg/ml)	5 aggregates	1 aggregate

[1]Colonies-clusters counted after seven days of incubation at 37°C.

trations of LPS and TPA, quantitative changes in number of colonies/clusters
were observed. It is apparent from Tables 1 and 2 that the number and size
of the colonies appearing in the agar is related to the concentration of
LPS and TPA which was added to the agar cultures. The interaction of TPA

Table 2. Titration of Synergistic Activity of LPS with TPA in
Inducing Clonal Growth of BM Cells in Soft Agar Cultures

Treatment of BM Cells in agar cultures	No. of Colonies-Clusters /10^5 BM cells[1]
TPA (10^{-6}M)	89 clusters
TPA (10^{-6}M) + LPS (50 µg/ml)	97 colonies
TPA (10^{-6}M) + LPS (25 µg/ml)	110 colonies
TPA (10^{-6}M) + LPS (10 µg/ml)	99 colonies
TPA (10^{-6}M) + LPS (1 µg/ml)	85 clusters
TPA (10^{-6}M) + LPS (0.1 µg/ml)	92 clusters

[1]Assayed on CBA/J BM cells. Colonies-clusters counted after
seven days of incubation at 37°C.

Table 3. Generation of CSF from BM Cultures Stimulated
with TPA and LPS

Treatment of BM cells[1]	No. of Colonies-Clusters /10^5 BM cells[2]
TPA + LPS	21 colonies
TPA	20 clusters
LPS	0

[1]CBA/J BM cells (10^6/ml) were incubated with TPA (10^{-6}M) and/or
(25 µg/ml) LPS for four hour after which the cells were washed
and incubated for additional 20 hours in growth medium.
Supernatants were used as a source of CSF.
[2]Assayed on C3H/HeJ BM cells.

with BM cells resulted in only a few cell divisions giving clusters of 50
or less cells while the simultaneous addition of LPS and TPA resulted in
colonies containing up to 1000 cells. The interaction of TPA with the BM
cells is specific since replacement of TPA by the non tumor promoting phorbol
ester, 4-O-Me-TPA, in combination with LPS did not induce colonies but rather
aggregates similar to the effect of the LPS alone. Also, the interaction
of LPS with BM cells is specific, since LPS had no effect when added together
with TPA to BM cells from C3H/HeJ mice.

Generation of CSF by BM Cells Stimulated with LPS and TPA

The data presented in Tables 1 and 2 indicate a synergistic effect of
LPS and TPA in inducing proliferation of committeed hemopoietic stem cells
into colonies of mature granulocytes/macrophages. To demonstrate that this
clonal growth is mediated by the release of CSF from BM cells stimulated
with LPS and TPA, supernatants from such cells were incorporated in the
hard agar layer on top of which fresh unstimulated C3H/HeJ BM cells were
cloned in soft agar. Table 3 shows that only supernatants from BM cultures
stimulated with LPS and TPA contained CSF activity as demonstrated by the
growth of colonies in the soft agar cultures supplemented with such supernatants. Moreover, since the BM cells which were used to detect CSF activity
in the supernatants were from C3H/HeJ mice, this would eliminate the possibility that residual LPS was responsible for the clonal growth of BM colonies.

Assaying CSF activity in soft agar cultures using BM cells as target
cells has some limitations. The amount of CSF to be assayed determines not
only the number of colonies but also their size. While the number of colonies
does not increase at high concentrations of CSF, the size of the colonies
increases. At low concentrations of CSF it is difficult to discriminate
between a colony and a cluster. Therefore, it would be advantageous to
quantitate CSF in a system which is independent of colony number or size
but dependent upon an objective measurement such as DNA synthesis. Recently,
we described a basophil/mast cell line, PT-18, whose proliferation is dependent on the presence of CSF (22,26). In addition, using this cell line to
quantitate CSF allows for the separation between the BM cell which generates
the CSF and the cells which are used for the assay of CSF.

When varying concentrations of LPS, ranging from 3-50 µg/ml and varying
concentrations of TPA ranging from 5 x 10^{-8} up to 10^{-6} were mixed to stimulate CBA/J BM cells to generate CSF, the combination of 12.5 µg/ml LPS and

Table 4. Effect of Combined Stimulation of Varying Concentrations of TPA and LPS on CSF-Generation by CBA/J BM Cells

LPS (μg/ml)	3.12	6.25	12.5	25.0	0.0
TPA (M)					
10^{-6}	72.4	98.2	129.8	126.1	8.2
5×10^{-7}	23.5	64.3	69.4	76.4	3.4
10^{-7}	8.1	20.6	25.2	21.7	1.4
5×10^{-8}	4.7	11.4	13.5	15.2	1.2
0	0.9	0.8	1.1	0.9	0.8

CSF Activity (Stimulation Index)[1]

[1]LPS and TPA were added to BM cells (5×10^{6}/ml) for four hours after which the cells were washed and incubated for additional 20 hours in growth medium. Supernatants were assayed for CSF activity on PT-18 cells.

10^{-6}M TPA was found to be optimal (Table 4). Replacing TPA with 4-0-Me-TPA had no effect when added with LPS to CBA/J BM cells. When LPS and TPA were added to C3H/HeJ BM cells, no CSF was generated. The results of these experiments are summarized in Table 5. These data confirm the specificity of the interaction of LPS and TPA with BM cells. To further elucidate the

Table 5. Effect of Combined Stimulation of Varying Concentrations of Phorbol Esters and LPS on CSF-Generation by CBA/J and C3H/HeJ BM Cells

LPS (μg/ml) Phorbol esters	6.25	12.5	25.0	50
CBA/J BM:				
TPA 10^{-6}M	98.2	129.8	126.1	NT
4-0-Me-TPA 10^{-6}M	0.5	0.5	0.7	0.6
C3H/HeJ BM:				
TPA 10^{-5}M	1.5	2.0	1.8	2.0
TPA 10^{-6}M	0.3	0.8	1.0	0.8

[1]LPS and phorbol esters were added to BM cells (5×10^{6}/ml) for four hours after which the cells were washed and incubated for additional 20 hours in growth medium. Supernatants were assayed for CSF activity on PT-18 cells.

Table 6. Effect of Combined Stimulation of Various Preparations of LPS and TPA on CSF-Generation by CBA/J and C3H/HeJ BM Cells

Preparations of LPS[1]	CSF Activity (Stimulation Index)	
	CBA/J	C3H/HeJ
S. abortus equi	158.7	3.1
S. minn. R595	176.2	5.7
Lipid A (E. coli Re)	170.6	4.2
Polysaccharide	5.3	3.4

[1] 10 µg/ml of each preparation was added together with 10^{-6}M TPA to BM cell (5×10^6/ml) for four hours after which the cells were washed and incubated for additional 20 hours in growth medium. Supernatants were assayed for CSF activity on PT-18 cells.

specificity of interaction of LPS with BM cells, we tested the effect of various products related to LPS, on the generation of CSF. The results (Table 6) demonstrate that the glycolipid from Salmonella minnesota R595, and the lipid A obtained by acid hydrolysis from Escherichia coli Re were as active as the intact LPS obtained from Salmonella abortus equi, while the polysaccharide moiety obtained by hydrolysis did not stimulate CSF production by BM cells. None of the preparations had activity when tested on BM cells from C3H/HeJ mice. These data reveal the specificity of the interaction of LPS with the BM cells.

Table 7. Generation of CSF by BM and Spleen Cells from Athymic (nu/nu) and Euthymic (nu/+) Mice Stimulated with LPS and TPA or with Con A

Treatment of Cells	CSF Activity (Stimulation Index)[1]	
	nu/nu	nu/+
BM Cells:		
TPA + LPS	208.8	184.1
TPA	5.8	6.0
LPS	1.6	1.6
medium	0.5	1.0
Spleen Cells:		
Con A	1.3	444.6

[1] TPA (10^{-6}M) and/or LPS (12.5 µg/ml) were added to BM cells (5×10^6/ml) for four hours after which cells were washed and incubated for additional 20 hours in growth medium. Spleen cells (5×10^6/ml) were incubated with Con A (5 µg/ml) for 24 hours. Supernatants were assayed for CSF activity on PT-18 cells.

Table 8. Generation of CSF by Fractionated BM Cells on Polystyrene Surfaces and on Polystyrene Surfaces Coated
with Goat Antibodies to Murine Immunoglobulins

BM Subpopulations	CSF Activity (Stimulation Index)
I Polystyrene Surface:[1]	
Unfractionated	167.4
Adherent	69.4
Non-adherent	82.9
Reconstituted	146.2
II Immunoglobulin Coated Surface:[2]	
Unfractionated	141.1
Adherent	157.4
Non-adherent	16.9
Reconstituted	155.2

[1] 2 ml BM cells (2.5×10^6/ml) in RPMI-1640 medium + 10% fetal
calf serum were incubated in a polystyrene plate (35 mm diameter) for 90 minutes at 37°C. Non-adherent cells were removed
by washing the plate with warm medium. Non-adherent cells
were centrifuged and resuspended in 1 ml DMEM + 25% horse serum.
1 ml of the same medium was also added to the plate containing
the adherent cells. For reconstitution 1 ml of non-adherent
cells was added back to the plate containing the adherent
cells. Each of the BM subpopulations was stimulated with
TPA (10^{-6}M) and LPS 12.5 µg/ml) for four hours after which
the cells were washed and incubated for an additional 20 hours
in growth medium. Supernatants were assayed for CSF activity
on PT-18 cells.

[2] Polystyrene plates (35 mm diameter) were incubated with 1 ml/
plate of goat anti-mouse immunoglobulin diluted in PBS (150 µg/
ml) for 60 minutes at 22°C. After incubation, the serum was
removed and the plates washed five times with 2 ml PBS. After
washing, the plates containing PBS (2 ml/plate) were incubated
overnight at 4°C. The PBS was removed and 2 ml BM cells ($2.5 \times
10^6$/ml) were incubated for 60 minutes at 22°C. Non-adherent
cells were removed by washing the plates with PBS. All further
steps were identical to those as with polystyrene plates.

Identification of the Cells in the BM Responding to LPS and TPA in
Generating CSF

Since bone marrow contains a heterogenous population of cells, we attempted to identify the types of cells which respond to LPS and TPA. First,
the effect of LPS and TPA on BM cells obtained from athymic mice (nu/nu)
and their heterozygous euthymic controls (nu/+) was compared. Table 7 shows
that BM cells from athymic mice generated similar amounts of CSF when compared with cells from euthymic mice. However, when spleen cells from these
two groups of mice were stimulated with a T cell mitogen (Con A), only cells
from the euthymic mice generated CSF. These data suggest that mature T
lymphocytes are not involved in the response of BM cells to LPS and TPA.
Similar results were obtained in a previous study (7) in which spleen cells

Table 9. Generation of CSF by BM Cells Stimulated by Subsequent or Simultaneous Addition of TPA and LPS

Treatment of BM Cells	CSF Activity (Stimulation Index)
TPA 2h + LPS 2h	36.4
LPS 2h + TPA 2h	15.1
TPA + LPS 2h	135.3
TPA + LPS 4h	155.2
TPA 2h	6.3
TPA 4h	5.1
LPS 2h	2.4
LPS 4h	3.3
Medium	1.2

[1]BM cells (5×10^6/ml) were incubated with TPA (10^{-6}M) and/or LPS
(12.5 µg/ml) for various time periods, as indicated in the table,
after which the cells were washed and incubated for additional
20 hours in growth medium. Supernatants were assayed for CSF
activity on PT-18 cells.

from athymic and euthymic mice generated CSF in response to LPS. These
results also explain why T cell mitogens like Con A and PHA were ineffective
with TPA in stimulating BM cells. Secondly, we separated BM cells into
adherent and non-adherent cells and tested each subpopulation for its capability to generate CSF (Table 8). Each subpopulation generated about 50%
of CSF when compared to the unfractionated or reconstituted BM cells. In
further experiments, polystyrene surfaces were coated with goat anti-mouse
immunoglobulin antibodies (14). When BM cells were separated into adherent
and non-adherent populations, the adherent cells had the capacity to generate
CSF almost as well as the unfractionated or the reconstituted (adherent and
non-adherent) BM cells (Table 8). Since we used intact antibodies and not
the Fab fraction it cannot be concluded that cells bearing Ig receptors
(B0lymphocytes) are the only cells responding to LPS and TPA. Other cells
bearing Fc receptors (macrophages and granulocytes) must also be considered
as possible sources for generating CSF in response to LPS and TPA.

Possible Explanations for the Requirement of Synergism Between LPS and TPA

In the next group of experiments we tried to elucidate the unique requirement for a cooperation between LPS and TPA to stimulate BM cells. LPS
alone can stimulate murine spleen cells but not murine BM cells to generate
CSF. The spleen contains a much more mature population of lymphoreticular
cells when compared to the BM. It is known from other studies that TPA has
the capacity to induce rapid maturation of cells of two human leukemic cell
lines, U-937 and HL-60, into mature macrophages (24). It is possible, therefore, that in our experiments TPA induces a rapid maturation of the BM cells
which in turn will respond to LPS in generating CSF. To test this hypothesis
we compared the generation of CSF by BM cells stimulated by TPA alone with
subsequent addition of LPS, to that of BM cells stimulated by a simultaneous
addition of LPS and TPA. The results of these experiments are shown in
Table 9. Optimal generation of CSF was only achieved when LPS and TPA were
added simultaneously. When TPA was added for two hours followed by LPS for
an additional two hours, or addition of these agents in the reverse order,
only small amounts of CSF were generated. These results suggest that TPA
does not induce maturation of BM cells followed by increased sensitivity to
LPS.

Table 10. Effect of Combined Stimulation of BM Cells by
IL-1 with TPA or with LPS on Generation of CSF

Treatment of BM Cells[1]	CSF Activity (Stimulation Index)
TPA + LPS	153.9
TPA + IL-1	8.1
LPS + IL-1	2.3
TPA	4.8
LPS	2.4
IL-1	2.4
Medium	1.2

[1] TPA 10^{-6}, LPS 12.5 µg/ml, IL-1 2 units/ml. BM cells (5 x 10^6/ml) were incubated for four hours with TPA and/or LPS and/or IL-1, after which the cells were washed and incubated for additional 20 hours in growth medium. Supernatants assayed for CSF activity on PT-18 cells.

In a previous study (7) we showed that LPS stimulates macrophages to release a soluble factor which in turn induces lymphocytes to generate CSF. This soluble factor could be a lymphocyte activating factor (Interleukin-1, IL-1). Both LPS and TPA stimulate monocytes to produce IL-1 (12,16). Therefore, we tested whether IL-1 is involved in the generation of CSF by BM cells stimulated with LPS and TPA. Purified IL-1 was added to BM cells either alone or in combination with TPA or LPS and the amount of CSF was determined. Table 10 shows that IL-1 alone or in combination with either TPA or LPS, had no effect on the generation of CSF, while the simultaneous addition of TPA and LPS had the usual stimulatory effect on generation of CSF. In addition, supernatants from BM cells, stimulated with LPS and TPA, contained no IL-1 activity.

LPS is known to stimulate macrophages to generate prostaglandins (PGE) which in turn have an inhibitory effect on the generation of CSF by LPS (13). TPA is also known to enhance the activity of CSF (11). Thus, it is possible that the cooperation between LPS and TPA is required to overcome an inhibition by PGE. To test such a possibility we stimulated BM cells with LPS in the presence of indomethacin, which inhibits the synthesis of PGE (17), and compared it to the effect of the simultaneous addition of LPS and TPA. Table 11 shows that addition of indomethacin to the stimulated BM cultures with LPS alone had no effect on the generation of CSF. Only a simultaneous addition of LPS and TPA was effective in stimulating the generation of CSF by the BM cells. In the presence of indomethacine an increase of 12% in CSF activity was observed.

Potential Mechanism of the Synergistic Effect of LPS and TPA

Recent studies (18,9) have suggested that the interaction of a wide variety of biologically active substances with their specific cell surface receptors is followed by an immediate breakdown of membrane inositol phospholipids associated with an increase in intracellular calcium. These biochemical events seem to mediate many physiological responses of cells. Two of the earliest products of the breakdown of inositol phospholipids which accumulate in the cell are 1,2-diacylglycerol and myo-inositol 1,4,5 triphosphate. The diacylglycerol activates the calcium dependent enzyme, protein kinase C, and the inositol triphosphate mobilizes intracellular calcium. Protein

Table 11. Effect of Indomethacin on Generation of CSF
by BM Cells

Treatment of BM Cells[1]	CSF Activity (Stimulation Index)
TPA + LPS	130.4
TPA + LPS + Indomethacin	155.2
LPS	5.3
LPS + Indomethacin	8.4

[1]LPS 12.5 µg/ml, TPA 10^{-6}M, Indomethacin 10^{-6}M. BM cells were
incubated with Indomethacin for four hours in the presence of
LPS and TPA, after which the cells were washed and incubated
for additional 20 hours in growth medium containing Indomethacin. Supernatants were assayed for CSF activity on PT-18
cells.

kinase C is now widely accepted to be the cellular target for TPA, which
by-passes the inositol phospholipid breakdown and interacts directly with
the enzyme. In view of these data we postulated that the cooperation between
LPS and TPA in stimulating BM cells to generate CSF could be linked to activation of protein kinase C by TPA and calcium mobilization by LPS (Figure
2). The interaction of LPS with the BM cells could be followed by inositol
phospholipid breakdown releasing inositol triphosphate. This in turn

Table 12. Replacement of LPS by the Calcium Ionophore A 23187
in Stimulating BM Cells to Generate CSF

Treatment of BM Cells[1]	CSF Activity (Stimulation Index)
TPA 10^{-6}M + A 23187 5×10^{-6}M	55.2
TPA 10^{-6}M + A 23187 10^{-6}M	98.3
TPA 10^{-6}M + A 23187 5×10^{-7}M	76.1
TPA 10^{-6}M + A 23187 10^{-7}M	11.4
A 23187 5×10^{-6}M	3.2
A 23187 10^{-6}M	2.1
A 23187 5×10^{-7}M	14.3
A 23187 10^{-7}M	16.1

[1]BM cells were incubated with TPA and A 23187 for four hours
after which the cells were washed and incubated for additional
20 hours in growth medium. Supernatants were assayed for CSF
activity on PT-18 cells.

mobilizes calcium required for the activation of protein kinase C. It is even possible that LPS may directly mobilize calcium. To examine these possibilities we tested whether calcium ionophores such as A 23187 can stimulate BM cells in the presence of TPA. No production of CSF was observed when BM cells were stimulated with LPS and A 23187. Further experiments are required to examine the interesting possibility that LPS exerts an influence on cellular activity by mobilizing calcium which is required for other metabolic processes.

SUMMARY

Stimulation of murine bone marrow (BM) cells with the tumor-promoting phorbol ester 12-0-tetradecanoyl-phorbol-13-acetate (TPA) induces the growth of clusters of granulocytes/macrophages in soft agar cultures. However, the simultaneous addition of lipopolysaccharide (LPS) and TPA results in the formation of large colonies of these cells in the soft agar cultures. The growth of these large colonies is mediated via the production of colony-stimulating factor (CSF) by the BM cells. CSF was quantitated by measuring the 3H-thymidine incorporation into the DNA of a basophil/mast cell line, PT-18, which is CSF dependent. Replacing TPA with its non-tumor promoting phorbol ester, 4-0-Methyl-TPA, abolished the production of CSF. LPS can be replaced by its lipid A moiety but not by the polysaccharide component in stimulating the production of CSF. Treatment of BM cells from the LPS-low responder mouse strain, C3H/HeJ, with LPS and TPA was ineffective in inducing CSF. To identify the cells which participate in generation of CSF, BM cells were separated into polystyrene adherent and non-adherent cells. Each of these two populations generated similar amounts of CSF which were about 50% of the amount generated by the unfractionated BM cells. However, when BM cells were separated on surfaces coated with goat anti-murine immunoglobulin antibodies, the adherent cells produced as much CSF as did the unfractionated BM cells, while the non-adherent cells generated only minute quantities of

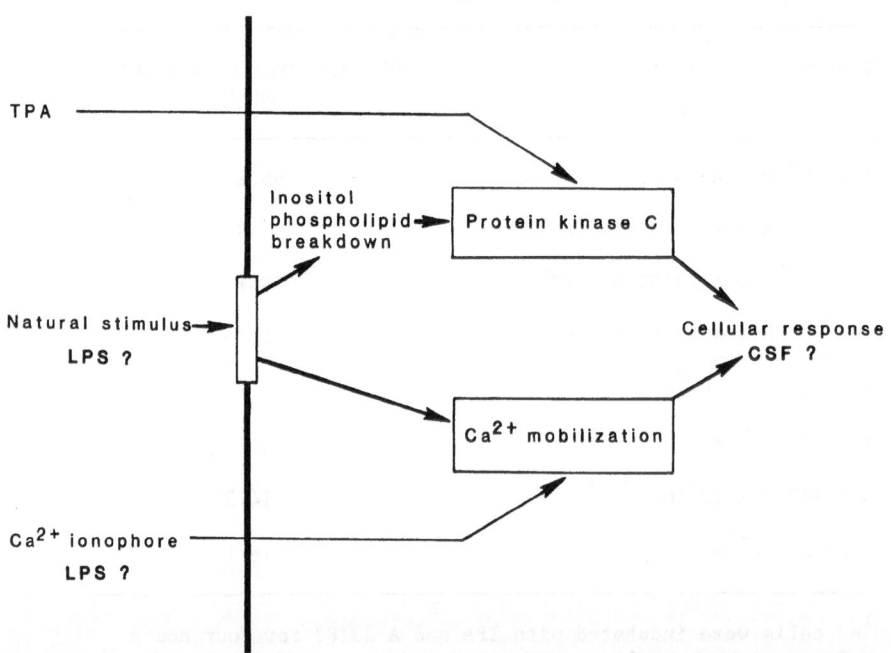

Figure 2. Synergistic roles of calcium and protein kinase C in cellular responses, and the possible relation of LPS to these events.

CSF. Additional experiments demonstrated that the cooperation between LPS and TPA was effective only when both agents were added simultaneously to the BM cells. Since the calcium ionophore A 23187 could replace LPS in cooperating with TPA in stimulating BM cells to generate CSF, a hypothesis was put forward that LPS could be involved in calcium mobilization which is required for many cellular activities and in the production of various soluble mediators.

REFERENCES

1. R. N. Apte and D. H. Pluznik, Control mechanisms of endotoxin and particulate material stimulation of hemopoietic colony forming cell differentiation, Exp. Hematol. 4:10 (1976).
2. R. N. Apte, C. Galanos, and D. H. Pluznik, Lipid A, the active part of bacterial endotoxins in inducing serum colony stimulating activity and proliferation of splenic granulocyte-macrophage progenitor cells, J. Cell. Physiol. 87:71 (1976).
3. R. N. Apte and D. H. Pluznik, Genetic control of lipopolysaccharide induced generation of serum colony-stimulating factor and proliferation of splenic granulocyte/macrophage precursor cells, J. Cell. Physiol. 89:313 (1976).
4. R. N. Apte, C. F. Hertogs, and D. H. Pluznik, Regulation of lipopolysaccharide-induced granulopoiesis and macrophage formation by lymphoid cells. I. Relationship between colony stimulating factor release and lymphocyte activation in vitro, J. Immunol. 118:1435 (1977).
5. R. N. Apte, O. Ascher, and D. H. Pluznik, Genetic analysis of generation of serum interferon by bacterial lipopolysaccharide, J. Immunol. 119:1899 (1977).
6. R. N. Apte, C. F. Hertogs, and D. H. Pluznik, Generation of colony-stimulating factor by purified macrophages and lymphocytes, J. Reticuloendothelial. Soc. 26:491 (1979).
7. R. N. Apte, C. F. Hertogs, and D. H. Pluznik, Regulation of lipopolysaccharide-induced granulopoiesis and macrophage formation by spleen cells. II. Macrophage-lymphocyte interactions in the process of generation of colony-stimulating factor, J. Immunol. 124:1223 (1980).
8. O. Ascher, R. N. Apte, and D. H. Pluznik, Generation of lipopolysaccharide-induced interferon in spleen cell cultures. I. Genetic analysis and cellular requirements, Immunogenetics 12:117 (1981).
9. M. J. Berridge and R. F. Irvine, Inositol triphosphate, a novel second messenger in cellular signal transduction, Nature 312:315 (1984).
10. E. Fibach, P. Marks, and R. A. Rifkind, Tumor promoters enhance myeloid and erythroid colony formation by normal mouse hemopoietic cells, Proc. Nat. Acad. Sci. USA 77:4152 (1980).
11. L. J. Guilbert, D. J. Nelson, J. A. Hamilton, and N. Williams, The nature of 12-0-tetradecanoyl phorbol-13-acetate (TPA)-stimulated hemopoiesis, colony stimulating factor (CSF) requirement for colony formation, and the effect of TPA on ^{125}I-CSF-1 binding to macrophages, J. Cell. Physiol. 115:276 (1983).
12. T. Krakauer, D. Mizel, and J. Oppenheim, Independent and synergistic thymocyte proliferative activities of PMA and IL-1, J. Immunol. 129:939 (1982).
13. J. I. Kurland, H. E. Broxmeyer, and M.A.S. Moore, Role of monocyte-macrophage-derived colony-stimulating factor and prostaglandin E in the positive and negative feedback control of myeloid stem cell proliferation, Blood 52:388 (1978).
14. M. G. Mage, L. L. McHugh, and T. L. Rothstein, Mouse lymphocytes with and without surface immunoglobulin: preparative scale separation in

polystyrene tissue culture dishes coated with specifically purified anti-immunoglobulin, J. Immunol. Methods 15:47 (1977).

15. D. Metcalf, Acute antigen-induced proliferation in vitro of bone marrow precursors of granulocytes and macrophages, Immunology 21:427 (1971).

16. S. B. Mizel, D. Rosenstreich, and J. J. Oppenheim, Phorbol myristic acetate stimulates LAF production by the macrophage cell line P388D1, Cell. Immunol. 40:230 (1978).

17. R. N. Moore, R. Urbaschek, L. M. Wahl, and S. E. Mergenhagen, Prostaglandin regulation of colony-stimulating factor production by lipopolysaccharide-stimulated murine leukocytes, Infec. Immun. 26:408 (1979).

18. Y. Nishizuka, The role of protein kinase C in cell surface signal transduction and tumor promotion, Nature 308:693 (1984).

19. A. Nowotny, U. H. Behling, F. Madani, A. M. Nowotny, P. H. Pham, C. F. Hertogs, and D. H. Pluznik, Studies on the optimal conditions of CSF generation by endotoxic LPS and its PS derivative in mice, J. Immunopharmacology 5:93 (1983).

20. D. H. Pluznik and L. Sachs, The cloning of normal "mast" cells in tissue culture, J. Cell. Physiol. 66:319 (1965).

21. D. H. Pluznik and L. Sachs, The induction of clones of normal "mast" cells by a substance from conditioned medium, Exp. Cell Res. 43:553 (1966).

22. D. H. Pluznik, N. S. Tare, M. M. Zatz, and A. L. Goldstein, A mast/basophil cell line dependent on colony stimulating factor, Exp. Hematol. 10:221 (1982).

23. P. J. Quesenberry, A. Morely, F. Stohlman, Jr., K. Rickard, D. Howard and M. Smith, Effect of endotoxin on granulopoiesis and colony-stimulating factor, N. Eng. J. Med. 286:227 (1972).

24. G. Rovera, D. Santoli, and C. Damsky, Human promyelocytic leukemia cells in culture differentiate into macrophage-like cells when treated with phorbol-diester, Proc. Natl. Acad. Sci USA 76:2779 (1979).

25. R. K. Stuart and T. A. Hamilton, Tumor-promoting phorbol esters stimulate hemopoietic colony formation in vitro, Science 208:402 (1980).

26. J. Y. Vanderhook, N. S. Tare, J. M. Bailey, A. L. Goldstein, and D. H. Pluznik, New role for 15-HETE: activator of leukotriene biosynthesis in PT-18 mast/basophil cells, J. Biol. Chem. 257:12191 (1982).

27. Z. Williams, C. F. Hertogs, and D. H. Pluznik, Use of mice tolerant to LPS, to demonstrate requirement of cooperation between macrophages and lymphocytes, to generate LPS-induced CSF in vivo, Infec. and Immun. 41:1 (1983).

SECTION V. MODULATION OF THE IMMUNE RESPONSE BY ENDOTOXINS

Modulation of Humoral and Cell-Mediated Immune Responses by a
Structurally Established Nontoxic Lipid A
Edgar Ribi, John L. Cantrell, Kuni Takayama, Hans O. Ribi,
Kent R. Myers, and Nilofer Qureshi

Endotoxin Membrane Protein Complexes as Immunomodulators
Kathryn Nixdorff and Sigrid Schell

Endotoxin Associated Proteins and Their Polyclonal and Adjuvant Activities
Barnet M. Sultzer, John P. Craig, and Raymond Castagna

Lipopolysaccharide-Induced Recurrence of Arthritis Initiated
by Peptidoglycan-Polysaccharide
Stephen A. Stimpson, Ronald E. Esser, William J. Cromartie, and
John H. Schwab

Antibiotics, Immunity, and Immunoenhancement
G. Pulverer, W. Roszkowski, H. L. Ko, K. Roszkowski, and J. Jeljaszewicz

The Effect of Selected Antibiotics on the Endogenous Intestinal Microflora
and its Consequences for Experimental Tumor Growth in BALB/c Mice
K. Roszkowski, H. L. Ko, D. van der Waaij, W. Roszkowski,
J. Jeljaszewicz, and G. Pulverer

Restoration of Antibody Responsiveness by Endotoxin in Retrovirus-
Immunodepressed Mice: Role of Macrophages
Mauro Bendinelli, Donatella Matteucci, Anna Maria Giangregorio,
and Pier Giulio Conaldi

Endotoxin and Polysaccharide Derivative Induced Enhanced Antibody
Formation in Leukemia Virus Infected Mice
Herman Friedman and Andor Szentivanyi

Clinical Relevance of Endotoxemia
James P. Nolan

MODULATION OF HUMORAL AND CELL-MEDIATED IMMUNE RESPONSES

BY A STRUCTURALLY ESTABLISHED NONTOXIC LIPID A

Edgar Ribi[1], John L. Cantrell[1], Kuni Takayama[2],
Hans O. Ribi[3], Kent R. Myers[1], and Nilofer Qureshi[2]

[1]Ribi ImmunoChem Research, Inc., Hamilton, Montana
[2]William S. Middleton Memorial Veterans Hospital, Madison
Wisconsin, [3]Department of Cell Biology, Stanford University
Stanford, California

In 1954 Westphal and his associates reported on the isolation of a
moiety of bacterial endotoxin which they liberated by means of hydrolysis
in dilute acetic or hydrochloric acid solutions (25). The water soluble
phase of the hydrolysis reaction contained a haptenic polysaccharide which
no longer retained the ability to stimulate physiological responses characteristic of the starting material (3,4). On the other hand, the hydrolysis
products extractable with organic solvents did retain some of the endotoxic
properties of the original substance, leading Westphal and coworkers to
postulate that the "endotoxic" activities of LPS were attributable to a
lipidic component, which they designated lipid A (5).

EARLY STUDIES ON THE TOXICITY OF LIPID A

The relationship between lipid A and toxicity was initially unclear,
in light of work we reported in 1961 where we showed that lipids isolated
from endotoxins by acid hydrolysis, even when well-dispersed with the aid
of detergents, exerted less than one percent of the toxicity of the parent
endotoxins from which they were derived (5). The conclusion we drew at the
time was that, if lipid A was in fact responsible for all of the biological
properties of endotoxin, including toxicity, then it must have been chemically altered by acid hydrolysis in such a way that its toxic nature is
severely diminished.

In a study to determine the rate at which toxicity decreased in relation
to the release of bound lipid during acid hydrolysis of a highly toxic
Salmonella enteritidis lipopolysaccharide (LPS), we found that the toxicity
was essentially eliminated prior to any significant liberation of lipid
from LPS. When data from assays measuring the lethality in mice and pyrogenicity in rabbits of acid-hydrolyzed endotoxins were expressed relative
to the potency of the original endotoxin, it became apparent that as little
as five minutes of hydrolysis with boiling 0.1 N acetic acid (100°C) was
sufficient to reduce endotoxin potency by 50 percent or more (Figure 1).
It was especially noteworthy that after 60 minutes of acid hydrolysis, when
the loss of potency became pronounced, very little lipid (0.2%) had been
released and the turbidity of the reaction mixture did not differ from that
of the zero time sample. After 90 minutes a slight increase in the turbidity

of the hydrolysate was observed and about 0.6 percent bound lipid had been set free. At this point in hydrolysis, the decline of the pyrogenicity was about 80-fold. Since complete hydrolysis of this endotoxin with boiling 1 N hydrochloric acid for 45 minutes yielded 6.4 percent organic solvent soluble lipid A, no more than one-tenth of this fraction could have been released by 90 minute hydrolysis with 0.1 N acetic acid. Therefore, we concluded that the structural feature(s) of endotoxin responsible for its toxicity is highly acid-labile and would not be present in lipid fractions isolated from acid hydrolysates. In retrospect, we now know that a different condition of hydrolysis would be necessary to release a lipid moiety from LPS that is as toxic as the original endotoxin.

PREPARATION OF TOXIC AND NONTOXIC LIPID A

With the above conclusion in mind, further work has led to a delineation of the structure-function relationship of toxic and nontoxic hydrolysis reaction products derived from endotoxin (8-d10,13,15,20-22). Figure 2 gives an outline of how the Re chemotype LPS from S. typhimurium was prepared, hydrolyzed to lipid A, and then purified by either thin layer or column chromatography. For these studies, we used the Re mutants of Salmonella whose endotoxins are deficient in polysaccharide and contain the KDO moiety (and possibly aminoarabinose and phosphoethanolamine) which forms the link between the polysaccharide moiety and the lipid A residue. These are referred to here as Re-glycolipids. We prepared the endotoxin from purified cell wall rather than the whole cells because much cleaner endotoxin could be obtained which facilitated the early work on the purification of LPS and lipid A.

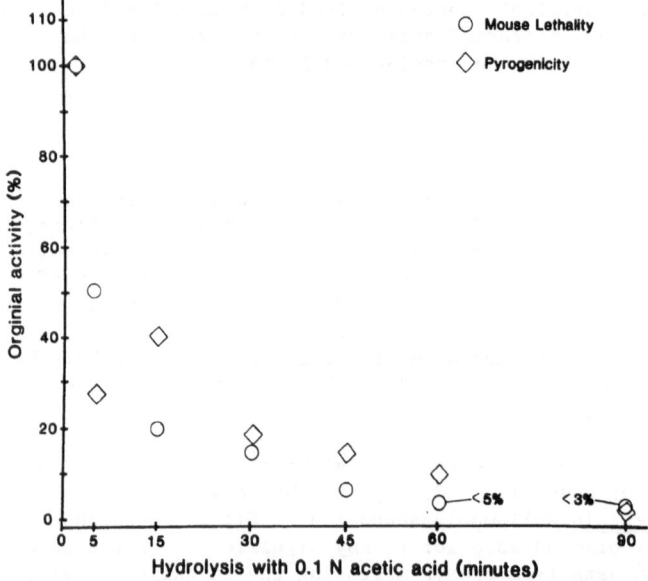

Figure 1. Kinetics of acid hydrolysis of lipopolysaccharide as measured by the decrease of lethal activity in mouse and pyrogenicity in rabbit. Lipopolysaccharide obtained from S. enteritidis was suspended in 0.1 N acetic acid and incubated at 100°C for 90 min. Aliquots were taken at various time intervals and the hydrolyzed lipopolysaccharide was assayed for the two biological activities. Unhydrolyzed lipopolysaccharide was set at 100% activity (5).

Table 1. Toxicity and Chemical Compositions of Refined Endotoxin
and Various Lipid A Preparations

Lipid A – Components from S. typhimurium	Chick Embryo Lethality CELD$_{50}$ (µg)[a]	Molar Ratio			
		GlcN	Phosphate	KDO	Fatty acids[b]
Refined endotoxin	0.0005	2	2	2	4
Diphosphoryl lipid A	0.0009	2	2	0	4
Monophosphoryl lipid A	10	2	1	0	4
"Free lipid A" Westphal method	0.9	-	-	-	-

[a] CELD$_{50}$ = 50% chick embryo lethal dose.
[b] Average number of moles of different fatty acids in nonfractionated mixtures of parent endotoxin and acid hydrolysis products.

Treatment of this purified Re glycolipid with a dilute solution of sodium acetate (pH 4.5, 100°C, 30 minutes), a procedure described by Rosner et al. (17), resulted in the selective removal of the KDO moiety. As determined by lethality in chick embryos, the resulting lipid A fraction retained the full toxicity of the parent endotoxin (Table 1). Because it contained two phosphate groups per mole of glucosamine disaccharide, this pH 4.5

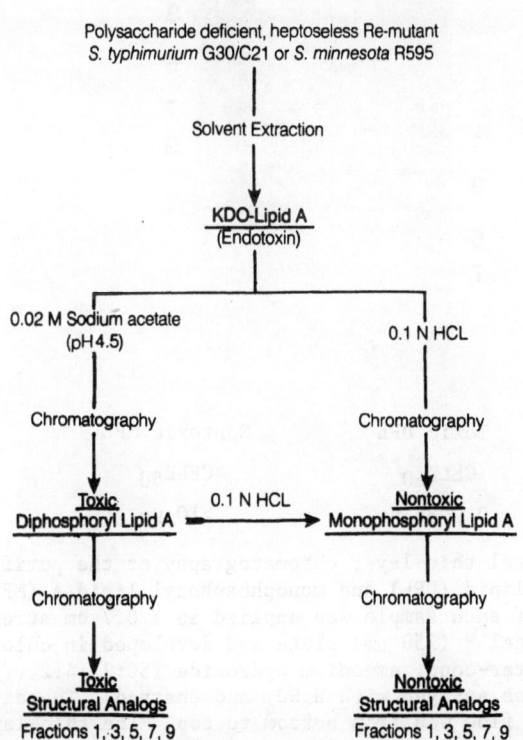

Figure 2. Scheme for the isolation of endotoxin from S. typhimurium, preparation of toxic and nontoxic lipid A, and the purification of the structural analogs.

product was called diphosphoryl lipid A (DPL). Treatment of DPL with 0.1 N hydrochloric acid resulted in selective removal of one-half of the phosphates without detectable changes in composition of fatty acids (13). The resulting compound, designated monophosphoryl lipid A (MPL), could then be separated from unreacted DPL by column chromatography. The purified MPL was found to be at least one-thousand-fold less toxic than DPL as determined by chick embryo lethality test (Table 1). MPL can also be obtained directly by treating the Re glycolipid with aqueous hydrochloric acid which removes both KDO and the labile phosphate group (Figure 2).

We now know that the toxicity of conventional "free lipid A's" liberated from endotoxin with mineral acid hydrolysis is determined by the amount of DPL remaining in the hydrolysis mixture. The presence of only 0.1% of DPL in the hydrolysate of Re mutant endotoxin would, for example, result in a $CELD_{50}$ value of 1 µg (Table 1).

STRUCTURAL BASIS FOR TOXICITY OF LIPID A

We found that both DPL and MPL are composed of a homologous series of components, as can be seen by the thin layer chromatography (TLC) patterns in Figure 3. Note that the nontoxic monophosphoryl components migrate more rapidly than the corresponding toxic diphosphoryl components. The components, designated TLC 1, 3, 5, 7 and 9, were isolated from one another by

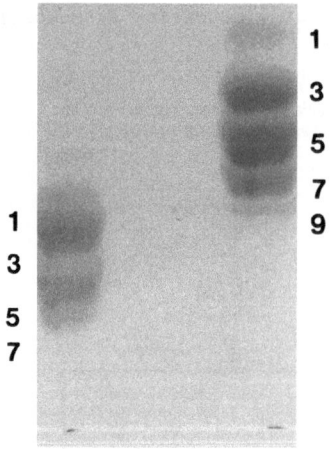

Toxic DPL   Nontoxic MPL

$CELD_{50}$ $CELD_{50}$

0.0088 µg   >10 µg

Figure 3.   Silica gel thin-layer chromatography of the purified diphosphoryl lipid (DPL) and monophosphoryl lipid A (MPL). About 40 µg of each sample was applied as a 0.7 cm streak on a silica gel H (250 µm) plate and developed in chloroform-methanol-water-conc. ammonium hydroxide (50:25:4:2,v/v). The plate was sprayed with H_2SO_4 and charred. The direction of solvent flow was from bottom to top. The thin-layer chromatography band numbers (TLC-1,-3,-5,-7, and -9) were assigned according to the structural similarity of the DPL-MPL pair as established by fast atom bombardment mass spectrometry (9,20).

Table 2. Molecular Weights, Toxicity and Fatty Acid Content of Isolated
          Pairs of Diphosphoryl Lipid A (DPL) and Monophosphoryl Lipid A
          (MPL) Homologs from S. typhimurium G30/C21

Chromatographic fraction number	DPL		MPL		Moles of fatty acid/ mole of sample			
	Molecular weight	$CELD_{50}$ a (μg)	Molecular weight b	$CELD_{50}$ a (μg)	$OH-C_{14}$	C_{12}	C_{14}	Total
3	1,797	0.0061	1,717	>10	4	1	1	6
5	1,587	0.0050	1,507	>10	4	1	0	5
7	1,361	0.0039	1,281	>10	3	1	0	4

a $CELD_{50}$ = 50% chick embryo lethal dose
b The difference between molecular weights of corresponding DPL and MPL pair
is 80 atomic mass units, which represents the loss of a phosphate group.
$OH-C_{14}$ = hydroxy myristate; C_{12} = laurate; C_{14} = myristate

silica gel preparative TLC or column chromatography. The free acid form of
the isolated TLC fractions were methylated with diazomethane and purified
to homogeneity by reverse phase high performance liquid chromatography
(HPLC). The exact structures of such HPLC-purified samples were then determined with the aid of chemical, mass spectrometry and nuclear magnetic resonance analysis (8-10,15,19,20). The homologs in each series differed in
number, type and position of ester-linked fatty acids (Table 2), i.e., TLC-3
having six, TLC-5 having five, and TLC-7 having four fatty acids. The molecular weight of each toxic homolog differed from its nontoxic counterpart by
80 atomic mass units, which is the mass of one phosphate group.

    The structure of the toxic TLC-3 lipid A homolog is shown in Figure 4
where R = H_2PO_3. TLC-3 from Re-mutants of both S. typhimurium and S.
minnesota consists of a glucosamine disaccharide backbone to which are linked
six long-chain fatty acid residues. Both the reducing and distal glucos-

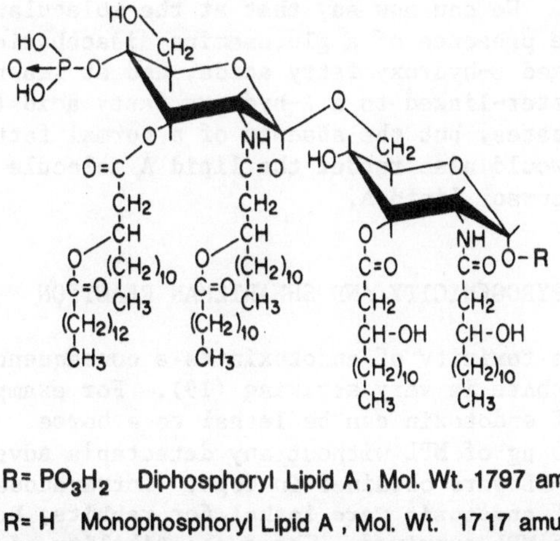

R = PO_3H_2 Diphosphoryl Lipid A , Mol. Wt. 1797 amu

R = H Monophosphoryl Lipid A , Mol. Wt. 1717 amu

Figure 4. Established structures of diphosphoryl lipid A (where R = H_2PO_3)
           and monophosphoryl lipid A (where R = H) obtained from the endo-
           toxin of S. typhimurium having the highest degree of acylation
           (the TLC-3 fractions).

411

Table 3. Effect of Structure of LPS and Lipid A on Pyrogenicity for
Rabbits and Lethality for Chick Embryos

Compound	Rabbit Pyrogenicity APD_{50}[a] (µg/kg)	Chick Embryo Lethality[b] $CELD_{50}$ (µg)
Endotoxin	0.0001 – 0.0003	0.0031
Diphosphoryl lipid A (unfractionated)	0.0005 – 0.001	0.0088
Monophosphoryl lipid A (unfractionated)	2–5	6.7
Monophosphoryl lipid A (purified, TLC-3)	>10	>20

[a]The approximate dose necessary to cause febrile response of >0.46 C in
50% of a test population.
[b]The test material was solubilized in pyrogen-free 0.15 M NaCl containing
0.5% triethylamine and diluted appropriately before intravenous injection
into 11-day old chick embryos.

amines contain two β-hydroxymyristates. In addition, a myristic acid and a
lauric acid are ester-linked to the β-hydroxyl groups of the distal end
hydroxymyristates. The toxic form of lipid A contains two phosphate groups,
one at each end of the disaccharide, whereas the detoxified form lacks the
acid-labile phosphate group at the reducing end of the molecule.

The TLC-3 homolog of both DPL and MPL from S. typhimurium represented
the lipid A with the highest degree of acylation (i.e., six acyl groups).
The lipid A from S. minnesota R595 (a TLC-1 fraction) was shown to contain
a homolog with seven acyl groups (8). The additional fatty acyl residue
was a palmitate which was ester-linked to a hydroxy-myristate at the reducing
end. Homologs with a total of five, four and three fatty acids are also
isolated from both Re mutant strains.

The factors responsible for the toxicity of endotoxins continue to
attract much interest. We can now say that at the molecular level, toxicity
appears to require the presence of a glucosamine disaccharide, two phosphates, two amide-linked β-hydroxy fatty acids, and at least one normal
fatty acid which is ester-linked to a β-hydroxy fatty acid (19). Thus, the
presence of two phosphates, but the absence of a normal fatty acid in an
acyloxyacyl linkage, would also render the lipid A molecule nontoxic as is
the case with the precursor lipid A.

TESTS FOR LETHALITY, PYROGENICITY AND SHWARTZMAN REACTION

The change in the toxicity of endotoxin as a consequence of removing
the reducing end phosphate is very striking (19). For example, a single
injection of 150 µg of endotoxin can be lethal to a horse. In contrast, a
horse was given 20,000 µg of MPL without any detectable adverse effects.
Similar results with MPL were obtained in dogs. Intravenous administration
of 1 to 10 µg doses of endotoxin were lethal for rabbits, however rabbits
receiving 15,000 µg of MPL survived. The susceptibility of humans to endotoxin is estimated to be similar to that for dogs and rabbits.

In addition to being considerably less toxic for rabbits, MPL also was
demonstrated to be less pyrogenic for these animals (Table 3). The simi-

larity of endotoxin-induced fever effects in man and rabbits suggests that
MPL is also less pyrogenic in man (24). Also noteworthy is the observation
that parent endotoxin and DPL are highly effective in provoking a dermal
Shwartzman reaction in rabbits, while MPL is about 200 times less reactive
(21). Figure 5 illustrates the pronounced differences in the reactions
provoked by toxic and nontoxic forms of lipid A.

RELATING TOXICITY OF LIPID A TO THE STATE OF AGGREGATION IN A SUSPENSION

Toxicity also appears to require that the endotoxin or lipid A molecules
be aggregated into particles of colloidal dimensions. The ability of these
lipopolysaccharides or glycolipids to form a wide variety of colloidal structures has been well established with the aid of electron microscopy (see
below) as well as other physical techniques. The biological relevance of
these colloidal particles was demonstrated in a paper published in 1966 in
which we reported that dissociation of colloidal endotoxic particles by
treatment with sodium deoxycholate resulted in a loss of toxicity and pyrogenicity (14). Removal of the detergent led simultaneously to reaggregation
of the endotoxin and recovery of its toxic properties.

The question immediately arises as to whether or not nontoxic MPL can
associate into particles of colloidal dimensions in the manner of the parent
endotoxin. In an effort to answer this question, we recently obtained electron micrographs of endotoxin, DPL, and MPL from an Re-mutant strain of S.
minnesota (R595). The micrographs reveal the presence of vesicles in preparations of all three glycolipids (Figure 6a-6d). For example, Figures 6a
and 6b are micrographs of parent endotoxin prepared from a solution of endotoxin in pure water and in dilute aqueous triethylamine (TEA), respectively.
The samples were negatively stained with one percent uranyl acetate. The
structures visible in these micrographs resemble collapsed liposomal vesi-

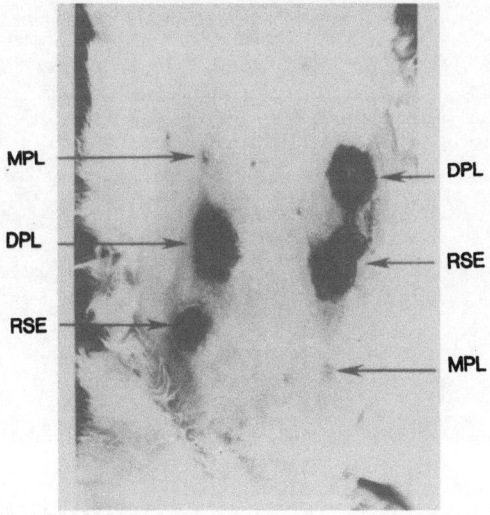

Figure 5.   A comparison of the responses of refined standard endotoxin
            (RSE), diphosphoryl lipid A (DPL), and monophosphoryl lipid A
            (MPL) in the dermal Shwartzman reaction in rabbit. The extent
            of hemorrhagic lesions produced at sites of intradermal injec-
            tion of 100 μg of MPL, 20 μg of DPL, and 20 μg of RSE were ob-
            served 18 hr after intravenous challenge with 40 μg of RSE.
            (A. Johnson and E. Ribi, unpublished results).

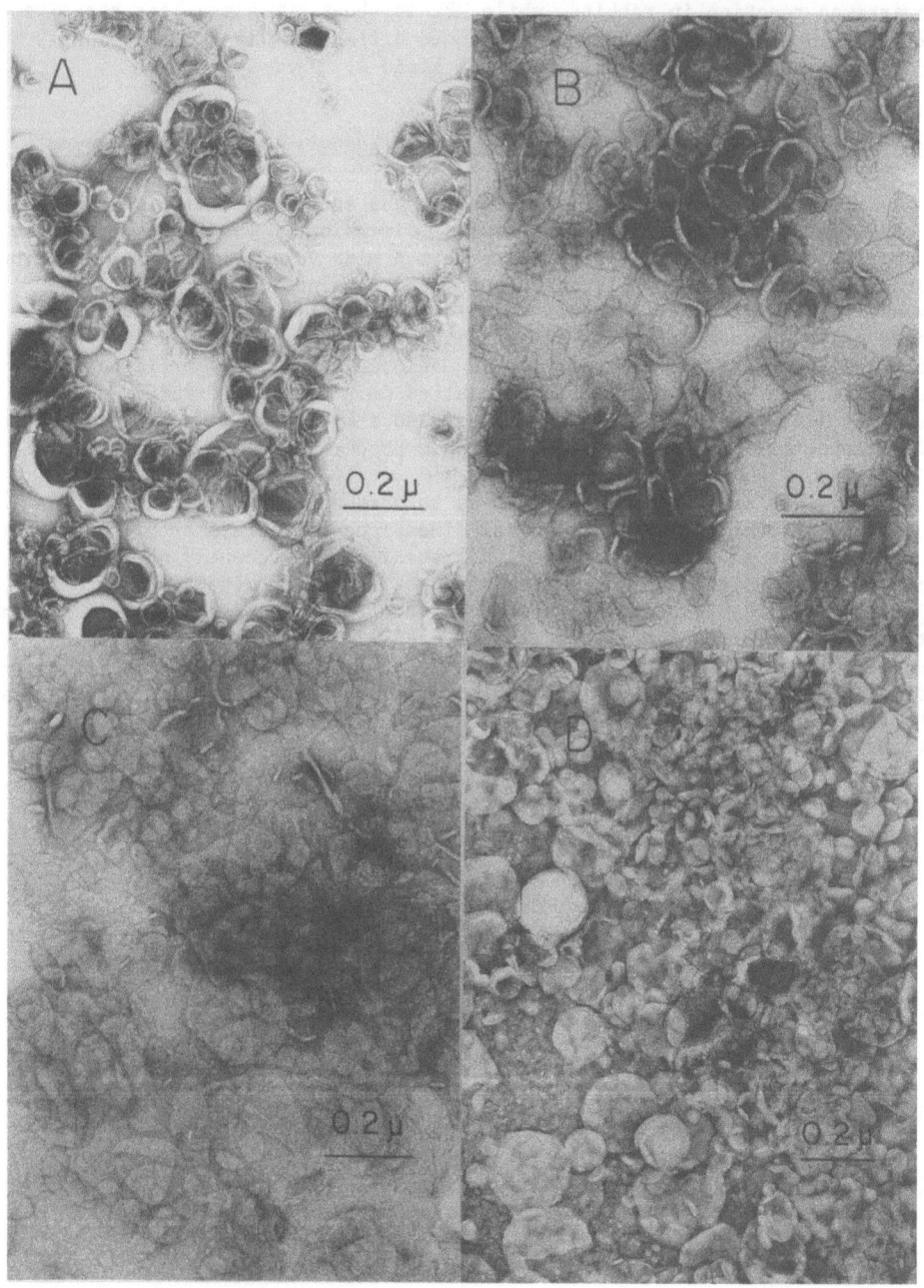

Figure 6. Electron micrographs of endotoxic glycolipids from Re mutant $\underline{S}$. minnesota (R595): a) refined standard endotoxin (RSE) dispersed in pure water; b) RSE dispersed in 0.005% triethylamine (TEA); c) diphosphoryl lipid A dispersed in 0.005% TEA; d) the TEA salt of monophosphoryl lipid A dispersed in 0.005% TEA. All samples were negatively stained with 1.0% uranyl acetate.

Table 4. Adjuvant Activities Elicited by Endotoxin, Toxic Diphosphoryl (DPL) and Nontoxic Monophosphoryl Lipid A (MPL)

Biological Activity	Endotoxin	DPL	MPL
A. B cell mitogen on mouse spleen (stimulation index)[a]	17.2	8.1	27.0
B. Activation of peritoneal murine MØ (phagocytosis of fluorescent beads, percent increase[b]	905	540	730
C. Chemiluminescence response of PEC of mice pretreated with 100 µg samples (CPM x 1000, 18 hr after Zymosan stimulation[c]) Tris-sucrose control: 275	480	1180	1220
D. Colony stimulating activity (CSA) Mouse serum CSA (No. CFU-GM stimul. with 50 µl serum)[d] Saline: 0	60 ± 13.2	–	64 ± 6.3
E. Stimulation of interleukin (IL-1) by human monocytes (CPM in lysates of monocytes)[a] none, 15,173	43,833	39,000	50,626
F. Protection of mice against x-irradiation with 600R (% survivors)[c] Diluent control: 20	75	–	75

[a] I. Gery and E. Ribi, unpublished results
[b] A. Johnson and E. Ribi, unpublished results
[c] N. Masihi and E. Ribi, unpublished results
[d] R. Urbaschek, B. Urbaschek and E. Ribi, unpublished results

cles. The vesicles in the sample prepared from the dilute aqueous solution of TEA (Figure 6b) were of relatively constant size while those in pure water are polydispersed (Figure 6a). The endotoxin vesicles appear to be somewhat rigid, which may be a consequence of the presence of the KDO moiety (compared with DPL and MPL, discussed below). These vesicles are very similar to those observed with Re-mutant endotoxins by Nixdorff et al. (7) and Amano and Fukushi (1). They differ, however, from the open bilayer structures which are typically observed with wild-type lipopolysaccharides (18). Figures 6C and 6d are micrographs of DPL and MPL, respectively. In both cases, the samples were dispersed in dilute aqueous TEA and were negatively stained with one percent uranyl acetate. Like the parent endotoxin, DPL is able to form vesicles, although they are less rigid in appearance. Finally, MPL is clearly revealed to form uni- and possibly multilamellar vesicles, ranging in size from a few hundred to several thousand angstroms in diameter. Thus, these micrographs demonstrate that loss of reducing-end phosphate group does not interfere with the ability of MPL to form vesicles.

Endotoxin, DPL and MPL are amphiphilic molecules and are therefore able to spontaneously form membranous structures when dispersed in water. The liposomal structures formed by these powerful adjuvants are in close analogy to the phospholipid arrangement found in biomembranes. The inter-

Table 5. ELISA Activity of Serum from (C57BL/10 x BALB/c) F_1 Mice
Given 50 μg Ovalbumin Alone or with Adjuvants

Adjuvant	Dose (μg)	Form	ELISA titers (days after immunization)	
			8	14
None	none	aqueous	200	400
MPL[a]	100	aqueous	1,600	6,400
CFA[b]	-	emulsion (50% oil)	800	1,600
IFA[c]	-	emulsion (50% oil)	400	1,600
MPL + TDM[d]	50 + 50	emulsion (2% oil)	6,400	102,400
MPL + TDM	25 + 50	emulsion (2% oil)	1,600	204,800
TDM	50	emulsion (2% oil)	400	12,800

[a] MPL = monophosphoryl (detoxified) lipid A from Salmonella minnesota R595
[b] CFA = comlete Freund's adjuvant
[c] IFA = incomplete Freund's adjuvant
[d] TDM = trehalose dimycolate (P3)

action of these "adjuvant membranes" with host cellular membranes may be a
fundamental requirement for elicitation of their biological activities.
This is a subject of current investigation in our laboratories.

IMMUNOLOGICAL PROPERTIES OF NONTOXIC LIPID A

MPL has been tested in parallel with DPL and endotoxin in a variety of
in vitro and in vivo systems in order to evaluate its immunological properties, and the results of some of these studies will be summarized in the
remainder of this chapter. Data in Table 4A show that MPL has strong mitogenic activity for spleen cells from Swiss outbred mice. At a concentration
of 50 μg/ml, the activity of MPL was higher than that of the other products
tested. We may speculate that at this concentration, DPL and endotoxin
elicit their toxic effects towards spleen cells. Similar results were obtained with spleen cells from BALB/c nude mice.

Arthur Johnson tested the capacity of MPL to induce phagocytosis of
latex beads by peritoneal macrophages from C58 mice (Table 4B) and found
that it was active in this regard as parent endotoxin and DPL. Another
measure of the degree of stimulation of phagocytosis activity by macrophages
is the extent of hydroxyl radical formation, which leads to the emission of
protons under certain conditions. This release of energy in the form of
light is termed chemiluminescence (CL). Masihi et al. (6) used a Limunol-
dependent Zymosan-induced CL assay to examine the effects of muramyl
dipeptide (MDP) and trehalose dimycolate (TDM) on the phagocytotic activity
of murine spleen cells. Data in Table 4C show the high response elicited
by 100 μg quantities of either DPL or MPL. Interestingly, the response
obtained with parent endotoxin in this test system was lower and appeared
to parallel the activity of 100 μg of MDP.

Table 6. Nontoxic Monophosphoryl Lipid A (MPL) Can Replace Endotoxin (ET) to Enhance the Tumor Regressive Potency of Mycobacterial Cell Wall Skeleton (CWS)

Material Associated with Oil Droplets Injected into Tumors[a]	Dose (µg)	Cured/Total	Percent Cured
ET	150	0/8	0
CWS	50	10/48	20
ET + CWS	50 + 50	8/8	100
	5 + 50	6/8	75
MPL	50	0/18	0
MPL + CWS	50 + 50	50/52	96
	25 + 25	8/10	90
	8 + 50	4/8	50

[a]Strain 2 guinea pigs bearing one-week-old line-10 tumors (8-10 mm in diameter) inoculated intralesionally with doses contained in 0.4 ml volumes.

Renate and Bernhard Urbaschek observed a significant effect of MPL on in vivo generation of the colony stimulating activity (CSA) (Table 4D). Sera of mice were assayed two hours after i.v. inoculation of 5 µg of either MPL or reference standard endotoxin. It is evident from the data in Table 4D that MPL and endotoxin exhibit comparable activity in this assay.

It has been reported that IL-1 production is stimulated by CSA, and this is in agreement with the finding of Igal Gery that MPL stimulates the production and release of IL-1 by human monocytes (Table 4E). However, when compared to the activity of parent endotoxin, it was less effective in stimulating the release of IL-1 into the medium (data not shown here). The latter finding may indicate that MPL is less damaging to monocytes than the toxic compounds. It is not yet known what "adjuvant" role each of the IL-1 pools plays--it could be that the intracellular fraction is of major importance, and that IL-1 release is of less significance.

The capacity of sera containing endotoxin to induce elevated levels of CSA and to stimulate granulopoiesis was shown to be correlated with its ability to increase nonspecific protection against lethal x-irradiation (23). Data in Table 4F supplied by the Urbaschek's show that 100 µg of MPL protected up to 75 percent of the mice when given i.v. 24 hours prior to exposure to 600 R whole body radiation. A dose of 10 µg of parent endotoxin was lethal for radiated mice (ten percent survivors). However, a dose of 1 µg of endotoxin was not lethal and protected 75 percent of the animals.

The ability of nontoxic MPL to enhance the formation of antibodies to nonrelated antigens is illustrated in Table 5. Addition of MPL to ovalbumin caused a 16-fold increase in the ELISA titer as compared to a four-fold increase using complete Freund's adjuvant (CFA) or incomplete Freund's adjuvant (IFA) on day 14. In addition, a striking synergistic response was observed when the antigen was incorporated together with MPL and mycobacterial trehalose dimycolate (TDM) into minute droplets of squalene and dispersed in a large volume of saline to a final concentration oil of one percent. Titers were increased by up to 512-fold by such preparations.

Table 7. Effect of Nontoxic Monophosphoryl Lipid A (MPL) and Trehalose Dimycolate (TDM) on Resistance of Mice to Aerogenic Influenza Virus Infection

Treatment[a]	Dose (µg)	Dead/Total (day 14)	% Survival
TDM	50	10/10	0
MPL	50	10/10	0
TDM + MPL	50 + 50	0/10	100
Oil droplet control	–	21/21	0

[a]Mice were treated i.v. three weeks prior to infection. Materials to be tested were incorporated into minute droplets of oil (squalane) and dispersed in saline (concentration of oil: 1%). N. Masihi, W. Brehmer and E. Ribi (unpublished results).

MPL also retained the ability of the parent endotoxin to synergistically enhance the ability of the mycobacterial cell wall skeleton (CWS) to regress line-10 tumors in syngeneic guinea pigs and to concomitantly establish systemic tumor immunity (Table 6). The adjuvant combination was applied as a one percent oil droplet in saline suspension. In contrast with endotoxin, however, MPL does not exhibit any lethal shock activity in the animals tested when administered with either CWS or muramyl dipeptide, a monomeric constituent of CWS (12).

Using another tumor model, Meth A induced fibrosarcoma in mice, Bloksma et al. (2) has made similar observations and confirmed our conclusion that the toxicity of endotoxin is not directly related to the therapeutic potential and that the combination of MDP and MPL, which in this case was administered without incorporation into oil droplets, was more effective than either of these two components alone.

Finally, a dramatic synergism was observed by Noel Masihi when TDM and MPL were used in combination to enhance resistance in mice against airborne infection with lethal doses of influenza virus, without specific antigen being incorporated (Table 7). Neither component alone was effective, but full protection was obtained by i.v. administration of a combination containing only 50 µg of each agent. In this test system, a similar synergism was observed between MDP and TDM (6). It would appear that when administered i.v., the adjuvant covered oil droplets which became lodged in the lung tissue, caused persistent accumulation of activated macrophages which served to nonspecifically kill the invading virus. During this process, specific immunity against the viral antigen may be augmented.

Stability, including resistance to intracellular degradation by macrophages, appears to be an important characteristic of successful vaccines. A similar mechanism may be involved in the development of cell-mediated tumor specific transplantation immunity in the line-10 tumor system upon intratumor injection of a one percent oil-in-water emulsion of CWS plus MPL.

In contrast to the Re-mutant glycolipid vesicles which were formed in the absence of oil, the physical make-up of the oil droplet adjuvant system is thought to consist of a glycolipid monolayer which covers the surface of the oil. The hydrophobic fatty acid residues of MPL and the TDM molecules would extend into the oil phase acting as anchors to stabilize the surface monolayer. The hydrophilic disaccharide heads would protrude into the water interface. Protein molecules or peptidic fragments derived from them are thought to be attached to the polar heads or embedded in between them. These and other molecules carrying antigenic determinants are incorporated into the adjuvant system during the preparation of the oil-in-water emulsion (11,16). In the case of nonspecific protection where the antigen is not added to the adjuvant system, we speculate that immune response stimulation may require the adjuvant to interact as a surface monolayer with the host cell plasma membrane. This situation might be similar to the innocent by-stander phenomenon of tumor regression by nonspecific stimulants (26).

The data reviewed here establish the structural relationship between the toxic DPL and the nontoxic MPL which are derived from the native LPS molecule. MPL retains many of the beneficial properties of the parent endotoxin, such as being able to stimulate humoral and cell-mediated immune responses, while being significantly less active in causing fever, shock and death in animals which are highly susceptible to the toxic effects of endotoxin. Moreover, MPL used in combination with chemically-defined adjuvants from mycobacteria acts synergistically in mediating tumor and anti-microbial immunity. The data from these studies strongly suggest that MPL has the potential for a wide range of clinical applications.

REFERENCES

1. K. Amano and K. Fukushi, Chemical and ultrastructural differences in endotoxic glycolipids from Salmonella minnesota Re mutant extracted with various solvent systems, Microbiol. Immunol. 28:135 (1984).
2. N. Bloksma, F.M.A. Hofhuis, and J.M.N. Willers, Endotoxin-induced antitumor activity in the mouse is highly potentiated by muramyl dipeptide, Cancer Lett. 23:159 (1984).
3. A. Boivin and L. Mesrobeanu, Recherches sur les antigenes somatiques et sur les endotoxines des bacteries. I. Considerations generales et expose des techniques utilisees, Rev. Immunol. 1:553 (1935).
4. G. G. Freeman, The preparation and properties of a specific polysaccharide from Bacterium typhosum Ty$_2$, Biochem. J. 36:340 (1942).
5. W. T. Haskins, M. Landy, K. C. Milner, and E. Ribi, Biological properties of parent endotoxins and lipoid fractions, with a kinetic study of acid-hydrolyzed endotoxin, J. Exp. Med. 114:665 (1961).
6. K. N. Masihi, W. Brehmer, I. Azuma, W. Lange, and S. Miller, Stimulation of chemiluminescence and resistance against aerogenic influenza virus infection by synthetic muramyl dipeptide combined with trehalose dimycolate, Infect. Immun. 43:233 (1984).
7. K. Nixdorff, J. Gmeiner, and H. H. Martin, Interaction of lipopolysaccharide with detergents and its possible role in the detergent resistance of the outer membrane of gram-negative bacteria, Biochim. Biophys. Acta. 510:87 (1978).
8. N. Qureshi, P. Mascagni, E. Ribi, and K. Takayama, Monophosphoryl lipid A obtained from lipopolysaccharides of Salmonella minnesota R595. Purification of the dimethyl derivative by high performance liquid chromatography and complete structural determination, J. Biol. Chem., in press (1985).
9. N. Qureshi, K. Takayama, D. Heller, and C. Fenselau, Position of ester groups in the lipid A backbone of lipopolysaccharides obtained from Salmonella typhimurium, J. Biol. Chem. 258:12947 (1983).
10. N. Qureshi, K. Takayama, and E. Ribi, Purification and structural determination of nontoxic lipid A obtained from lipopolysaccharide of Salmonella typhimurium, J. Biol. Chem. 257:11808 (1982).
11. E. Ribi, Structure-function relationship of bacterial adjuvants, in: "Proceedings of Advances in Carriers and Adjuvants for Veterinary Biologics Symposium," Ames, Iowa, in press (1985).

12. E. Ribi, Beneficial modification of the endotoxin molecule, J. Biol. Resp. Mod. 3:1 (1984).

13. E. Ribi, K. Amano, J. L. Cantrell, S. M. Schwartzman, R. Parker, and K. Takayama, Preparation and antitumor activity of nontoxic lipid A, Cancer Immunol. Immunother. 12:91 (1982).

14. E. Ribi, R. L. Anacker, R. Brown, W. T. Haskins, B. Malmgren, K. C. Milner, and J. A. Rudbach, Reaction of endotoxin and surfactants. I. Physical and biological properties of endotoxin treated with sodium deoxycholate, J. Bacteriol. 92: 1493 (1966).

15. E. Ribi, J. L. Cantrell, K. Takayama, N. Qureshi, J. Peterson, and H. O. Ribi, Lipid A and immunotherapy, Rev. Infect. Dis. 6:567 (1984).

16. E. Ribi, D. L. Granger, K. C. Milner, K. Yamamoto, S. M. Strain, R. Parker, R. W. Smith, W. Brehmer, and I. Azuma, Induction of resistance to tuberculosis in mice with defined components of mycobacteria and with some unrelated materials, Immunology 46:297 (1982).

17. M. R. Rosner, J. Y. Tang, I. Barzilay, and H. G. Khorana, Structure of the lipopolysaccharide from an Escherichia coli heptoseless mutant, J. Biol. Chem. 254:5906 (1979).

18. J. W. Shands, Jr., The physical structure of bacterial lipopolysaccharides, in: "Microbial Toxins. IV. Bacterial Endotoxins," G. Weinbaum, S. Kadis, S. J. Ajl, eds., Academic Press, New York (1971).

19. K. Takayama, N. Qureshi, E. Ribi, and J. L. Cantrell, Separation and characterization of toxic and nontoxic forms of lipid A, Rev. Infect. Dis. 6:439 (1984).

20. K. Takayama, N. Qureshi, and P. Mascagni, Complete structure of lipid A obtained from the lipopolysaccharides of the heptoseless mutant of Salmonella typhimurium, J. Biol. Chem. 258:128901 (1983).

21. K. Takayama, N. Qureshi, C.R.H. Raetz, E. Ribi, J. Peterson, J. L. Cantrell, F. C. Pearson, J. Wiggins, and A. G. Johnson, Influence of fine structure of lipid A on Limulus amebocyte lysate clotting and toxic activities, Infect. Immun. 45:350 (1984).

22. K. Takayama, E. Ribi, and J. L. Cantrell, Isolation of a nontoxic lipid A fraction containing tumor regression activity, Cancer. Res. 41:2654 (1981).

23. R. Urbaschek, Effect of bacterial products on granulopoiesis, in: "Macrophages and Lymphocytes," Part B, M. R. Escobar and H. Friedman, eds., Plenum Publishing, New York (1980).

24. G. J. Vosika, C. Barr, and D. Gilbertson, Phase one study of intravenous modified lipid A, Cancer Immunol. Immunother. 18:107 (1984).

25. O. Westphal and O. Luderitz, Chemische Erforschung von Lipopolysacchariden gramnegativer Bakterien, Angew. Chem. 66:407 (1954).

26. B. Zbar, E. Ribi, M. T. Kelly, D. Granger, C. Evans, and H. J. Rapp, Immunologic approaches to the treatment of human cancer based on a guinea pig model, Cancer. Immunol. Immunother. 1:127 (1976).

ENDOTOXIN MEMBRANE PROTEIN COMPLEXES AS IMMUNOMODULATORS*

Kathryn Nixdorff and Sigrid Schell

Institute für Mikrobiologie
Technische Hochschule Darmstadt
Darmstadt, Federal Republic of Germany

INTRODUCTION

The surface antigens on a bacterial cell certainly play an essential role in the defense reactions of the host to an infection. The bacterium is recognized and attacked by the immune system of the host primarily at these cell surface structures.

The outer membrane of gram-negative bacteria is a mosaic of components including lipopolysaccharide (LPS), a few major proteins and phospholipids, which are complexed very tightly with one another in the native outer membrane structure. Since the immune system of the host surely encounters complexes of these outer membrane components in the course of an infection with gram-negative bacteria, the effects of such complexes on the immune response should be considered. In what ways various outer membrane components can modulate the immune responses to antigens contained in the outer membrane structure is a question that has received relatively little attention.

We have addressed this question by studying the modulation of immune responses in a model system consisting of defined complexes produced from isolated, purified components of the outer membrane of Proteus mirabilis, which serve as modulators (LPS, phospholipids, proteins) and as immunogens (LPS, proteins).

These investigations have shown that outer membrane components have a profound effect in modulating both the strength and the character of the antibody-producing cell responses in mice to LPS. In this regard, the immune responses to LPS were altered from a primarily IgM-type response with very little production of IgG to a much stronger response that was mainly IgG in character when LPS was incorporated into bacterial membrane phospholipid vesicles (23) or complexed with major outer membrane proteins (6,7). In addition, a selective induction of either IgG1 or IgG2 subclasses of antibody-producing cells was effected by complexing LPS with different modulators, which was apparently dependent upon the physical character of the modulator (6).

---

* This chapter is dedicated to Dr. Otto Lüderitz on the occasion of his 65th birthday, March 2, 1985.

The present chapter summarizes some of our previous results and presents an extension of these studies.

MATERIALS AND METHODS

Bacterial Strains

Proteus mirabilis strains D52, 19 and VI were obtained from Prof. Dr. H. H. Martin, Institut für Mikrobiologie, Technische Hochschule Darmstadt, Darmstadt, FRG. These bacteria were cultivated as previously described (10).

Extraction and Purification of LPS

Lipopolysaccharides were extracted from P. mirabilis strains D52 and VI with phenol/water (26) and purified according to Gmeiner (2). To remove contaminating proteins, these preparations were re-extracted with phenol/water. The resulting extracts were free of protein according to amino acid analyses (see below).

Extraction of Phospholipids

Phospholipids were extracted from whole cells of P. mirabilis 19 with chloroform: methanol according to Gmeiner and Martin (3). These extracts contained approximately 80% phosphatidyl ethanolamine, 12% phosphatidyl glycerol, four percent diphosphatidyl glycerol, one percent lysophospholipids and two percent "neutral lipids" (3).

Isolation of M_r 39,000 Protein

A major outer membrane protein with a molecular weight of approximately 39,000 (M_r 39,000 protein) was isolated from purified cell walls of P. mirabilis 19 and purified according to methods described previously (6,18). Contaminating LPS was difficult to remove. After extensive purification, some batches were free from detectable amounts of LPS (<0.085%) while other batches contained up to 0.3% LPS according to gas-liquid chromatography (see below). The contaminating LPS did not affect the modulating capacity of the proteins.

Isolation of Lipoprotein

Free lipoprotein with a molecular weight of approximately 7,300 was isolated from cell walls of P. mirabilis 19 and purified as described previously (6). This protein was free of detectable LPS according to gas-liquid chromatography (see below).

Analytical Methods

Protein content was determined by a modification of the Lowry method (9). Amino acid analysis of LPS extracts was performed after hydrolysis in 4.0 N HCl for 16 hours at 100°C. Amounts of phospholipids and LPS were determined by gas-liquid chromatography of fatty acids (3). Sodium dodecyl sulfate polyacrylamide gel electrophoresis of proteins was performed in slab gels as previously described (18).

Experimental Animals

Specific pathogen-free NMRI mice were obtained from Charles River Wiga, Sulzfeld, FRG. C57B1/6JHan and nu/nu NMRI mice were obtained from

Zentralinstitut für Versuchstierzucht, Hannover, FRG. All mice were seven weeks old at the time of primary immunization.

Immunization of Mice

LPS alone or mixtures of LPS with phospholipids or proteins were sonicated in buffer solution as previously described (6). Mice were injected intraperitoneally with 0.2 ml of the sonicated solutions. A primary injection was given on day 0 and a secondary injection on day 14. Per injection, mice received either 25 μg LPS, 25 μg LPS plus 300 μg phospholipids, or 25 μg LPS plus 12.5 μg protein.

In Vitro Cell Culture System

Spleen cells from C57BL/6JHan mice were cultivated and stimulated to antibody production in the Mishell-Dutton system (13,14). Briefly, washed, single spleen cells at a density of 3×10^6 in 500 μl RPMI 1640 medium plus 5% fetal calf serum (FCS) were placed in wells of Coster cell culture dishes (well size 16mm). Antigen contained in 100 μl of RPMI medium plus 5% FCS was added to the wells. The dishes were placed in a Bellco gas chamber (TecNoMara, Huttenberg bei Giessen, FRG), and air was exchanged with a gas mixture of 7% O_2, 10% CO_2 and 83% N_2. The dishes were placed on a rocker platform and incubated with gentle rocking at 37°C for four days. Two drops of nutrient cocktail (14) were added to each well daily.

Assay of Antibody-Producing Cells

The IgM and the IgG antibody-producing cell responses to LPS were measured in the hemolytic plaque test (5), using a modification of the microscope slide assay (15) with sheep red blood cells (SRBC) sensitized with alkali-treated LPS (1,24) as previously described (6). For the measurement of IgG subclass responses, indirect plaques were developed with specific rabbit anti-mouse IgG1, IgG2$_{ab}$, or IgG3 serum (Nordic Laboratories, Tilburg, NL). The specificity of these IgG subclass antisera has been documented (6).

RESULTS

Modulation of the IgG Subclass Responses to LPS by Outer Membrane Components of P. Mirabilis

Table 1 presents the results of the modulating effects of various components of the outer membrane of P. mirabilis on the response to LPS from this organism.

The outer membrane components had little effect in modulating the IgM responses to LPS. In contrast, IgG responses were greatly enhanced. Enhancement of the primary IgG responses (up to day 14) could be observed, but effects were most pronounced on the secondary responses, which reached a peak in our system on day 19, or five days after the second injection. The IgG responses on day 19 to LPS alone were relatively weak and the numbers of plaque-forming cells (PFC) were mainly distributed in the IgG1 (40% of total IgG) and the IgG2 (44% of total IgG) subclasses. When LPS was complexed with phospholipids, both the strength and the IgG subclass distribution were altered. In this case, the greatest enhancing effect (adjuvant factor of 105) was seen on the IgG1 subclass of PFC (82 percent of total IgG). On the other hand, when LPS was complexed with the M_r 39,000 protein, the greatest enhancing effect (adjuvant factor of 43) was seen on the IgG2 subclass of PFC (62% of total IgG). With lipoprotein as modulator, predominantly IgG1 PFC were induced.

Table 1. IgM and IgG Subclass Responses in Mice to <u>Proteus Mirabilis</u> D52 Lipopolysaccharide (LPS) Alone and in Combination with Phospholipids, M_r 39,000 Protein and Lipoprotein[a]

Immunogen[b]	Antibody Type	PFC/10^6 Spleen Cells[c] on Day				% IgG Subclass of Total IgG on Day 19	Adjuvant Factor on Day 19
		4	14	18	19		
LPS	IgM	32 ± 4	7 ± 1	41 ± 7	10 ± 1	–	–
	IgG1	0	3 ± 1	4 ± 2	17 ± 3	40	1
	IgG2	0	1 ± 1	14 ± 3	19 ± 5	44	1
	IgG3	0	2 ± 1	9 ± 3	7 ± 1	16	1
LPS + Phospholipid	IgM	55 ± 9	16 ± 2	52 ± 7	26 ± 15	–	–
	IgG1	0	69 ± 22	529 ± 20	1793 ± 507	82	105
	IgG2	0	5 ± 2	82 ± 11	248 ± 12	11	13
	IgG3	0	7 ± 1	81 ± 18	147 ± 21	7	21
LPS + M_r 39,000 Protein	IgM	141 ± 5	5 ± 1	52 ± 25	31 ± 5	–	–
	IgG1	0	38 ± 3	240 ± 35	337 ± 75	25	20
	IgG2	0	59 ± 1	378 ± 71	818 ± 133	62	43
	IgG3	0	3 ± 3	66 ± 22	168 ± 49	13	24
LPS + Lipoprotein	IgM	49 ± 2	5 ± 1	41 ± 6	12 ± 1	–	–
	IgG1	n.d.	n.d.	302 ± 36	369 ± 64	59	22
	IgG2	n.d.	n.d.	67 ± 38	145 ± 21	23	8
	IgG3	n.d.	n.d.	34 ± 10	113 ± 24	18	16

[a] Data from Karch, Gmeiner and Nixdorff (6).
[b] Dosages per injection were 25 µg LPS, 300 µg phospholipids and 12.5 µg proteins. Mice received a primary injection on day 0 and a secondary injection on day 14.
[c] Geometric means ± standard errors of plaque-forming cells (PFC). Values are the results of three to five separate experiments. Responses were measured against P. mirabilis D52 LPS coupled to SRBC

n.d. = not determined

Table 2. IgM and IgG Subclass Responses in Mice to Proteus mirabilis D52
Lipopolysaccharide (LPS) in Combination with Bovine Serum
Albumin (BSA) or Methylated BSA (meth. BSA)[a]

Immunogen[b]	Antibody Type	PFC/10^6 Spleen Cells[c] on Day 19	% IgG Subclass of Total IgG	Adjuvant Factor
LPS + BSA	IgM	8 ± 4	--	--
	IgG1	177 ± 44	32	10
	IgG2	322 ± 43	60	17
	IgG3	42 ± 14	8	6
LPS + meth. BSA	IgM	16 ± 6	--	--
	IgG1	270 ± 68	69	16
	IgG2	52 ± 12	13	3
	IgG3	72 ± 24	18	10

[a] Data taken from Karch, Gmeiner and Nixdorff (6).
[b] Dosages per injection were 25 µg LPS and 12.5 µg proteins. Mice received a primary injection on day 0 and a secondary injection on day 14.
[c] Geometric means ± standard errors of plaque-forming cells (PFC). Values are the results of three to four separate experiments. Responses were measured against P. mirabilis D52 LPS coupled to SRBC.

Thus, phospholipids and the lipoprotein had the greatest effect on the induction of IgG1 PFC while the M_r 39,000 protein had the greatest effect on the induction of IgG2-producing cells, although significant enhancing effects by all modulators were observed in all three IgG subclasses.

It should be noted that the IgM responses to these immunogenic preparations are relatively non-specific, while the IgG responses are highly specific for the LPS used as immunogen. We could not detect IgG PFC directed against uncoupled SRBC or SRBC coupled with heterologous LPS at any time during the course of immunization with these LPS-modulator complexes (6,7,23).

Modulation of the IgG Subclass Responses to LPS by Bovine Serum Albumin and Methylated Bovine Serum Albumin

Phospholipids and the lipoprotein have known hydrophobic properties. Although the M_r 39,000 protein is an outer membrane protein, it has strong hydrophilic properties. It forms hydrophilic pores in model membrane phospholipid vesicles, whereas lipoprotein does not (18), and it is easily solubilized in aqueous buffers after brief sonication. It therefore appeared as if the ability to selectively enhance the production of either IgG1 PFC or IgG2 PFC might reside in the physical character of the modulator.

To investigate this possibility further, we mixed LPS with either bovine serum albumin (BSA), which is relatively hydrophobic, and used methylated BSA (meth. BSA), which is more hydrophobic, and used these preparations as model immunogens. The results are presented in Table 2. LPS-BSA preparations induced predominantly IgG2 PFC specific for LPS, while LPS-meth. BSA preparations induced mainly IgG1 PFC. These results further support the concept that the selective induction of IgG1 or IgG2 antibody-producing cells is determined by the physical character of the immunogenic preparation.

Subsequent investigations showed this to be true as well for the structure of the LPS used as immunogen without adjuvant. LPS containing short O-polysaccharide chains induced predominantly IgG1 and IgG2 antibody-producing cells, while LPS containing long O-polysaccharide chains induced mainly IgG2 and IgG3 antibody-producing cells. However, regardless of the type of LPS employed, phospholipids and proteins used as modulators effected the same alterations characteristic for the particular modulator (6).

It was also determined that LPS and modulators had to be complexed in order to obtain the observed alterations in the responses to LPS (6).

Differential Effects of Modulators on the Primary and Secondary Stimulations

Reports in the literature indicate that T cells can augment the IgG responses to thymus-independent antigens. In this regard, T cells were involved in the enhancement of IgG2a and responses to trinitrophenyl (TNP)-Ficoll (16,17) or IgG1 and IgG2 polyclonal responses to LPS (11,12). Because IgG1 and IgG2 responses to LPS were greatly enhanced by the modulators in our studies, and because LPS and modulators had to be complexed to obtain these effects, an investigation was undertaken to determine whether these modulators were acting as carriers in the classic immunological sense. In this case, a strong secondary (IgG) response to LPS would occur only if the same modulator were used for the primary and secondary stimulations (15,19). This would in turn imply that cooperation between carrier (modulator)-specific T cells and hapten (LPS)-specific B cells is necessary in order to effect the induction of an LPS-specific secondary IgG response (21).

The result of this investigation (8) was that the modulators were not acting as carriers in the classic immunological sense. Undiminished secondary IgG responses to LPS were obtained, regardless of which adjuvant was used for the secondary injection.

However, the character of the immune response (IgG subclass distribution) was determined by the modulator used for the primary stimulus (Table 3). In the first experiment in Table 3 mice were given a primary injection of the LPS-M_r 39,000 protein complex and a secondary injection of the LPS-lipoprotein complex. A strong enhancement of the secondary responses was observed. The IgG subclass distribution was, however, characteristic of the effects of the M_r 39,000 protein as modulator. In the second experiment mice were given a primary injection of the LPS-lipoprotein complex and a secondary injection of the LPS-M_r 39,000 protein complex. The IgG subclass distribution in this case was characteristic of the effects of lipoprotein as modulator.

These results indicated that the modulator used for the primary stimulus determined the number and subclass of LPS-specific IgG memory cells induced. The same or a different modulator used for the secondary stimulus effected only the differentiation of these memory cells to antibody-producing cells.

The last experiment in Table 3 demonstrates the specificity of the responses for LPS. The strong secondary IgG responses observed after primary and secondary stimulation with P. mirabilis D52 LPS complexed with phospholipids (Table 1) did not occur if P. mirabilis 19 LPS was used for the secondary stimulation. Thus, the induction of the IgG memory cells by the first injection and the differentiation of these cells to antibody producing cells by the second injection were both LPS-specific processes.

Requirement for T Cells

Although no modulator-specific T cells seem to be involved in the selective enhancing effects observed, T cells apparently do play a role in the

Table 3. Secondary IgG Responses of Mice to Lipopolysaccharide (LPS) from Proteus Mirabilis D52 After a Primary Stimulus (Day 0) with One LPS-Adjuvant Preparation and a Secondary Stimulus (Day 14) with Another LPS-Adjuvant Preparation[a]

Immunogen[b]	Antibody Type	PFC/10^6 Spleen Cells[c] on Day 19	% IgG Subclass of Total IgG
LPS D52 + M$_r$ 39,000 Protein (Primary)	IgG1	200 ± 56	22
LPS D52 + Lipoprotein (Secondary)	IgG2	655 ± 182	70
	IgG3	73 ± 24	8
LPS D52 + Lipoprotein (Primary)	IgG1	444 ± 18	59
LPS D52 + M$_r$ 39,000 Protein (Secondary)	IgG2	240 ± 20	32
	IgG3	65 ± 17	9
LPS D52 + Phospholipids (Primary)	IgG1	62 ± 2	63
LPS 19 + Phospholipids (Secondary)	IgG2	25 ± 2	26
	IgG3	11 ± 1	11

[a]Data taken from Karch and Nixdorff (8).

[b]Dosages per injection were 25 µg LPS from either P. mirabilis D52 or P. mirabilis 19, 300 µg phospholipids and 12.5 µg proteins.

[c]Geometric means ± standard errors of plaque-forming cells (PFC) from three separate experiments. Responses were measured against P. mirabilis D52 LPS coupled to SRBC.

427

Table 4. Comparison of the Secondary IgM and IgG Subclass Responses in Normal and in nu/nu NMRI Mice to Lipopolysaccharide (LPS) After Injection with a Mixture of LPS and M_r 39,000 Protein from the Outer Membrane of _Proteus mirabilis_[a]

Mice	Antibody Type	PFC/10^6 Spleen Cells on Day 18[b]	% IgG Subclass of Total IgG
Normal NMRI	IgM	200 ± 24	--
	IgG1	450 ± 162	37
	IgG2	622 ± 247	51
	IgG3	157 ± 17	12
nu/nu NMRI	IgM	173 ± 37	--
	IgG1	9 ± 3	26
	IgG2	20 ± 4	57
	IgG3	6 ± 10	17

[a] Dosages per injection were 25 µg LPS and 12.5 µg M_r 39,000 protein. Mice received a primary injection on day 0 and a secondary injection on day 14.
[b] Geometric means ± standard errors from two separate experiments. Responses were measured against _P. mirabilis_ LPS coupled to SRBC.

IgG responses to LPS-protein complexes as shown by the results presented in Table 4. In this case we have compared the responses to the complex of LPS with the M_r 39,000 protein in normal and in nu/nu NMRI mice. The results show that the IgM responses to this immunogen in normal and in nu/nu mice were of the same strength. In contrast, the IgG responses in the nu/nu mice were greatly diminished in all IgG subclasses of PFC. The strength of the IgG subclasses responses in nu/nu mice was very similar to that of the IgG subclass responses in normal NMRI mice to LPS alone, without modulator (Table 1).

Secondary In Vitro Responses to the LPS-M_r 39,000 Protein Complex

Our studies have shown that a strong secondary IgG response to LPS was obtained only when a modulator complexed with LPS was used as immunogen for both the primary and the secondary injections (8). Because, on the other hand, the first stimulus determines the IgG character of the response, we are interested in testing whether T cells are required at the time of the secondary stimulus for the differentiation of IgG memory cells induced by the primary stimulus. In this regard, Schuler et al. (25) recently showed that although T cells are required for the production of IgG memory cells induced by primary injection of mice with dextran, the differentiation of these IgG memory cells to antibody-producing cells by a second injection is T cell independent.

We are presently carrying out such studies in an in vitro system where T cells can be eliminated before giving spleen cells the secondary stimulus. For these investigations, homozygous inbred strains of mice must be used.

It was therefore necessary to determine if the responses to our immunogens in inbred mice are the same as those obtained in the NMRI mice which we have consistently employed for our investigations to date. In this regard, we have compared the secondry IgG responses to the LPS-M_r 39,000 protein complex in NMRI and in C57BL/6 mice. The results presented in

Table 5. Comparison of the Secondary IgG Subclass Response in NMRI and in C57BL/6 Mice to Lipopolysaccharide (LPS) After Injection with a Mixture of LPS and M_r 39,000 Protein from the Outer Membrane of Proteus mirabilis[a]

Mouse Strain	Antibody Type	PFC/10^6 Spleen Cells on Day 18[b]	% IgG Subclass of Total IgG
NMRI	IgG1	430 ± 143	27
	IgG2	932 ± 383	58
	IgG3	248 ± 126	15
C57BL/6	IgG1	324 ± 61	19
	IgG2	1222 ± 346	72
	IgG3	152 ± 40	9

[a]Dosages per injection were 25 µg LPS and 12.5 µg M_r 39,000 protein. Mice received a primary injection on day 0 and a secondary injection on day 14.
[b]Geometric means ± standard errors from three separate experiments. Responses were measured against P. mirabilis LPS coupled to SRBC.

Table 5 show that the IgG subclass responses in these two strains of mice are indeed comparable. We could detect no significant differences either in the strength or in the IgG subclass distribution of the responses.

For testing the secondary in vitro responses to the LPS-M_r 39,000 protein complex, we used the Mishell-Dutton culture system (13,14) described in Materials and Methods. In our hands, this system gave very good primary responses of C57BL/6 mouse spleen cells to SRBC. On an average, we obtained 4622 ± 843 IgM PFC per culture, as compared with 845 ± 122 IgM PFC for the control cultures, which received no antigen.

Using this system, we have now obtained preliminary results of the secondary in vitro responses to LPS alone and to LPS complexed with the M_r 39,000 protein, which are presented in Table 6.

We immunized one group of mice in vivo on day 0 with LPS alone. On day 14, we removed the spleens of these mice, placed single cell suspensions in culture and restimulated these cells with LPS. Control cultures that received no antigen were included. We immunized another group of mice in vivo with the LPS-M_r 39,000 protein complex, and restimulated the spleen cells of these mice with the LPS-protein complex in vitro. We also included control cultures which received no antigen.

The results in Table 6 show that the IgG responses to LPS alone were relatively weak. When no secondary stimulus was given, no IgG response could be detected. Complexing LPS with the M_r 39,000 protein effected a distinct enhancement of the IgG responses. However, the distribution of IgG subclasses of PFC was not typical of the response normally obtained with the LPS-protein complex in vivo (compare with Table 5). Nevertheless, there was a characteristic tendency for enhanced induction of IgG2 PFC (adjuvant factor 9.8) after secondary stimulation in vitro. When no secondary stimulus with the LPS-protein complex was given in vitro, IgG PFC were detectable, but there was no adjuvant effect (adjuvant factors < 1). In this case, it seems likely that IgG memory cells were induced by the primary stimulus in vivo, and a small number of these cells could be differentiated

Table 6. Secondary In Vitro IgM and IgG Subclass Responses of Spleen Cells from C57BL/6 Mice Given a Primary Injection In Vivo of Either Isolated Lipopolysaccharide (LPS) or a Mixture of LPS with M_r 39,000 Protein from the Outer Membrane of Proteus mirabilis

Immunogen[a] (Primary, In Vivo)	Immunogen[b] (Secondary, In Vitro)	Antibody Type	PFC/Culture on Day 18[c]	% IgG Subclass of Total IgG	Adjuvant Factor
LPS	LPS	IgM	273 ± 67	--	--
		IgG1	48 ± 47	47	1
		IgG2	23 ± 22	23	1
		IgG3	30 ± 30	30	1
LPS	--	IgM	262 ± 150	--	--
		IgG1	0	--	--
		IgG2	0	--	--
		IgG3	0	--	--
LPS + M_r 39,000 Protein	LPS + M_r 39,000 Protein	IgM	705 ± 358	--	--
		IgG1	438 ± 35	51	9.1
		IgG2	225 ± 19	26	9.8
		IgG3	198 ± 75	23	6.6
LPS + M_r 39,000 Protein	--	IgM	186 ± 99	--	--
		IgG1	47 ± 74	56	0.98
		IgG2	9 ± 8	11	0.39
		IgG3	28 ± 27	33	0.93

[a] Dosages for the primary injection in vivo on day 0 were 1 μg LPS and 0.5 μg M_r 39,000 protein.

[b] Dosages for the secondary injection in vitro on day 14 were 0.0001 g LPS and 0.0005 g M_r 39,000 protein.

[c] Responses were measured against P. mirabilis LPS coupled to SRBC. Values are reported as geometric means ± standard errors from two separate experiments.

into antibody-producing cells by cultivation in the presence of FCS. Under these conditions, predominantly IgG1 and IgG3 antibody-producing cells were stimulated. Taking these results into consideration, the secondary stimulus with the LPS-protein complex in vitro shows a definite tendency indeed toward a selective induction of IgG2-producing cells. However, this system is obviously not optimal.

Lack of a strong characteristic enhancement of IgG2 responses in vitro may be due to the restricted amounts of immunogen employed (1 µg LPS + 0.5 µg protein in vivo; 0.0001 µg LPS + 0.00005 µg protein in vitro), which in preliminary tests appeared to be optimal for this system. This restriction may possibly be dictated by toxic effects of LPS and/or induction of T suppressor cells. Before determining the T cell dependence of the secondary responses to LPS-protein complex, we are presently attempting to alter the culture conditions and the ratio of the amounts of LPS to protein in order to obtain optimal results.

DISCUSSION

Our results show a characteristic enhanced, selective induction of particular subclasses of IgG antibody-producing cells specific for LPS effected by complexing LPS with bacterial outer membrane components used as modulators of the immune response. T cells are involved in this selective modulation, and these T cells are apparently not specific for the modulators. The induction of IgG memory cells effected by a primary stimulus with these complexed and the differentiation of the memory cells to antibody-producing cells effected by a secondary stimulus are both LPS-specific processes.

Another pertinent observation is that LPS molecules of different structure also effect a selective induction of particular subclasses of IgG antibody-producing cells. LPS molecules having short O-polysaccharide chains show a tendency to induce IgG1 and IgG2 PFC while LPS molecules possessing long O-polysaccharide chains induce mainly IgG2 and IgG3 PFC (6). However, regardless of the type of LPS employed as antigen, phospholipids and proteins used as modulators effect the selective induction characteristic for the particular modulator (6).

Taken together, these results suggest a change in the physical character or perhaps the conformation of the LPS molecule occurs upon complex formation with modulators, which in some manner not only allows enhanced IgG production but also determines the IgG subclass induced. Our studies do not distinguish between a directed isotype switch or a selective induction of B cells already expressing a particular isotype of immunoglobulin on the membrane surface. (11,16,22).

Our work in the future will be concerned with delineating the roles T cells, macrophages and their products play in the selective induction of particular subclasses of IgG-producing cells. We feel that we have a well-defined system that should be helpful in these studies. For example, it would be of interest to determine what roles T helper vs. T suppressor cells play in this system, and whether specific T cell factors such as the B cell differentiation factor BCDFγ, which apparently effects a selective switch to IgG1-producing B cells in response to LPS (20), play a role in our system.

ACKNOWLEDGMENTS

This work was supported by the Deutsche Forschungsgemeinschaft.

431

REFERENCES

1. F. Bub, P. Bieker, H. H. Martin, and K. Nixdorff, Immunological characterization of two major proteins isolated from the outer membrane
of Proteus mirabilis, Infect. Immun. 27:315 (1980).
2. J. Gmeiner, The isolation of two different lipopolysaccharide fractions
from various Proteus mirabilis strains, Eur. J. Biochem. 58:621
(1975).
3. J. Gmeiner and H. H. Martin, Phospholipids and lipopolysaccharide in
Proteus mirabilis and its stable protoplast L-form. Difference in
content and fatty acid composition, Eur. J. Biochem. 67:487 (1976).
4. P. C. Isakson, E. Pure, E. S. Vitetta, and P. H. Krammer, T cell
derived B cell differentiation factor(s). Effect on the isotype
switch of murine B cells, J. Exp. Med. 155:734 (1982).
5. N. K. Jerne and A. A. Nordin, Plaque formation in agar by single
antibody-producing cells, Science 140:405 (1963).
6. H. Karch, J. Gmeiner, and K. Nixdorff, Alteration of the immunoglobulin
G subclass responses in mice to lipopolysaccharide: effect of nonbacterial proteins and bacterial membrane phospholipids or outer
membrane proteins of Proteus mirabilis Infect. Immun. 40:157 (1983).
7. H. Karch and K. Nixdorff, Antibody-producing cell responses to an isolated outer membrane protein and to complexes of this antigen with
lipopolysaccharide or with vesicles of phospholipids from Proteus
mirabilis, Infect. Immun. 31:862 (1981).
8. H. Karch and K. Nixdorff, Modulation of the IgG subclass responses to
lipopolysaccharide by bacterial membrane components: differential
adjuvant effects produced by primary and secondary stimulation, J.
Immunol. 131:6 (1983).
9. M.A.K. Markwell, S. M. Haar, L. L. Bieker, and W. E. Tolbert, A
modification of the Lowry procedure to simplify protein determination
in membrane and lipoprotein samples, Anal. Biochem. 87:206 (1978).
10. H. H. Martin, Composition of the mucopolymer in cell walls of the un-
stable and stable L-form of Proteus mirabilis, J. Gen. Microbiol.
36:441 (1964).
11. C. Martinez-Alonso, A. Coutinho, and A. A. Augustin, Immunoglobulin
C-gene expression. I. The commitment to IgG subclass of secretory
cells is determined by the quality of the non-specific stimuli, Eur.
J. Immunol. 10:698 (1980).
12. J. P. McKearn, J. W. Paslay, J. Slack, C. Baum, and J. M. Davie, B cell
subsets and differential response to mitogens, Immunol. Rev. 64:5
(1982).
13. R. I. Mishell and D. W. Dutton, Immunization of dissociated spleen cell
cultures from normal mice, J. Exp. Med. 126:423 (1967).
14. B. B. Mishell and R. I. Mishell, Primary immunization in suspension
cultures, in: "Selected Methods in Cellular Immunology," B. B.
Mishell and S. M. Shiigi, eds., W. H. Freeman and Company, San
Francisco (1980).
15. N. A. Mitchison, The carrier effect in the secondary response to
hapten-protein conjugates. I. Measurement of the effect with transferred cells and objection to the local environment hypothesis, Eur.
J. Immunol. 1:10 (1971).
16. P.K.A. Mongini, W. E. Paul, and E. S. Metcalf, T cell regulation of
immunoglobulin class expression in the antibody response to trinitro-
phenyl-Ficoll. Evidence for T cell enhancement of the immunoglobulin
class switch, J. Exp. Med. 155:884 (1982).
17. P.K.A. Mongini, K. E. Stein, and W. E. Paul, T cell regulation of the
IgG subclass antibody production in response to T-independent antigens, J. Exp. Med. 153:1 (1981).

18. K. Nixdorff, H. Fitzer, J. Gmeiner, and H. H. Martin, Reconstitution of model membranes from phospholipid and outer membrane proteins of Proteus mirabilis. Role of proteins in the formation of hydrophilic pores and protection of membranes against detergent, Eur. J. Biochem. 81:63 (1977).

19. Z. Ovary and B. Benacerraf, Immunological specificity of the secondary response with dinitrophenylated proteins, Proc. Soc. Exp. Biol. Med. 114 (1963).

20. E. Pure, P. C. Isakson, J. W. Kappler, P. Marrack, P. H. Krammer, and E. S. Vitetta, T cell-derived B cell growth and differentiation factors. Dichotomy between the responsiveness of B cells from adult and neonatal mice, J. Exp. Med. 157:600 (1983).

21. M. C. Raff, Role of thymus-derived lymphocytes in the secondary humoral immune response in mice, Nature 226:1257 (1970).

22. Y. J. Rosenberg, Isotype-specific T cell regulation of immunoglubulin expression, Immunol. Rev. 67:33 (1982).

23. E. Ruttkowski and K. Nixdorff, Qualitative and quantitative changes in the antibody-producing cell responses to lipopolysaccharide induced after incorporation of the antigen into bacterial membrane phospholipid vesicles, J. Immunol. 124:2548 (1980).

24. S. Schlecht and O. Westphal, Über die Herstellung von Antiseren gegen die somatischen (O-) Antigene von Salmonellen. II. Mitteilung: Untersuchungen über Hämagglutinintiter, Zentralbl. Bakteriol. Parasitenkd. Infektionskr. Hyg. Abt. I. Orig. 205:487 (1967).

25. W. Schuler, A. Schuler, and E. Kölsch, Immune response against the T-independent antigen $\alpha(1 \longrightarrow 3)$ dextran. II. Occurrence of B_γ memory cells in the course of immunization with the native polysaccharide is T cell dependent, Eur. J. Immunol. 14:578 (1984).

26. O. Westphal, O. Lüderitz, F. Bister, Über die Extraktion von Bakterien mit phenol/wasser, Z. Naturforsch. Teil B. 7:148 (1952).

ENDOTOXIN ASSOCIATED PROTEINS AND THEIR POLYCLONAL AND ADJUVANT ACTIVITIES

Barnet M. Sultzer, John P. Craig, and Raymond Castagna

State University of New York, Downstate Medical Center
Department of Microbiology and Immunology
Brooklyn, New York

INTRODUCTION

The study and analysis of the structure and function of gram-negative bacterial outer membrane proteins in recent years has attracted increasing attention. This interest has developed because of two reasons. First, in the case of several pathogenic organisms, some of these proteins may act as antigens which induce protective immune responses in the host and, second many of the major proteins act as matrix components, ion channels or receptors of various types, all of which are essential to the life of these organisms. Of additional interest is the more recent finding that certain outer membrane proteins have been found to be immunobiologically active in ways other than as specific antigens. This discovery was the outgrowth of experiments on the lymphocytes of the C3H/HeJ mouse which genetically is hyporesponsive to the pathophysiological effects of the outer membrane lipopolysaccharide endotoxin (LPS). The lymphocytes of these mice are deficient in their response to the polyclonal activation of LPS; however, the stimulation of such lymphocytes by selected outer membrane proteins, which are closely associated with but nevertheless separable from the LPS, is essentially normal (18).

The proteins which co-extract with the LPS by the tricholoracetic acid (TCA) method (19), represent a select number of outer membrane polypeptides. In the case of Salmonella sp. and Escherichia coli, these polypeptides consist of the porins and some lower molecular weight components but exclude the major low molecular lipoprotein found free in the outer membrane and also linked to the peptidoglycan (4,5). When these polypeptides were dissociated from the LPS but examined as a protein complex, which we have referred to as endotoxin associated protein (EP), the first activity measured was mitogenicity for mouse lymphocytes. EP activates B cells of the C3H/HeJ mouse to proliferate. In general, EP has been found to be more active than LPS as a mitogen for mouse lymphocytes which are LPS responsive and, furthermore, activates lymphocytes of other species which are poorly reactive to LPS including human peripheral blood lymphocytes (6,7). Additionally, EP has been found to be active in a variety of immunobiological ways as summarized in Table 1. Polyclonal antibody production, the activation of macrophages to be cytotoxic as well as the induction of interferon are other examples of the nonspecific activating properties of EP. But the EP from Salmonella typhimurium also has been found to act as a protective antigen(s) for mouse typhoid in strains of mice which are "normally" susceptible but not hypersusceptible to this pathogen (2,3).

435

Table 1. Immunobiological Activities of Outer Membrane Proteins
Associated with Endotoxin from Gram-negative Bacteria

	Reference
Mitogenicity	
1. Mouse B Lymphocytes	(19)
2. Guinea Pig Spleen Cells	(6)
3. Rabbit Spleen Cells	(6)
4. Rat Spleen Cells	(6)
5. Human Peripheral Blood Lymphocytes	(7)
Nonspecific Antibody Production (PCA)	
1. Mouse B Lymphocytes	(6,19)
2. Human B Lymphocytes	(7)
Macrophage Activation	
1. Cytotoxicity for Tumor Cells (in vitro)	(18)
2. Prostaglandin Release (in vitro)	(18)
Interferon Induction	
1. Mouse (in vivo)	
2. Mouse Spleen Cells	(18)
Protective Antigens	
1. Mouse Typhoid (Salmonella typhimurium)	(2,3)
2. Pertussis in Mice (?)	

In other studies, the so-called lipid A associated protein (LAP), which is similar to but does not have the same polypeptide composition of EP, has been shown to increase glucose utilization in mouse fibroblasts and can cause the release of histamine and serotonin from rat peritoneal mast cells without cytolysis taking place (11). On the other hand, the EP produced by Sipe et al. has been reported to initiate acute phase serum amyloid A in C3H/HeJ mice and to stimulate IL-1 production from macrophages (17).

In view of the varied and potent immunostimulatory properties of EP, we considered the proposition that these selected outer membrane polypeptides could act as immunomodulating agents and thereby enhance or perhaps suppress the immune response to other antigens. For this purpose, we adopted both in vivo and in vitro model systems to measure the immune response to cholera enterotoxin and sheep erythrocytes. EP preparations from Salmonella typhi, Vibrio cholerae and Bordetella pertussis were used for comparison purposes.

MATERIALS AND METHODS

Animals

CF-1 mice were obtained from Charles River Breeding Laboratories (Wilmington, MA). CBA/J and C3H/HeJ mice were obtained from Jackson Laboratories (Bar Harbor, ME). Female mice were used at two to four months of age and were maintained on water and Purina mouse chow ad lib. Male rabbits (6 lbs.) were obtained from Marland Farms (Wayne, NJ).

Antigens and Adjuvants

Purified cholera enterotoxin was obtained from List Biological Laboratories, Inc. (Campbell, CA). Procholeragenoid (Pcg) was prepared by heating the purified toxin for 30 min. at 65°C. Cholera toxin concentrated filtrate from V. cholerae 569B was prepared by the method of Kateley et al. (9) and used for coating sheep erythrocytes. EP was prepared from S. typhi O-901, V. cholerae 569B and B. pertussis 3779B-114 cultures as previously described (5,19). Three day shaker flask cultures were used for S. typhi and V. cholerae, whereas static 72 hr. cultures of B. pertussis were grown according to the method of Morse and Morse (12). Stock solutions of each EP at 1 mg/ml were prepared and stored at -20°C. Protein-free LPS was recovered from the aqueous phase of the phenol-water fractionation of the tricholoracetic acid extracted endotoxin of B. pertussis (5).

Lymphocyte Cultures

Single cell suspensions were prepared from spleens, and 1×10^6 cells per ml were cultured in RPMI-1640 medium supplemented with 2.5% fetal calf serum (Gibco Laboratories, Grand Island, NY), 100 U of penicillin per ml and 100 µg of streptomycin per ml as previously described (6,19). DNA synthesis was measured by incorporation of [³H]-thymidine. Twenty-four hours before harvesting the cells, one µCi of [³H]-thymidine was added to each culture tube. The cells were collected on fiber glass filters in a Titertek cell harvester (Flow Laboratories, Rockville, MD). The dried filters were counted by means of liquid scintillation (Beckman LS-250). All cultures were done in triplicate and replicate values were within ± 10% of the mean. The stimulation index (S.I.) is equal to the mean cpm of the stimulated culture divided by the mean cpm of the control culture.

Plaque Forming Cell Assay

Direct plaques representing IgM antibody producing cells were measured in a modified Jerne system by coating sheep erythrocytes with cholera enterotoxin concentrated filtrate. Spleen cells were cultures at a concentration of 10×10^6 cells per ml in supplemented RPMI-1640. One ml of the suspension was incubated in 35 mm plastic petri dishes at 37°C in an atmosphere of 83% N_2, 10% CO_2 and 7% O_2 with or without the antigen procholeragenoid and/or EP added at the initiation of the culture. All cultures were done in triplicate and fed daily with 0.1 ml of culture medium. For determining PFC's against cholera toxin-coated sheep erythrocytes, the mean number of plaques obtained were subtracted from the mean number of plaques obtained with uncoated sheep erythrocytes. For polyclonal antibody experiments, the spleen cell cultures were incubated with or without EP or LPS alone. The plaque assay was done after 72 hrs of culture using trinitrophenylated (TNP) coated sheep erythrocytes as the substrate (16).

Mouse Protection Test

In order to assess the adjuvant properties of EP, CF-1 mice were immunized by intraperitoneal injection of a single dose of cholera toxoid, with or without a single dose of EP then challenged intravenously 42 days later with a dose of approximately 3 LD_{50} of cholera toxin. Three LD_{50} of cholera toxin kills all mice within two to three days. The 50% protective dose (PD_{50}) of glutaraldehyde cholera toxoid has been found to be 0.1 to 3 µg of toxoid protein. The relative enhancing effect of the adjuvants was estimated by comparing PD_{50} values. The LD_{50} and PD_{50} values were calculated by the method of Reed and Muench (15) using six mice per group and two to three fold dilutions of the cholera toxoid. The 95% limits of confidence of these values were determined by the method of Pizzi (14).

Antitoxin Measurements

The rabbit intracutaneous assay of cholera enterotoxin induced vascular permeability was used to measure the level of antitoxin antibodies in the serum of mice taken 24 to 48 hrs. before the intravenous challenge of cholera enterotoxin (1). One unit of antitoxin neutralizes 40 ng of cholera enterotoxin.

Gel Electrophoresis

Analytical polyacrylamide gel electrophoresis (PAGE) was performed with the discontinuous sodium dodecyl sulfate (SDS) buffer system of Laemmli (10) using slab gels as previously described (5). The stacking gel was 3% acrylamide and the separating gel was 12%. The samples were dissolved in an SDS and 2-mercaptoethanol buffer and run at 25 and 50 µg. The slab gel was run overnight against a constant voltage (30V). Gels were stained with 0.1% Coomassie brilliant blue R-250 (Bio-Rad Laboratories) in 40% methanol and 7.5% acetic acid for one hour at 37°C and destained with 7% acetic acid.

For preparative SDS-PAGE, the slab gel (15 cm x 24 cm) was prepared at a thickness of 6 mm using 12% acrylamide as a separating gel and 3% acrylamide as the stacking gel. A 10 mg sample was prepared and run as described above. After the completion of the run, the marker lanes on each side of the gel were cut off and stained with Coomassie brilliant blue. Using the stained marker lanes to delineate the major polypeptide bands, the appropriate gel bands were sliced from the slab, minced and lyophilized. The dried gel slices were then ground and extracted in the running buffer followed by extensive dialysis with ten changes of 20 volumes of distilled water. The extracts were then redried and extracted three times in 90% acetone to remove

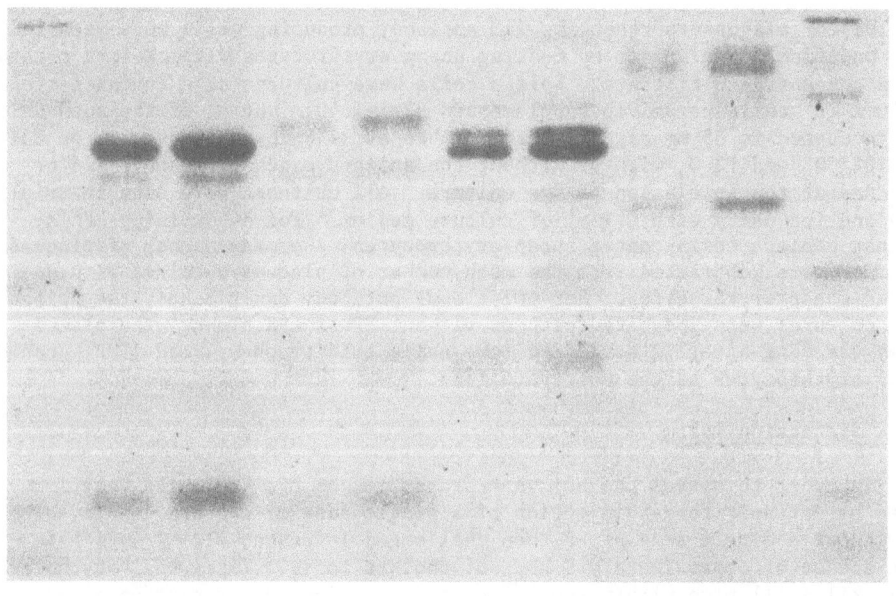

Figure 1. Analytical SDS-PAGE. The molecular weight markers from top to bottom (lanes 1 and 10) are globulin, 160K; albumin, 68K; ovalbumin, 45K; chymotrypsinogen, 25K; myoglobulin, 17.2K; and cytochrome C, 12-13K. Lanes 2 and 3 are S. typhi EP, lanes 4 and 5 are Serratia marcescens EP, lanes 6 and 7, V. cholerae EP, and lanes 8 and 9 are B. pertussis EP.

Table 2. Mitogenic Activity of the Polypeptide Components of
Salmonella typhimurium EP for C3H/HeJ Spleen Cells

Fraction (µg/ml)		Mean Net CPM of ^{3}H-Thymidine[a] Incorporation (S.I.)[b]	
Control		2,908	
EP 12082	(1)	35,509	(12.2)
	(10)	55,147	(18.9)
32-35K	(1)	4,284	(1.5)
	(10)	10,054	(3.5)
20K	(1)	11,536	(3.9)
	(10)	24,181	(8.3)
17K	(1)	22,103	(7.6)
	(10)	28,457	(9.8)

[a]All cultures were harvested after 48 hours of incubation. The
mean of triplicate cultures did not vary more than ± 10%.
[b]S.I. = stimulation index, see Methods.

residual SDS. The protein precipitates were then washed three times with
non-pyrogenic distilled water, relyophilized and subsequently tested for
mitogenic activity.

RESULTS

The polypeptides separated from the TCA extracted LPS by the hot-phenol
method (5) produce various profiles on SDS-PAGE when they are derived from
enteric and non-enteric gram-negative bacteria. As shown in Figure 1, the
polypeptides from Salmonella typhi and Serratia marcescens are very similar,
whereas those from Vibrio cholerae and Bordetella pertussis vary considerably. All have polypeptides between 30 and 35 kilodaltons but noticeably
different are the higher molecular weight polypeptides of Bordetella pertussis of approximately 65 and 68 kilodaltons.

To assess the activity of each component, we have separated the polypeptides of Salmonella typhimurium which are similar to those of S. typhi
by preparative SDS-PAGE. As shown in Table 2, each of the major polypeptides
stimulated C3H/HeJ splenic lymphocytes, although none exceeded the undissociated complex from which the components were derived. Indeed, the sum of
the quantitative activity of the [^{3}H]-thymidine incorporation at each concentration of the 17,20 and 32-35 kilodalton fractions approximated the
values obtained with the whole EP; however, sufficient trials have not been
done to conclude that each component proportionately contributes to the
total activity as expressed in this particular data. Nevertheless, it is
clear that the EP complex is the most active of all the individual polypeptides analyzed. Consequently, in all of the experiments described below,
we used the undissociated FP from several organisms to study the adjuvant
effect on selected antigens.

For this purpose, we have an adopted mouse protection test that was
developed to measure the immune response to cholera enterotoxin after

immunization with cholera toxoid. CF-1 mice were given intraperitoneal injections in graded three-fold doses of cholera toxoid with or without selected constant doses of EP from four different organisms. Six weeks later, the mice were challenged intravenously with approximately 3 LD_{50} of cholera toxin. As depicted in Figure 2, the PD_{50} for the reference cholera toxoid varied from one trial to another in part as a function of the variation in the LD_{50} of the cholera toxin challenge dose. Nevertheless, in each instance where EP was administered at the time of the cholera toxoid immunization, the µg of cholera toxoid needed for the PD_{50} was reduced. The most significant enhancement of the protection provided by cholera toxoid can be seen in those mice which received <u>Salmonella typhi</u> and <u>Bordetella pertussis</u> EP. The factor of enhancement ranged from 5.2 to 16.2 with <u>S. typhi</u> EP, whereas the increased protection produced by <u>B. pertussis</u> EP varied from 5.3 to 27.0 times that obtained when cholera toxoid was given alone. The EP's from <u>S. typhimurium</u> and <u>V. cholerae</u> in these trials were relatively less effective but an adjuvant effect can be clearly observed. In addition, in another trial both EP and purified LPS from <u>S. typhi</u> were compared. As shown in Figure 3, a similar level of enhancement was obtained; however, there was not significant synergistic or additive effects when both agents were administered together.

In the adjuvant experiments just described, serum samples were obtained from the mice 24 to 48 hours before the intravenous challenge with cholera toxin. From these samples toxin neutralizing antibody levels (IgG) were determined by the rabbit intracutaneous method of cholera toxin induced vascular permeability (1). The results of these experiments are shown in

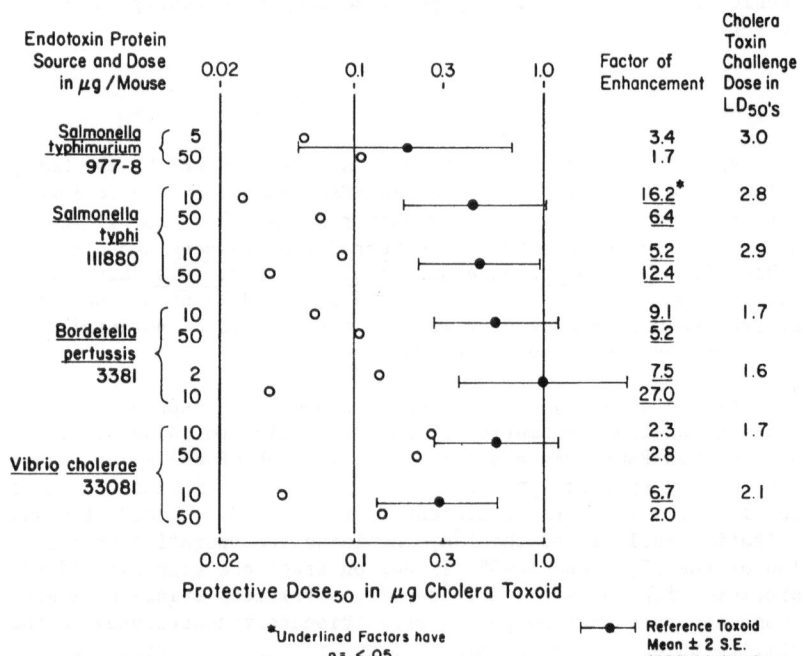

Figure 2. Adjuvant effects of endotoxin protein on immunogenicity of cholera toxoid in CF-1 female mice. The closed circles are the mean PD_{50} in µg of the reference cholera toxoid. The open circles are the PD_{50}s in µg of cholera toxoid when the indicated doses of EP were given together with the cholera toxoid.

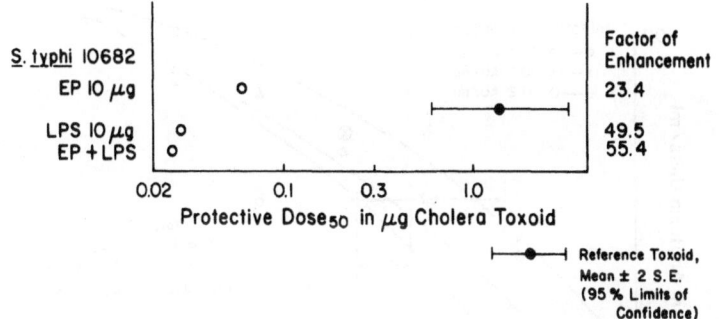

Figure 3. Adjuvant effect of EP and LPS from <u>Salmonella</u> <u>typhi</u> on immunogenicity of cholera toxoid in CF-1 female mice. A comparison
of the adjuvant effect of EP and LPS from <u>S</u>. <u>typhi</u> on the
immunogenicity of cholera toxoid in CF-1 female mice. See
Figure 2 for the notations.

Figures 4 through 7. The serum antitoxin units are plotted against the
cholera toxoid dose and the dose response curves were drawn by inspection.
In each instance, the antitoxin levels evoked by a given dose of toxoid was
increased when EP was given along with the toxoid. This is shown by a left-
ward displacement of the dose response curve. The actual fold increase in
the immunogenicity of the cholera toxoid provided by the adjuvant effect of
the EP is indicated on each graph. Furthermore, by extrapolating from the
PD_{50} dose of the toxoid on the vertical axis, we can estimate the serum
antitoxin levels associated with 50% protection in each challenge. In the
case of the <u>V</u>. <u>cholerae</u> EP, 50% protection was associated with 10.5 to 11.2
AU/ml as compared to 14.5 AU/ml for the control toxoid alone (Figure 3).
For the <u>S</u>. <u>typhi</u> EP, 3 to 5.9 AU/ml yielded a PD_{50} as compared to 9.3 AU/ml
for the control toxoid. The most dramatic enhancement in immunogenicity
was obtained with the <u>B</u>. <u>pertussis</u> EP where a 7.5 to 27-fold increase was
obtained and 7.6 to 10.0 AU/ml provided 50% protection as compared to 9.3
AU/ml in the toxoid control.

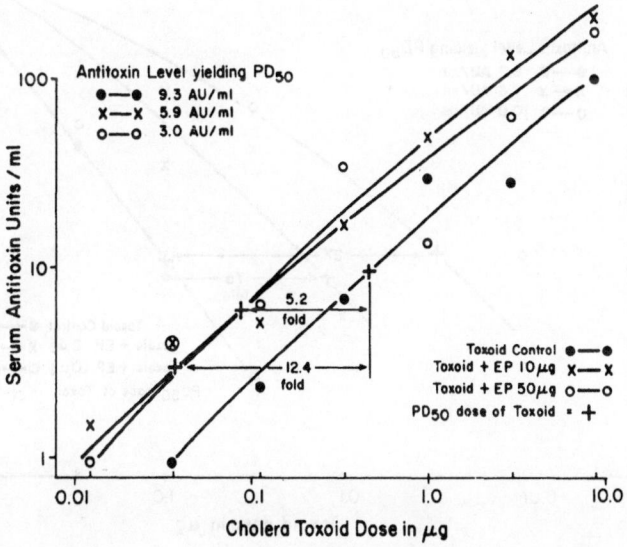

Figure 4. The relationship between serum antitoxin levels and protection
against cholera toxin challenge in CF-1 mice immunized with and
without <u>S</u>. <u>typhi</u> endotoxin protein.

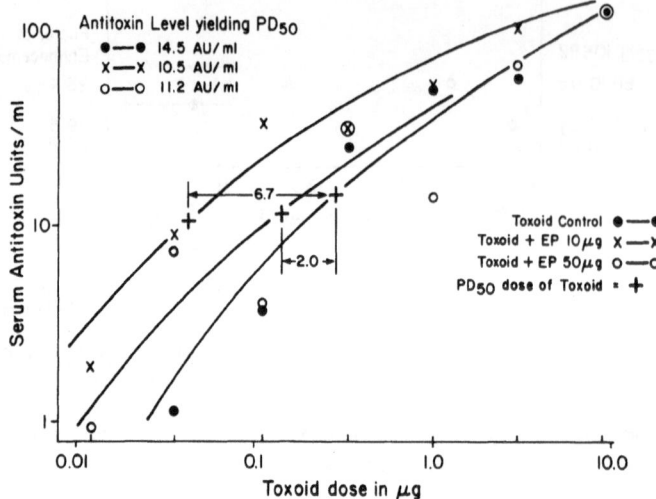

Figure 5. The relationship between serum antitoxin levels and protection
against cholera in CF-1 mice immunized with cholera toxoid with
and without <u>V</u>. cholerae endotoxin protein.

From these results, it is clear that protection is directly related to
the availability of circulating antitoxin antibodies. Moreover, a mean
antitoxin level of about 9 AU/ml correlates with comparable PD_{50} values in
all groups. Consequently, it is not necessary to invoke any other factor
other than the ability to increase antitoxin antibody as an explanation of
the adjuvant effect of EP upon cholera toxoid.

From the results obtained in vivo, it was reasonable to expect that EP
could adjuvant antibody production in cultured lymphocytes. At the same

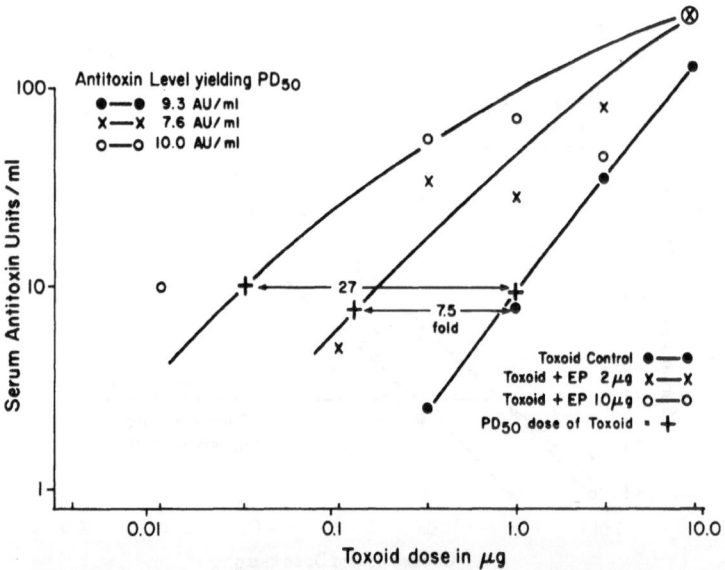

Figure 6. Relationship between serum antitoxin levels and protection
against cholera toxin challenge in CF-1 mice immunized with
cholera toxoid with and without <u>B</u>. <u>pertussis</u> (3381) endotoxin
protein.

Table 3. Number of PFC/10^6 Viable Cells Against Cholera Enterotoxin-Coated Erythrocytes After Various Times of Incubation of Normal CBA/J Spleen Cells with Procholeragenoid (Pcg) and Endotoxin Protein (EP) from <u>Salmonella typhi</u> 0901

| Days of Culture | Mean Number of Specific CT-PFC/10^6 Cells[a] | |
	Pcg[b]	Pcg + EP[b]
3	179	252
5	106	637
6	113	721
7	43	1242

[a]Variation about the mean did not exceed ±10%.
[b]Pcg + 1 µg/ml; EP = 10 µg/ml

time the adjuvanticity could be examined from the standpoint of the need for accessory cells and the role of polyclonal activation in the adjuvant process. As a corollary of the cholera toxoid protection experiments, there-fore, CBA/J splenic lymphocyte cultures were incubated with procholeragenoid, S. typhi EP, and the combination of procholeragenoid and EP for various periods of time followed by a direct plaque forming cell (PFC) assay against cholera enterotoxin coated sheep erythrocytes. The procholeragenoid antigen was obtained by heating purified cholera toxin for 30 minutes at 65°C. As shown in Table 3, procholeragenoid effectively stimulated specific PFC's (IgM) for cholera toxin coated sheep erythrocytes from three to six days of culture. This response, however, was significantly enhanced five to seven days after the initiation of the cultures by the initial addition of S. typhi EP. Similar results were obtained with CF-1 mice (data not shown) which had been used for the in vivo protection experiments.

In view of the potent adjuvant activity of the B. pertussis EP (PEP) in vivo, we examined this polypeptide complex from two aspects. Although the composition of the PEP differed from that obtained from other gram-

Table 4. Stimulation of DNA Synthesis in Mouse Splenic Lymphocytes by Pertussis Endotoxin Protein (PEP)

| Stimulant (µg/ml) | Mean Net CPM of ^{3}H-Thymidine Incorporation[a] (±SE) | | | | | |
	CBA/J		C3H/HeJ		CF-1	
Control	1,397	(±77)	3,569	(±15)	2,128	(±150)
PEP (1)	26,453	(±1,035)	26,917	(±987)	34,957	(±3,182)
PEP (10)	39,306	(±1,136)	31,450	(±2,360)	36,295	(±1,567)
PEP (50)	33,216	(±5,383)	27,843	(±838)	30,040	(±646)
LPS (1)	21,973	(±748)	2,132	(±55)	17,164	(±1,185)
LPS (10)	21,880	(±281)	2,423	(±72)	19,405	(±647)
LPS (50)	20,690	(±1,134)	2,969	(±280)	11,319	(±52)

[a]Cultures harvested after 48 hr incubation. (±SE) = Standard error of mean.

Table 5. Effect of Macrophage Depletion on Polyclonal Activation of
C3H/HeJ Splenic Lymphocyte by Pertussis Endotoxin Protein (PEP)

Stimulant (μg/ml)	Mean Number PFC/10^6 Viable Cells[a]
Normal Spleen Cells	
None	12
LPS (1)	4
LPS (10)	2
PEP (1)	119
PEP (10)	694
Macrophage Depleted Spleen Cells[b]	
None	2
LPS (1)	5
LPS (10)	0
PEP (1)	33
PEP (10)	173

[a]Mouse spleen cells were cultured for 72 hours and then plaqued against
TNP/SRBC. Variation about the mean did not exceed ±10%.
[b]Macrophages were removed by the carbonyl iron method (2x) with less than
1% nonspecific esterase staining cells remaining.

negative organisms, the possibility existed that it would have immunobiological properties similar to the S. typhi EP and stimulate polyclonal activation. As shown in Table 4, PEP is a potent mitogen for cultured mouse
splenic lymphocytes including LPS low-responder C3H/HeJ cells. The LPS
extracted from B. pertussis and purified to the extent of having less than
1% protein is somewhat less active than PEP and as expected cannot activate
DNA synthesis in C3H/HeJ lymphocytes. In addition PEP activated C3H/HeJ
spleen cells to make IgM antibody against TNP-SRBC in contrast to pertussis
LPS which was ineffective (Table 4). However, this polyclonal effect was
diminished as much as 75% when the spleen cells were depleted of macrophages
suggesting a role for macrophages as an accessory cell but not necessarily
as an absolute requirement (Table 4).

Table 6. Enhancement of PFC to Cholera Enterotoxin-Coated Sheep
Erythrocytes by Pertussis EP in CF-1 Spleen Cell Cultures

Stimulant (μg/ml)	Mean Nos. of Specific CT-PFC/10^6 Cells Pcg[a] Antigen at			
	0.0	0.1	1.0	10.0 (μg/ml)
None	0	0	4	0
Pertussis LPS (10)	38	ND	42	48
Pertussis EP (10)	35	52	63	194

[a]Pcg = procholeragenoid, see Materials and Methods.

Table 7. Increased PFC Response to Sheep Erythrocytes Stimulated by
LPS or EP in CF-1 Mouse Spleen Cell Cultures

Stimulant (10 µg/ml)	Mean Nos. of PFC/10^6 Cells at 5 Days of Culture	
	No SRBC	With SRBC
None	2	17
S. typhi LPS	37	493
S. typhi EP	20	308
B. pertussis LPS	108	453
B. pertussis EP	23	865

These results, as well as the effects of PEP in vivo, clearly suggested that this material would be active as an adjuvant for cultured lymphocytes. As expected PEP enhanced specific direct PFC to cholera enterotoxin coated sheep erythrocytes in CF-1 mouse spleen cell cultures, whereas the pertussis LPS in this experiment was ineffective (Table 6). Finally, we have tested PEP in comparison to S. typhi EP and purified LPS preparations from both organisms as an adjuvant for a classical T-dependent antigen, sheep erythrocytes. As depicted in Table 7, PEP is superior to pertussis LPS and S. typhi LPS or EP in enhancing the IgM response of CF-1 splenic lymphocytes to this antigen.

DISCUSSION

The major conclusion we can derive from these studies is that selected outer membrane proteins of various gram-negative bacteria, particularly those associated with the lipopolysaccharide in the outer membrane are potent immunomodulators. From our preliminary separation experiments with S. typhimurium it appears that the polypeptide components of the EP complex are active in stimulating B lymphocytes to proliferate, given that the undissociated EP has been shown to be a B-cell mitogen and to activate polyclonal antibody synthesis. If mitogenicity is a hallmark of the other activities ascribed to EP, we would assume that each of these separated polypeptides would also be active as adjuvants, but this remains to be tested.

In any event, the EP complex as such appears to be most active as a polyclonal activator and has been used in vivo and in vitro to establish that both IgM and IgG antibody production can be enhanced by the simultaneous presentation of EP and the antigen in question to either mouse or its lymphocytes in culture.

Of more than passing interest is the fact that B. pertussis EP is a potent adjuvant and probably contributes to the long known adjuvant effect of the whole cell vaccine previously attributed solely to the endotoxin. Whether these same proteins may serve as protective immunogens against B. pertussis infection is currently under study in our laboratory.

Since at least portions of these outer membrane polypeptides in situ may interact with host cells under natural conditions of infection, it is also conceivable that they may alter the immune response to various antigens from the very same organisms. Indeed, Karch and Nixdorff (8) have shown that certain outer membrane proteins complexed to the lipopolysaccharide from _Proteus mirabilis_ can modulate the antibody response to LPS in mice converting this response from predominantly IgM to IgG antibody. Consequently, it would seem reasonable to conclude that the maturation and differentiation of B-lymphocytes in the host in response to gram-negative bacterial antigens is dependent, at least in part, on the immunomodulating components of the outer membranes of these organisms as they interact with these cells directly or indirectly through the activation of accessory immunocompetent cells.

ACKNOWLEDGMENTS

The technical assistance of Lynne Kilbourne and Arthur Buonaspina is gratefully acknowledged, as well as the assistance of Janice Howard for preparation of the manuscript.

This work was supported in part by the Office of Naval Research Contract N0014-84-K-0693.

REFERENCES

1. J. P. Craig, Cholera toxins, in: "Microbial Toxins, A Comprehensive Treatise, Volume IIA Bacterial Protein Toxins," S. Kadis, T. C. Montie, andk S. T. Ajl, eds., Academic Press, New York (1971).
2. T. K. Eisenstein, L. M. Killar, and B. M. Sultzer, Immunity to Infection with _Salmonella typhimurium_: mouse-strain difference in vaccine- and serum-mediated protection, J. Inf. Dis. 150:425 (1984).
3. T. K. Eisenstein and B. M. Sultzer, _Salmonella_ vaccines: protection by endotoxin protein and lipopolysaccharide in two different mouse strains, in: "Bacterial Vaccines," J. Robbins, ed., Decker, New York (1982).
4. R. G. Goldman, D. White, and L. Lieve, Identification of outer membrane proteins, including known lymphocyte mitogens as the endotoxin protein of _Escherichia coli_ O111, J. Immunol. 127:1290 (1981).
5. G. W. Goodman and B. M. Sultzer, Characterization of the chemical and physical properties of a novel B-lymphocyte activator, endotoxin protein, Inf. and Immun. 24:685 (1979).
6. G. W. Goodman and B. M. Sultzer, Further studies on the activation of lymphocytes by endotoxin protein, J. Immunol. 122:1329 (1979).
7. G. W. Goodman and B. M. Sultzer, Endotoxin protein is a mitogen and polyclonal activator of human B lymphocytes, J. Exp. Med. 149:713 (1979).
8. H. Karch and H. Nixdorff, Antibody-producing cell responses to an isolated outer membrane protein and to complexes of this antigen with lipopolysaccharide or with vesicles of phospholipids from _Proteus mirabilis_, Inf. Immun. 31:862 (1981).
9. J. R. Kateley, S. F. Lyons, and H. Friedman, Immunocyte response to a purified bacterial toxin (choleragen) and toxoid: cytokinetics, immunoglobulin class, and specificity, J. Immunol. 112:1452 (1974).
10. U. K. Laemmli, Cleavage of structural proteins during the assembly of the head of bacteriophage T4, _Nature_ (London) 227:680 (1970).
11. D. C. Morrison and S. J. Betz, Chemical and biological properties of a protein-rich fraction of bacterial lipopolysaccharide. II. The in vitro peritoneal mast cell response, J. Immunol. 119:1790 (1977).
12. S. I. Morse and J. H. Morse, Isolation and properties of the leukocytosis- and lymphocytosis-promoting factor of _Bordetella pertussis_, J. Exp. Med. 143:1483 (1976).
13. B. S. Nilsson, B. M. Sultzer, and W. W. Bullock, Purified protein derivative of tuberculin induced immunoglobulin production in normal mouse spleen cells, J. Exp. Med. 137 (1973).

14. M. Pizzi, Sampling variation of the fifty percent endpoint, determined by the Reed Muench (Behrens) method, Human Biology 22:151 (1950).

15. L. J. Reed and H. Muench, A simple method of estimating fifty percent endpoints, Amer. J. Hyg. 27:493 (1938).

16. M. B. Rittenberg and K. L. Pratt, Antitrinitrophenyl (TNP) plaque assay: primary response of BALB/c mice to soluble and particulate immunogen, Proc. Soc. Exp. Biol. Med. 132:575 (1969).

17. J. D. Sipe, M. Johns, A. DeMaria, A. S. Cohen, and W. R. McCabe, Acute phase serum amyloid A response in endotoxin non-responder mice: a sensitive and specific effect of endotoxin associated protein, Fed. Proc. 43:1452 (1984).

18. B. M. Sultzer, Lymphocyte activation by endotoxin and endotoxin protein: the role of the C3H/HeJ mouse, in: "Beneficial Effects of Endotoxins," A. Nowotny, ed., Plenum Press, New York (1983).

19. B. M. Sultzer and G. W. Goodman, Endotoxin protein: a B-cell mitogen and polyclonal activator of C3H/HeJ lymphocytes, J. Exp. Med. 144:821 (1976).

LIPOPOLYSACCHARIDE-INDUCED RECURRENCE OF ARTHRITIS INITIATED BY PEPTIDOGLYCAN-POLYSACCHARIDE

Stephen A. Stimpson, Ronald E. Esser, William J. Cromartie, and John H. Schwab

Department of Microbiology, School of Medicine
University of North Carolina
Chapel Hill, North Carolina

INTRODUCTION

Lipopolysaccharide (LPS) and covalent complexes of peptidoglycan and polysaccharide (PG-PS) are the major toxic components of bacterial cell walls. The lipid A and peptidoglycan moieties are responsible for much of the biological activity of LPS and PG-PS, respectively. Both are ubiquitous in the environment and have many biological properties in common, including polyclonal activation of lymphoid cells, complement activation, mitogenicity, macrophage activation, and adjuvanticity (3,8,10,11). In spite of these similarities, the interaction or complementation of the phlogistic properties of these toxic polymers in inflammation has received little attention.

The present study focuses on the ability of systemic injection of LPS to induce a recurrence of arthritis in rat ankle joints which had been previously inflamed by intraarticular (ia) injection of PG-PS (5). The effects of various types and doses of LPS, and the histological features of LPS-induced exacerbation of joint inflammation, are described.

MATERIALS AND METHODS

Purification of Streptococcus pyogenes Group A PG-PS

S. pyogenes (Group A, strain D-58) cell walls were prepared from whole cells broken by shaking with glass beads. Cell walls were extracted three times with 2% sodium dodecyl sulfate in phosphate-buffered saline (PBS) at 56°C for two hours each time, followed by extensive washing with PBS and distilled water to yield large PG-PS polymers (5). These were sonicated to generate a range of smaller fragments, which were further fractionated by differential centrifugation (6). PG-PS fragments which sedimented at 100,000 x g, but not at 10,000 x g, were used in this study.

Source of Endotoxin

LPS isolated from Escherichia coli 0111:B4 and Salmonella typhimurium wild type LT-2, and endotoxic glycolipid from S. typhimurium G30/C21 Re mutant, were obtained from Ribi Immunochem Research, Inc., Hamilton, MT. S. typhimurium W LPS was obtained from Difco Laboratories, Detroit, MI.

Table 1. Recurrence of Arthritis Induced by Systemic Injection of Various Types and Doses of Endotoxin[a]

Type of Endotoxin Injected	Dose	PG-PS-injected (right) Ankle		PBS-injected (left) control ankle	
		Incidence[b]	Severity[c]	Incidence	Severity
S. typhimurium LT-2 wild type LPS	100 μg Exp 1	7/7	2.2±0.3 p=0.001[d]	5/7	1.1±0.3
	50 μg Exp 1	6/6	2.5±0.3 p<0.005	4/6	1.1±0.2
	Exp 2	8/8	2.9±0.1 p<0.001	6/8	1.2±0.3
	25 μg Exp 1	8/8	2.3±0.2 p<0.001	4/8	0.6±0.3
	10 μg Exp 1	8/8	2.4±0.1 p<0.001	3/8	0.4±0.3
	Exp 2	7/8	2.0±0.4 p<0.001	2/8	0.2±0.1
	3 μg Exp 2	5/8	1.3±0.3 p<0.005	0/8	0.1±0.1[f]
	1 μg Exp 2	2/8	0.8±0.3	0/8	0
	0.3 μg Exp 2	1/8	0.5±0.2	0/8	0
	0.1 μg Exp 2	0/8	0.5±0.2[f]	0/8	0
S. typhimurium W wild type LPS	50 μg	6/6 (72 h)[e]	2.0±0.3	0/8	0
S. typhimurium G30/C21 Re glycolipid	50 μg	5/8	1.3±0.2	0/8	0
E. coli 0111:B4 LPS	100 μg	8/8	2.3±0.2 p<0.001	2/8	0.6±0.2

[a]On day 0 the right ankle of each rat received an ia injection of 5 μg of S. pyogenes PG-PS in 10 μl of PBS. The left ankle joint received an ia injection of 10 μl of PBS alone. Three weeks after ia injection, each rat received iv injection of indicated type and dose of endotoxin in 0.5 ml PBS.
[b]Number of ankles in which joint score increased by at least one unit 48 h after injection of endotoxin/total number of ankles injected.
[c]Mean joint score (±1 standard error) 48 h after injection of endotoxin, including those joints which are negative. Immediately prior to injection of endotoxin, the mean joint score of PG-PS-injected joints was never more than 0.4±0.2; PBS-injected control joints showed no apparent inflammation.
[d]Significance of the difference between the mean joint score of PG-PS-injected vs. PBS-injected joints, by Student's t-test.
[e]Joint inflammation induced by this type of LPS peaked at 72 h, instead of 48 h as seen in response to other types of endotoxin.
[f]Although some joints in this group were slightly inflamed, none met the criteria indicated in b.

Induction of Arthritis

Female Sprague-Dawley rats six to eight weeks old (Zivic-Miller Laboratories, Allison Park, PA) were used. PG-PS (5 μg) was injected ia into the right ankle (through the Achilles tendon just above the calcaneal process) using a 25-gauge needle adapted to a micropipette. The left ankle was injected with sterile PBS as a control (5). Nineteen to 22 days later, endotoxin in 0.5 ml PBS was injected iv into each rat. The dose of LPS varied as described below for each experiment. The severity of inflammation was scored grossly on a scale of zero (no apparent inflammation) to four (severe inflammation) based on erythema and edema of the periarticular tissues and enlargement of the joints (4).

Histology

Ankle joints, fixed in 10% formaline and decalcified, were embedded in paraffin, sectioned, and stained with hematoxylin and eosin. Each stained section was examined and scored for the following changes in the tissues of the ankle and adjacent tarsal joints: (a) presence of edema, infiltration of inflammatory cells, and extravasation of fibrin and red blood cells (acute exudative reaction); (b) hyperplasia of synovial lining cells, and increased numbers of fibroblasts and increased vascularity in the synovial stroma (proliferative reaction); (c) presence of diffuse and focal collections of

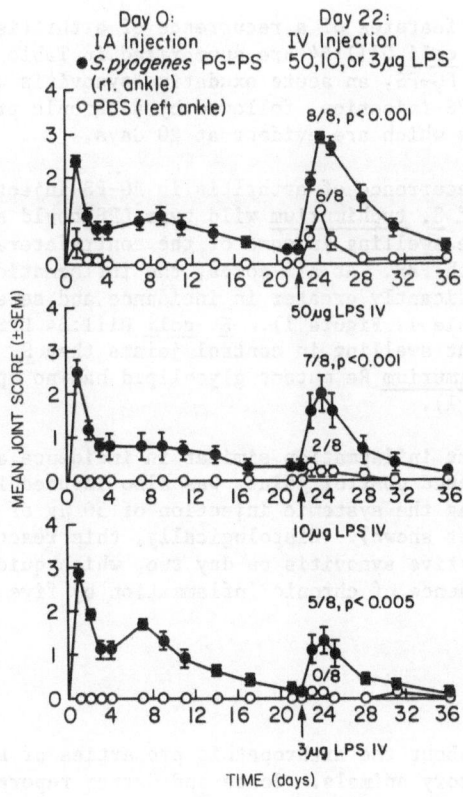

Figure 1. Course of arthritis induced by the ia injection of S. pyogenes PG-PS (5 μg) and recurrence of arthritis induced by three different systemic doses of S. typhimurium LT-2 wild type LPS. See Table 1 for experimental details. The incidence of joint inflammation and the significance of the difference between the mean joint score of PG-PS-injected vs. PBS-injected joints 48 hrs after LPS injection is indicated.

lymphocytes and macrophages (chronic inflammatory reaction); and (d) scarring, and fibrosis of the synovial stroma. Each characteristic was scored on a scale of zero to ++++.

RESULTS

A systemic injection of S. typhimurium wild type LPS induced a recurrence of arthritis in joints which had received an ia injection of S. pyogenes PG-PS 21 days earlier (Table 1). The severity of the inflammation increased to a peak at 48 hours, and then gradually subsided (Figure 1). The smallest dose of S. typhimurium wild type LPS which induced a recurrence of inflammation in 100% of the PG-PS-injected joints was 10 µg per rat, and a recurrence could be induced with as little as 300 ng (Table 1). Other studies (5) showed that no recurrences were induced in joints inflamed by the ia injection of LPS three weeks previous to the systemic LPS injection (data not shown here).

E. coli 0111:B4 LPS and S. typhimurium Re mutant glycolipid were also active in this model (Table 1), suggesting that lipid A might be the active moiety of LPS, and that this arthropathic activity might be widespread among endotoxins in nature. The Re mutant glycolipid appears to be less active than wild type LPS on a dry weight basis.

The histological features of a recurrence of arthritis induced by systemic injection of E. coli 0111:B4 are summarized in Table 2. In joints previously exposed to PG-PS, an acute exudative synovitis was present two days after systemic LPS injection, followed by a chronic proliferative and infiltrative synovitis which are evident at 20 days.

In addition to recurrence of arthritis in PG-PS-injected test joints, high systemic doses of S. typhimurium wild type LPS could also elicit a transient acute tissue swelling in some of the contralateral control joints initially injected with PBS. At all doses, the inflammation induced in test joints was significantly greater in incidence and severity than that in control joints (Table 1, Figure 1). E. coli 0111:B4 LPS appeared to induce less acute joint swelling in control joints than S. typhimurium wild type LPS, and S. typhimurium Re mutant glycolipid had no apparent effect on control joints (Table 1).

An acute transient inflammation similar in incidence and severity to that seen in PBS-injected control joints was also induced in ankle joints of naive rats following the systemic injection of 50 µg of S. typhimurium wild type LPS (data not shown). Histologically, this reaction was characterized by an acute exudative synovitis on day two, which quickly resolved with little or no evidence of chronic inflammation by five days after LPS injection.

CONCLUSIONS

Little is known about the arthropathic properties of LPS upon systemic injection into laboratory animals. Jones and Carter reported in 1957 that somatic antigen from Shigella paradysenteriae Type Z induced a transient acute synovitis in guinea pigs (7). Synovitis was also seen with lethal endotoxemia in rabbits (1). The present study in rats supports the idea that the systemic administration of a high dose of LPS induces only a transient acute joint inflammation.

We also show that a systemic injection of LPS, even in low doses, induces a recurrence of arthritis in joints previously injured by exposure

Table 2. Histological Features of LPS-Induced Recurrence of Arthritis[a]

Days After IV injection of LPS	Joint	Mean Joint Score[b]	Rat	Histological Features[c] of Individual Joints				
				Acute Exudative	Proliferative	Chronic Inflammatory	Erosive	Fibrosis and Scarring
2	Left	0.6±.2	R2	0	0	0	0	0
			R5	±	±	0	±	0
	Right	2.3±0.2	R2	++++	++	0	±	0
			R5	++	++	0	±	0
9	Left	0	R1	0	±	±	±	0
			R3	0	0	0	0	0
			R6	0	0	0	0	0
	Right	0.2±0.1	R1	0	±	+++	±	++
			R3	±	±	++++	0	++
			R6	±	++++	++++	+	++
20	Left	0	R4	0	0	0	0	0
			R7	0	0	0	0	0
			R8	0	0	0	0	0
	Right	0	R4	+	+++	++++	±	+
			R7	0	0	+	0	+
			R8	0	0	+++	0	++

[a]On day 0, the right ankle joint of each of eight rats was given an ia injection of 5 μg S. pyogenes PG-PS in 10 μl PBS. The left ankle received PBS alone. Nineteen days later each rat was given an iv injection of 100 μg of E. coli 0111:B4 LPS. On days 2, 9, and 20 following LPS injection, two or three rats were sacrificed and the ankle joints collected for histological examination.
[b]Mean joint score, ± 1 standard error, of all rats including those rats sacrificed for histology.
[c]The histological features included in the evaluation of each tissue section are described in Methods.

to S. pyogenes PG-PS. This reaction is significantly more severe and greater in incidence and duration than that seen in joints of naive rats or control joints previously injected with PBS.

S. typhimurium Re mutant glycolipid also induces a recurrence of arthritis in joints previously exposed to PG-PS, but has no apparent affect on PBS-injected control joints, suggesting that lipid A is the active moiety, and that the polysaccharide portion might be necessary for the induction of acute inflammation in naive or control joints.

We hypothesize that LPS, although usually associated with acute toxic reactions, might play a role in chronic arthritis through the exacerbation and perpetuation of inflammation initiated by certain peptidoglycan-containing polymers.

The mechanism by which LPS induces a recurrence of arthritis in joints previously exposed to PG-PS is not known, but may involve the shared biological activities of lipid A and peptidoglycan. Important in this regard may be the synergistic interaction of muramyl dipeptide and endotoxin observed by Ribi et al. in an in vivo model of antitumor activity (9) and by Butler and Friedman in in vitro adjuvanticity assays (2).

REFERENCES

1. S. Aoki and K. Ikuta, Immunopathological studies on experimental arthritis-lesions in synovial tissue during endotoxemia, Bull. Osaka Med. Sch. 14:99 (1968).

2. R. C. Butler and H. Friedman, Restoration of leukemia cell immune responses by bacterial products, in: "Current Chemotherapy and Infectious Disease," J. D. Nelson and C. Grassi, eds., American Society for Microbiology, Washington, D. C. (1980).

3. C. Chetty and J. H. Schwab, Endotoxin-like products of gram-positive bacteria, in: "Handbook of Endotoxin, Volume 1, Chemistry of Endotoxin," E. Th. Rietschel, ed., Elsevier Science Publishers, Amsterdam (1984).

4. W. J. Cromartie, J. G. Craddock, J. H. Schwab, S. K. Anderle, and C.H. Yang, Arthritis in rats after systemic injection of streptococcal cells or cell walls, J. Exp. Med. 146:1585 (1977).

5. R. E. Esser, S. A. Stimpson, W. J. Cromartie and J. H. Schwab, Reactivation of cell wall-induced arthritis by homologous and heterologous cell wall polymers, submitted for publication (1985).

6. A. Fox, R. R. Brown, S. K. Anderle, C. Chetty, W. J. Cromartie, H. Gooder, and J. H. Schwab, Arthropathic properties related to the molecular weight of peptidoglycan-polysaccharide polymers of streptococcal cell walls, Infect. Immun. 35:1003 (1982).

7. R. S. Jones and Y. Carter, Experimental arthritis. I. Morphologic alterations in the guinea pig after the parenteral injection of bacterial extracts, AMA Arch. Path. 63:472 (1957).

8. A. Nowotny, In search of the active sites in endotoxins, in: "Beneficial Effects of Endotoxins," A. Nowotny, ed., Plenum Press, New York (1983).

9. E. Ribi, R. Parker, S. M. Strain, Y. Mizuno, A. Nowotny, K. B. Von-Eschen, J. L. Cantrell, C. A. McLaughlin, K. M. Hwang, and M. B. Goren, Peptides as requirement for immunotherapy of the guinea pig line-10 tumor with endotoxins, Cancer Immunol. Immunother. 7:43 (1979).

10. E.Th. Rietschel, I. Schade, M. Jensen, H. W. Wollenwever, O. Luderitz, and S. G. Greisman, Bacterial endotoxins: chemical structure, biological activity and role in septicaemia, Scand. J. Infect. Dis. Supp 31:8 (1982).

11. D.E.S. Stewart-Tull, The immunological activities of bacterial peptidoglycans, Ann. Rev. Microbiol. 34:311 (1980).

ANTIBIOTICS, IMMUNITY AND IMMUNOENHANCEMENT

G. Pulverer[1], W. Roszkowski[2], H. L. Ko[1], K. Roszkowski[2], and J. Jeljaszewicz[3]

Institute of Hygiene, University of Cologne, West Germany[1]
Institute of Lung Diseases, Warsaw, Poland[2]
National Institute of Hygiene, Warsaw, Poland[3]

INTRODUCTION

In a previous paper (3) we have shown that a seven day chemotherapy can induce in BALB/c mice pronounced suppression of the immunity system. The most dramatic and long lasting immunosuppressive effect was caused by mezlocillin. Furthermore we could demonstrate that mezlocillin therapy modified the growth of an experimental murine tumor (1). The immunomodulatory effects of mezlocillin are obviously associated with significant changes in the endogenous intestinal microflora of BALB/c mice (2). The present report deals with the results of immunochemotherapy in BALB/c mice, using mezlocillin as a chemotherapeutic drug and a cell preparation of the strong immunomodulatory strain Propionibacterium granulosum KP-45 (4).

MATERIALS AND METHODS

Inbred male BALB/c mice eight to ten weeks old received subcutaneous injections of mezlocillin (Bayer AG) twice a day at ten hour intervals in a daily dose of 300 mg/kg body weight. After a three day mezlocillin therapy animals as well as controls (treated with phosphate buffer solution (PBS) instead of mezlocillin) were killed for the evaluation of their spleen weight as well as for the preparation of spleen cells. One mg of Concanavalin A was used for the stimulation of spleen cell proliferation (3). The spleen cell suspension was also analyzed for Ig-positive and theta antigen-positive cells using direct immunofluorescence. The evaluation of mouse sarcoma growth was done according to the techniques described previously (1). Sarcoma L-1 tumor was used for these experiments which arose spontaneously in the lung of a BALB/c mouse and was maintained in this species of mice. Doses of 1×10^5 viable L-1 tumor cells were given by s.c injections to mice. Propionibacterium granulosum (P.g.) and/or mezlocillin-therapy was started the same day. Fourteen days thereafter mice were killed and the local tumor growth was evaluated. P.g. KP-45 was administered over three days in the form of a suspension of killed microorganisms (4) either intraperitoneally (1 mg of dry mass per mouse) or orally (5 mg of dry mass per mouse).

Table 1. Relative Spleen Weight of BALB/c Mice Treated with Mezlocillin or with Mezlocillin and P. granulosum KP-45 (P.g.)

Experimental Group	Relative Spleen Weight** Mean ± SD	% of Control
Control	0.42 ± 0.08	100
Mezlocillin	0.28 ± 0.08*	67
Mezlocillin + P.g. i.p. 1 mg	0.87 ± 0.19*	207
Mezlocillin + P.g. orally 5 mg	0.53 ± 0.14	126

SD = standard deviation; * = p < 0.01
** = relative spleen weight = $\dfrac{\text{spleen weight}}{\text{body weight}}$ x 100

RESULTS

BALB/c mice treated for three consecutive days with mezlocillin showed a significant reduction of their spleen weight (Table 1). When mezlocillin therapy was combined with P.g. KP-45 either intraperitoneally or orally, the spleen weight reduction could be completely prevented and even reversed to a positive stimulatory effect. Mezlocillin therapy caused a significant reduction of the total number of spleen cells, not influencing the relative ratios of B- and T-lymphocytes (Table 2). P.g. application reversed the mezlocillin effect and the total number of spleen lymphocytes increased significantly.

Table 3 shows the experimental data concerning a stimulation of the spleen cell proliferation by concanavalin A. Mezlocillin caused a marked

Table 2. Composition of Spleen Cells from BALB/c Mice Treated with Mezlocillin or with Mezlocillin and P. granulosum KP-45 (P.g.)

Experimental Group	Total Number of cells x 10^6 mean ± SD	% of Ig Positive Cells mean ± SD	% of Theta Positive Cells mean ± SD
Control	148 ± 17	38.9 ± 2.5	41.3 ± 3.8
Mezlocillin	25 ± 8*	35.6 ± 4.3	37.8 ± 6.2
Mezlocillin + P.g. i.p. 1 mg	283 ± 56*	46.2 ± 7.3	41.1 ± 4.2
Mezlocillin + P.g. orally 5 mg	174 ± 34	41.8 ± 4.8	39.2 ± 5.1

SD = standard deviation; * = p < 0.01

Table 3. Proliferation of Spleen Cells from BALB/c Mice Treated with
Mezlocillin or with Mezlocillin and P. Granulosum KP-45 (P.g.)

Experimental Group	Concanavalin A cpm/10^6 spleen cells x 10^3 mean ± SD	% of Control
Control	27.9 ± 3.6	100
Mezlocillin	6.4 ± 1.9*	23
Mezlocillin + P.g. i.p. 1 mg	24.3 ± 3.2	87
Mezlocillin + P.g. orally 5 mg	31.2 ± 6.1	112

*$p < 0.01$; SD = standard deviation

reduction of the proliferative ability of spleen cells whereas P.g. was able
to abolish this mezlocillin effect.

In Table 4 the results of a three day mezlocillin and combined mezlo-
cillin-P.g. immunochemotherapy on the local growth of sarcoma L-1 are summarized. P.g. therapy was able to reverse the marked anti-tumor effect of
mezlocillin.

DISCUSSION

In our previous investigations we demonstrated that mezlocillin induces
in BALB/c mice a marked suppression of the cellular and humoral immune responses (3) as well as a modification of the transplantable sarcoma L-1
tumor growth (1). The data of the present study show that mezlocillin causes
a significant atrophy of the spleen, the total number of spleen cells being
reduced but the relative B and T-cell ratios not changed. The spleen cell
proliferative reactivity is suppressed, too. The mechanisms involved are
not yet clear, a possible role of the endogenous intestinal flora for the
mezlocillin-effect has been discussed (2). The simultaneous application of

Table 4. Local Sarcoma L-1 Tumor Growth in BALB/c Mice Treated with
Mezlocillin or with Mezlocillin and P. granulosum KP-45 (P.g.)

Experimental Group	Tumor Weight in mg mean ± SD	% of Control
Control	482 ± 96	100
Mezlocillin	220 ± 63*	46
Mezlocillin + P.g. i.p. 1 mg	386 ± 112	80
Mezlocillin + P.g. orally 5 mg	415 ± 108	86

*$p < 0.01$; SD = standard deviation

a cell preparation of <u>Propionibacterium granulosum</u> KP-45, a very potent
biological immunomodifier (4), is able to abolish completely these immunosuppressive effects of mezlocillin. Furthermore, P.g. therapy reversed the
indirect antitumor effect of mezlocillin. No explanation for these surprising effects can yet be presented, more extended experiments are needed.

REFERENCES

1. K. Roszkowski, H. L. Ko, W. Roszkowski, J. Jeljaszewicz, and G.
   Pulverer, Effects of cefotaxime, clindamycin, mezlocillin, and
   piperacillin on mouse sarcoma L-1 tumor, <u>Cancer Immunol</u>. <u>Immunother</u>.
   18:164 (1984).
2. K. Roszkowski, H. L. Ko, D. van der Waaij, W. Roszkowski, J. Jeljaszewicz, and G. Pulverer, The effect of selected antibiotics on the
   endogenous intestinal microflora and its consequences for experimental tumor growth in mice, <u>in</u>: "The Immunobiology and Immunopharmacology of Bacterial Endotoxins. Basic and Clinical Aspects," A.
   Szentivanyi and H. Friedman, eds., Plenum Publishing Corporation,
   New York (1986).
3. W. Roszkowski, H. L. Ko, K. Roszkowski, J. Jeljaszewicz, and G.
   Pulverer, Antibiotics and immunomodulation: effects of cefotaxime,
   amikacin, mezlocillin, piperacillin, and clindamycin, <u>Med</u>.
   <u>Microbiol</u>. <u>Immunol</u>. 173:279 (1985).
4. S. Szmigielski, W. Roszkowski, K. Roszkowski, H. L. Ko, J. Jeljaszewicz,
   and G. Pulverer, Experimental immunostimulation by propionibacteria,
   <u>in</u>: "Bacteria and Cancer," J. Jeljaszewicz, G. Pulverer, and W.
   Roszkowski, eds., Academic Press, New York (1982).

THE EFFECT OF SELECTED ANTIBIOTICS ON THE ENDOGENOUS INTESTINAL MICROFLORA

AND ITS CONSEQUENCES FOR EXPERIMENTAL TUMOR GROWTH IN BALB/C MICE

K. Roszkowski[1], H. L. Ko[2], D. van der Waaij[3], W. Roszkowski[1]
J. Jeljaszewicz[4], and G. Pulverer[2]

Institute of Lung Diseases, Warsaw, Poland[1]
Institute of Hygiene, University of Cologne, West Germany[2]
University Hospital, Groningen, The Netherlands[3]
National Institute of Hygiene, Warsaw, Poland[4]

INTRODUCTION

In previous studies we have shown that some antibiotics may influence the growth of experimental tumors. Concerning rifampicin an enhancing effect on the tumor development was combined with the impairment of some parameters of antitumor immunity (6). A marked suppression of the antitumor immunity was also observed in mezlocillin treated mice (5). However, this antibiotic caused a surprising antitumor effect when given during early tumor growth. Because a direct cytotoxic activity of mezlocillin could not be demonstrated, indirect mechanisms have been suggested for this mezlocillin effect. The present paper deals with the influence of mezlocillin and other selected antibiotics on the endogenous intestinal flora of BALB/c mice and on the local growth of sarcoma L-1 tumor.

MATERIALS AND METHODS

Inbred male BALB/c mice eight to ten weeks old have been used for all experiments. Antibiotics were given twice a day (s.c.) at ten hour intervals over a period of ten days in the following doses: mezlocillin, 300 mg/kg body weight; piperacillin, 260 mg/kg body weight; cefotaxime, 100 mg/kg body weight; gentamicin, 4 mg/kg body weight; clindamycin, 30 mg/kg body weight. Sarcoma L-1 tumor was used which arose spontaneously in the lung of a BALB/c mouse and was maintained in this mouse species. Doses of 1×10^5 viable L-1 tumor cells were injected subcutaneously to mice the same day when antibiotic therapy was started. All animals were killed the day after completing antibiotic therapy, and the cecum was isolated for the evaluation of weight and endogenous intestinal flora. The procedures of cultivation, isolation and identification of bacteria are in accordance with previously described methods (7). Local L-1 tumor weight was recorded eleven days after implantation (5).

Some modifications of experiments have been performed: mezlocillin therapy in a daily dose of 300 mg/kg body weight was given for three, seven and ten days, respectively. Mezlocillin was administered not only subcutaneously, but also orally (daily dose, 150 mg/kg body weight). A three day

Table 1. Influence of 10 Day Antibiotic Treatment on the Endogenous Intestinal Flora and on the Local Tumor Growth in BALB/c Mice

Antibiotic Used	Log # of Bacteria/1g feces Mean ± SD			Relative Cecal Wt. mean ± SD	Tumor Wt. in mg mean ± SD
	Entero-bacteriaceae	S. faecalis	S. viridans		
Control	4.9 ± 0.9	5.5 ± 1.2	5.2 ± 1.6	2.7 ± 0.2	710 ± 149
Cefotaxime	0.0 ± 0.0*	6.2 ± 1.2	6.4 ± 1.8	2.9 ± 0.3	639 ± 127
Clindamycin	5.3 ± 1.1	5.7 ± 0.9	6.1 ± 1.9	3.0 ± 0.3	660 ± 136
Gentamicin	4.4 ± 0.8	6.1 ± 1.7	5.8 ± 1.3	2.5 ± 0.1	723 ± 142
Mezlocillin	0.0 ± 0.0*	0.0 ± 0.0*	0.0 ± 0.0*	5.6 ± 0.7*	350 ± 81*
Piperacillin	3.5 ± 1.4	0.0 ± 0.0*	0.0 ± 0.0*	3.6 ± 0.4	611 ± 164

*$p < 0.01$; SD = standard deviation

therapy was used for this comparison. Half of the animals were killed the day after finishing the mezlocillin therapy for the evaluation of the intestinal flora; the remaining mice were killed 14 days after tumor implantation for recording the local tumor weight. When investigating the dynamics of the mezlocillin effect mice were killed 4 hours, 24 hours and 72 hours after the first mezlocillin injection as well as one, three and seven days after finishing the three day mezlocillin treatment. In these animals the concentration of E. coli endotoxin in the cecal content was determined also using the limulus amoebocyte lysate test (4).

RESULTS

As shown in Table 1, mezlocillin, piperacillin and cefotaxime showed a marked influence on the endogenous intestinal flora of BALB/c mice after a

Table 2. Effect of Mezlocillin on the Endogenous Intestinal Flora and Tumor Growth in BALB/C Mice in Relation to the Duration of Treatment

Experimental Group	Log # of Bacteria/1g Feces Mean ± SD			Relative Cecal Wt. mean ± SD	Tumor Wt. in mg mean ± SD
	Entero-bacteriaceae	S. faecalis	S. viridans		
Control	4.9 ± 0.9	5.5 ± 1.2	5.2 ± 1.6	2.7 ± 0.2	719 ± 149
Mezlocillin					
10 days	0.0 ± 0.0*	0.0 ± 0.0*	0.0 ± 0.0*	5.6 ± 0.7*	350 ± 81*
7 days	0.0 ± 0.0*	0.0 ± 0.0*	0.0 ± 0.0*	5.4 ± 0.8*	381 ± 94*
3 days	0.0 ± 0.0*	0.0 ± 0.0*	0.0 ± 0.0*	3.5 ± 0.3*	320 ± 118*

*$p < 0.01$; SD = standard deviation

Table 3. Comparison of the Effect of Mezlocillin on the Endogenous
Intestinal Flora and the Local Tumor Growth after 3 Days
Systemic or Oral Treatment

Route of Administration and dose	Log # of Bacteria/1g feces Mean ± SD			Relative Cecal Wt. mean ± SD	Tumor Wt. in mg Mean ± SD
	Entero-bacteriaceae	S. faecalis	S. viridans		
Control	4.3 ± 0.7	5.1 ± 1.3	4.6 ± 0.8	2.4 ± 0.2	854 ± 173
Mezlocillin 300 mg/kg b.w. subcutaneously	0.0 ± 0.0*	0.0 ± 0.0*	0.0 ± 0.0*	4.6 ± 0.6*	432 ± 106*
Mezlocillin 150 mg/kg b.w. orally	0.0 ± 0.0*	0.0 ± 0.0*	0.0 ± 0.0*	5.2 ± 0.8*	331 ± 68*

*p < 0.01; SD = standard deviation

ten day treatment. No changes were observed after therapy with clindamycin
or gentamicin. Cefotaxime caused a complete eradication of Enterobacteriaceae, piperacillin on the other hand produced the absence of aerobic gram-
positive cocci. Mezlocillin was the only drug examined which eliminated
gram-negative as well as gram-positive aerobes. Another surprising mezlocillin effect after a ten day therapy was that the relative weight of the
cecum was significantly increased, whereas the local L-1 tumor growth was
significantly inhibited.

Table 2 demonstrates that a three day therapy of mezlocillin is already
sufficient for causing all effects mentioned above.

In next experiments it was investigated if oral therapy of mezlocillin
is as effective as systemic application in influencing the intestinal flora
of BALB/c mice and the local growth of L-1 tumor. For the oral treatment a
daily mezlocillin dose of 150 mg/kg body weight was chosen because we found
that 300 mg/kg body weight mezlocillin given subcutaneously is responsible
for exactly the same cecal mezlocillin concentration as achieved after 150
mg/kg body weight given orally. On the other hand, it must be stated that
we could not find any detectable mezlocillin serum level when this drug was
given orally. The data in Table 3 prove that oral and systemic administration of mezlocillin cause comparable effects.

Table 4 summarizes the dynamics of the changes of the intestinal flora
during and after mezlocillin therapy. As soon as four hours after the first
mezlocillin injection Enterobacteriaceae and S. viridans disappeared. Anaerobic bacteria could no longer be cultivated after 24 hours mezlocillin
treatment and S. faecalis was not present three days after the beginning of
mezlocillin therapy. Twenty-four hours after the end of mezlocillin application all aerobes and anaerobes were still absent in the intestine followed
by a slow recovery in the subsequent days. Even seven days after finishing
chemotherapy the number of anaerobes in the intestine was definitely lower
in comparison to non-treated controls. The concentration of E. coli endotoxin in cecal content showed a decreasing tendency through the mezlocillin
therapy period. A significant enhancement of E. coli endotoxin could be
seen during the recovery phase, i.e., three to seven days after the last
mezlocillin injection.

Table 4. Changes of the Endogenous Intestinal Flora and E. coli Endotoxin Concentration in Cecum During and After Mezlocillin Treatment

	Non-treated Control	During Treatment Time from Start			After Treatment Time from Completion		
		4 h	24 h	3 days	24 h	3 days	7 days
Log no. of bacteria/1g feces mean ± SD							
Enterobacteriaceae	4.1±1.1	0.0±0.0*	0.0±0.0*	0.0±0.0*	0.0±0.0*	2.3±1.9	6.9±1.3
S. faecalis	5.8±1.6	3.8±0.6	3.3±0.4	0.0±0.0*	0.0±0.0*	6.9±1.2	6.8±1.1
S. viridans	5.3±1.2	0.0±0.0*	0.0±0.0*	0.0±0.0*	0.0±0.0*	5.8±1.4	4.6±1.8
Evaluation of anaerobic bacteria growth	++++	+++	-	-	-	+	+++
E. coli endotoxin concentration in cecum in EU/ml; mean ± SD	0.7±0.2	1.3±0.4*	0.2±0.1	0.0±0.0	0.0±0.0	5.6±1.7*	7.8±3.1

* $p < 0.1$
SD = standard deviation
- = no growth
+ - ++++ = relative growth intensity

DISCUSSION

Mezlocillin levels in murine serum after parenteral administration of 300 mg/kg body weight daily are relatively low in comparison to the known situation in humans. The elimination of mezlocillin from mouse serum is relatively fast also (5). On the other hand, mezlocillin concentrations in the digestive tract of mice are rather high and long-lasting. These facts may explain the strong mezlocillin effects observed on the endogenous intestinal flora of BALB/c mice. The shortest therapy time necessary for the complete elimination of bacterial intestinal flora from the mouse cecum has been demonstrated to be three days. This mezlocillin treatment time has also been sufficient for a marked inhibition of the local growth of sarcoma L-1 tumor in BALB/c mice.

We have now demonstrated that oral application of mezlocillin over three days is inducing the same effects, namely elimination of intestinal flora and inhibition of local tumor growth. Because of no detectable mezlocillin levels in serum it seems that there exists no direct antitumor activity of this drug in vivo. An indirect mechanism of this local tumor inhibition is more likely and we speculate that there exist some correlations between intestinal flora inhibition and antitumor effects of mezlocillin. We do not think that E. coli endotoxin is involved in this effect (although it is known that endotoxin possesses antitumor activity), because local tumor growth was already inhibited just after finishing mezlocillin therapy when E. coli endotoxin was not yet detectable in the cecum.

It was observed that bacteria normally growing in the digestive tract can influence the proliferation of intestinal epithelium (1) and can penetrate the abdominal lymph nodes (2). The relevance of these phenomena are not known, but one can speculate that the presence of normal endogenous bacterial flora in the cecum of BALB/c mice may be a potent stimulus for the proliferation of normal nontransformed cells (proven in the case of plasma cells) (3) as well as for the proliferation of certain malignant cells.

REFERENCES

1. G. D. Abrams, H. Bauer, and H. Sprinz, Influence of the normal flora on mucosal morphology and cellular renewal in the ileum. A comparison of germfree and conventional mice, Lab. Invest. 12:335 (1963).
2. R. D. Berg and A. W. Garlington, Translocation of certain endogenous bacteria from the intestinal tract to the mesenteric lymph nodes and other organs in gonobiotic mouse model, Infect. Immun. 23:403 (1979).
3. P. A. Crabbe, H. Bazin, H. Eyssen, and J. F. Heremans, The normal microbial flora as a major stimulus for proliferation of plasma cells synthesizing IgA in the gut, Int. Arch. Allergy 34:362 (1968).
4. A. Gardi and G. R. Arpagaus, Improved microtechnique for endotoxin assay by the limulus amebocyte lysate test, Analyt. Biochem. 109:382 (1980).
5. K. Roszkowski, H. L. Ko, W. Roszkowski, J. Jeljaszewicz, and G. Pulverer, Effects of cefotaxime, clindamycin, mezlocillin, and piperacillin on mouse sarcoma L-1 tumor, Cancer Immunol. Immunother. 18:164 (1984a).
6. W. Roszkowski, R. Lipinsaka, K. Roszkowski, J. Jeljaszewicz, and G. Pulverer, Rifampicin-induced suppression of antitumor immunity, Med. Microbiol. Immunol. 172:197 (1984b).
7. D. Van der Waaij, J. M. Berghuis-DeVries, and J.E.C. Lekkerkerk, Colonization resistance of the digestive tract of mice during systemic antibiotic treatment, J. Hyg. 70:605 (1972).

RESTORATION OF ANTIBODY RESPONSIVENESS BY ENDOTOXIN IN RETROVIRUS-

IMMUNODEPRESSED MICE: ROLE OF MACROPHAGES

Mauro Bendinelli, Donatella Matteucci,
Anna Maria Giangregorio, and Pier Giulio Conaldi

Institute of Epidemiology, Hygiene and Virology
University of Pisa
Pisa, Italy

INTRODUCTION

Exogenous retroviruses are the cause of severe immunodeficiency syndromes in humans and animals (4). Studies in our laboratory have focused on elucidating the mechanisms underlying the immunodeficiency induced in mice by retroviruses of the Friend leukemia complex (FLC). The pathologies produced by the two viral components of this complex are considerably different. Genetically susceptible mice inoculated with the entire FLC develop a rapidly progressing erythroblastosis of the spleen followed by erythroleukemia within a few weeks postinfection--a series of changes thought to be initiated by the rapidly transforming replication-defective spleen focusforming virus. In contrast, the slow transforming replication-competent helper virus (F-MuLV), when injected alone into the same strains of mice, causes leukemias of varied histotype after a latency of several months, the only change noticeable during the long incubation period being a slight hyperplasia of the spleen (11,32).

Yet, both viral preparations cause profound derangements in the immunological responsiveness of the infected hosts. Thus, for example, the only major difference observed in the effects on the splenic antibody response to a standard antigen such as sheep red cells (SRC) is that the suppression caused by FLC progresses until the animals die (in most instances due to rupture of the spleen), while that caused by F-MuLV peaks within ten to 15 days and then subsides (but the responsiveness remains at about half the normal level throughout life). Indeed, it has been suggested that F-MuLV, which represents 90-99% of the infectivity found in FLC preparations, plays an important role in the early immunodepressive properties of the viral complex (7).

In any case, the parallel use of the two viral preparations has afforded us with the opportunity of comparing the immunodepressive effects of two retroviruses that are structurally and antigenically identical, but substantially different in other important aspects including the rapidity of neoplasia induction. This has, for example, facilitated the discrimination between virus and tumor-induced immunodepressive changes and helped in identifying a major functional deficit at the level of the antigen-presenting accessory cells (5).

465

Bacterial endotoxins are powerful inducers of adaptive biological responses (26). Extensive investigations have shown that the administration
of lipopolysaccharide (LPS) modifies immune functions in many different
ways. For example, the production of antibody in vitro can be either augmented or suppressed depending on the dose, the timing relative to immunization, the density of responder cells, the strain of donor mice, etc. (20).
Such changes are generally believed to result from the direct interaction
of LPS with various classes of immunocompetent cells and/or through indirect
mechanisms. Thus, enhancement of the humoral immune response of normal
mice has been attributed to the sum of the direct polyclonal stimulation of
B cells, the activation of macrophages, and the induction of soluble helper
factors (29).

The data dealing with possible effects of bacterial endotoxins on
infection and tumorigenesis by leukemia viruses are few. Existing evidence,
is, however, indicative that LPS can substantially influence host response
to retroviruses. For example, inoculation of mice with LPS extracted from
several gram-negative bacteria was seen to augment the spleen focus-forming
efficiency of FLC without affecting the eventual tumor-dependent lethality.
The effect was influenced by varying the timing of LPS inoculation as well
as by the genetic constitution of the host, and appeared to be primarily
related to the lipid moiety of LPS and was attributed to an expansion of
the cellular targets for viral replication and transformation (31). In
other studies, in vitro treatment with LPS increased the replication of
exogenous leukemia viruses, including F-MuLV, and the expression of endogenous retroviruses by murine lymphoid cells (19,28,30). This activity was
attributed to an enhanced permissiveness of B cells consequent to their
polyclonal stimulation. On the other hand, there are also indications that
retroviruses modify host responses to bacterial endotoxin. For example,
mice infected with leukemia viruses exhibited a markedly increased sensitivity to the lethal action of LPS (8).

The present report focuses on the effects of endotoxin on the immunological reactivity of mice severely immunodepressed by FLC or F-MuLV. We
first review findings from our and other groups that provide the rationale
for the subsequent experiments. Briefly, these results have established
that in infected mice the blastogenic and antibody responses to LPS are
relatively spared and that LPS administration can also restore antibody
response to other antigens. We proceed to show that in vitro such beneficial
effects are produced also by doses of LPS which do not effect appreciable
polyclonal activation of lymphocytes, and then examine the role of macrophage
activation and macrophage-released soluble factors in the phenomenon.

RESULTS

Antibody Responsiveness to LPS

Humoral responses are especially prone to suppression by FLC viruses
and LPS is one of the several antigens which have been shown to elicit depressed antibody responses in infected mice. Yet, a constant feature of
the LPS-specific antibody response is that it is relatively spared as compared to those elicited by other antigens. For example, in BALB/c mice
immunized one week after F-MuLV infection the peak LPS-specific plaque-
forming cell (PFC) response in the spleen averaged 50% of control or more,
while that directed to SRC was less than 10%. Similarly, in leukemia-resistant C57BL/6 mice injected with FLC the anti-SRC PFC response was suppressed
by over 70% but that against LPS remained virtually unchanged (3).

In the virus-mouse strain combinations which lead to the early development of erythroblastosis and leukemia (such as BALB/c mice inoculated with

466

Table 1. Effect of FLC and F-MuLV Infections on the
Blastogenic Response of Spleen Cells to LPS

Virus Inoculation[§]	Spleen Weight*	Anti-SRC PFC/Spleen†	Blastogenic Activity – LPS	+ LPS**
none	123	100	4,348 ± 360	38,227 ± 1,240
FLC, day – 5	218	13	5,187 ± 101	40,043 ± 1,702
FLC, day – 10	716	5	3,221 ± 1,105	29,677 ± 2,765
F-MuLV, day – 5	159	18	6,345 ± 277	37,065 ± 2,056
F-MuLV, day – 10	186	31	7,343 ± 652	43,698 ± 812

[§] In this and all the following experiments, two-month old inbred BALB/c mice from our own colony were infected intravenously (iv) with 10^3 PFU of FLC or with $10^{2.5}$ PFU of F-MuLV. The viral stocks used were LDH virus-free and were prepared as described (7).

* Average weight in mg.

† PFC were assayed four days after iv immunization with 2×10^7 SRC. The results are expressed as percent of control. Means ± S.E. of controls: $102 \pm 31 \times 10^3$ PFC/spleen (5 mice/group).

**The LPS used in this and all the following experiments was a preparation extracted from E. coli 0127: B8 by the Westphal technique (Difco, Detroit Michigan), stored at -30°C in stock solution at 1 mg/ml in sterile saline and freshly diluted in medium with the help of a brief sonication. For the blastogenic assay 10^6 spleen cells were incubated with or without 50 µg/ml LPS for three days. Tritiated thymidine was added 16 hr before harvesting. The results are expressed as mean ± S.E. cpm/well (4 mice/group; 3 wells/mouse).

FLC), the suppression of the anti-LPS response was pronounced but delayed, suggesting that it might depend at least in part on the rapid proliferation of pathologic cells that occurs in the spleen of these mice. Dracott et al., who have examined the response of BALB/c mice to LPS in the very early stages of FLC infection, have reached similar conclusions (15). The considerable resistance of LPS-specific antibody responsiveness to suppression by viruses of the FLC markedly contrasts with findings showing that antibody production to other T-independent antigens such as pneumococcus polysaccharide and TNP-ficoll is effectively suppressed by infection (15; unpublished results).

Blastogenic Response to LPS

The proliferative response of spleen cells to LPS has also proved relatively resistant to infection. For example, in the experiment presented in Table 1 splenocytes of F-MuLV-infected BALB/c mice showed no reduction in thymidine uptake, and in mice infected with FLC the loss was moderate and late as compared to that exhibited by humoral responses. The remarkable resistance of LPS-induced blastogenesis to suppression by FLC has been noted also by Dracott et al. (15) and by Genovesi et al. (17).

Adjuvanticity of LPS in Vivo

LPS is a powerful adjuvant when given at selected times relative to the immunizing antigen (26). It seemed therefore possible that the

Table 2. Effect of LPS on the Antibody Response in Vitro of Spleen Cells from Infected Mice.

Responder cultures§	Exp. 1 (LPS: 2 µg/ml)	Exp. 2 (LPS: 0.2 µg/ml)	Exp. 3 (Alk-LPS*: 0.2 µg/ml)
Normal	7,503 ± 242	947 ± 231	13,816 ± 2,306
+ LPS	17,510 ± 515 (2.3)	1,745 ± 302 (1.8)	13,420 ± 250 (1.0)
FLC-infected†	459 ± 161	235 ± 96	4,840 ± 234
+ LPS	4,259 ± 280 (9.3)	2,530 ± 267 (10.8)	7,517 ± 1,034 (1.5)
F-MuLV-infected†	459 ± 96	327 ± 67	2,270 ± 253
+ LPS	9,826 ± 231 (21.4)	3,460 ± 365 (10.6)	2,075 ± 467 (0.9)

§In these and in all the following experiments spleen cells from four to six BALB/c mice were immunized with SRC as described (6), except that 1.5×10^6 SRC were used (5 wells/group). The indicated doses of LPS or Alk-LPS were added at the initiation of the cultures. Direct PFC assayed 5 days later and expressed as mean PFC per culture ± S.E. (in parenthesis, the stimulation index).

*Alk-LPS was prepared according to Niwa (25). It was amitogenic at concentration up to 100 µg/ml.

†In this and all the following experiments, mice were infected eight days before sacrifice.

Table 3. Effect of Polymyxin B (PB) on the Ability of LPS to Enhance the Antibody Response in Vitro of Spleen Cells from Infected Mice

| Materials Added | Responder Cultures | |
	Normal	F-MuLV-Infected
None	4,200 ± 651	500 ± 153
LPS (0.2 µg/ml)	7,533 ± 353 (1.8)	3,400 ± 473 (6.9)
PB§ (5 µg/ml)		600 ± 20 (1.2)
LPS (0.2 µg/ml) + PB (5 µg/ml)		1,067 ± 426 (2.1)

§LPS was incubated with PB (Sigma, St. Louis, MO) at room temperature for one hour prior to addition to responder cultures.

relative resistance of the LPS-specific PFC response to depression was due to the intrinsic adjuvanticity of LPS. The fact that the antibody response to delipidized LPS (Alk-LPS) had proved more susceptible to suppression than that elicited by the intact molecule supported this concept.

In experiments designed to test such possibility, a wide range of LPS concentrations given at the same time as the antigen enhanced the antibody response to SRC much more dramatically in infected than in control animals. For example, the administration of 30 µg of LPS induced a 24-fold increase in the anti-SRC PFC response of FLC-infected C57BL/6 mice as compared to a four-fold increase in the uninfected controls (3). The adjuvanticity of LPS has been shown to depend on the dose of antigen used for immunization (29). However, the augmented adjuvanticity of LPS observed in infected mice did not appear to be due to a shift in the optimal dose of antigen since essentially similar results were observed using varied doses of SRC (unpublished results).

Adjuvanticity of LPS in Vitro

When the above experiments were repeated by exposing spleen cells to the direct action of LPS in vitro the difference between the adjuvant activities observed in infected and control animals was even more striking. The administration of LPS enhanced the numbers of PFC produced by cultures of infected BALB/c mice immunized with optimal doses of SRC ten to 20-fold, while in the uninfected counterparts the enhancement rarely exceeded a factor of two. In addition, dose-response experiments showed that submitogenic doses of LPS (0.2 µg/ml) sufficed to significantly increase the response of infected spleen cells. In similar conditions, Alk-LPS had no effects (Table 2). Furthermore, when similar experiments were done in the presence of polymixin B (PB), an antibiotic known to neutralize many biological activities of LPS (27), the potentiating effect was markedly reduced (Table 3). These results indicated that most probably potentiation of infected spleen cell responses was due to the lipid moiety of the endotoxin molecule.

Since in most instances the antibody response of infected cultures stimulated with LPS was as high or even higher than that of control cultures given the same dose of LPS, it seems appropriate to define the effects of LPS on the antibody responsiveness of infected mice as "reconstitutive" or "beneficial".

Role of Macrophages in the Beneficial Effects of LPS

Results in the previous section had shown that doses of LPS which did not induce appreciable mitogenic activation of lymphocytes from normal or infected mice were beneficial to the responsiveness of infected spleen cells. This suggested that the effects of LPS might not be directly on B cells but instead be mediated via other cell type(s). There is solid evidence that impairment of macrophages and other accessory cells plays a key part in the genesis of retrovirus-induced immunodeficiencies (2). On the other hand, macrophages have long been known to respond to LPS in several ways, and are currently attributed an increasingly crucial role in the adjuvanticity of LPS as studied in normal mice (23,24). Thus, we investigated whether these cells might be involved in the beneficial effects exerted by LPS on the responsiveness of infected cells.

In a first series of experiments we looked at the effects of LPS on the ability of normal macrophages to reconstitute the response of infected spleen cells. Previous studies had shown that, proteose peptone-elicited peritoneal macrophages are very active in doing so. As few as 10^5 such cells restored the antibody response of 10^7 infected spleen cells to normal levels. In contrast, resident peritoneal and spleen macrophages had very little activity, if any (6). However, when preincubated for two hours in the presence of LPS, these cells caused a substantial increase in the PFC response of infected spleen cells (Table 4). In previous experiments, intraperitoneal inoculation of donor mice with LPS had caused little or no change in such behavior of peritoneal macrophages harvested three to four days later (6). At present, we have no explanation for this discrepancy, apart from the complex dose and timing relationships which reportedly regulate macrophage responses to LPS, including surface Ia expression and antigen presentation (34).

A more direct indication for the involvement of macrophage was obtained in cross-recombination experiments designed to evaluate the ability of macrophage-rich adherent cells derived from infected spleens to restore the responsiveness of macrophage-depleted nonadherent cells from normal spleens. Consistent with previous results (6), adherent cells from F-MuLV-infected spleens did not cooperate normally with nonadherent spleen cells of normal mice. However, pretreatment with LPS for two hours in vitro completely reversed this deficit (Experiment 1 in Table 5). Interestingly, LPS-treated infected macrophages showed a high efficiency also when the responder cultures contained PB at a dose which should have blocked the action of any residual LPS (Experiment 2 in Table 5). It must be mentioned, however, that attempts to reproduce these results using cells derived from FLC-infected mice failed. This is tentatively attributed to the fact that spleen infiltration by neoplastic cells in such mice can considerably hinder the cell separation procedures.

Role of Macrophage-Produced Factors

Previous studies by Butler et al. had shown that factors released by LPS-stimulated lymphoreticular cells enhance the response of FLC-infected spleen cells (10). In preliminary experiments we confirmed and extended these findings. Supernatants conditioned from LPS-stimulated uninfected spleen cells potentiated the PFC response of FLC- or F-MuLV-infected responder cultures much more effectively than that of control cells. Interestingly, on many occasions the restorative activity of the supernatant was even higher than that achieved by the direct addition of LPS. In contrast, culture fluids from unstimulated or Alk-LPS-stimulated cells had no activity or were suppressive (data not shown).

Table 4. Effect of Pretreatment with LPS on the Ability of Normal Spleen and Resident Peritoneal Macrophages to Restore the Antibody Responsiveness of Spleen Cells from Infected Mice

Responder Cultures	Normal Macrophages Added§		
	None	Peritoneal	Peritoneal LPS-treated
Exp. 1			
Normal	5,083 ± 501	5,720 ± 640 (1.1)	6,450 ± 519 (1.3)
FLC-infected	819 ± 37	805 ± 25 (1.0)	1,687 ± 127 (2.1)
F-MuLV-infected	435 ± 144	510 ± 188 (1.2)	2,055 ± 120 (4.7)
	None	Splenic	Splenic LPS-treated
Exp. 2			
Normal	2,920 ± 162	3,270 ± 334 (1.1)	1,260 ± 349 (0.5)
FLC-infected	1,047 ± 124	830 ± 153 (0.8)	2,400 ± 96 (2.3)

§Macrophages were separated by adherence as described (6). Ninty-five percent or more of the adherent cells resembled macrophages morphologically, phagocytozed latex beads, were FC‡, Thy-1⁻, and Lyt-2⁻. Peritoneal (5 x 10^5) or splenic macrophages (10^6) from normal BALB/c mice were incubated with or without 10 µg/ml LPS for two hours, washed thoroughly three times, and added to 10^7 responder cells at the initiation of culture.

Table 5. Effect of Pretreatment with LPS on the Ability of Adherent Spleen Cells from Infected Mice to Cooperate with Nonadherent Spleen Cells of Normal Mice

Responder Cultures[§]				
Normal nonadherent	Normal Adherent	F-MuLV-infected adherent	Exp. 1	Exp. 2[*]
10^7	--	--	295 ± 41	418 ± 55
--	10^6	--	20 ± 0	50 ± 2
10^7	10^6	--	3,840 ± 231	5,720 ± 418
--	--	10^6	100 ± 30	50 ± 0
10^7	--	10^6	612 ± 56	918 ± 101
10^7	--	10^6 + LPS**	4,276 ± 424	5,180 ± 703

[§] Spleen cells from normal and 7-day infected BALB/c mice were separated into adherent and nonadherent fractions and recombined as indicated.
[*] In experiment 2 the responder cultures contained polymyxin B (5 µg/ml).
**The cells were incubated with LPS (10 µg/ml) for two hours and then washed thoroughly before mixing with nonadherent cells.

Macrophage-produced factors appear to be important determinants of the adjuvant action exerted by LPS in normal mice (13,18,33). We, therefore sought to establish whether similar factors are involved in the ability of LPS-conditioned supernatants to restore the responsiveness of infected spleen cells. As shown by Table 6, supernatants generated under similar conditions of LPS stimulation by 10^6 95% pure splenic macrophages or by 10^7 unfractionated spleen cells were equally active. Also in these experiments the potentiation of infected responder cells was much higher than that of the uninfected counterparts. Again, culture fluids generated in similar conditions except for the absence of LPS had no effect or were inhibitory (data not shown).

Although the supernatants produced by LPS-stimulated macrophages were active at dilutions such that the amount of LPS present was supposedly too low to exert detectable effects, it was still possible that their action was due to carryover of LPS or to macrophage-processsed forms of LPS. Reportedly, LPS may be retained in biologically active form by reticuloendothelial cells for prolonged periods of time (9) and be processed by macrophages in such a way that its mitogenicity is greatly increased (16). However, the activity of macrophage-generated supernatants was only marginally reduced by incubation with PB for one hour before addition to the responder cultures (Table 7), thus excluding the possibility that it was solely due to the residual LPS. This conclusion was corroborated by experiments in which LPS-resistant C3H/HeJ mice were used for assaying the supernatants (data not shown).

Effect of Exogenously Added IL-1

IL-1 is presently considered a key factor in the ability of macrophages to support antibody synthesis in cultures of normal murine lymphoid cells (14) and several studies have shown that the soluble mediator secreted by LPS-treated macrophages which augments the antibody response of uninfected

Table 6. Effect of Supernatants from LPS-treated Normal Lymphoid Cells on the Antibody Response of Spleen Cells from Infected Mice

Cell Source	Supernatant[§] Final Concentration	Responder Cultures Normal	FLC-infected	F-MuLV-infected
Total spleen (10^7/ml)	10%	0.9*	11.2	18.7
	1%	1.1	6.4	ND
Adherent spleen (10^6/ml)	10%	0.5	8.0	12.4
	1%	1.6	7.3	6.7
Peritoneal, resident (10^7/ml)	1%	2.2	15.0	9.6

[§]Supernatants were generated by incubating the indicated cells from normal mice in RPMI 1640 medium supplemented with 10% fetal calf serum and 10 µg/ml LPS in a humidified environment of 6% CO_2 in air. After one (peritoneal cells) or five days (spleen cells) at 37°C, the supernatants were clarified by centrifugation, dialyzed, filtered through 22 µm filters (Millipore, Bedford, MA), and frozen at 75°C in small aliquots. The supernatants were added to the responder cells at the initiation of the cultures.
*Ratio between numbers of SRC-specific PFC produced in the presence and in absence of the indicated supernatants.

spleen cells in vitro is most likely IL-1 (13,18,33). To verify the possibility that the beneficial effects of LPS-conditioned supernatants on the antibody responsiveness of infected spleen cells were similarly due to this mediator, we studied the effect of adding two different preparations of purified IL-1 to infected responder cultures. The results of two representative experiments are shown in Table 8. Over a wide range of concentrations IL-1 had no appreciable effects on the PFC response of infected spleen cells.

IL-1 Production by Macrophages of Infected Mice

The results in the previous section seemed to exclude that LPS-conditioned macrophage supernatants were active because they were capable of

Table 7. Effect of Polymyxin B (PB) on the Ability of Supernatant from LPS-treated Macrophages to Enhance the Antibody Response in Vitro of Spleen Cells from Infected Mice

Materials Added	Responder Cultures Normal	F-MuLV-infected
None	6,033 ± 651	1,100 ± 326
Supernatant 1%		3,900 ± 651 (3.5)
Supernatant 1% + PB		3,100 ± 252 (2.8)
PB alone		867 ± 133 (0.8)

[§]Supernatant was incubated with 5 µg/ml PB at room temperature for one hour prior to addition to responder cultures.

473

Table 8. Effect of Two Preparations of Purified Interleukin 1 (IL-1) on the Antibody Response of Spleen Cells from Infected Mice

Units of IL-1 Added	Responder Cultures Normal	FLC-infected	F-MuLV-infected
Exp. 1[§]			
--	18.700 ± 3.089	4,987 ± 481	2,449 ± 130
0.25	15,070 ± 1,655 (0.8)	5,133 ± 481 (1.0)	2,486 ± 518 (1.0)
0.50	11,000 ± 1,103 (0.6)	5,324 ± 217 (1.1)	1,775 ± 128 (0.7)
1.0	18,260 ± 1,324 (1.0)	8,727 ± 265 (1.8)	2,295 ± 124 (0.9)
2.0	11,165 ± 275 (0.6)	3,549 ± 426 (0.7)	2,442 ± 322 (1.0)
Exp. 2*			
Control fluid	12,700 ± 603	2,300 ± 158	4,417 ± 501
5	15,267 ± 1,303 (1.2)	1,833 ± 203 (0.8)	2,767 ± 372 (0.6)
10	18,733 ± 546 (1.5)	2,233 ± 376 (1.0)	ND
15	11,467 ± 353 (0.9)	2,800 ± 231 (1.2)	3,700 ± 451 (0.8)

[§]Human IL-1 purified from S. albus-stimulated human monocytes by immuno-absorption chromatography (specific activity: >100 U/ml) was purchased from Koch-Light (Haverhill, UK). The indicated amounts were added at the initiation of the cultures.
*Recombinant mouse IL-1 (specific activity: >10^6) was obtained from Hoffman-LaRoche (Nutley, NJ). Control fluid consisted of nonrecombinant bacterial extract in 5M guanidine-HCl solvent.

compensating a defect in IL-1 synthesis by infected cells. In these experiments, we examined this aspect further by comparing the levels of IL-1 produced by adherent spleen cells derived from normal and infected mice under conditions similar to those used to produce the LPS-conditioned supernatants. IL-1 activity was measured by the thymocyte proliferation assay (21). The proliferation of thymocytes induced by the supernatants generated from FLC- or F-MuLV-infected cells was either normal or increased (Figure 1). Separate experiments excluded that the enhanced thymocyte proliferation was due to costimulation by viral products present in the former supernatants (data not shown). It was therefore, concluded that FLC- and F-MuLV -infected macrophages have no deficits in the production of IL-1. This is in keeping with findings showing that human monocytes exposed to retroviruses in vitro produce normal levels of IL-1 (12). On the other hand, increased production of IL-1 has been documented in at least two other conditions associated with reduced immunoresponsiveness (1,22).

CONCLUSIONS

The experiments presented herein confirm the beneficial effects of LPS on the antibody responsiveness of mice infected with viruses of the FLC and demonstrate that macrophages are deeply involved in the phenomenon. First, macrophage-rich cell populations derived from normal mice, which were by themselves incapable of reversing the immunological deficit of infected spleen cell cultures, became capable of doing so once pretreated with LPS. Second, LPS restored the ability of F-MuLV infected spleen macrophages to help nonadherent cells from normal mice in the generation of an antibody response. Third, supernatants conditioned from LPS-stimulated cells which were over 95% pure macrophages proved very effective in enhancing the responsiveness of infected cells.

Owing to the varied effects of LPS on macrophage activities (24), establishment of the precise macrophage function(s) responsible for restoring infected spleen cell responsiveness is far from easy. A distinct possibility is that LPS acts by reversing the defect of antigen presentation produced by FLC and F-MuLV (6). It has been reported that LPS induces macrophages to express more Ia antigen, whose importance in antigen presentation is well documented (34). Possible influences on antigen uptake and processing could also play some role in the improvement of macrophage accessory functions brought about by LPS. On the other hand, the finding that supernatants of LPS-stimulated macrophages are as active or even more active than LPS, suggests a major role for the secretory functions of these cells.

The best characterized soluble product of macrophages active on immunocompetent cells is IL-1. However, repeated attempts to reproduce the effects of the LPS-conditioned supernatants using purified IL-1 failed. Moreover, infected spleen macrophages produced normal or enhanced levels of this important mediator following in vitro stimulation with LPS. Thus, it seems likely that other soluble products of LPS-stimulated macrophages, by themselves or in concert with IL-1, mediate the restorative activity of LPS. Their nature and mode of action is currently under investigation in our laboratory.

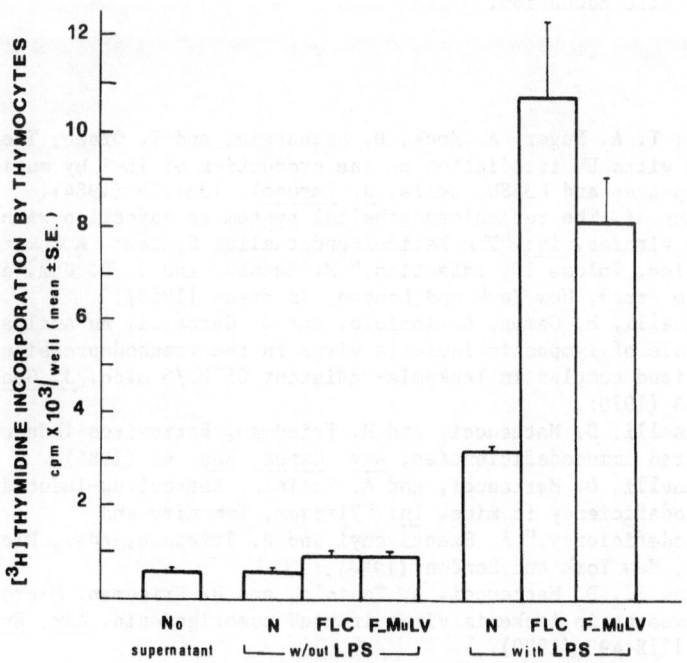

Figure 1. Effect of FLC and F-MuLV infections on the production of interleukin 1 (IL 1) by spleen macrophages following stimulation in vitro with LPS. Adherent spleen cells were prepared and stimulated with LPS as described in Table 6. The culture fluids were assayed for IL 1 activity by an established method (21).

At present, it is also impossible to come to any firm conclusion on whether the beneficial effects of LPS are solely the result of its action on macrophages. It seems doubtful that a nonspecific agent such as endotoxin is capable of selectively eliminating a single aspect of the cell defects responsible for the impaired responsiveness of infected mice. Clearly, one or more additional mechanisms might be operative. A resonable possibility is, for example, that in the infected mouse the feedback mechanisms that supposedly regulate the adjuvanticity of LPS in normal mice (20) are either not triggered or not working properly. Further studies are needed to evaluate this and other possibilities.

A constant feature of our results is that FLC-infected spleen cells proved amenable to reconstitution by LPS or LPS-conditioned supernatants as promptly as those of F-MuLV-infected spleen cells. This probably reflects the fact that the experiments were done at day eight postinfection, when the pathologic changes caused by FLC in the spleen are still limited. In any case, the finding confirms that the mechanisms whereby the two viruses immunosuppress are, at least in the initial stages of infection, very similar. Collectively, the present data also indicate that the manifestations of immunodeficiency caused by retroviruses are partially reversible by relatively simple compounds and reemphasize the centrality of accessory cells in their genesis.

ACKNOWLEDGMENTS

The skilled assistance of Giulietta Cerretini and Luciana Montagnani is deeply appreciated. This work was supported by grants from the Italian National Research Council, Special Project "Oncology," and from the Ministry of Public Education.

REFERENCES

1. J. Ansel, T. A. Luger, A. Kock, D. Hochstein, and I. Green, The effect of in vitro UV irradiation on the production of IL-1 by murine macrophages and P388D$_1$ cells, J. Immunol. 133:135 (1984).
2. M. Bendinelli, The reticuloendothelial system in infection with RNA tumor viruses, in: "The Reticuloendothelial System: A Comprehensive Treatise, Volume 10, Infection," M. Escobar and J. P. Utz, eds., Plenum Press, New York and London, in press (1985).
3. M. Bendinelli, M. Campa, A. Toniolo, and C. Garzelli, An analysis of the role of lymphatic leukemia virus in the immunodepression exerted by Friend complex in leukemia-resistant C57BL/6 mice, J. Gen. Virol. 39:243 (1978).
4. M. Bendinelli, D. Matteucci, and H. Friedman, Retrovirus-induced acquired immunodeficiencies, Adv. Cancer Res. 45 (1985).
5. M. Bendinelli, D. Matteucci, and A. Toniolo, Retrovirus-induced immunodeficiency in mice, in: "Viruses, Immunity and Immunodeficiency," A. Szentivanyi and H. Friedman, eds., Plenum Press, New York and London (1986).
6. M. Bendinelli, D. Matteucci, A. Toniolo, and H. Friedman, Macrophage involvement in leukemia virus-induced tumorigenesis, Adv. Exp. Biol. Med. 121B:493 (1980).
7. M. Bendinelli and L. Nardini, Immunodepression by Rowson-Parr virus in mice. 1. Growth curves of Rowson-Parr virus and immunological relationships with Friend virus, Inf. Immun. 7:152 (1973).
8. S. S. Boggs and G. N. Schartz, Increased lethality after endotoxin in old or leukemic AKR mice, Proc. Soc. Exp. Biol. Med. 157:424 (1978).
9. S. Britton, T. Wepsic, and G. Moller, Persistence of immunogenicity of two complex antigens retained in vivo, Immunology 14:491 (1968).

10. R. C. Butler, J. M. Frier, M. S. Chapekar, M. O. Graham, and H. Friedman, Role of antibody response helper factors in immunosuppressive effects of Friend leukemia virus, Infec. Immun. 39:1260 (1983).

11. B. Chesebro, J. L. Portis, K. Wehrly, and J. Nishio, Effect of murine host genotype on MCF virus expression, latency and leukemia cell type of leukemias induced by Friend murine leukemia helper virus, Virology 128:221 (1983).

12. E. A. Copelan, J. J. Rinehart, M. Lewis, L. Mathes, R. Olsen, and A. Sagone, The mechanism of retrovirus suppression of human T cell proliferation in vitro, J. Immunol. 131:2017 (1983).

13. J. L. Curtis and A. A. Nordin, Primary in vitro plaque-forming cell response to DAGG-ficoll: LPS-induced enhancement mediated by interleukin-1, Immunology 52:711 (1984).

14. C. A. Dinarello, Interleukin-1, Rev. Inf. Dis. 6:51 (1984).

15. B. N. Dracott, N. Wedderburn and M. J. Doenhoff, The immunodepressive effect of Friend virus. IV. Effects on spleen B lymphocytes, Immunology 34:679 (1978).

16. R. L. Duncan and D. C. Morrison, The fate of E. coli lipopolysaccharide after uptake of E. coli by murine macrophages in vitro, J. Immunol. 132:1416 (1984).

17. E. V. Genovesi, D. Livnat, and J. J. Collins, Immunotherapy of murine leukemia. VII. Prevention of Friend leukemia virus-induced immunosuppression by passive serum therapy, Int. J. Cancer 30:609 (1982).

18. M. K. Hoffman, S. B. Mizel, and J. A. Hirst, IL-1 requirement for B cell activation revealed by use of adult serum, J. Immunol. 133:2566 (1984).

19. D. D. Isaak, R. M. Miceli, and J. P. Lake, Target cell heterogeneity in murine leukemia virus infection. III. Identification of susceptible Lyt 1$^+$ and resistant Lyt 2$^+$ T cell subsets following in vitro infection with Friend murine leukemia virus, Cell. Immunol. 88:464 (1984).

20. S. Koenig and M. K. Hoffman, Bacterial lipopolysaccharide activates suppressor B lymphocytes, Proc. Natl. Acad. Sci. 76:4608 (1979).

21. L. B. Lachman, M. P. Hacker, G. T. Blyden, and R. E. Handschumacher, Preparation of lymphocyte activating factor from continuous murine macrophage cell lines, Cell. Immunol. 34:416 (1977).

22. R. Lelchuk, G. Rose, and J. H. L. Playfair, Changes in the capacity of macrophages and T cells to produce interleukins during murine malaria infection, Cell. Immunol. 84:253 (1984).

23. R. N. Moore, K. J. Goodrum, and L. J. Berry, Mediation of endotoxin effect by macrophages, J. Reticuloendothelial Soc. 19:187 (1976).

24. D. C. Morrison and J. L. Ryan, Bacterial endotoxins and host immune responses, Adv. Immunol. 28:293 (1979).

25. M. Niwa, K. C. Milner, E. Ribi, and J. A. Rudbach, Alteration of physical, chemical, and biological properties of endotoxin by treatment with mild alkali, J. Bacteriol. 97: 1069 (1969).

26. A. Nowotny, ed., "Beneficial Effects of Endotoxins," Plenum Press, New York and London (1983).

27. D. Rifkind and J. D. Palmer, Neutralization of endotoxin toxicity in chick embryos by antibiotics, J. Bacteriol. 92:815 (1966).

28. N. H. Ruddle, M. Y. K. Armstrong, and F. F. Richards, Replication of murine leukemia virus in bone marrow-derived lymphocytes, Proc. Natl. Acad. Sci. 73:3714 (1976).

29. I. J. T. Seppala and O. Makela, Adjuvant effect of bacterial LPS and/or alum precipitation in responses to polysaccharide and protein antigens, Immunology 53:827 (1984).

30. G. Schumann and C. Moroni, Mitogen induction of murine C-type viruses. I. Analysis of lymphoid cell subpopulations, J. Immunol. 116:1145 (1976).

31. R. A. Steeves and I. Grundke-Iqbal, Bacterial lipopolysaccharides as helper factors for Friend spleen focus-forming virus in mice, J. Natl. Cancer Inst. 56:541 (1976).

32. R. Weiss, N. Teich, H. Varmus, and J. Coffin, eds., "Molecular Biology of Tumor Viruses: RNA Tumor Viruses," Second Edition, Cold Spring Harbor Laboratory, Cold Spring Harbor, New York (1982).

33. D. D. Wood, P. M. Cameron, M. T. Poe, and C. A. Morris, Resolution of a factor that enhances the antibody response of T cell depleted murine splenocytes from several other monocyte products, Cell. Immunol. 21:88 (1976).

34. H. K. Ziegler, L. K. Staffileno, and P. Wentworth, Modulation of macrophage Ia-expression by lipopolysaccharide. I. Induction of Ia expression in vivo, J. Immunol. 133:1825 (1984).

ENDOTOXIN AND POLYSACCHARIDE DERIVATIVE INDUCED ENHANCED ANTIBODY FORMATION IN LEUKEMIA VIRUS INFECTED MICE

Herman Friedman and Andor Szentivanyi

Departments of Medical Microbiology and Immunology
and Pharmacology and Therapeutics
University of South Florida College of Medicine
Tampa, Florida

Endotoxins constitute a major component of the cell walls of gram-negative bacteria and are ubiquitously found in many microbial species (2,3,12,15). The molecules have been studied in great detail over the last few decades in terms of their chemical, physiologic and pharmacologic activities (11,13,20). They are also known to be potent modulators of immune responses, either enhancing or suppressing a wide variety of antibody or cell mediated activities (18-20). Endotoxin containing bacterial products have also been used as antitumor agents, presumably because they enhance nonspecific antitumor responses (14,17,19). It is now recognized that the Coley serum used over 60 to 70 years ago had effects because of the presence of bacterial endotoxin components and other biologically active cell wall materials in the preparations. In this context it has once again become popular in some instances to use bacterial components in attempts to induce antitumor immunity. Some investigators believe that treatment of cancer patients with bacterial products might stimulate nonspecific resistance to a tumor by activation of macrophages or other nonspecific effectors of immunity. Alternatively, such bacterial components might induce the production of soluble mediators of immunity, including interleukins and interferons (20,21).

Exposure of cells of the immune system to endotoxins in conjunction with various antigens results in dramatic modification of specific immune response which would not occur otherwise. For example, studies in this laboratory have shown that bacterial endotoxins are potent immunostimulators and may also reverse immunologic tolerance not only to soluble nonbacterial antigens such as serum proteins but also to bacterial antigens. Similar studies have been performed with the lower molecular weight Lipid A-free polysaccharide derivatives from endotoxins (6,9). These nontoxic polysaccharide-rich materials, when injected into experimental animals or added to spleen cell cultures, may influence the antibody response similar to intact endotoxin. Both endotoxins and polysaccharide derivatives can stimulate mediator release from murine lymphoid cells, including interleukins and interferons (10,16). As described in the present chapter, these materials, i.e., both intact endotoxin and polysaccharide-rich derivative, were also found to have similar immunoenhancing activity for spleen cells from leukemia virus infected mice. Furthermore, they had the ability to either partially or completely moderate leukemogenesis. These bacterial products, when added to spleen cultures from normal animals, induced a soluble immunoenhancing factor, presumably interleukin, which enhanced the antibody response not

only of normal spleen cells but also of leukemic cells. In contrast, these
materials failed to induce similar immunoenhancing factors by spleen cells
from leukemic animals, although the endotoxin itself had some immunostimulatory activity for spleen cells from the leukemic mice.

GENERAL EXPERIMENTAL APPROACHES

For these studies the murine model system was utilized with the Friend
leukemia virus (FLV) to examine immunomodulation by bacterial endotoxin and
nontoxic polysaccharide derivatives (7,8). FLV is a retrovirus which induces
in susceptible mice a rapid erythroleukemia resulting in marked splenomegaly,
hepatomegaly and blood cell dyscrasia, resulting in death of the animal
within three to five weeks, depending on the initial dose of virus and route
of administration. Infection with this virus also results in a marked
immunosuppression, due to direct effects of the virus on antibody forming B
lymphocytes and their precursors, although effects on T cells, especially
helper T cells, have also been noted. Deleterious effects on macrophages
may also be induced. Induction of leukemogenesis as well as immunosuppression was found to be modulated by bacterial endotoxin administration. For
this purpose, lipopolysaccharide (LPS) was obtained from Serratia marcescens
by the hot acid extraction procedure as described previously (3,12). The
LPS containing extract contained less than one percent nitrogen, but was
rich in the Lipid A moiety and polysaccharide. The endotoxin had the usual
characteristics associated with LPS, including the ability to stimulate
enhanced antibody formation in vivo and in vitro, induction of Shwartzman
reactivity in rabbits, pyrogenicity, etc. The nontoxic polysaccharide derivative was obtained by acid hydrolysis of the intact endotoxin exactly
as described previously (6,13). Control E. coli endotoxin preparations
were obtained commercially from Difco Laboratories, Detroit, MI.

The adjuvant activity of the preparations was determined by injecting
graded quantities of LPS or polysaccharide-rich preparations intraperitoneally (IP) into groups of five to ten BALB/c mice approximately six to eight
weeks of age, followed by challenge immunization at various times by IP
injection of graded amounts of sheep erythrocytes (SRBC) or E. coli vaccine.
The antibody response was determined at various times thereafter by obtaining
individual blood specimens by retroorbital venous plexus puncture. Micro-
hemagglutination and hemolysin tests were performed using 0.25 ml volumes
of serial two-fold dilutions of serum to determine anti-SRBC antibody.
Anti-E. coli antibody was determined by bacterial microagglutination with a
suspension of 10^8 heat killed E. coli as the test antigen. Groups of mice
were sacrificed at various times after immunization, their spleens obtained
and dispersed single cell suspensions prepared by teasing with needles and
forceps in sterile Medium 199 containing sterile fetal calf serum. The
number of individual hemolytic antibody plaque forming cells (PFC) per cell
suspension was determined by the standard hemolytic plaque assay using a
microprocedure with untreated SRBC for direct 19S PFC to SRBC or E. coli
LPS treated SRBC for the bacterial PFC. The number of individual PFCs was
calculated from the number of plaques for three or more cell cultures and
calculated per million spleen cells tested or per cell suspension. In addition, a completely in vitro culture system was used in which 5×10^7 viable
spleen cells, either from normal or virus infected animals, were cultured
in microtiter plates in 0.25 ml volumes of medium to which were added
graded concentrations of either LPS or polysaccharide derivative and sheep
RBCs as the antigen (14,15). The number of hemolytic anti-E. coli PFCs per
well was determined at various times thereafter by the microplaque assay
exactly as for the in vivo studies.

In order to determine whether soluble mediators were involved in immunomodulation, cell-free culture supernatants were obtained from spleen cells

Table 1. Antibody Responses of FLV Infected Mice to Sheep
Erythrocytes and E. coli Vaccine

Time in Days after Infection[a]	Spleen Weight (mg)[b]	Antibody Response (PFC)[c] SRBC	E. coli
None (control)	118 ± 16	975 ± 16	1130 ± 129
+ 1	121 ± 14	630 ± 56	850 ± 138
+ 3	136 ± 22	410 ± 16	576 ± 79
+ 7	186 ± 45	130 ± 20	143 ± 20
+ 10	215 ± 36	72 ± 17	87 ± 15
+ 15	876 ± 130	8 ± 2	30 ± 5
+ 40	1250 ± 260	<5	<10

[a]Groups of mice infected i.p. with 100 LD_{50} FLV on day indicated before
challenge immunization with 4×10^8 SRBC or 10^7 killed E. coli.
[b]Average spleen weight in mg for 5 mice per group at time of testing for
antibody response.
[c]Average number of PFC to SRBC or E. coli per 10^6 spleen cells 5 days after
challenge immunization.

incubated with graded concentrations of either LPS or polysaccharide derivative and these were added to suspensions of normal spleen cells immunized
in vitro with SRBC (5,7). The culture supernatants were characterized by
treatment with enzymes or heat. In addition, interferon activity was determined in culture supernatants by assessing the ability of graded dilutions
to inhibit the cytolytic activity of vesicular stomatitiss virus against
Vero cells in culture. Interleukin 1 (IL-1) activity was assessed by determining the ability of cell-free supernatants to enhance the blastogenic
responses of normal C3H/HeJ thymocytes in vitro exactly as described (5).
Interleukin 2 (IL-2) activity was determined in the culture supernatants by
the ability to increase proliferation of IL-2 dependent CTLL cells in vitro.

The Friend leukemia virus preparation used in these studies was free
of lactic dehydrogenase or lymphocytic choriomeningitis virus. Homogenates
of clarified spleen cell extracts from two to three week infected leukemic
mice were used as the stock leukemia virus preparation. Injection of graded
amounts of the stock virus into susceptible BALB/c mice resulted in marked
splenomegaly and death within three to four weeks.

EXPERIMENTAL RESULTS

Mice injected with FLV showed rapid development of splenomegaly and a
concomitant decrease in their antibody responsiveness to the T-dependent
antigen sheep erythrocytes as well as to the T-independent antigen E. coli
(Table 1). It is noteworthy that although very little if any significant
splenomegaly was detectable during the first few days after infection, there
was still a marked decrease in the antibody response per spleen of mice
immunized with either the SRBC or the E. coli. One week after infection,
when splenomegaly became evident, there was an 80 to 90% or greater depression of the expected antibody response. By the second or third week after
infection the animals were essentially anergic and showed marked leukemogenesis.

Table 2. Susceptibility of FLV Infected Mice to LPS Induced Toxicity

Time in Days After Infection[a]	Dose of FLV for Infection (ID_{50})[b]			
	0	10.0	50.0	100.0
None (control)	0/5	0/5	0/5	0/6
+ 3	0/3	0/4	0/4	2/5
+ 7	0/3	0/3	1/5	2/4
+ 15	0/5	1/4	3/4	4/4
+ 20	0/5	2/4	4/4	--

[a]Groups of mice infected at time indicated with tabulated dose of FLV.
[b]Number of mice surviving/number challenged with 100 µg LPS after 48 hrs.

Mice infected with FLV also showed increased susceptibility to LPS-induced toxicity, either by the Serratia or unrelated E. coli preparations. As is apparent in Table 2, mice infected with FLV showed increased suscepti-bility to LPS as a function of time after infection and also of the dose of the virus. An inoculum of 50 to 100 ID_{50} FLV resulted in a slight to moderate splenomegaly on day seven and lower doses of virus had essentially no effect on spleen size at that time. Nevertheless, susceptibility to lethality by endotoxin was quite evident in the 50-100 ID_{50} FLV infected mice on day seven. By the second week after infection even the mice given the 10 ID_{50} FLV dose, culminating in death of all mice within four to six weeks, exhibited enhanced susceptibility to endotoxemia. This increased susceptibility to endotoxin appeared to limit the potential use of LPS for treatment of leukemic animals. Even if there was a therapeutic effect by LPS, the leukemic state would probably have made the animals quite suscep-tible to the endotoxin. Nevertheless, as shown in Table 3, a lower amount of endotoxin, i.e., 20 µg LPS, resulted in longer survival time of FLV infected animals.

Table 3. Effect of Endotoxin Preparations on Resistance of Mice to FLV

Day of Treatment[a]	Survival of FLV Infected Mice[b]	
	LPS	PS
- 15	5/7 (48 ± 8)	6/8 (52 ± 4)
- 10	1/8 (37 ± 6)	4/6 (48 ± 5)
- 5	4/5 (42 ± 5)	4/7 (47 ± 5)
- 2	6/9 (49 ± 4)	6/8 (49 ± 6)
0	5/7 (56 ± 5)	4/5 (40 ± 3)
+ 5	3/5 (39 ± 3)	1/6 (35 ± 3)

[a]Groups of mice injected on indicated day with LPS or PS (20 µg) relative to day of infection with 100 LD_{50} FLV.
[b]Number of mice surviving/number mice injected after 30 days (mean sur-vival time after 30 days); control untreated mice showed only a mean survival time of 35 ± 5 days.

Table 4. Effect of LPS on Antibody Response of Spleen Cells
from Normal or FLV Infected Mice

Dose per Mouse (μg)	PFC/10^6 Spleen Cells[b]	
	Normal Mice	FLV Infected[c]
None (control)	836 ± 57	120 ± 18
1.0	892 ± 130	146 ± 26
10.0	2261 ± 375	973 ± 240
20.0	2640 ± 280	1860 ± 340
50.0	1976 ± 310	1810 ± 270

[a]Indicated dose of LPS injected i.p. into mice at time of challenge
immunization with 4 x 10^6 SRBC.
[b]Average number of PFC/10^6 spleen cells for 3-4 mice 4 days after
challenge immunization.
[c]Mice injected seven days earlier with 100 LD_{50} FLV.

The polysaccharide (PS), which also had immunomodulatory effects in
both normal and leukemic animals, produced a prolongation of survival of
FLV-infected animals (Table 2). This was observed when the PS was injected
simultaneously with the leukemia virus. However, as shown in Table 3, the
time of administration of LPS or PS derivative markedly influenced the resistance of the mice to FLV infection. Enhanced resistance occurred when the
LPS was given about two weeks before challenge with FLV. Increased survival
also occurred when the mice were given LPS about the time of infection or
two to five days before. In contrast, mice given FLV ten days after injection with LPS showed greater leukemogenesis, suggesting an increased susceptibility. This did not occur when the PS was given ten days before challenge
with FLV. Greater resistance to FLV was evident in the mice exposed to the
virus either at the time of treatment with PS or even when given as long as
two weeks after treatment.

Table 5. Effect of PS on Antibody Response of Spleen Cells from
Normal or FLV Infected Mice

Dose per Mouse (μg)[a]	PFC/10^6 Spleen Cells[b]	
	Normal Mice	FLV Infected[c]
None (control)	760 ± 49	138 ± 27
1.0	932 ± 72	159 ± 42
10.0	1540 ± 132	765 ± 180
20.0	1960 ± 275	1636 ± 285
50.0	2150 ± 310	1930 ± 310

[a]Indicated dose of PS injected i/p. into mice at time of challenge
immunization with 4 x 10^8 SRBC.
[b]Average number of PFC/10^6 spleen cells for 3-4 mice 4 days after
challenge immunization.
[c]Mice injected 7 days earlier with 100 LD_{50} FLV.

Table 6. Effect of LPS or PS on In Vitro Antibody Response of Mouse Spleen Cells to Sheep Erythrocytes

Dose per culture[a] (μg/culture)	Antibody Response[b]			
	LPS Treatment		PS Treatment	
	Normal	FLV Infected[c]	Normal	FLV Infected[c]
None (control)	835 ± 65	82 ± 10	796 ± 48	97 ± 22
1.0	973 ± 42	120 ± 14	810 ± 60	186 ± 38
10.0	1730 ± 240	862 ± 38	1530 ± 184	673 ± 120
20.0	2970 ± 290	1470 ± 140	2650 ± 286	1310 ± 265
40.0	3100 ± 270	1630 ± 195	2970 ± 265	1830 ± 193

[a] Indicated dose of LPS or PS added to cultures of 5×10^6 spleen cells from normal or FLV-infected mice.
[b] PFC per 10^6 spleen cells for 3-4 cultures per group 5 days after in vitro immunization with 2×10^6 SRBC.
[c] Mice injected 10 days earlier with 100 LD_{50} FLV.

In order to examine the mechanisms involved, experiments were performed concerning effects of LPS or PS on antibody responsiveness of infected mice. Spleen cells were obtained from mice infected with FLV one week earlier and either untreated or treated at the time of infection with LPS or PS. As seen in Table 4, the LPS caused a dose-related stimulation of the PFC response of spleen cells from normal mice as well as those from FLV infected mice. Although the PS-rich material had no toxic effect with the dosages used, nevertheless it induced essentially the same immunoenhancement by spleen cells from normal or FLV infected mice (Table 5). The PS was essentially equivalent to LPS in stimulating enhanced antibody responses in vivo by spleen cells from normal and leukemic mice.

The effects of LPS or PS injected into mice IP at the time of challenge immunization with sheep RBCs could have been due to direct activation of lymphoid or non-lymphoid cells in vivo, as well as alterations in the pattern of traffic of cells in the animals. Thus it was of interest to determine whether the LPS, as well as PS, could affect lymphoid cells from either normal or FLV-infected animals cultured in vitro with LPS or PS. As shown in Table 6, relatively similar results were obtained with both bacterial preparations. For example, graded amounts of LPS or PS, when added to spleen cells from normal or FLV-infected animals, resulted in similarly enhanced antibody responses. Both bacterial products resulted in about the same stimulation of the PFC response of normal and FLV-infected spleen cells in vitro at a dose of 20-40 μg per culture. Even 10 μg had a stimulatory effect. Enhanced PFC responses of FLV-infected spleen cells treated in vitro with these bacterial products were seen with these doses. Although the maximum response was lower than that observed with spleen cells from normal animals, it is important to note that the number of PFCs by untreated spleen cells from leukemic mice was extremely low, i.e., usually less than approximately 10 to 15% of the control spleen cell response. Thus the LPS or PS resulted in even greater enhancement of the response of FLV-infected spleen cells in terms of percent increase or stimulation index.

Various other substances have been reported to have effects on the antibody response of leukemic animals. For example, BCG has been often used to study the effects of nonspecific stimulation of the immune system on resistance to leukemogenesis, both in experimental animals and man.

Treatment[a]	Response of Mice[b]			
	Control		FLV Infected	
	Spleen Weight	PFC/10^6 Spleen Cells	Spleen Weight	PFC/10^6 Spleen Cells
None (control)	119 ± 16	768 ± 45	878 ± 138	113 ± 32
LPS alone	158 ± 32	1836 ± 120	490 ± 156	960 ± 86
PS alone	130 ± 18	1640 ± 110	526 ± 130	865 ± 72
BCG alone	199 ± 52	2140 ± 156	310 ± 116	580 ± 40
MDP alone	138 ± 22	1809 ± 210	638 ± 82	618 ± 83
LPS + BCG	189 ± 252	2965 ± 320	432 ± 72	1180 ± 130
LPS + MDP	171 ± 60	2760 ± 410	398 ± 63	1045 ± 72
PS + BCG	188 ± 252	2530 ± 270	420 ± 72	976 ± 68
PS + MDP	156 ± 40	2640 ± 210	486 ± 120	1132 ± 72

[a]Groups of 5-6 mice injected i.p. with indicated stimulator on same day
as injection i.p. with 100 LD_{50} FLV; LPS, PS or MDP injected in saline at
a dose of 25 μg/mouse or 1.0 mg BCG i.p. at time of infection.
[b]Average response for 3-4 mice per group 10 days after infection with FLV.

Muramyldipeptide (MDP), a small molecular weight derivative of cell walls
of bacteria such as mycobacteria and also present in other microorganisms,
appears to contain most of the immunostimulatory or adjuvantic properties
of the whole bacterial cell wall. MDP is readily synthesized and has been
used extensively in experimental studies concerning adjuvanticity. As is
apparent in Table 7, MDP alone had some immunostimulatory activity for FLV-
infected mice, as it did for normal mice. For these experiments an optimal
dose of sheep red cells was used to stimulate the immune response of spleen
cells from normal mice an presumably this was an optimal dose for stimulating
spleen cells from FLV-infected mice. The MDP caused a marked adjuvantic
effect for these spleen cells. LPS alone also had good stimulatory effects.
Treatment of the mice at the time of infection with both LPS and MDP resulted
in an even greater stimulation for normal mice and also had a strong effect
on the FLV-infected splenocytes. The polysaccharide derivative, when given
to cultures from normal or FLV-infected mice, together with MDP, resulted
in essentially the same response as occurred when LPS and MDP were used.
It seems apparent that a small molecular weight synthetic derivative of
bacterial cell walls contains much of the adjuvantic effect attributed to
whole cell wall adjuvants. Treatment of mice with MDP and the polysaccharide
derivative resulted in essentially the same immunostimulatory activity as
that induced by less defined bacterial products such as BCG and intact LPS.

In an attempt to determine the mechanisms involved in such immunostimulation, experiments were performed with culture supernatants from normal or
FLV-infected splenocytes treated with the bacterial products. As shown in
Table 8, supernatants from normal mouse spleen cells treated in vitro for
24 hours with LPS had the ability to increase the antibody responsiveness
of spleen cells from normal mice. The polysaccharide had similar immunostimulatory activity and induced the same level of antibody helper activity in

Table 8. Effect of LPS or PS Induced Cell Free Culture Supernatants on
PFC Response of Spleen Cells from Normal or FLV-Infected Mice

| Culture Supernatant Addition[a] | Normal | PFC/10^6 Spleen Cells[b] | | |
		Percent of Control	FLV Infected[c]	Percent of Control
None	912 ± 38	--	189 ± 15	--
LPS Induced 1: 5	1820 ± 125	199	1630 ± 48	862
1:10	1938 ± 210	213	1710 ± 75	799
1:20	1263 ± 196	135	730 ± 36	386
PS Induced 1: 5	1973 ± 138	216	1430 ± 40	757
1:10	1494 ± 173	164	1270 ± 93	672
1:20	1130 ± 210	124	1020 ± 65	540

[a]Indicated dilution of 24 hr cell free culture supernatant from LPS or PS
stimulated normal mouse spleen cell cultures added to cultures of 5 x 10^6
spleen cells from normal or FLV-infected mice.
[b]Average number of PFC 5 days after in vitro immunization of cultures with
2 x 10^6 SRBC.
[c]Mice injected 10 days earlier with 100 LD_{50} FLV.

spleen cell cultures. These supernatants, i.e., those induced by LPS or PS
in normal spleen cell cultures, had the ability to enhance the antibody
responsiveness of FLV-infected splenocytes.

Previous studies had shown that the ability of leukemic splenocytes to
generate antibody helper activity in response to LPS or PS was markedly
diminished. Thus it appeared that the immunoenhancing activity of LPS as
well as the polysaccharide derivative was related to release of soluble
factors which had antibody helper activity for both normal and FLV-infected
splenocytes. The factor(s) appeared to be immunoenhancing cytokines such
as IL-1 or interferon. Characterization of the supernatants indicated that
spleen cells from normal mice, when incubated either with LPS or the polysaccharide derivative, contained interferon, mainly of the gamma type, as
well as IL-1. The FLV-infected splenocytes, in contrast, had reduced levels
of interferon and IL-1 activity, but under some circumstances had higher
amounts. These experiments are being continued.

DISCUSSION AND CONCLUSIONS

Recent studies concerning the immunomodulatory effects of bacterial
endotoxins have revealed that these microbial products exhibit a wide variety
of activities on the immune response system (2,3,5,7,8,12). It is widely
accepted, as indicated in the chapters of this volume, as well as in many
reviews and publications, that endotoxin derived from gram-negative bacteria
can either enhance or suppress the immune response to a wide variety of B
cell or T cell dependent antigens depending upon the dose and the type of
antigen and the time interval between administration of antigen and endotoxin. Many models have been used in this regard, including antibody formation in vivo as well as immune parameters in vitro utilizing newer immuno-
biologic and culture methods.

Table 9. Induction of Antibody Helper Activity by LPS or PS in Cultures of
Spleen Cells from Normal or FLV-Infected Mice

Cell Type Tested[a]	PFC/10^6 Spleen Cells[b]			
	LPS Treated		PS Treated	
	Normal	FLV-Infected	Normal	FLV-Infected
None	930 ± 48	92 ± 18	--	--
Spleen	2760 ± 348	1430 ± 138	2630 ± 280	1260 ± 48
Adherent Spleen	3450 ± 460	1590 ± 230	2970 ± 270	1410 ± 156
Nonadherent Spleen	1140 ± 68	136 ± 26	1030 ± 76	140 ± 29
PE Cells	2560 ± 130	1510 ± 32	2435 ± 322	1370 ± 448

[a]Cultures of 5×10^6 spleen cells or separated adherent or nonadherent cells
or peritoneal cells, immunized in vitro with 4×10^6 SRBC.
[b]Average PFC response, ± SE, for 3-4 cultures per group 5 days after in vitro
culture and treatment with 0.1 ml cell free supernatant for normal cultures
incubated for 24 hrs with 20.0 µg LPS or PS or, as control, medium alone.
[c]Donor mice infected 10 days earlier with 100 ID_{50} FLV.

Previous studies in this laboratory have demonstrated that not only
intact bacterial endotoxin but also the lipid-free polysaccharide-enriched
derivative could enhance antibody responses to sheep red blood cells, as
well as to other antigens (4). There is evidence from the results presented
in this chapter that graded doses of either the lipopolysaccharide or the
polysaccharide-rich derivative could enhance antibody response of mouse
spleen cells after immunization with sheep erythrocytes in vivo as well as
after in vitro challenge with the same antigen. It is apparent that the
polysaccharide-rich derivative free of the lipid A moiety is as effective
as the intact endotoxin in enhancing antibody formation to sheep erythrocytes
in the mouse model system. On the other hand, the polysaccharide rich fraction which is not toxic (greater than 1000 µg can be administered to mice
without toxicity) did not cause increased mortality in mice infected with a
leukemogenic virus which induces acquired immune deficiency.

Previous studies in this and other laboratories had indicated that
retroviruses in mice suppress immune responses. Studies in this laboratory
have focused attention on the effects of Friend leukemia virus infection on
the nature and mechanism of how the virus diverts B lymphocytes and their
precursors from the pathway of antibody formation to the pathway of leukemogenesis. Although initially it was thought that FLV directly infects B
lymphocyte precursors, studies inthis and other laboratories have shown
that FLV infection is a very complex situation and that both B and T lymphocytes may be affected. Studies in this laboratory, as well as by Bendinelli
and others, have also shown that macrophages can be affected by Friend leukemia virus infection (1). The summation of all these effects, however, is
that the antibody-forming response to a T cell dependent antigen such as
sheep erythrocytes is markedly diminished within a few days after infection
of mice. With a relatively small dose of FLV, which does not cause death
until 30-40 days later, the suppression of the sheep red blood cell response
is one of the first indicators of suppression of immunity and the first
indicator of infection.

As shown previously and in the present chapter, intact endotoxin can
reverse either partially or even markedly immunosuppression induced by FLV.

The endotoxin itself, when given to FLV-infected mice, resulted in excessive mortality, especially when used at a dose which would normally not cause endotoxemia in normal mice. Thus it appears that the Friend leukemia virus infection makes mice more susceptible to endotoxin induced lethality. Although in vitro studies have shown that the endotoxin can reverse significantly the immune suppression of splenocytes from leukemia virus infected mice, it seems unlikely that endotoxin per se could be utilized completely safely in vivo. This is true of many bacterial products which contain endotoxin or even a toxic component of a microorganism which may either directly affect antibody forming cells or be an inadvertent contaminant or component of the microbe without direct effects in restoring antibody formation. On the other hand, the polysaccharide-rich derivative devoid of toxicity was found to cause essentially no increased mortality in FLV-infected animals. Furthermore, the polysaccharide derivative had essentially the same ability to increase antibody forming activity of spleen cells, both in vivo and in vitro, from FLV infected animals.

Earlier studies in this laboratory, as well as other laboratories, have shown that LPS affects immune responses by stimulating interleukin production by macrophages (4). Thus it was of interest to determine whether the polysaccharide-rich derivative functioned in a similar manner. Indeed, as shown in this study, the polysaccharide derivative and the intact endotoxin had essentially equivalent ability to induce and/or stimulate enhanced immune response in vitro, regardless of the dose of antigen and/or time of addition to cultures, whether derived from normal mice or FLV-infected mice. Furthermore, serum obtained from normal mice injected two hours earlier with either the intact LPS or the polysaccharide-rich derivative had similar immunoenhancing activity for other spleen cell cultures, either normal or leukemic after immunization in vitro with SRBC.

In contrast, the LPS or polysaccharide-rich extract, when injected into leukemic mice, induced relatively little immunoenhancing factor in the serum and also had little if any ability to induce leukemic spleen cells to produce the immunoenhancing factor (i.e., probably interleukins). Cultures from normal mice exposed either to the endotoxin or polysaccharide-rich derivative resulted in supernatants which had immunoenhancing activity for both normal spleen cells as well as FLV suppressed splenocytes. These results indicate that the polysaccharide-rich derivative is active in stimulating immunoenhancing soluble factors, presumably interleukins, in normal splenocytes.

There have been no similar reported results to date showing that Lipid A per se or a derivative of Lipid A synthetic or natural, has similar activities. Thus, the lipid-free polysaccharide induced immunoenhancement appears due to the polysaccharide component itself. The stimulation of intermediate substances such as interleukins would account for much if not all the immunoenhancement induced by both endotoxin and the purified polysaccharide derivatives.

It is of interest that immunosuppressed spleen cell preparations when exposed either directly to the polysaccharide derivative or the intact LPS, or even to small quantities of serum from normal mice treated two hours earlier with these preparations, responded in an enhanced manner to challenge immunization with sheep erythrocytes. This suggests that even though immunosuppressed spleen cell cultures have few immunocompetent cells capable of producing detectable antibody to sheep erythrocytes, following incubation with these bacterial preparations and/or induced soluble cell-free factors or serum components stimulated by these materials, increased numbers of antibody forming cells appeared. This suggests that the effects are due to enhancement of cellular differentiation or clonal expansion. Interleukins have such effects in a wide variety of model systems and it seems likely

that the polysaccharide derivatives, as well as the intact endotoxin, has
the ability to stimulate interleukin activity in supernatants of spleen
cell cultures from normal mice. It is also of interest, at least in the
experiments in this laboratory, that such immunoenhancing activity by splenocytes from leukemic mice are depressed.

The polysaccharide derivative was equivalent to the intact endotoxin
in stimulating antibody responsiveness of control and FLV-infected mice,
and, moreover, the polysaccharide extract also collaborated with BCG or MDP
in inducing enhanced antibody responses of normal as well as leukemic mouse
spleen cells. Since the polysaccharide derivative is much smaller than the
intact endotoxin and is relatively purified, and certainly does not contain
the toxic component of the LPS, and since MDP has been synthesized or purified from mycobacterium and shown to have the same or equivalent adjuvanticity as the intact cell wall extracts of mycobacteria such as BCG, it is
apparent that both of these smaller molecular weight materials may be of
potential use in stimulating antibody formation in leukemia virus infected
individuals with acquired immunodeficiency.

SUMMARY

Endotoxins from gram-negative bacteria, as well as the nontoxic polysaccharide enriched derivative, were studied in terms of the nature and
mechanism of immunoenhancement in vivo and in vitro with lymphoid cells
from normal and Friend leukemia virus-infected mice which exhibit acquired
immunodeficiency. Whereas the intact endotoxin showed increased toxicity
for mice infected with the leukemia virus, the polysaccharide-rich derivative did not. However, both the intact endotoxin and the polysaccharide-
rich derivative were immunoenhancing for antibody-forming cells to sheep
erythrocytes by both normal and leukemia virus-infected splenocytes. The
polysaccharide-rich fraction, devoid of Lipid A induced toxicity, showed
both enhancing activity for antibody formation and protective activity
against leukemia virus infection similar to the intact endotoxin but did
not have similar toxic activity. The polysaccharide-rich extract was also
equivalent to intact endotoxin in stimulating serum immunoenhancing factors
in normal mice given BCG and/or inducing immunoenhancing factor-containing
supernatants in spleen cell cultures from normal mice. The Lipid A free
polysaccharide-rich derivative also augmented the activity of BCG or even
purified synthetic MDP in increasing antibody formation by normal and leukemic splenocytes and also stimulated cell-free factors which are immunoenhancing. Further studies concerning the nontoxic polyssacharide-rich
derivative of endotoxin as an immunoadjuvant in acquired immunodeficiency
induced by retroviruses are warranted.

REFERENCES

1. M. Bendinelli, P. Matteuci, and H. Friedman, Retrovirus-induced
   acquired immunodeficiencies, Ad. Canc. Res. 46:78 (1986).
2. L. J. Berry, The mediation of endotoxemia effects, in: "Toxins," P.
   Rosenberg, ed., Pergamon Press, New York (1978).
3. A. Boivin, J. Masrobenu, and L. Mesrobenu, Extration d'un complexe
   toxique et antigenique a partier du bacille d'aertrycke, C.R. Soc.
   Biol. 114:307 (1933).
4. R. C. Butler, H. Friedman, and A. Nowotny, Restoration of depressed
   antibody responses of leukemic splenocytes treated with LPS-induced
   factors, Adv. Exp. Med. Biol. 121:315 (1980).
5. R. C. Butler, H. Friedman, S. Specter, and T. K. Eisenstein, Indication
   of immunoenhancing factors for murine splenocyte cultures by

Salmonella typhosa ribosome and RNA extracts, Infect. Immun. 32:1123 (1981).

6. S. J. Frank, S. Specter, A. Nowotny, and H. Friedman, Immunocyte stimulation in vitro by nontoxic bacterial lipopolysaccharide derivatives, J. Immunol. 119:855 (1977).

7. H. Friedman and S. Specter, Virus induced immunomodulation, in: "Immunopharmacology," L. Chedid and J. Hadden, eds., Pergamon Press, New York (1980).

8. H. Friedman, S. Specter, and M. Bendinelli, Viruses and the immune response, in: "Bacterial and Viral Inhibition and Modulation of Host Defenses," Academic Press, London (1984).

9. H. Friedman, S. Specter, and R. C. Butler, Stimulation of immunomodulatory factors by bacterial endotoxin and non-toxic polysaccharide, in: "Beneficial Effects of Endotoxins," A. Nowotny, ed., Plenum Press, New York (1983).

10. L. L. Lachman, Interleukin 1 release from LPS-stimulated mononuclear phagocytes, in: "Beneficial Effects of Endotoxins," A. Nowotny, ed., Plenum Press (1983).

11. H. C. Nauts, The beneficial effects of bacterial injections on host resistance to cancer, N.Y. Cancer Res. Inst. Monograph 8 (1946).

12. A. Nowotny, Endotoxoid preparations, Nature 197:721 (1963).

13. A. Nowotny, S. Golub, and B. Key, Fate and effect of endotoxin derivatives in tumor-bearing mice, Proc. Soc. Exp. Biol. Med. 132:26 (1971).

14. M. D. Prager, C. M. Ludden, W. J. Landy, J. P. Allison, and G. B. Kelto, Endotoxin stimulated immune response to modified lymphoma cells, J. Natl. Canc. Inst. 75:773 (1975).

15. O. Westphal, O. Luderitz, E. G. Rietschell, and C. Galanos, Bacterial lipopolysaccharides and Lipid A component: some historical and some current aspects, Biochem. Soc. Trans. 9:191 (1980).

16. D. D. Wood and P. M. Cameron, The relationship between bacterial endotoxin and B cell activation factor, J. Immunol. 121:53 (1978).

17. C. Yang and A. Youndra, Effect of endotoxin on tumor resistance in mice, Infect. Immun. 9:95 (1974).

18. H. Friedman, T. W. Klein, and A. Szentivanyi, eds., "Immunomodulation by Bacteria and Their Products," Plenum Press, New York (1981).

19. T. W. Klein, S. Specter, H. Friedman, and A. Szentivanyi, eds., "Biological Response Modifiers in Human Oncology and Immunology," Plenum Press, New York (1983).

20. A. Szentivanyi, E. Middleton, J. F. Williams, and H. Friedman, Effect of microbial agents on the immune network and associated pharmacologic reactivities, in: "Allergy: Principles and Practice," Second Edition, E. Middleton, C. E. Reed, and E. F. Ellis, eds., The C.V. Mosby Company, St. Louis, Missouri (1983).

21. A. Szentivanyi and H. Friedman, eds., "Viruses, Immunity and Immunodeficiency," Plenum Press, New York (1986).

CLINICAL RELEVANCE OF ENDOTOXEMIA

James P. Nolan

Department of Medicine
State University of New York at Buffalo
School of Medicine
Buffalo, New York

Endotoxins are in the outer cell wall of gram-negative bacteria. While endotoxins contain protein, the toxic portion is composed of lipopolysaccharides (LPS) and the terms are often used interchangeably. The innermost portion is lipid A and this moiety demonstrates all the biological effects of this potent material, and is similar in structure among most endotoxins. Endotoxin is formed by death of bacteria as well as by vesicle formation and release by the living bacterium.

As can be seen in Figure 1, the biological effects of this material are extremely varied and affect many systems. Profound vascular effects occur early after administration and are related to the release of catecholamines and other vasoactive peptides as well as direct endothelial injury with the leakage of protein-rich material from the intravascular compartment. Metabolically, endotoxemia can lead to profound hypoglycemia and to hyperlipidemia. Hematologic actions of LPS include an initial leukopenia related to margination of polymorphonuclear leukocytes on endothelial surfaces, sludging of red blood cells and the clinically important disseminated intravascular coagulation often seen in septic states. Long known to have the potent immunologic effects depicted, any interpretation of immunologic abnormalities in clinical conditions must take into account the presence or absence of endotoxemia if a correct interpretation of the abnormality is to be made. There is increasing interest in how cellular injury may be caused clinically in states associated with endotoxemia. Because of a large body of work on the C3H/HeJ endotoxin resistant mouse, it is believed that many if not all the biological effects of LPS on the host are mediated through the monocyte-macrophage system (28). Controversy exists, however, as to whether endotoxin has direct cellular effects independent of ischemia. Against a direct effect is the observation that uncoupling of mitochondrial function in hepatocytes is not identified during the early phase of endotoxemia but that a reduction in nutrient blood flow does occur in this early phase (3). On the other hand, endotoxin administration had an inhibitory effect on the adenylate cyclase enzyme system in dog liver plasma membranes both in vivo and in vitro (21). Lipid peroxidation has been shown to be induced by endotoxin in experimental animals (39,61), and LPS has been shown to change membrane lipid fluidity by phospholipase A activation (25). Regardless of whether LPS has direct toxic effects on cells or whether these effects are all mediated, it is obvious that both the individual resistance

491

of the host to these toxins as well as the duration and amount of the exposure are critically important.

Despite our wide knowledge in experimental animals of the potent effects
of purified endotoxins, and despite the detection of endotoxemia in a number
of clinical conditions, its clinical significance remains suggestive but
inconclusive. This assessment applies to both endotoxemia associated with
presence of gram-negative bacteria in the blood, and to the ever increasing
number of conditions where endotoxins are found in the circulation without
concomitant sepsis. The LPS in these conditions is assumed to be of enteric
origin. The reasons why the clinical relevance of these circulatory endotoxins has been questioned are varied. Detection of endotoxins in the systemic circulation has involved cumbersome biologic assays. Until recently,
the Limulus lysate assay has become the standard for measurement, and has
been made increasingly sensitive through the use of a chromogenic substrate
(12). The specificity, however, has been questioned, and some studies using
this technique have failed to confirm observations made by others (51). A
new immunoradiometric assay for lipid A may obviate some of these difficulties (38). Another difficulty may lie with interpreting biologic effects
in animals with that found in humans. Animal studies have used purified
LPS preparations but this form is unlikely to exist in clinical endotoxemia
since the LPS arising in the gut is composed of much cruder parts of cell
walls with less defined activities. The state of endotoxins in the circulation may have a profound effect on their activity and some forms may be
non-toxic or complexed with specific immunoglobulins (16). Lastly, non-specific tolerance may vary according to the individual or the accompanying
disease state altering toxicity of any given amount of this material.

In this chapter we will review and evaluate current evidence that endotoxins are present and may be relevant in human disease. It must always be
remembered, however, that it is often difficult to decide whether endotoxins
in the circulation contribute to the disease process itself or simply represent an associated epiphenomenon.

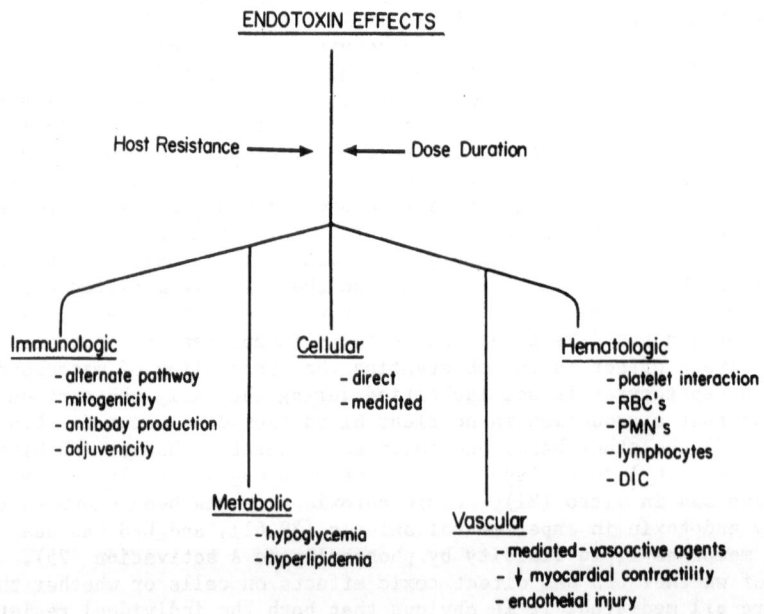

Figure 1. The multiple biological effects of bacterial endotoxin.

It has been observed for many years that infection with gram-negative organisms leads to states of endotoxemia. Further, the profound shock that frequently accompanies sepsis with these organisms can be reproduced in experimental animals by injecting purified LPS from the outer cell wall of these bacteria without the necessity of injecting the living organism. Anti-sera with reactivity against a number of endotoxins from gram-negative bacteria also protect animals against the lethality of bacterial sepsis as well (53). All these observations would suggest a key role for LPS in the clinical manifestations of infection with these bacteria. Certain contra-dictory evidence as to the role of LPS exist here as well. Despite the fact that LPS preparations have almost identical biologic effects in animals, infections with gram-negative organisms in humans do not behave identically. Typhoid fever has a different clinical spectrum than a Pseudomonas septicemia and E. coli bacteremia may be dissimilar from both.

There is little question that in sepsis, gram-negative bacteria release endotoxin. Of interest is the fact that antibiotics are not as effective in these states as in sepsis associated with gram-positive organisms. In an excellent study Shenep and Morgan showed a 10-2,000 fold increase in circulating endotoxin in animals treated with antibiotics despite decreasing levels of bacteremia (50). While it is assumed that these liberated endo-toxins detected by the Limulus assay come from bacteria in the bloodstream, this conclusion may not be correct. Cuevas and Fine calculated the amount of LPS found in rabbits with bacteremia as being far greater than that likely to be liberated by the bacteria present (11). It was their contention that the hypotension associated with gram-negative bacteremia promotes increased absorption of endotoxins from the large intra-intestinal pool. If such a mechanism of increased gut absorption does occur in sepsis, then the finding of Limulus positivity in septic states associated with the non-endotoxin forming Pneumococcus and Staphylococcus may not represent false positives as alleged (51). Further, if intra-intestinal endotoxin plays a role in gram-negative bacterial infections, therapy directed toward this extra-vas-cular pool may be indicated as well.

Endotoxemic States Without Bacteremia

The concept that endotoxemia may exist in the absence of circulating bacteria is one that is gaining in acceptance, but still is unfamiliar to many clinicians. Basically, the origin of the endotoxins in these clinical states is from the cell walls contained in large quantities in the intestine. Either by the mechanism of increased absorption through a break in the usually efficient mucosal barrier, or by a failure of hepatic removal and detoxification mechanisms, LPS gains entry into the systemic circulation.

While the most intensive and convincing demonstration of the clinical importance of endotoxemia lies in liver disease, studies implicating endo-toxins as key pathophysiological mediators will be examined in other condi-tions as well before concentrating on its role in hepatic injury.

1. Acute Renal Failure   Bacterial infections particularly those asso-ciated with gram-negative organisms often cause vasospastic renal failure. This type of injury leads to necrosis of the tubules and severe oliguria or anuria. In most cases, a preceding episode of hypotension can be documented and the entire clinical picture of cortical ischemia can be produced in experimental animals by the injection of LPS. Acute renal failure occurs clinically after hypotension associated with trauma or hemorrhage and where sepsis is not present. The renal hemodynamics in these settings are identi-cal to the vasospastic cortical changes seen with sepsis, leading Wardle to suggest that enteric endotoxin reaching the systemic circulation may play a

role in these renal injuries (57). A classic experimental model of acute vasospastic renal failure is the intramuscular injection of glycerol in the rat. In this model, endotoxin tolerant rats are resistant to the lesion and endotoxemia can be demonstrated by the lead acetate enhancement assay (37). Clinically, endotoxemia has been demonstrated by the Limulus assay in patients with acute and chronic renal failure (58). An interesting suggestion has been made that the acute renal failure induced by aminoglycoside antibiotics may in certain cases be the result of the sudden release of endotoxins from the killed circulating bacteria (50). Whether endotoxins can indeed be implicated as a common denominator in many types of acute renal failure remains unproven; but if true, might lead to new approaches to prevention and therapy.

2. Adult Respiratory Distress Syndrome   Like Acute Renal Failure, similar clinical settings underlie the development of the Adult Respiratory Distress Syndrome (ARDS). Frequently in the Intensive Care Unit, patients suffering from sepsis or shock will develop an alveolar capillary leak syndrome. This leakage of protein-rich material into the alveolar space leads to the typical "white-out" appearance on chest x-ray and to severe and progressive hypoxia. This type of pulmonary edema is not ordinarily associated with an increased pulmonary artery pressure, but rather is due to capillary endothelial damage. Endotoxins are known to cause capillary disruption and lead to such exudation (9). Indeed a classic model for the production of ARDS in animals is the injection of LPS (45). It is attractive again to involve endotoxins as a common denominator in a number of causes of this disorder. Severe trauma and prolonged shock from any cause might promote increased LPS absorption from the gut, and might impair the ordinarily efficient Kupffer cell removal of endotoxins by the liver. Once past the liver, the lungs become the target organ. But as with acute renal failure, endotoxins while able to mimic the clinical picture, have not been demonstrated conclusively to play this unifying pathophysiological role.

3. Neonatal Endotoxemia   Immaturity has been shown to increase the intestinal absorption of the macromolecules similar in size to endotoxins (56). Scherfele and his colleagues from Vancouver first demonstrated with the Limulus assay systemic endotoxemia in prematures with small bowel bacterial overgrowth or with necrotizing enterocolitis (48). In an extension of that study, these investigators found spontaneous endotoxemia to be common in neonates less than 1500 grams who were ill enough to be admitted to an intensive care unit. Endotoxin-like activity was found in 28 of 47 such infants and in only two of 58 separate episodes could be attributed to sepsis (49). Of importance, endotoxemia in this group was uncommon before oral feeding when little or no Limulus positivity was detected in stools. With feeding, stool positivity for endotoxin rose sharply as did the systemic presence of LPS. Whether such sick infants also had a decreased capacity for the hepatic removal of LPS was not examined but seems quite possible. No direct toxic effects of the endotoxin were observed in the positive group, but it seems possible that this spontaneous endotoxemia might add to their already significant problems. Since endotoxin is also a potent immunostimulator, it is also conceivable that small doses in the circulation might have advantages to the host defenses.

4. Gastrointestinal Disease   Inflammatory bowel disease would seem to have a classic potential for leading to increased endotoxin absorption through an inflamed and damaged mucosa. A number of studies have examined the problem with somewhat conflicting results. A study by Kruis and his associates found significantly elevated lipid A antibody titers in patients with Crohn's disease, but titers in patients with ulcerative colitis were not different than controls (24). Simultaneous determinations for circulating endotoxins using the Limulus lysate assay were negative in both groups. On the other hand, a group from Hanover, West Germany found marked endotox-

emia in 12 patients suffering acute relapses of both Crohn's disease and ulcerative colitis (59). These investigators used whole gut irrigation to markedly reduce the endotoxin load. With this irrigation, plasma endotoxin levels dropped rapidly, serum iron levels rose and in febrile patients, the temperature returned to normal. Thus the suggestion is made that absorbed endotoxins may play a crucial pathophysiological role in the clinical picture of inflammatory bowel disease. In a study performed at the time of colectomy for severe, uncontrolled ulcerative colitis, portal bacteremia was found very infrequently, but systemic endotoxemia was common (41). The authors felt that this endotoxemia was likely the result of the bowel manipulation because mesenteric blood was negative for LPS prior to bowel mobilization. Since disseminated intravascular coagulation (D.I.C.) occurs after colectomy, these investigators felt endotoxemia was the likely cause.

Bacterial overgrowth with the potential of a marked increase in the endotoxin pool occurs in blind loop syndromes and after jejunoileal (J-I) bypass surgery. In dogs with a J-I bypass, antibiotic treatment to reduce both aerobic and anaerobic bacteria protected the animals against liver injury when compared to non-treated controls (40). These authors speculated that endotoxins might damage an already compromised liver. However, moving to the clinical arena, patients given doxycycline hyclate after by-pass surgery showed no difference in the incidence of liver abnormalities when compared to randomized, non-treated groups post-operatively (62). As we shall see in the next section, endotoxins may be prominently involved in alcoholic liver injury, and the hepatic lesions following J-I bypass closely resemble alcoholic hepatitis.

5. Liver Disease Both experimentally and clinically, a pathogenic role for absorbed enteric endotoxins has been most convincingly made in the area of liver injury. Much of the work and interest in this relationship has come from laboratories in Europe, and it is surprising how unfamiliar many American gastroenterologists are with this concept. While the evidence of a relationship is strong, a clear clinical correlation between the presence of systemic endotoxemia and the clinical manifestations remains suggestive but not conclusively demonstrated. The hypothesis that is still being extensively tested (Figure 2) states:

A. that portal vein endotoxemia is a normal condition;

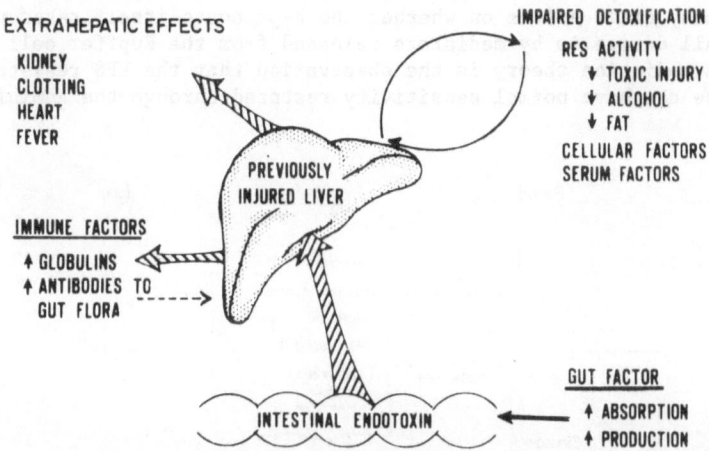

Figure 2. Depiction of hypothesis suggesting a key role for endotoxin as a co-factor in toxic liver injury.

B. that the liver, particularly the fixed macrophages (Kupffer cells) normally remove these small quantities of LPS;

C. that injury by a number of agents is to these lining cells of the sinusoids, impairing the ability to detoxify enterically derived endotoxins. Damage by toxins such as alcohol to the gut mucosa may also increase LPS absorption; and

D. that this increase in LPS sensitivity (in some cases up to 1,000 fold) leads to further liver damage and by spillover into the general circulation cause extra-hepatic signs and symptoms (29).

Historically, a link between the intestinal flora and liver injury in experimental animals has been known for decades. In 1954, it was demonstrated that rats fed a necrogenic diet did not develop liver injury if they were reared in germ-free environment (26). When micronodular cirrhosis was produced in rats by a choline deficient diet, the administration of a non-absorbable antibiotic markedly retarded the development of the process (46). Broitman and his colleagues in 1964 demonstrated that it was endotoxin rather than the intact bacteria that was crucial for the development of the liver injury (6). Simply by adding purified E. coli 026 LPS to the drinking water of the animals, the protective effect of the neomycin on choline deficiency cirrhosis was negated. Further, choline deficiency reduces the lethal dose in rats to endotoxin challenge by ten fold and significant transaminase elevations occur at small doses that cause no chemical injury in pair-fed controls (30). In rabbits, a marked synergism occurs between endotoxin and CCl_4. In vitro, the impressive capacity of liver homogenates to inactivate the toxic effects of LPS is completely lost if the animal is pretreated with carbon tetrachloride (15), and the LD_{50} to endotoxin is 600 fold less in guinea pigs given a sublethal dose of CCl_4 prior to challenge (17).

Kupffer cells seem to be the major removers of LPS from the circulation and their injury or blockade leads to a marked decrease in the removal of circulating endotoxin and a marked increase in their toxic effects. As early as 1947, Beeson demonstrated that endotoxin tolerance could be abolished by loading the reticuloendothelial system (RES) with particulate matter (5). Since then, alcohol, CCl_4 and fatty infiltration associated with choline deficiency has been shown to depress RES function and impair LPS removal (2,35,36). The mechanism by which endotoxin eventually damages the hepatocyte is not entirely clear but possible mechanisms are depicted in Figure 3. Controversy still exists on whether the hepatocyte itself takes up LPS or whether all damage is by mediators released from the Kupffer cell. In favor of the mediation theory is the observation that the LPS resistant C3H/HeJ mouse can have normal sensitivity restored through the administra-

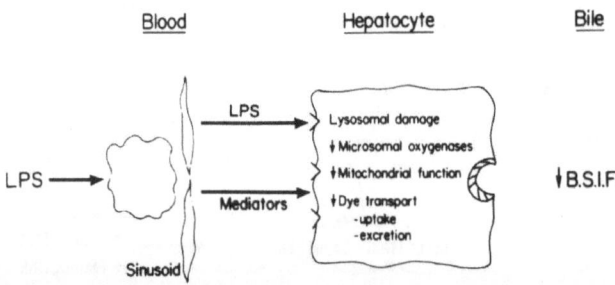

Figure 3. Mechanisms by which LPS can cause direct or mediated liver cell damage.

tion of macrophages from a normally responsive strain (28). Wisse and his associates found that an injected Cr[51]-labelled LPS dose could be found almost exclusively in the non-parenchymal liver cells (42), but others have found significant amounts in hepatocytes and the type and form (smooth or rough mutants) were important determinants of the distribution (27). It is likely that receptors to LPS as well as to lipid A do exist on the hepatocyte membrane, and the events depicted intracellularly in Figure 3 are well documented.

In experimental animals, a role for intestinal endotoxin in liver injury is strongly suggested in a number of different models. In chronic liver injury and cirrhosis due to choline deficiency, the role of LPS as described by Broitman (6), has been described earlier in this chapter. Three models of acute liver injury have been extensively examined for a similar relationship. Carbon tetrachloride (CCl$_4$) is believed to cause hepatic injury by lipoperoxidation of the cell membrane. While evidence is extensive that such an event occurs, measures to reduce the toxicity of LPS significantly ameliorate the injury. Rats made tolerant over nine days to bacterial endotoxin show significantly less biochemical and histologic damage to CCl$_4$ challenge than non-tolerant controls (32). Polymyxin B is an antibiotic that disrupts LPS and renders it non-toxic. Gentamicin has a similar antibacterial spectrum but has no effect on LPS. When animals are challenged with CCl$_4$ after pretreatment with polymyxin B, significant histologic and biochemical reduction in hepatic toxicity is evident compared to the gentamicin pre-treated group (34). Further pretreatment with small doses of CCl$_4$ significantly prolongs the clearance of a test dose of E. coli endotoxin as measured by an immunoradiometric assay (36). Massive hepatic necrosis induced by D-galactosamine has had endotoxins of intestinal origin implicated in the lesion, although such claims have been disputed by others. In an elegant series of experiments diagramed in Figure 4, the Wurzburg group disputed the explanation that hepatocellular death was caused solely by an interference with protein synthesis (22). These investigators noted mast cell degranulation as the initial event followed by histaminemia and colonic edema. This colonic injury led to increased endotoxin absorption and endotoxicosis depending on whether the RES was stimulated or unstimulated. Complement activation then occurred with resultant liver cell necrosis. By simply performing a colectomy five days before galactosamine administration, no hepatic necrosis occurred despite an identical drop in the level of uracil nucleotides in the hepatocytes. The role of endotoxin in this injury is

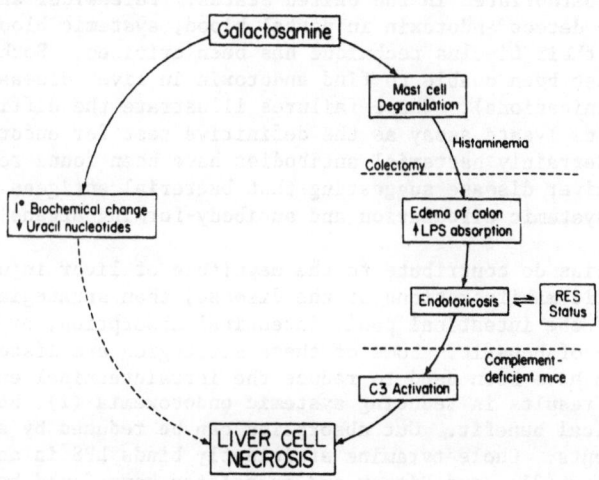

Figure 4. Schema for a role of enteric endotoxin in the galactosamine injury.

supported by another study that demonstrates galactosamine induced sensitization to the lethal effects of endotoxin in rabbits, rats and mice (19).
Where a marker LPS was instilled into the intestine of rabbits challenged
with galactosamine, the endotoxin appeared promptly in the systemic circulation in significant amounts, although small amounts appeared after anesthesia
alone in controls (8). Rasenach and his associates, dispute the importance
of endotoxin in galactosamine injury because galactosamine caused enzyme
leakage in the isolated liver preparation and the addition of exogenous LPS
did not increase the enzyme release (44). Another interesting animal model
that suggests enteric endotoxin as a critical co-factor for injury is that
of murine hepatitis caused by the toxic protein of Frog Virus 3. This virus
which does not replicate in mammals causes early and severe death of sinusoidal lining cells followed by hepatocytolysis. Colectomy does not prevent
the damage to the Kupffer and endothelial cells but does protect against
hepatocyte necrosis. Polymyxin B protects against the injury as does the
induction of tolerance (23).

What evidence exists that endotoxemia is relevant to human liver disease. A large number of studies have been previously reviewed that demonstrate Limulus lysate positivity in a variety of acute and chronic liver
diseases (33). A correlation exists between the presence of endotoxin in
the circulation and clinical manifestations known to be caused by LPS.
Whether these relationships are cause and effect or rather an epiphenomenon
is difficult to establish. Wilkinson and his associates in 1974 demonstrated
an almost perfect correlation between an impairment in renal function and
systemic endotoxemia by the Limulus lysate assay (60). Later studies by
others correlated endotoxemia and diminished creatinine clearance in cirrhosis, obstructive jaundice and acute viral hepatitis (4,10,55). Even these
observations have been questioned. Gatta and his colleagues tested 57 patients with alcoholic cirrhosis and found no significant difference in the
frequency of endotoxemia between patients with impaired renal blood flow or
decreased effective renal plasma flow (20). In a large number of patients
with acute and chronic liver disease, Limulus positivity has correlated
with the presence or absence of endotoxemia with survival at six months; 48
percent of those dying as compared to only 17 percent without (52). After
the peritoneovenous shunts (LeVeen) of ascites into the systemic circulation,
coagulopathy not infrequently results. Since ascitic fluid may contain
endotoxins, some have felt that LPS is responsible for activating the clotting system although this has been disputed (43). While endotoxemia still
is commonly detected in liver disease, it is of interest that this finding
has not been substantiated in the United States. Fulenwider and the Atlanta
group failed to detect endotoxin in portal blood, systemic blood or ascites
(18), although their Limulus technique has been critized. Both Gans and
DiLuzio have also been unable to find endotoxin in liver disease patients
(personal communications). These failures illustrate the difficulties in
using the Limulus lysate assay as the definitive test for endotoxins in
body fluids. Certainly bacterial antibodies have been found regularly in
patients with liver disease suggesting that bacterial antigens must gain
access to the systemic circulation and antibody-forming organs (54).

If endotoxins do contribute to the magnitude of liver injury and to
the extra-hepatic manifestations of the disease, then strategies for modifying the size of the intestinal pool, intestinal absorption, or host defenses
to LPS might be of benefit. Some of these strategies are listed in Table
1. Antibiotics have been used to reduce the intraintestinal endotoxin pool
with excellent results in reducing systemic endotoxemia (1), but with less
clear cut clinical benefit. Gut absorption can be reduced by a number of
LPS binding agents. Cholestyramine efficiently binds LPS in animals and
prevents toxicity (31), and Ditter and associates have found betonite to be
the single best oral adsorbent in preventing enterically induced endotoxemia

498

Table 1. Strategies to Modify Endotoxin Toxicity in Liver Disease

Decrease Intraintestinal LPS
- paramonycin
- neomycin

Decrease Gut Absorption
- cholestyramine
- lactulose
- betonite
- charcoal
- bile salts

Increase Resistance
- induce tolerance
- immunization (active and passive) to core endotoxins
- lysosomal stabilizers (steroids, flavinoids)

Systemic Agents
- colistin
- "anti-endotoxins"

(13). Bile salts disrupt and inactivate LPS and the post-operative renal failure related to endotoxemia in patients with obstructive jaundice can be prevented by the oral administration of bile salts (7). Host resistance to endotoxin could be heightened by inducing LPS tolerance or by the use of passively administered IgM antibodies to core LPS mutants such as are now used in treating gram-negative sepsis (53). Polymyxin B would seem promising as a therapy to inactivate endotoxins but its toxicity limits its usefulness. Colistin has similar anti-endotoxic properties but is less toxic and has been shown to prevent endotoxin induced cholestasis in dogs (14). The flavinoids such as cyanidonol-3 protect in rats against both endotoxin and galactosamine liver injury and may be another promising approach (47).

Whether a cause and effect relationship exists between enteric endotoxin and liver injury after alcohol or toxins has not been proven. Nevertheless, accumulating evidence of such a relationship is consistent with the hypothesis. Better methods are necessary to measure and quantitate endotoxemia in human disease. The state of LPS tolerance of the human host is obviously important but is difficult to determine. It is likely, however, that the state and form of defense against endotoxins are critical to the host and may well determine the extent of clinical injury. If endotoxins from the gut prove not to be an important co-factor in toxic liver injury, then the search for alternative explanations of the protective effects of tolerance, antiendotoxins, and colectomy in experimental animals should lead to a better understanding of mechanisms involved in hepatic necrosis.

ACKNOWLEDGMENTS

This work was supported in part by USPHS Grant 5 R01 AM 25130 and the Margaret Duffy and Robert Cameron Troup Memorial Fund of the Buffalo General Hospital.

REFERENCES

1. Y. Adachi, M. Enomoto, M. Adachi, et al., Enteric coated polymyxin B in the treatment of hyperammonemia and endotoxemia in liver disease, Gastroent. Japan 17:550 (1982).

2. M. V. Ali and J. P. Nolan, Serial assessment of reticuloendothelial function in experimentally induced nutritional cirrhosis, Lab Invest 21:184 (1969).

3. E. F. Asher, N. R. Garrison, D. J. Ratcliffe, and D. E. Fry, Endotoxin, cellular function, and nutrient blood flow, Arch. Surg. 118:441 (1983).

4. M. E. Bailey, Endotoxin bile salts and renal function in obstructive jaundice, Brit. J. Surg. 63:774 (1976).

5. P. B. Beeson, Tolerance to bacterial pyrogens. II. Role of the reticulo-endothelial system, J. Exp. Med. 86:39 (1947).

6. S. A. Broitman, L. S. Gottlieb, and N. Zamcheck, Influence of neomycin and ingested endotoxin in the pathogenesis of choline deficiency cirrhosis in the adult rat, J. Exp. Med. 119:633 (1964).

7. C. J. Cahill, Prevention of post-operative renal failure in patient with obstructive jaundice--the role of bile salts, Brit. J. Surg. 70:590 (1983).

8. D. S. Camara, J. A. Caruana, K. A. Schwartz, et al., D-galactosamine liver injury: absorption of endotoxin and protective effect of small bowel resection in rabbits, Proc. Soc. Exp. Biol. Med. 172:255 (1983).

9. S. Chien, D. G. Sinclair, R. J. Dellenback, et al., Effect of endotoxin on capillary permeability to macromolecules, Am. J. Surg. 207:518 (1964).

10. C. Clemente, J. Bosch, J. Rodes, et al., Functional renal failure and hemorrhogic gastritis associated with endotoxemia in cirrhosis, Gut 18:556 (1977).

11. P. Cuevas and J. Fine, Route of absorption of endotoxin from the intestine in non-septic shock, J. Reticuloendothel. Soc. 11:535 (1972).

12. B. Ditter, K. P. Becker, R. Urbaschek, and B. Urbaschek, Quantitativer Endotoxin - Nachweis. Arzneim. Forsch., Drug Res. 33:681 (1983).

13. B. Ditter, R. Urbaschek, and B. Urbaschek, Ability of various absorbents to bind endotoxins in vitro and to prevent orally induced endotoxemia in mice, Gastroenterol. 84:1547 (1983).

14. P. Escartin, J. Rodrigues-Montes, V. Cuervas-Mons, et al., Effect of colistin on reduction of biliary flow induced by endotoxin of E. coli, Dig. Dis. Sci. 27:875 (1982).

15. W. E. Farrar, M. Edison, and T. H. Kant, Susceptibility of rabbits to pyrogenic and lethal effects of endotoxin after acute liver injury, Proc. Soc. Exp. Biol. Med. 128:711 (1968).

16. P. C. Fink and K. D. Schultze, The polyethylene glycol precipitation technique and particle counting immunoassay for detection of circulatory immune complex material in liver cirrhosis and septicemia, J. Lab. Clin. Med. 90:852 (1982).

17. S. B. Formal, H. E. Noyes, and H. Schneider, Experimental Shigella infections. III. Sensitivity of normal, starved and carbon tetrachloride treated guinea pigs to endotoxin, Proc. Soc. Exp. Biol. Med. 103:415 (1960).

18. J. T. Fulenwider, C. Sibley, S. F. Stein, et al., Endotoxemia of cirrhosis; an observation not substantiated, Gastroenterol. 78:1001 (1980).

19. C. Galanos, M. S. Freudenberg, and W. Rutter, Galactosamine induced sensitization to the lethal effects of endotoxin, Proc. Natl. Acad. Sci. USA 76:939 (1979).

20. A. Gatta, L. Milani, C. Merkel, et al., Lack of correlation between endotoxemia and renal hypoperfusion in cirrhotics without overt renal failure, Eur. J. Clin. Invest. 12:417 (1982).

21. S. Ghosh and M. Liu, Decrease in adenylate cyclase activity in dog livers during endotoxic shock, Am. J. Physiol. 245:R737 (1983).

22. M. Grun, H. Liehr, and U. Rasanach, Significance of endotoxemia in experimental "galactosamine hepatitis" in the rat, Acta. Hepatogast. 23:64 (1976).

23. J. R. Gut, S. Schmitt, A. Bingen, et al., Probable role of endogenous endotoxins in hepatocytolysis during murine hepatitis caused by frog virus 3, J. Infect. Dis. 149 (1984).

24. W. Kruis, P. Schussler, M. Weinzierl, C. Galanos, and J. Eisenburg, Circulating lipid A antibodies despite absence of systemic endotoxemia in patients with Crohn's disease, Dig. Dis. Sci. 29:502 (1984).

25. M. Liu, S. Ghosh, and Y. Yang, Change in membrane lipid fluidity induced by phospholipase A activation: A mechanism of endotoxin shock, Life Sci. 33:1995 (1983).

26. T. D. Luckey, J. A. Reyniers, P. Gyorgy, et al., Germ-free animals and liver necrosis, Ann. N.Y. Acad. Sci. 57:932 (1954).

27. S. K. Maitra, A. Rachmileuwitz, D. Eberle, et al., The hepatocellular uptake and excretion of endotoxin in the rat, Hepatology 1:401 (1981).

28. D. C. Morrison and R. J. Ulevitch, The effects of bacterial endotoxins on host mediation systems, Am. J. Pathol. 93:527 (1978).

29. J. P. Nolan, The role of endotoxins in liver injury, Gastroenterol. 69:1346 (1975).

30. J. P. Nolan and M. V. Ali, Endotoxin and the liver. I. Toxicity in rats with choline deficient fatty livers, Proc. Soc. Exp. Biol. Med. 129:29 (1968).

31. J. P. Nolan and M. V. Ali, Effect of cholestyamine on endotoxin toxicity and absorption, Am. J. Dig. Dis. 17:161 (1972).

32. J. P. Nolan and M. V. Ali, Endotoxin and liver. II. Effect of tolerance on carbon tetrachloride induced injury, J. Med. 4:28 (1973).

33. J. P. Nolan and D. S. Camara, Endotoxin, sinusoidal cells and liver injury, in: "Progress in Liver Disease, Volume VII," H. Popper, and F. Schaffner, eds., Grune and Stratton, Inc. (1982).

34. J. P. Nolan and A. I. Leibowitz, Endotoxin and the liver. III. Modification of acute carbon tetrachloride injury by polymyxin B--an antiendotoxin, Gastroenterol. 75:445 (1978).

35. J. P. Nolan, A. I. Leibowitz, and A. O. Vladutiu, Influence of alcohol on Kupffer cell function and possible significance in liver injury, in: "The Reticuloendothelial System and The Pathogenesis of Liver Disease," H. Liehr and M. Grun, eds., Elsevier/North Holland, Amsterdam (1980a).

36. J. P. Nolan, A. I. Leibowitz, and A. O. Vladutiu, Influence of carbon tetrachloride on circulatory endotoxin after the exogeneous administration of endotoxin in rats, Proc. Soc. Exp. Biol. Med. 165:453 (1980b).

37. J. P. Nolan, R. C. Venuto, and G. S. Goldmann, Role of endotoxin in glycerol-induced renal failure in the rat, Clin. Sci. Mol. Med. 54:615 (1978).

38. J. P. Nolan, A. O. Vladutiu, D. M. Moreno, S. A. Cohen, and D. S. Camara, Immunoradiometric assay of lipid A: A test for detecting and quantitating endotoxins of various origins, J. Immunol. Meth. 55:63 (1982).

39. R. Ogawa, T. Morita, F. Kunimoto, and T. Fujita, Changes in hepatic lipoperoxide concentration in endotoxemic rats, Circ. Shock 9:369 (1982).

40. P. J. O'Leary, J. W. Maher, J. I. Hollenbeck, et al., Pathogenesis of hepatic failure after obesity bypass, Surg. Forum 23:356 (1974).

41. K. R. Palmer, B. I. Duerden, and C. D. Holdsworth, Bacteriological and endotoxin studies in cases of ulcerative colitis submitted to surgery, Gut 21:851 (1980).

42. D. P. Praaning-VanDalen, A. Brouwer, and D. L. Knook, Clearance capacity of rat liver, Kupffer, endothelial, and parenchymal cells, Gastroenterol. 81:1036 (1981).

43. M. Ragni, J. Lewis, and J. Spero, Ascites-induced LeVeen shunt coagulopathy, Ann. Surg. 193:91 (1983).
44. J. Rasenach, A. K. Koch, J. Nowack, et al., Hepatotoxicity of D-glactosamine in the isolated perfused rat liver, Exp. Mol. Pathol. 32:264 (1980).
45. J. T. Reeves, and R. F. GroverBlockade of acute hypoxic pulmonary hypertension by endotoxin, J. Appl. Physiol. 36:328 (1974).
46. A. M. Rutenburg, E. Sonneblick, I. Koven, et al., The role of intestinal bacteria in the development of dietary cirrhosis in rats, J. Exp. Med. 106:1 (1957).
47. D. Scevola, E. Magliulo, G. Barbarini, et al., Possible anti-endotoxin activity of cyanidanol-3 in experimental hepatitis in the rat, Hepato-gastroenterol. 29:178 (1982).
48. D. W. Scherfele, P. Melton, and V. Whitechelo, Evaluation of the Limulus test for endotoxemia in neonates with suspected sepsis, J. Pediatr. 98:899 (1981).
49. D. W. Scherfele, E. Olsen, S. Fussell, and N. Pedray, Spontaneous endotoxemia in premature infant: correlations with oral feeding and bowel dysfunction, J. Pediatr. Gastroenterol. Nutr. 4:67 (1985).
50. J. L. Shenep and K. A. Morgan, Kinetics of endotoxin release during antibiotic therapy for experimental gram-negative bacterial sepsis, J. Inf. Dis. 150:380 (1984).
51. R. J. Stumacher, M. J. Kovnat, and W. R. McCabe, Limitations of the usefulness of the Limulus assay for endotoxin, New Eng. J. Med. 288:1261 (1973).
52. K. Tarao, K. So, T. Moroi, et al., Detection of endotoxin in plasma and ascites of patients with cirrhosis; its clinical significance, Gastroenterol. 73:539 (1977).
53. N. Teng, H. Kaplan, J. Herbert, C. Moore, H. Douglas, A. Wunderlich, and A. Braude, Protection against gram-negative bacteremia and endotoxemia with human monoclonal IgM antibodies, Proc. Natl. Acad. Sci. USA 82:1790 (1985).
54. U. Turunen, M. Maikamaki, V. V. Valtoneu, et al., High titers of enterobacterial common antigen antibodies in patient with alcoholic cirrhosis, Gut 22:849 (1981).
55. A. C. VanVilet, H. C. Maas, and J.H.P. Wilson, The effect of portasystemic shunting and RES function on endotoxemia in human liver disease, in: "The Reticuloendothelial System and Pathogenesis of Liver Disease," H. Liehr, and M. Grun, eds., Elsevier/North Holland, Amsterdam (1980).
56. W. A. Walker, R. Cornell, L. M. Davenport, et al., Macromolecular absorption mechanism of horseradish peroxidase uptake in adult and neonatal rat intestine, J. Cell. Biol. 54:195 (1972).
57. E. N. Wardle, Endotoxemia and the pathogenesis of acute renal failure, Quart. J. Med. 44:389 (1975).
58. E. N. Wardle, Acute renal failure in the 1980's: the importance of septic shock and of endotoxemia, Nephron. 30:193 (1982).
59. W. Wellmann, P. C. Fink, and F. W. Schmidt, Whole-gut irrigation as antiendotoxinaemic therapy in inflammatory bowel disease, Hepato-gastroenterol. 31:91 (1984).
60. S. P. Wilkinson, V. Arroyo, B. G. Gazzard, et al., Relationship of renal impairment and hemorrhogic diathesis to endotoxemia in fulminant hepatic failure, Lancet 1:521 (1974).
61. T. Yoshikawa, M. Murakami, Y. Furukawa, H. Kato, S. Takemura, and M. Kondo, Lipid peroxidation and experimental disseminated intravascular coagulation in rats induced by endotoxin, Thromb. Haemostas. (Stuttgart) 49:214 (1983).
62. R. L. Yost, M. C. Duerson, W. L. Russel, et al., Doxycycline in the prevention of hepatic dysfunction: an evaluation of its use following jejunoileal bypass in humans, Arch. Surg. 114:931 (1979).

CONTRIBUTORS

APICELLA, Michael A., Ph.D., Division of Infectious Diseases, Department of
    Medicine, State University of New York at Buffalo School of Medicine,
    462 Grider Street, Buffalo, New York 14215

BAGBY, Gregory J., Department of Physiology, Louisiana State University
    Medical Center, 1901 Perdido Street, New Orleans, Louisiana 70112-1393

BECKER, Jeanne, Department of Medical Microbiology and Immunology,
    University of South Florida, College of Medicine, 12901 N. 30th
    Street, Box 10, Tampa, Florida 33612

BENDINELLI, Mauro, M.D., Ph.D., Institute of Epidemiology, Hygiene and
    Virology, University of Pisa, I-5600 Pisa, Italy

BERRY, L. Joe, Department of Microbiology, University of Texas, Austin,
    Texas 78712

BLANCHARD, D. K., Department of Medical Microbiology and Immunology, Univer-
    sity of South Florida, College of Medicine, 12901 N. 30th Street,
    Box 10, Tampa, Florida 33612

BROWN, Thomas A., University of Alabama School of Medicine, University of
    Alabama in Birmingham, University Station, Birmingham, Alabama 35294

BRUMFIELD, Brent A., B.S., Department of Physiology, Louisiana State
    University Medical Center, 1901 Perdido Street, New Orleans, Louisiana
    70112-1393

BUTLER, R. Christopher, Ph.D., Department of Medical Microbiology and Immun-
    ology, The Arlington Hospital, 1701 North George Mason Drive, Arlington,
    Virginia 22205

CANTRELL, John L., Ribi ImmunoChem Research, Inc., P. O. Box 1409, Hamilton,
    Montana 59840

CASTAGNA, Raymond, Department of Microbiology and Immunology, State Univer-
    sity of New York, Downstate Medical Center, 450 Clarkson Avenue,
    Box 44, Brooklyn, New York 11203

CHAPEKAR, Mrunal S., Department of Medical Microbiology and Immunology,
    The Arlington Hospital, 1701 North George Mason Drive, Arlington,
    Virginia 22205

CHEDID, Louis, M.D., Ph.D., Institute Pasteur, Immunotherapie Experimentale,
    28 Rue du Dr. Roux, 75724 Paris, Cedex 15, France

COFFEY, Ronald G., Ph.D., Department of Pharmacology and Therapeutics, University of South Florida, College of Medicine, 12901 N. 30th Street, Box 9, Tampa, Florida 33612

COLWELL, Dawn E., University of Alabama School of Medicine, University of Alabama in Birmingham, University Station, Birmingham, Alabama 35294

CONALDI, Pier Giulio, Institute of Epidemiology, Hygiene and Virology, University of Pisa, I-56100 Pisa, Italy

CRAIG, John P., Department of Microbiology and Immunology, State University of New York, Downstate Medical Center, 450 Clarkson Avenue, Box 44, Brooklyn, New York 11203

CROMARTIE, William J., Department of Microbiology, University of North Carolina School of Medicine, Chapel Hill, North Carolina 27514

CULLOR, J. S., Department of Veterinary Pathology, School of Veterinary Medicine, University of California, Davis, California 95616

DUNCAN, Jr., Robert L., Ph.D., Department of Oral Biology, Emory University School of Medicine, 1462 Clifton Road, N.E., Atlanta, GA 30322

DUSAPIN, Karen R., M.S., Department of Pharmacology and Experimental Therapeutics, Louisiana State University Medical Center, 1901 Perdido Street, New Orleans, Louisiana 70112-1393

EISENSTEIN, Toby K., Ph.D., Temple University School of Medicine, Department of Microbiology and Immunology, Philadelphia, Pennsylvania 19140

ESSER, Ronald E., Department of Microbiology, University of North Carolina School of Medicine, Chapel Hill, North Carolina 27514

FENWICK, Brad W., D.V.M., M.S., Department of Veterinary Pathology, School of Veterinary Medicine, University of California, Davis, California 95616

FILKINS, James P., Ph.D., Department of Physiology, Loyola University of Chicago, Stritch School of Medicine, 2160 South First Avenue, Maywood, Illinois 60153

FRASCH, Carl E., Office of Biologics Research and Review, Food and Drug Administration, 8800 Rockville Pike, Bethesda, Maryland 20205

FRIEDMAN, Herman, Ph.D., Department of Medical Microbiology and Immunology, University of South Florida, College of Medicine, 12901 N. 30th Street, Box 10, Tampa, Florida 33612

FRIER, Jeri M., Department of Medical Microbiology and Immunology, The Arlington Hospital, 1701 North George Mason Drive, Arlington, Virginia 22205

GADKE, P., Institute for Medical Microbiology, Free University of Berlin, Berlin, Federal Republic of Germany

GALANOS, Chris, Max Planck Institut für Immunbiologie, D-7800 Freiburg, Federal Republic of Germany

GALY, Ann, Program of Immunopharmacology, Department of Internal Medicine, University of South Florida, College of Medicine, 12901 N. 30th Street, Box 19, Tampa, Florida 33612

GIANGREGORIO, Anna Maria, Institute of Epidemiology, Hygiene and Virology, University of Pisa, I-56100 Pisa, Italy

GOODMAN, S. A., University of Kansas Medical Center, Microbiology Department, 39th and Rainbow Boulevard, Kansas City, Kansas 66103

GRASSO, Robert J., Ph.D., Department of Medical Microbiology and Immunology, University of South Florida, College of Medicine, 12901 N. 30th Street, Box 10, Tampa, Florida 33612

HADDEN, Elba M., Ph.D., Program of Immunopharmacology, Department of Internal Medicine, University of South Florida, College of Medicine, 12901 N. 30th Street, Box 19, Tampa, Florida 33612

HADDEN, John W., M.D., Program of Immunopharmacology, Department of Internal Medicine, University of South Florida, College of Medicine, 12901 N. 30th Street, Box 19, Tampa, Florida 33612

HAHN, H., Institute for Medical Microbiology, Free University of Berlin, Berlin, Federal Republic of Germany

JAHNSEN, Mark, Cetus Immune Research Labs, 3400 West Bayshore Road, Palo Alto, California 94303

JELJASZEWICZ, J., National Institute of Hygiene, Warsaw, Poland

KELLY, A., Department of Veterinary Medicine, University of California, Davis, California, 95616

KILLAR, Loran M., Yale University School of Medicine, 333 Cedar Street, New Haven, Connecticut 06510

KLEIN, Thomas W., Ph.D., Department of Medical Microbilogy and Immunology, University of South Florida, College of Medicine, 12901 N. 30th Street, Box 10, Tampa, Florida 33612

KO, H. L., Institute of Hygiene, University of Cologne, Cologne, West Germany

KOOPMAN, William J., University of Alabama School of Medicine, University of Alabama in Birmingham, University Station, Birmingham, Alabama 35294

LANG, Charles H., Ph.D., Department of Physiology, Louisiana State University Medical Center, 1901 Perdido Street, New Orleans, Louisiana 70112-1393

LARRICK, James W., M.D., Ph.D., Cetus Immune Research Labs, 3400 West Bayshore Road, Palo Alto, California 94303

LATHAM, Patricia S., M.D., Department of Medicine, Division of Gastroenterology, University of Maryland Hospital, Baltimore, Maryland 21201

LÜDERITZ, Otto, M.D., Max Planck Institut für Immunbiologie, D-7800 Freiburg, Federal Republic of Germany

LUX, F., Institute for Medical Microbiology, Free University of Berlin, Berlin, Federal Republic of Germany

MADONNA, Gary S., Ph.D., Department of Microbiology, Uniformed Services University of the Health Sciences, F. Edward Hebert School of Medicine, 4301 Jones Bridge Road, Bethesda, Maryland 20814-4799

MANNEL, Daniela N., German Cancer Research Center, Heidelberg, Federal Republic of Germany

MATTEUCCI, Donatella, Institute of Epidemiology, Hygiene, and Virology, University of Pisa, I-56100 Pisa, Italy

MCDONOUGH, Kathleen H., Ph.D., Department of Physiology, Louisiana State University Medical Center, 1901 Perdido Street, New Orleans, Louisiana 70112-1393

MCGHEE, Jerry R., Ph.D. Department of Microbiology, University of Alabama School of Medicine, University of Alabama in Birmingham, University Station, Birmingham, Alabama 35294

MCGROARTY, Estelle J., Department of Biochemistry, Michigan State University, East Lansing, Michigan 48824

MELCHERS, Fritz, M.D., Basel Institute for Immunology, Grenzacherstrasse 487, Postfach, CH-4005 Basel, Switzerland

MERGENHAGEN, Stephan E., Cellular Immunology Section, Laboratory of Microbiology and Immunology, National Institute of Dental Research, National Institutes of Health, Bethesda, Maryland 20205

MESTECKY, Jiri, University of Alabama School of Medicine, University of Alabama in Birmingham, University Station, Birmingham, Alabama 35294

MICHALEK, Suzanne M., Ph.D., Department of Microbiology, University of Alabama School of Medicine, University of Alabama in Birmingham, University Station, Birmingham, Alabama 35294

MORRISON, David C., Ph.D., Department of Microbiology, University of Kansas Medical Center, 39th and Rainbow Boulevard, Kansas City, Kansas 66103

MYERS, Kent R., Ribi ImmunoChem Research, Inc., P. O. Box 1409, Hamilton, Montana 59840

NAKANO, Masayasu, M.D., Department of Microbiology, Jichi Medical School, Tochigiken 329-04, Japan

NITTA, Toshimasa, Department of Bacteriology, Tohoku Dental University, Koriyama 963, Japan

NIXDORFF, Kathryn, Institut für Mikrobiologie, Technische Hochschule Darmstadt, Schnittspahnstr. 9, D-6100 Darmstadt, Federal Republic of Germany

NOLAN, James P., M.D., Department of Medicine, State University of New York at Buffalo, School of Medicine, Erie County Medical Center, 462 Grider Street, Buffalo, New York 14215

NOWOTNY, Alois, Ph.D., University of Pennsylvania, Medical Center for Oral Health Research, School of Dental Medicine, 40th and Spruce Streets, Philadelphia, Pennsylvania 19104

OLANDER, H. J., Department of Veterinary Pathology, School of Veterinary Medicine, University of California, Davis, California 95616

OSBURN, B. I., Department of Veterinary Pathology, School of Veterinary
   Medicine, University of California, Davis, California 95616

PARANT, F., Institute Pasteur, Immunotherapie Experimentale, 28 Rue de Dr.
   Roux, 75724 Paris, Cedex 15, France

PARANT, M., Institute Pasteur, Immunotherapie Experimentale, 28 Rue de Dr.
   Roux, 75724 Paris, Cedex 15, France

PELZER, K., Department Veterinary Pathology, School of Veterinary Medicine,
   University of California, Davis, California 95616

PETERSON, Arnold A., Department of Biophysics, Michigan State University,
   East Lansing, Michigan 48824

PHILLIPS, Marshall, National Animal Disease Center, Ames, Iowa 50010

PICK, Edgar, M.D., Ph.D., Department of Human Microbiology, Sackler School
   of Medicine, Tel-Aviv University, Ramat-Aviv, Tel-Aviv 69978, Israel

PLUZNIK, Dov H., Ph.D., Cellular Immunology Section, Laboratory of Microbiology and Immunology, National Institute of Dental Research, National
   Institutes of Health, Bethesda, Maryland 20205

PULVERER, G., M.D., Institute of Hygiene, University of Cologne, Cologne,
   West Germany

QURESHI, Nilofer, William S. Middleton Memorial Veterans Hospital, Madison,
   Wisconsin 53705

RENDA, Pamela B., Department of Microbiology, University of Pittsburgh,
   School of Medicine, Pittsburgh, Pennsylvania 15261

RIBI, Edgar, Ribi ImmunoChem Research, Inc., P. O. Box 1409, Hamilton,
   Montana 59840

RIBI, Hans O., Department of Cell Biology, Stanford University, Stanford,
   California 94305

RIETSCHEL, Ernst Th., Forschungsinstitut Borstel, D-2061, Borstel, Federal
   Republic of Germany

RIVEAU, Gilles J., Institute Pasteur, Immunotherapie Experimentale, 28 Rue
   du Dr. Roux, 75724 Paris, Cedex 15, France

ROBIE, Norman W., Ph.D., Department of Pharmacology and Experimental
   Therapeutics, Louisiana State University Medical Center, 1901 Perdido
   Street, New Orleans, Louisiana 70112-1393

RODLOFF, A. C., Ph.D., Institute for Medical Microbiology, Free University
   of Berlin, Berlin, Federal Republic of Germany

ROSZKOWSKI, K., Institute of Lung Diseases, Warsaw, Poland

ROSZKOWSKI, W., Institute of Lung Diseases, Warsaw, Poland

SALVIN, Samuel B., Department of Microbiology, University of Pittsburgh,
   School of Medicine, Pittsburgh, Pennsylvania 15261

SCHELL, Sigrid, Institut für Mikrobiologie, Technische Hochschule Darmstadt,
   Schnittspahnstr. 9, D-6100 Darmstadt, Federal Republic of Germany

SCHWAB, John H., Department of Microbiology, University of North Carolina School of Medicine, Chapel Hill, North Carolina 27514

SENYK, George, Cetus Immune Research Labs, 3400 West Bayshore Road, Palo Alto, California 94303

SEPELAK, Susan B., Department of Pathology, Baltimore Veteran's Administration Hospital, Baltimore, Maryland

SHEPHERD, Raymond E., Ph.D., Department of Physiology, Louisiana State University Medical Center, 1901 Perdido Street, New Orleans, Louisiana 70112-1393

SINKOVICS, Joseph G., M.D., Community Cancer Center, St. Joseph's Hospital, 3001 W. Buffalo Avenue, Tampa, Florida 33607

SMITH, B. P., Department of Veterinary Pathology, School of Veterinary Medicine, University of California, Davis, California 95616

SPALDING, David M., University of Alabama School of Medicine, University of Alabama in Birmingham, University Station, Birmingham, Alabama 35294

SPITZER, John J., M.D., Department of Physiology, Louisiana State University Medical Center, 1901 Perdido Street, New Orleans, Louisiana 70112-1393

SPITZER, Judy A., Ph.D., Department of Physiology, Louisiana State University Medical Center, 1901 Perdido Street, New Orleans, Louisiana 70112-1393

STEWART II, William E., Ph.D., Department of Medical Microbiology and Immunology, University of South Florida, College of Medicine, 12901 N. 30th Street, Box 10, Tampa, Florida 33612

STIMPSON, Stephen A., Department of Microbiology, University of North Carolina School of Medicine, Chapel Hill, North Carolina 27514

SULTZER, Barnet M., Department of Microbiology and Immunology, State University of New York, Downstate Medical Center 450 Clarkson Avenue, Brooklyn, New York 11203

SZENTIVANYI, Andor, M.D., Departments of Pharmacology and Therapeutics and Internal Medicine, University of South Florida, College of Medicine, 12901 N. 30th Street, Box 2, Tampa, Florida 33612

TAKAYAMA, Kuni, William S. Middleton Memorial Veterans Hospital, Madison, Wisconsin 53705

TESH, Vernon, Department of Microbiology and Immunology, Emory University School of Medicine, 1462 Clifton Road, Atlanta, Georgia 30322

TOURAINE, J. L., Program of Immunopharmacology, Department of Internal Medicine, University of South Florida, College of Medicine, 12901 N. 30th Street, Box 19, Tampa, Florida 33612

TSAI, Chao-Ming, Ph.D., Office of Biologics Research and Review, Bacterial Polysaccharides Branch, Center for Drugs and Biologics, Food and Drug Administration, 8800 Rockville Pike, Bethesda, Maryland 20205

URBASCHEK, Bernhard, M.D., Department of Immunology and Serology, Institute for Hygiene and Medical Microbiology, Klinikum Mannheim, University of Heidelberg, 6800 Mannheim, Free Republic of Germany

URBASCHEK, Renate, M.D., Department of Immunology and Serology, Institute for Hygiene and Medical Microbiology, Klinikum Mannheim, University of Heidelberg, 6800 Mannheim, Free Republic of Germany

VAN DER WAAIJ, D., University Hospital, Groningen, The Netherlands

VOGEL, Stefanie N., Ph.D., Department of Microbiology, Uniformed Services University of the Health Science, F. Edward Hebert School of Medicine, 4301 Jones Bridge Road, Bethesda, Maryland 20814-4799

VUKAJLOVICH, S. W., University of Kansas Medical Center, Department of Microbiology, 39th and Rainbow Boulevard, Kansas City, Kansas 66103

WATSON, Karen, Cetus Immune Research Labs, 3400 West Bayshore Road, Palo Alto, California 94303

WEISS, Stefan, Cetus Immune Research Labs, 3400 West Bayshore Road, Palo Alto, California 94303

WESTPHAL, Otto, Max Planck Institut für Immunbiologie, D-7800 Freiburg, Federal Republic of Germany

WILLIAMS, Joseph F., Ph.D., Department of Pharmacology and Therapeutics, University of South Florida, College of Medicine, 12901 N. 30th Street, Box 9, Tampa, Florida 33612

WILLIAMSON, Shane I., University of Alabama School of Medicine, University of Alabama in Birmingham, University Station, Birmingham, Alabama 35294

WITEK-JANUSEK, Linda, Ph.D., Department of Physiology, Loyola University of Chicago, Stritch School of Medicine, 2160 South First Avenue, Maywood, Illinois 60153

WOLLENWEBER, H-W., Department of Microbiology, University of Kansas Medical Center, 39th and Rainbow Boulevard, Kansas City, Kansas 66103

YELICH, Michael R., Ph.D., Department of Physiology, Loyola University of Chicago, Stritch School of Medicine, 2160 South First Avenue, Maywood, Illinois 60153

Mouse (continued)
  typhoid, 25
  vaccinated, 242
Murabutide, 203, 205, 209
Muramyl dipeptide, 167, 369, 371,
  485
Muramyl peptide, 203-210
  anti-Klebsiella activity, 203-210
  toxicity, 203
Murine leukemia virus, see Leukemia
  virus
Mycobacterium bovis, see BCG
Myocyte, 174-175, 180
Myoglobinuria, 272

NADPH cytochrome c reductase, 159
NADPH oxidase, 340, 341, 346, 347,
  349
Naloxone, 25
Neisseria gonorrhoeae, 3, 57-64,
  90, 93
  N. lactamicae, 93
  N. meningitidis, 90-93
Neutropenia, induced, 257, 258
Neutrophil, 257-263, 341-342
Newcastle disease virus, 294
Norepinephrine, 137-138, 141-143,
  176-177
Nucleotide, cyclic, 187-198

Ovalbumin, 320
Oxygen radical production, 339-351
  see Burst, oxidative
Oxidase, mixed function - ,
  159-169
  and endotoxin, 159-169

Pasteurella multocida, 93, 206,
  207, 211
Peptide, opioid, 270
Peptidoglycan and arthritis,
  449-454
Peyer's patches and lipopoly-
  saccharide, 329-338
Phagocytosis, 281-290
  index, phagocytic, 283-287
Phenol/water extraction method for
  endotoxin, 19
Phenoxybenzamine, 274
Phentolamine, 274
Phorbol ester, 391-404
Phorbol myristate acetate, 194,
  340, 341
Phosphoenolpyruvate carboxykinase,
  23, 28, 107-108
Phospholipase
  A$_2$, 341
  C, 341
Phosphorylase
  B, 320
  kinase, 179

Phytohemagglutinin, 190, 356
Pig pneumonia, contagious, 233
  protection by core lipopoly-
    saccharide of E. coli,
    233-237
Piperacillin, 459-463
Plaque, hemolytic, forming cell
  assay, 370
Platelet count and hydroxyurea, 261
Polyacrylamide gel electrophoresis,
  see SDS-PAGE
Polyinosinic-polycytidylic acid, 294
Polymyxin B, 355, 356, 469, 473, 498
Polyuria, inappropriate, 272
Porin, 435
Procholeragenoid, 437, 443
Propionibacterium granulosum,
  455-458
Propranolol, 274
Prostaglandin, 191, 194, 269, 270,
  400, 436
Protein
  endotoxin-associated, 435-447
  immunobiology, listed, 436
  immunostimulation, 436
  lipid A-associated, 436
  mitogenicity, 435
  of outer membrane, bacterial,
    Gram-negative, 345-347
Protein kinase, 194, 400-402
Proteus mirabilis, 421-430, 446
Prothymocyte, 190, 191
Prothymosin, 292
Pseudomonas sp., 269
  bacteremia, 22
  sepsis, 274
  vaccine, 271
  P. aeruginosa, 78
Pump, osmotic, 133
Pyrogen, 16, 26, 190, 206, 208
Pyrogenicity, 14, 26, 203, 211, 372

Rabbit, 15, 18, 22, 26, 207, 272,
  294, 372, 412-413, 436, 438,
  452
Radioprotection, 221-225
Ra-reactive factor (RaRF), 211
Rat, 25, 121, 133-149, 151-158, 163,
  173-185, 436, 449-454
Receptor
  blockers, 274
  stimulators, 274
Renal failure, acute, 272
  and bacterial infection, 493-494
Resistance to infection, 17-18
Respiratory
  control index (RCI), 144, 145
  distress syndrome in adult, 494
Reticuloendothelial system, 112-114,
  126-127